Volumes already published

Volume 1: Trees I (1986)
Volume 2: Crops I (1986)
Volume 3: Potato (1987)
Volume 4: Medicinal and Aromatic Plants I (1988)
Volume 5: Trees II (1989)
Volume 6: Crops II (1988)
Volume 7: Medicinal and Aromatic Plants II (1989)
Volume 8: Plant Protoplasts and Genetic Engineering I (1989)
Volume 9: Plant Protoplasts and Genetic Engineering II (1989)
Volume 10: Legumes and Oilseed Crops I (1990)
Volume 11: Somaclonal Variation in Crop Improvement I (1990)
Volume 12: Haploids in Crop Improvement I (1990)
Volume 13: Wheat (1990)
Volume 15: Medicinal and Aromatic Plants III (1991)

Volumes in preparation

Volume 14: Rice (1991)
Volume 16: Trees III (1991)
Volume 17: High-Tech and Micropropagation I
Volume 18: High-Tech and Micropropagation II

Biotechnology in Agriculture and Forestry 15

Medicinal and Aromatic Plants III

Edited by Y.P.S. Bajaj

With 208 Figures

Springer-Verlag
Berlin Heidelberg New York
London Paris Tokyo
Hong Kong Barcelona
Budapest

Professor Dr. Y. P. S. Bajaj
A-137
New Friends Colony
New Delhi 110065, India

ISBN-13: 978-3-642-84073-9 e-ISBN-13: 978-3-642-84071-5
DOI: 10.1007/ 978-3-642-84071-5

Softcover reprint of the hardcover 1st edition 1991

Typesetting: International Typesetters Inc., Makati, Philippines
31/3145(3011)-543210 – Printed on acid-free paper

Dedicated to
Sarabjit, Bhupinder, and Sonia Bawa

Preface

The use of in vitro technology in the production of medicinal compounds through plant cell cultures is gaining momentum. The recent commercial production of shikonin in cell cultures of *Lithospermum* has especially attracted attention of the pharmaceutical industry. The advances made in the micropropagation, production of haploids, genetic transformation through *Agrobacterium,* the selection of high-yielding somaclones, and the cryopreservation of cell cultures of medicinal plants are significant and have far-reaching implications for commercial exploitation. These new developments need to be highlighted and brought to the attention of workers dealing with medicinal, aromatic and other plants of industrial importance. A series of books on the biotechnology of medicinal and aromatic plants are therefore being compiled to provide a survey of the literature focusing on recent information and the state of the art.

This volume, Medicinal and Aromatic Plants III, like the previous two (Medicinal and Aromatic Plants I and II, published in 1988 and 1989, respectively) is unique in its approach. It comprises 29 chapters dealing with the distribution, economic importance, conventional propagation, review of tissue culture studies, and the in vitro production of important medicinal and pharmaceutical compounds in various species of *Atropa, Ageratina, Ailanthus, Aconitum, Apium, Aloe, Akebia, Bidens, Carthamus, Carum, Chamomilla, Citrus, Cymbopogon, Dysoma, Euphorbia, Fritillaria, Glycyrrhiza, Lavandula, Nigella, Pelargonium, Perilla, Podophyllum, Rosa, Scutellaria, Securinega, Solanum, Swertia, Symphytum*, and *Syringa.* The potential role of biotechnology for industrial production is discussed. This book is tailored to the needs of advanced students, teachers and research workers in the fields of plant biotechnology and bioengineering, pharmacy, botany, and phytochemistry.

New Delhi, January 1991

Professor Y. P. S. Bajaj
Series Editor

Contents

IX *Carthamus tinctorius* L. (Safflower):
Production of Vitamin E in Cell Cultures
T. FURUYA and T. YOSHIKAWA (With 3 Figures)

X *Chamomilla recutita* (L.) Rauschert (Camomile):
In Vitro Culture and the Production of Secondary Metabolites
J. REICHLING and R. BEIDERBECK (With 7 Figures)

XI *Carum carvi* L. (Caraway): In Vitro Culture, Embryogenesis, and the Production of Aromatic Compounds
M. FURMANOWA, D. SOWINSKA, and A. PIETROSIUK
(With 8 Figures)

XII *Citrus* spp.: In Vitro Culture and the Production of Naringin and Limonin
R. L. MANSELL and C. A. McINTOSH (With 7 Figures)

XXVIII *Symphytum officinale* (Comfrey): In Vitro Culture, Regeneration, and Biogenesis of Pyrrolizidine Alkaloids
H.J. HUIZING and J.H. SIETSMA (With 12 Figures)

XXIX *Syringa vulgaris* L. (Common Lilac): In Vitro Culture and the Occurrence and Biosynthesis of Phenylpropanoid Glycosides
C.C.S. CHAPPLE and B.E. ELLIS (With 4 Figures)

List of Contributors

BAJAJ, Y. P. S., A-137, New Friends Colony, New Delhi 110065, India

BEIDERBECK, R., Botanisches Institut der Universität, Im Neuenheimer Feld 360, W-6900 Heidelberg, FRG

CALVO, M. C., Departament de Biologia Vegetal, Facultat de Farmàcia, Universitat de València, Avda. Blasco Ibàñez, 13, 46010 València, Spain

CASO, O., Departamento de Ciencias Biologicas, Facultad de Farmacia y Bioquimica, Universidad de Buenos Aires, (1113) Junin 956, Buenos Aires, Argentina

CASTORENA SANCHEZ, I., Department of Agricultural Plant Biology, Genetics Section, Via Matteotti 1/B, 56100 Pisa, Italy

CAVALLINI, A., Department of Agricultural Plant Biology, Genetics Section, Via Matteotti 1/B, 56100 Pisa, Italy

CHANG, W.-C., Institute of Botany, Academia Sinica, Taipei, Taiwan 11529, China

CHAPPLE, C. C. S., MSU-DOE Plant Research Laboratory, Michigan State University, East Lansing, MI 48824-1312, USA

CHARLWOOD, B. V., Plant Cell and Molecular Sciences Group, Division of Biosphere Sciences, King's College London, Campden Hill Road, London W8 7AH, United Kingdom

CHARLWOOD, K. A., School of Chemistry, Thames Polytechnic, Wellington Street, Woolwich, London SE18 6PF, United Kingdom

COLLIN, H. A., Department of Genetics and Microbiology, University of Liverpool, Liverpool L69 3BX, United Kingdom

DESMAREST, P., Centre de Recherches "Pernod-Ricard", 120, Av. du Maréchal Foch, 94003 Creteil, France

DOEBEL, P., Akademie der Wissenschaften, Institut für Biotechnologie, Permoserstraße 15, O-7050 Leipzig, FRG

DU MANOIR, J., Centre de Recherches "Pernod-Ricard", 120, Av. du Maréchal Foch, 94003 Creteil, France

EDY, A. M., Laboratoire de Botanique, Faculté des Sciences Pharmaceutiques, 31, Allées Jules Guesde, 31000 Toulouse, France

ELLIS, B. E., The University of British Columbia, Department of Plant Science, Suite 248, 2357 Main Mall, Vancouver, B.C., Canada V6T 2A2

FUJII, Y., Takasaki Research Laboratory, 219 Iwahana, Takasaki, Gunma, 370-12, Japan

FURMANOWA, M., Department of Biology and Pharmaceutical Botany, Medical Academy, 02-097 Warsaw, Banacha 1, Poland

FURUYA, T., School of Pharmaceutical Sciences, Kitasato University, Minato-ku, Tokyo 108, Japan

GIULIETTI, A. M., Biotecnologia – Facultad de Farmacia y Bioquimica, Universidad de Buenos Aires, (1113) Junin 956, Buenos Aires, Argentina

HATANO, K., Cooperate R and I, Plant Technology Institute, Ube Industries Ltd., 1978-5 Ogushi, Ubeshi, Yamaguchiken 755, Japan

HENRY, M., Laboratoire de Botanique, Faculté des Sciences Pharmaceutiques, 31, Allées Jules Guesde, 31000 Toulouse, France

HUIZING, H.-J., Agrotechnological Research Institute, ATO, P.O. Box 17, 6700 AA Wageningen, The Netherlands

IDE, A., Faculty of Agriculture, Yamaguchi University, 1677-1 Yoshida, Yamaguchi 753, Japan

IKUTA, A., The Research Institute for Biosciences, Science University of Tokyo, 2669 Yamazaki, Noda City, Chiba 278, Japan

ISAAC, S., Department of Genetics and Microbiology, University of Liverpool, Liverpool L69 3BX, United Kingdom

ISHIKURA, N., Department of Biological Science, Faculty of Science, Kumamoto University, 2-Chome, Kurokami, Kumamoto-Shi 860, Japan

JAGADISHCHANDRA, K. S., Department of Botany, University of Mysore, Manasagangotri, Mysore-570 066, India

MANSELL, R. L., Department of Biology, University of South Florida, Tampa, FL 336201, USA

McINTOSH, C. A., Department of Biology, University of South Florida, Tampa, FL 336201, USA

MIURA, H., Faculty of Pharmaceutical Sciences, Nagasaki University, 1–14, Bunkyo-machi, Nagasaki 852, Japan

NATALI, L., Department of Agricultural Plant Biology, Genetics Section, Via Matteotti 1/B, 56100 Pisa, Italy

NIGRA, H. M., Biotecnologia – Facultad de Farmacia y Bioquimica, Universidad de Buenos Aires, (1113) Junin 956, Buenos Aires, Argentina

NISHIOKA, I., Faculty of Pharmaceutical Sciences, Kyushu University, Higashiku, Fukuoka 812, Japan

NORTON, R. A., Department of Botany, University of British Columbia, Vancouver, B.C., Canada, V6T 2B1. (Present address: USDA, Agricultural Research Service, Plant Physiology Research Unit, P.O. Box 5677, Athens, GA 30613, USA)

PIETROSIUK, A., Department of Biology and Pharmaceutical Botany, Medical Academy, 02-097 Warsaw, Banacha 1, Poland

PROKSCH, P., Institut für Pharmazeutische Biologie, Technische Universität, Mendelssohnstraße 1, W-3300 Braunschweig, FRG

REICHLING, J., Institut für Pharmazeutische Biologie, Im Neuenheimer Feld 364, W-6900 Heidelberg, FRG

ROBERTS, A. V., Plant Biotechnology Research Unit, Polytechnic of East London, London E15 4LZ, United Kingdom

ROBERTS, M. F., Department of Pharmacognosy, The School of Pharmacy, London University, 29–39 Brunswick Square, London WC1N 1AX, United Kingdom

SCHMAUDER, H.-P., Friedrich-Schiller-Universität Jena, Biologisch-Pharmazeutische Fakultät, Institut für Mikrobiologie, Neugasse 23, O-6900 Jena, FRG

SEGURA, J., Departament de Biologia Vegetal, Facultat de Farmàcia, Universitat de València, Avgda. Blasco Ibàñez, 13, 46010 València, Spain

SHORT, K. C., Faculty of Science, Nottingham Polytechnic, Nottingham NG1 4BU, United Kingdom

SHOYAMA, Y., Faculty of Pharmaceutical Sciences, Kyushu University, Higashiku, Fukuoka 812, Japan

SIETSMA, J. H., Department of Molecular Plant Biology, University of Groningen, Biological Centre, 9751 NN Haren, The Netherlands

SIMOLA, L. K., Department of Botany, University of Helsinki, Unioninkatu 44, 00170 Helsinki, Finland

SOWÍNSKA, D., Department of Biology and Pharmaceutical Botany, Medical Academy, 02-097 Warsaw, Banacha 1, Poland

SREENATH, H. L., Department of Botany, University of Mysore, Manasagangotri, Mysore-570 066, India. (Present address: Tissue Culture Division, Central Coffee Research Institute, Coffee Research Station-577 177, Chikmagalur District, Karnataka, India)

SUN, C. S., Institute of Botany, Academia Sinica, Beijing, China

WANG, D. Y., China National Rice Research Institute, Hangzhou, China

YAMAMOTO, H., School of Pharmacy, Hokuriku University, Kanagawa-Machi, Kanazawa 920-11, Japan

YAMAMOTO, Y., Research Center, Nippon Paint Co., Ltd., Neyagawa, Osaka 572, Japan

YOSHIKAWA, T., School of Pharmaceutical Sciences, Kitasato University, Minato-ku, Tokyo 108, Japan

I *Atropa belladonna* L.: In Vitro Culture, Regeneration of Plants, Cryopreservation, and the Production of Tropane Alkaloids

Y.P.S. Bajaj[1] and L.K. Simola[2]

1 Introduction

Atropa belladonna L., commonly known as deadly nightshade, or belladonna belongs to the family Solanaceae. *Atropa* is a small genus native throughout middle and southern Europe and extending to central and western Asia up to the Himalayas. *A. belladonna* L., the economically most important species, is an erect perennial herb which may grow up to 5 ft in height. It has purple flowers (to distinguish it from the yellow-flowered *A. acuminata*) and black berries with many small seeds. This plant grows in damp or shady places, mainly in the mountains, and is attacked by various species of *Gonocephalum*, *Agrotis*, and *Archips*, which cause defoliation (Mathur and Anand 1964).

It is generally propagated through seed in the nurseries, and later transferred to the field. The leaves (folium belladonnae) are mainly used as a source for medicinal drugs, mainly atropine (dl-hyoscyamine), with scopolamine as a minor component.

Preparations of belladonna are known to have been used for many centuries. During the time of the Roman Empire and in the Middle Ages this plant was frequently used to produce marked dilatation of the pupils. Tropane alkaloids have been used in domestic medicine, for example, for inducing sleep and for toothache. Some of them have hallusinogenic effects. *A. belladonna* is known as a highly toxic species, and it has been employed as a poison since early classic times (Weiner 1985).

Tropane alkaloids (Figs. 1, 2) are the most important naturally occurring antimuscarinic drugs found especially in some solanaceous genera (*Atropa, Anisodus, Datura, Duboisia,* and *Hyoscyamus*). This class of drugs inhibits the action of acetylcholine on autonomic effectors innervated by postganglionic cholinergic nerves as well as on smooth muscles that lack cholinergic innervation. Since they antagonize the muscarinic action of acetylcholine they are known as antimuscarinic agents. The best-known member of this class of drugs is dl-hyoscyamine (Weiner 1985). Parasympathetic neuroeffector junctions in different organs are not all equally sensitive to these drugs. With small doses salivary and bronchial secretion and sweating are depressed. With large doses the pupil dilates and vagal effects on the heart are blocked so that the heart beat rate is increased.

The main pharmacological uses of hyoscyamine are as premedication for anesthesia, in gastrointestinal disorders, and in ophthalmology. Scopolamine

[1] Former Professor of Tissue Culture, Punjab Agricultural University, Ludhiana, India. Present Address: A-137 New Friends Colony, New Delhi 110065, India
[2] Department of Botany, University of Helsinki, Unioninkatu 44, 00170 Helsinki, Finland

Biotechnology in Agriculture and Forestry, Vol. 15
Medicinal and Aromatic Plants III (ed. by Y.P.S. Bajaj)

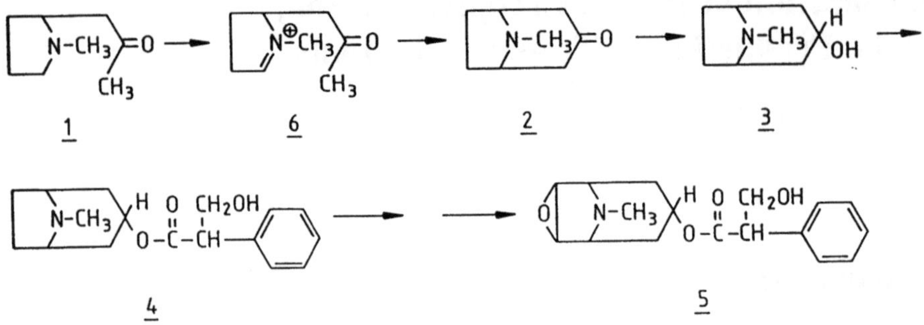

Fig. 1. Final stages in the biosynthesis of hyoscyamine and scopolamine. *1* Hygrine; *6* dehydrohygrine; *2* tropinone; *3* tropanol; *4* hyoscyamine; *5* scopolamine. (Simola et al. 1989)

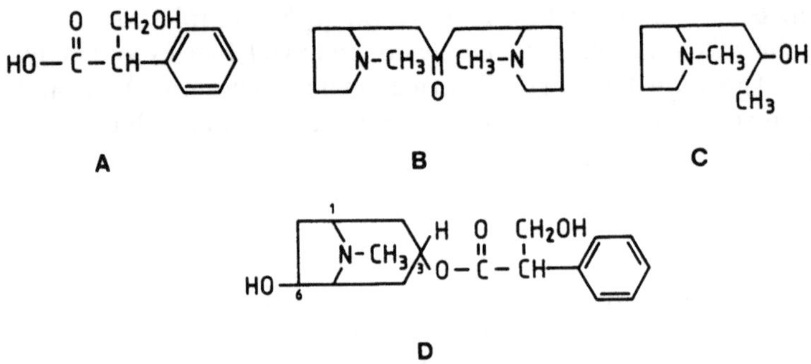

Fig. 2A-D. Structures of some alkaloid precursors and alkaloids. **A** Tropic acid. **B** Cuscohygrine. **C** Hygroline. **D** 6-Hydroxyhyoscyamine

prevents motion sickness, although some side effects are common (drowsiness and dry mouth). Many semi-synthetic and a large number of synthetic antimuscarinic compounds have been prepared with the objective of depressing gastric secretion without antimuscarinic effects on other organs (Weiner 1985). The new drugs have, however, few advantages over the naturally occurring alkaloids and their derivatives. Atropine is an effective medicine against some choline esterase inhibiting nerve gases (Weiner 1985).

Because of the immense importance of this plant as a source of medicines, there has been active research, employing biotechnological methods, to improve the yield of tropane alkaloids. Therefore in this chapter various aspects of tissue culture, micropropagation of plants, cryopreservation of cell cultures, and the in vitro production of tropane alkaloids are summarized.

2 In Vitro Culture Studies

Atropa belladonna, because of its high morphogenetic potential, has been extensively studied in vitro as a model system. Complete plants have been regenerated from callus cultures, excised anthers/pollen, isolated/fused protoplasts, and from the cryopreserved cultures. Moreover, significant literature has accumulated on the in vitro production of alkaloids in callus and suspension cultures and on nitrogen metabolism. Recent observations on the higher production of tropane alkaloids in vitro by "hairy roots" transformed through *Agrobacterium rhizogenes* offer yet another biotechnological approach for harvesting tropane alkaloids in this plant. The literature on various aspects of in vitro studies on *A. belladonna* is reviewed here (Tables 1, 2).

Table 1. In vitro culture studies on *Atropa belladonna* (see also Table 2)

Culture system	Observations/results	Reference
Callus/suspension cultures	Established callus and suspension cultures: differentiation of roots and shoots, regeneration of plants	Raj Bhandary et al. (1969)
	Formation of embryo-like structures from root callus	Thomas and Street (1970)
	Callus, buds, plants	Zenkteler (1971a)
	Effect of auxins on several enzymes in suspension culture	Simola and Sopanen (1971)
	Organogenesis and embryogenesis in root, callus, and suspension cultures	Thomas and Street (1972) Konar et al. (1972)
	Ultrastructural studies and changes in enzymes activity during differentiation of roots	Simola (1972) Simola (1973a)
	Development of chloroplasts in stem callus	Simola (1973b)
	Plants from haploid and diploid callus	Eapen et al. (1978)
	Effect of methionine on callus growth and ethylene production	Vakkari (1980)
	Effect of nitrogen sources and amino acids on callus growth and nitrogen metabolism	Salonen (1980, 1984); Salonen and Simola (1977, 1989)
	Phenylalanine transport in suspension cultures	Nyman and Simola (1988)
	Feeding experiments with precursors of tropane alkaloids	Simola et al. (1989)
Anther culture/ haploid production	Haploid plants and the study of stability in cell cultures	Zenkteler (1971b) Narayanaswamy and George (1972)
	Plants showed various level of ploidy ranging from haploids to hexaploids	Rashid and Street (1973, 1974) Bajaj et al. (1978)
Pollen mother cells	Repeated cell division	Bajaj (1974)
Pollen tetrads	Additional microspore formation	Bajaj (1975)

(*to be continued*)

Table 1. (*Continued*)

Culture system	Observations/results	Reference
Protoplast culture/ somatic hybridization	Embryos from cell suspension-derived protoplasts	Gosch et al. (1975)
	Plants from haploid mesophyll protoplasts showing genetic variability	Bajaj et al. (1978)
	Somatic hybridization with *Datura innoxia*	Krumbiegel and Schieder (1979, 1981)
	Somatic hybridization with *Nicotiana plumbaginifolia*	Gleba et al. (1988)
Cryopreservation	Cell suspension survived	Nag and Street (1975)
	Pollen embryos formed plants	Bajaj (1976, 1977, 1978)
	Mesophyll protoplasts formed embryos, plants	Bajaj (1988)
Transformation	Excised roots infected with *Agrobacterium rhizogenes*. Hairy root cultures with alkaloid production	Kamada et al. (1986) Jung and Tepfer (1987), Knopp et al. (1988)
In vitro production of alkaloids	See Table 2	

2.1 Callus/Suspension Culture and the Regeneration of Plants

Micropropagation of medicinal plants is of importance for mass production of high-yielding clones (see Bajaj et al. 1988). In *A. belladonna* there are three modes of regeneration, (1) organogenesis in callus cultures, (2) regeneration of buds on explants, and (3) somatic embryogenesis. The last two modes are preferred as they enable genetic stability of the stocks.

In their study of organogenesis, Raj Bhandary et al. (1969) also observed plant formation. Organogenesis and embryogenesis were described in detail in callus and suspension cultures of root origin (Thomas and Street 1970, 1972; Konar et al. 1972). Later, buds and eventual plant formation were obtained from excised leaves and root segments (Zenkteler 1971a).

Eapen et al. (1978) studied the morphogenetic ability of callus cultures and obtained haploid and diploid plants. The haploid cultures retained their organogenetic capacity longer than the diploid cultures. Likewise callus raised from protoplasts (Gosch et al. 1975; Bajaj et al. 1978) and anthers (Bajaj et al. 1978) underwent somatic embryogenesis and regenerated complete plants (see Sects. 2.3, 2.4 for details).

2.2 Developmental Physiology, Nitrogen Metabolism, and Ultrastructural Studies

Simola and her colleagues have conducted several studies on physiological and ultrastructural aspects of calli and suspension cultures. The original aim of the project was to use suspension cultures of *A. belladonna* as a model system for the regulation of differentiation with special emphasis on the changes in enzyme activities and ultrastructure during rhizogenesis. Higher activities for several enzymes were detected in undifferentiated cultures (medium with 2 mg/l NAA (1-naphthylacetic acid)), but not in cultures demonstrating root development (addition of 4 mg/l NOA (1-naphthoxyacetic acid) or lack of auxin). The latter two cultures were very similar in enzyme constitution (Simola and Sopanen 1971). In subsequent experiments the roots were dissected from the root-bearing aggregates (clumps). The roots and cell aggregates from the same suspension culture were found to have qualitatively dissimilar activities for several enzymes of primary metabolism (e.g., alanine aminopeptidase, glutamate, oxaloacetate transaminase, and ribonuclease). The undifferentiated aggregates had enzyme activities that were comparable to the clumps from which the roots were dissected (Simola 1973a).

Fine structure of calli and suspension cultures of *A. belladonna* has been compared with those of mesophyll cells of intact plants (Simola 1972, 1973b). Cytopathological features (lobed nuclei and cup-shaped organelles) were observed in cultured cells. Osmiophilic material interpreted to represent alkaloids was found as a network in the small vacuoles of the regenerated root tips. Corresponding deposits were not observed in undifferentiated parts of the suspension cultures (Simola 1972).

Due to the fact that alkaloid synthesis was very low or totally repressed in undifferentiated suspension cultures of *A. belladonna*, but organogenesis was stimulated by reduced nitrogen (Thomas and Street 1970, 1972), much work has been done in order to improve both growth and organogenesis in callus cultures (Salonen 1980, 1984). Formulation of a favorable but comparatively inexpensive nutrient medium was one of the problems to be resolved. The callus cultures of *A. belladonna* could be grown on NO_3^- and on combinations of NH_4^+ and NO_3^- in the ratios of 1:1 and 1:2 (5–35 mM N). Both forms were equally utilized up to 15 mM nitrogen, when growth, total nitrogen, and accumulation of NH_4^+ and NO_3^- were observed (Salonen 1984). At higher concentrations of nitrogen, combinations of NH_4^+ and NO_3^- became more suitable, producing higher total and organic nitrogen levels (about 3% of dry weight) compared with those of NO_3^--grown cultures (about 2% of dry weight).

The nitrogen requirements of the callus of *A. belladonna* could be met by some amino acids, but several dipeptides were poor nitrogen sources (Salonen and Simola 1977; Salonen 1980). Glutamate, aspartate, and their amides as well as proline could be used as effectively as inorganic nitrogen (2.5–7.5 mM N) whereas the end products (e.g., arginine, ornithine, methionine, and threonine) of the biosynthesis of glutamate- and aspartate-derived amino acids when used as the sole nitrogen sources were generally growth-retarding, often when in combination with inorganic nitrogen. However, ornithine (0.5 mM) in combination with glutamate or aspartate (5 mM) produced good growth (Salonen 1980).

Callus and suspension cultures of *A. belladonna* were also used as a model system for nitrogen metabolism of higher plants with special interest in amino acid precursors of tropane alkaloids. Repression of key enzymes of nitrogen assimilation by rather low levels of these amino acids might be problematic if biotechnological applications are planned. No repression of nitrogen reductase activity or growth inhibition were found in suspension cultures after feeding with glutamic acid or proline (2.5 mM) or their combinations with a low level of ornithine (0.25 mM) (Salonen and Simola 1989).

2.3 Anther/Pollen Culture and Production of Haploids

In vitro production of haploid plants enables easy detection of mutations, and early release of improved cultivars (see Bajaj 1990). *A. belladonna* anthers have been successfully cultured by various workers, and cell cultures and plants of various ploidy level produced (see Table 1). The possibility of the culture of pollen mother cells and pollen tetrads of *A. belladonna* was also explored by Bajaj (1974, 1975).

The pollen mother cells enclosed by the thick callose walls exhibited two types of behavior in culture. They either formed six to eight additional microspores, or they underwent repeated divisions to form a multicellular globular embryo-like structure. The frequency of these structures was increased if young anthers were subjected to low temperature (4°C for 48 h) before culturing (Bajaj 1974). Like the pollen mother cells, pollen tetrads of *A. belladonna* in culture also undergo divisions to form five to eight additional microspores, some microspores containing two or three nuclei. Even after 8 months in culture, the extra microspores were observed to be highly cytoplasmic and healthy (Bajaj 1975). This method offers a unique tool for the study of the ontogeny of pollen. The in vitro induction of repeated divisions also raises questions about the essential role of the tapetum during early microsporogenesis.

The excised anthers of *A. belladonna* cultured by various workers (see Table 1, Fig. 3A-C) resulted in the formation of haploid plants. Bajaj et al. (1978) cultured anthers of *A. belladonna* cv. *lutea* Döll at the uninucleate pollen stage in agar as well as in liquid media. The number of plants produced was increased when the anthers grown in liquid medium (for 4–5 weeks) were cut open and reinoculated on the same medium. This could be due to direct exposure of pollen embryos to the medium, where competition among them is reduced as compared to the ones enclosed in the anther. The anther-derived plants were transferred to pots (Fig. 3D,E) and allowed to grow further in the greenhouse. Plants of various ploidy were obtained, ranging from haploids (20%), diploids (70%), to triploids, hexaploids, and aneuploids (Fig. 3F-H). The regeneration of mixed population of plants of various ploidy, however, makes anther culture unfit for obtaining exclusively haploids; nevertheless, it can be employed for introducing genetic variation in plants.

2.4 Protoplast Culture and Somatic Hybridization

Complete *A. belladonna* plants have been regenerated from the cell suspension-derived protoplasts (Gosch et al. 1975) and mesophyll protoplasts (Bajaj et al. 1978),

Fig. 3A-H. Regeneration of plants of various ploidy from excised anthers of *Atropa belladonna*. **A** Anthers at the beginning of culture. **B** and **C** same, 7 and 9 weeks after culture showing the emergence of plantlets; **D** and **E** haploid (**D**) and diploid (**E**) anther-derived plants 10 weeks after transfer to pots (total time 20 weeks after anther culture). **F-H** root tip squashes from anther-derived plants showing various ploidy levels. (Bajaj et al. 1978)

and somatic hybrids have been obtained (Krumbiegel and Schieder 1979, 1981; Gleba et al. 1988).

Protoplasts isolated from actively growing cell suspensions were induced to divide repeatedly. An optimal protoplast yield of up to 80% was obtained in 4–5 h by treating cell suspensions with an enzyme mixture of Cellulase R 10 (1%) and Macerozyme R 10 (0.5%) in 0.6 M sorbitol at 30°C. The protoplasts cultured at a density of 6×10^4/ml in a modified MS liquid medium supplemented with NAA (2 mg/l), kinetin (0.1 mg/l) and 0.5 M sorbitol, and incubated in the dark at 28°C regenerated cell walls within 48 h. They underwent first division in 3–4 days and formed cell clumps and colonies in 10 days, which, when plated on an agar-solidified medium, developed into masses of calli. After transfer to an auxin-free liquid medium these calli underwent embryogenesis within the next 2 weeks and eventually developed into plantlets (Gosch et al. 1975). This work was extended to the mesophyll protoplasts obtained from leaves of anther-derived plantlets (Fig. 4A; Bajaj et al. 1978). For isolation, an enzyme mixture of 1.5% Cellulase and 0.3% Macerozyme gave an optimal protoplast yield of 1.3×10^6/g leaf material; however, in subsequent experiments the technique was refined and the amount of enzyme was considerably reduced to 0.5% Cellulase and 0.1% Macerozyme. These protoplasts isolated in low enzyme level (Fig. 4B) showed better growth response.

First division was noticed after 7 days of culture (Fig. 4C-E), and within 2 weeks small clumps of cells (Fig. 4F,G) were formed. At this stage dividing protoplasts and cell colonies were plated on agar medium of the same composition, and after 2–3 weeks colonies of callus were formed (Fig. 4H). When transferred to an auxin-free medium, within 2 weeks the callus started to differentiate, i.e., about 7–8 weeks from the time of incubation of protoplasts to the differentiation of plantlets (Fig. 4I; Bajaj et al. 1978).

The cultures exhibited three modes of morphogenesis, i.e., rhizogenesis, shoot-bud differentiation, and embryogenesis. Callus mostly differentiated to form only roots. In cultures where roots once started to initiate, the growth of callus was slowed down, and no shoot initiation was observed, while in others, on the agar medium shoots started to appear without forming any roots. In liquid cultures sometimes in addition to rhizogenesis the formation of embryos was also observed (Fig. 5A-C).

The enzymatically isolated mesophyll protoplasts subjected to freezing (-196°C) in the presence of cryoprotectants (DMSO, sucrose, mannitol) also retained their morphogenetic potential. A maximum of 44% cultures underwent somatic embryogenesis and organogenesis to form complete plants (Bajaj 1988).

The successful isolation and culture of protoplasts of *A. belladonna* resulted in studies on somatic hybridization by protoplast fusion. Krumbiegel and Schieder (1979, 1981) fused protoplasts of a tetraploid (4n = 48) chlorophyll-deficient mutant of *Datura innoxia* with the diploid (2n = 72) green protoplasts of *A. belladonna*, and selected 13 somatic hybrids. The chromosome number of these hybrids varied from 84 to 175. This work was successfully followed by Gleba et al. (1988), who reported somatic hybrids between *Nicotiana plumbaginifolia* and *A. belladonna*. These studies on somatic hybridization are very valuable if hybrid cell lines or plants capable of producing higher amounts of tropanes are isolated.

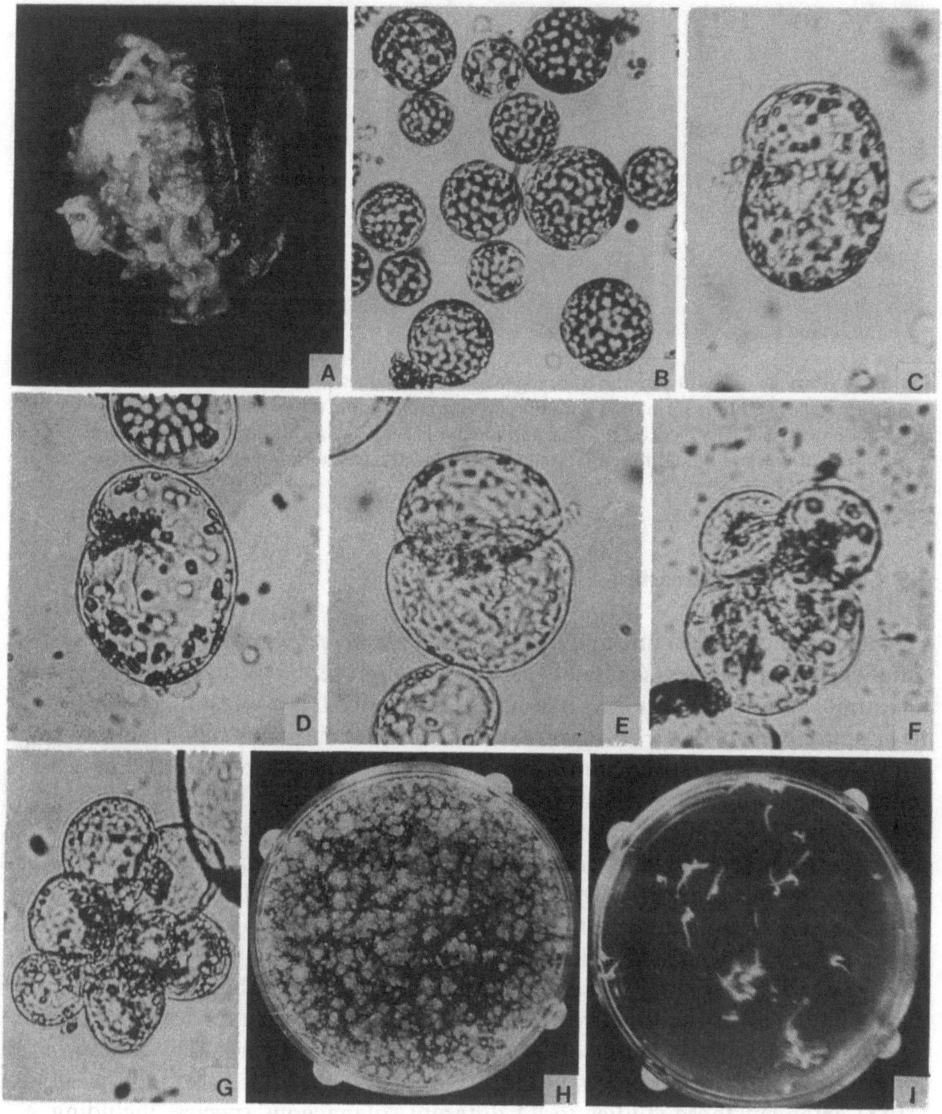

Fig. 4A-I. Isolation and culture of haploid mesophyll protoplasts of *Atropa belladonna*. **A** 5-week-old culture of an excised anther undergoing androgenesis (the leaves from the haploid plants thus raised were employed for the isolation of protoplasts). **B** Mesophyll protoplasts obtained from the peeled leaves incubated in an enzyme mixture containing cellulase (1.5%) and macerozyme (0.3%) in mannitol. **C-E** Various stages in the elongation and division of an isolated protoplast cultured in a liquid medium 5 days after isolation. **F,G** One- and 2-week-old cluster and colony of callus cells obtained from protoplasts. **H** Masses of calli obtained from cell colony (**F,G**) 3 weeks after plating on agar-solidified medium containing NAA. **I** Differentiation of embryoids and plantlets from calli after transferring to an auxin-free medium. (Bajaj et al. 1978)

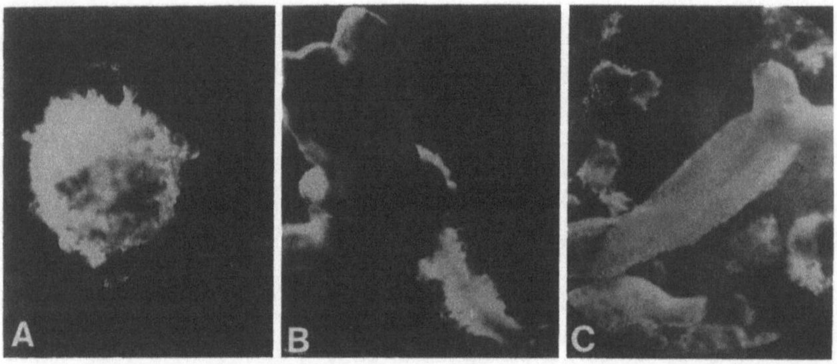

Fig. 5A-C. Various stages of somatic embryogenesis in protoplast-derived callus of *Atropa belladonna*. **A** An early stage of embryo formation in a callus transferred to liquid differentiation medium (NAA-free MS supplemented with 8% coconut water and kinetin 0.1 mg/l. **B,C** Maturing embryos from 2- and 6-week-old cultures in liquid differentiation medium; note the regeneration of a root in **B**. (Gosch et al. 1975)

2.5 Genetic Transformation

Genetic transformation through *Agrobacterium tumefaciens* has been reported in a number of plant species (see Bajaj 1989); however, recently the use of *A. rhizogenes* has attracted much attention as it induces hairy roots. It is now established that the Ri plasmids present in *A. rhizogenes* cause the transformation of cells by introducing T-DNA of the Ri plasmid into the genomic DNA of the plant cell. On a hormone-free medium these transformed cells form hairy roots. Kamada et al. (1986) induced hairy roots by inoculating stems of *A. belladonna* with *A. rhizogenes.* The axenic cultures of these hairy roots deriving from the stems proliferated 60-fold as based on the initial fresh weight after 1 month of culture. The amount of hyoscyamine and scopolamine in these cultures was the same or higher than in the normal plants grown in the field.

Jung and Tepfer (1987) observed considerable increase in biomass and alkaloid accumulation in root cultures by Ri T-DNA. Transformation improved root growth in agitated flasks by a factor of 24 in *A. belladonna,* and alkaloid content was improved in transformed roots vs. normal roots by 1.6-fold. Later, Knopp et al. (1988) extended these studies to 24 different solanaceous species, including *A. belladonna,* and reported that some of the root clones produced tropane alkaloids at levels similar to roots of the corresponding intact plants.

Genetic transformation is an area of research which is likely to yield dividends in terms of increased and efficient in vitro production of secondary metabolites not only in *A. belladonna* but in medicinal plants in general.

2.6 Cryopreservation of Cultures

Cryopreservation of in vitro cultures of medicinal and alkaloid-producing plants is an area of tremendous interest from the point of view of the pharmaceutical industry, and the subject has been reviewed by Bajaj (1988a). The significance of cryopreservation is threefold:

1. *Retention of Biosynthetic Potential in Cell Cultures.* In many cell cultures the amount of the desired metabolite that is formed is too little, or there is a decrease in biosynthetic capacity during serial subculture and transfer; thus cells are needed in large quantities. The cultures therefore have to be collected and stored until sufficient quantities are available for extraction purposes. The cultures preserved in liquid nitrogen (-196°C) retain their biosynthetic potential, therefore this method would be ideal for storage.
2. *Genetic Stability in High-Yielding Somaclones.* The cell cultures on periodical transfer undergo genetic erosions resulting in changes of varied nature. Most of these variation may not be of any use and therefore are discarded; however, sometimes mutant cell lines or high-yielding somaclones can be induced and isolated. Since the prime interest is to conserve the clones, for which there is no method that prevents genetic deterioration, cryopreservation holds great promise.
3. *Conservation of Germplasm.* There is an urgent need to conserve the germplasm of rare, elite, and endangered plants. Since entire plants can be regenerated from frozen cultures, cryopreservation will enable conservation of germplasm on a long-term basis.

In *A. belladonna*, complete plants have been regenerated from cryopreserved pollen embryos (Bajaj 1977), and from isolated protoplasts (Bajaj 1988). Following is the summary of published work:

2.6.1 Cryopreservation of Anthers and Pollen Embryos

The cryopreservation of anthers/pollen embryos would facilitate: mass production of haploids, maintenance of genetic stability of haploid cultures and conservation of germplasm.

The ampules containing suspensions of pollen embryos, or the androgenic anthers in cryoprotectant solution are subjected to freezing by different methods. After freezing the material is thawed by immediately immersing the ampule in warm water at 35–40°C. The survival/viability was judged by staining with fluorescein diacetate, or by the ability to grow, for which the retrieved material is washed with the culture medium, and recultured either in drops or plated on an agar medium. The ability to withstand freezing and the extent of survival of the frozen pollen embryos depends on their size and stage of development.

Freshly excised whole anthers of *A. belladonna* after sudden freezing and rapid thawing at 37°C occasionally survived and callused. The subsequent experiments were therefore conducted on the 4-week-old cultures of anthers undergoing androgenesis. The transversally cut halves of such anthers subjected to freezing continued androgenesis; however, better results were obtained with pollen-derived embryos. The globular embryos (Fig. 6) withstood freezing better than the late heart-shaped and fully differentiated embryos (Bajaj 1978). In cultures, the pol-

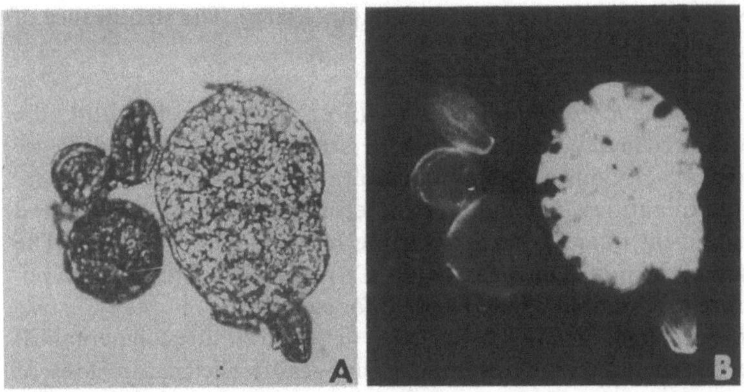

Fig. 6A,B. Survival of pollen embryos of *Atropa belladonna* frozen at -196°C in the presence of 5% DMSO (+4% sucrose in the medium): **A** Photographed in tungsten light. **B** Same, photographed in ultraviolet light; note the fluorescence of pollen embryos in **B** after staining with fluorescein diacetate. (Bajaj 1976)

len-derived embryos showed a lag phase proportional to their age and the extent of freeze injury. The frozen pollen-derived embryos eventually developed complete plants.

2.6.2 Cryopreservation of Protoplasts

The protoplasts differ from cell suspensions in two important aspects, (1) absence of cell wall, and (2) lack of cytoplasmic connections (plasmodesmata) with the adjacent cells. Thus protoplasts, by not being interdependent, are in fact a better system to study the mechanism of cryopreservation at the cellular level. The protoplasts, like the freeze preservation of cell suspension, can be employed for the storage and retention of biosynthetic potential and for genetic manipulation studies. There is no substantial literature on the freeze preservation of protoplasts of medicinal plants. Of the three alkaloid-producing plants studied (*Atropa, Datura, Nicotiana*), *A. belladonna* gave the best response (Bajaj 1988b). The enzymatically isolated mesophyll protoplasts were subjected to freezing in the presence of mannitol, sucrose, and dimethyl sulfoxide (DMSO), and subsequently thawed at 37°C. Initially the protoplasts lay quiescent for 3–5 days as compared to controls. After that they started to elongate, the chloroplasts became scanty and yellowish, and reorganized themselves at the periphery. The first division and the subsequent colony formation were invariably delayed; however, once they regenerated walls they looked like the control protoplasts. A maximum of 44% cultures showed differentiation of plantlets in callus. The morphogenetic events were not impaired and cultures showed both patterns of differentiation, as in control, embryogenesis, and organogenesis. The plants thus regenerated, when transferred to pots, continued to grow to form normal flowers and fruits. The seeds obtained from these plants were fertile, and in turn showed normal germination (Bajaj 1988b).

2.7 In Vitro Production of Tropane Alkaloids

The occurrence and chemistry of the tropane alkaloids have been recently reviewed by Lounasmaa (1988), and quite new aspects in their biosynthesis have been reported by Leete (1990). Earlier, the alkaloid composition of *Atropa belladonna* was considered to be relatively simple: the main constituents in intact plants being hyoscyamine, its N-oxide, scopolamine, and cuscohygrine (Phillipson and Handa 1976; Wilms et al. 1977). However, reinvestigation of roots of this species has resulted in the identification of many minor components (e.g., hygrine, hygroline, 6-hydroxyhyoscyamine) (Hartmann et al. 1986). The shoots revealed only seven alkaloids. Hygrines were completely absent but norhyoscyamine was frequently found in the aboveground parts of the plant. Many of these compounds occur as minor components or traces, and were not detected in cultured roots or suspension cultures.

Hyoscyamine and scopolamine (hyoscine) (Fig. 1), are esters formed by tropic acid (Fig. 2) and complex organic bases, either tropanol (tropine) or scopanol (scopine), respectively (Lounasmaa 1988). Although *A. belladonna* is considered one of our best-known medicinal plants, the biosynthetic pathway of tropane alkaloids is still somewhat obscure in this species.

The in vitro culture system has therefore been employed to understand more fully some of these mechanisms. There is a considerable literature on the production of tropane alkaloids in plant cell cultures, and the subject has recently been discussed (see Petri and Bajaj 1989). The work on *A. belladonna* is reviewed and summarized here (Table 2).

Studies on the in vitro production of alkaloids in *A. belladonna* were initiated in 1957 by West and Mika, who reported synthesis of atropine (hyoscyamine) in root callus, but did not detect any in stem and leaf callus. Excised root cultures of *A. belladonna* cultivar *lutea* contain hyoscyamine, scopolamine, tropine, and cuscohygrine (Figs. 1,2; Raj Bhandary et al. 1969). The roots in question, when subcultured every 21 days for more than 2 years and tested for alkaloid production, gave no positive reactions for hyoscyamine or scopolamine. However, they contained substances corresponding, in two-dimensional chromatography and color reaction, with Dragendorff's reagent to cuscohygrine, hygrine, tetramethylputrescine, tropine, and other Dragendorff-positive unidentified compounds (Thomas and Street 1970).

The conditions under which callus cultures of *A. belladonna* initiate roots and shoots and also which segments of their cultured excised roots can give rise to callus have been described (Raj Bhandary et al. 1969). Subsequently, several experiments were undertaken with the objective of inducing alkaloid production in cell suspension cultures of this species. The calli and suspension cultures were initiated from these cultured roots (Thomas and Street 1970). Aggregates from suspension cultures in SSM medium (with NAA) did not contain Dragendorff-positive compounds. Aggregates from cultures without NAA varied in alkaloid composition according to the frequency of the root formation.

In some cases Dragendorff-positive compounds corresponding in Rf to cuscohygrine, (Fig. 2), tropine (tropanol) and tetramethylputrescine (tetramethyldiaminobutane) could be detected, but in other cases, where more marked

Table 2. In vitro production of alkaloids in *Atropa belladonna*

Culture system employed	Observations/results	Reference
Root culture, callus	Synthesis of atropine[a] in root callus, no alkaloids were detected in stem and leaf callus	West and Mika (1957)
Root cultures of seedling and callus origin, shoots of callus origin	Atropine, scopolamine, tropine and cuscohygrine detected	Raj Bhandary et al. (1969)
Root-bearing aggregates, organ-free aggregates from suspension cultures with root formation	Seven Dragendorff-positive compounds including atropine and scopolamine	Thomas and Street (1970)
Callus	Tropine, atropine, and scopolamine detected	Eapen et al. (1978)
Callus cultures (stem origin), cell clumps, green embryoids	Quantitative estimation of total tropane alkaloids	Lindsey and Yeoman (1983)
Callus with roots, redifferentiated roots, plantlets	Quantitative estimation of total tropane alkaloids	Hamilton et al. (1986)
Root cultures	13 alkaloids from root cultures, their pattern was almost identical with that of roots from the intact plants	Hartmann et al. (1986)
Infected roots with *Agrobacterium rhizogenes*	Amount of atropine and scopolamine was higher in the hairy roots	Kamada et al. (1986)
Shoot cultures	Tropane alkaloids	Benjamin et al. (1987)
Root cultures infected with *A. rhizogenes*	Atropine, scopolamine, cuscohygrine detected	Jung and Tepfer (1987)
	Atropine and scopolamine detected	Knopp et al. (1988)
Callus culture	Studied the effect of various factors on alkaloid production in calli, synthesis of atropine in stem callus	Simola et al. (1988)
Cell suspension	Feeding experiments with precursors of tropane alkaloids (hygrine, tropinone, and tropic acid)	Simola et al. (1989)

[a] atropine = dl-hyoscyamine

organogenesis was reported, up to seven Dragendorff-positive compounds including hyoscyamine and scopolamine were found. Eluates of the bands at the Rf's of hyoscyamine and scopolamine gave IR spectra corresponding to these alkaloids (Thomas and Street 1970). A similar chromatographic picture was observed with extracts of aggregates from cultures without NAA containing either tropic acid or NOA. When organ-free aggregates from cultures showing profuse root formation were analyzed, the seven Dragendorff-positive compounds (including hyoscyamine) could be detected although they were present in smaller amounts than in the aggregates bearing visible roots.

Suspension cultures of *A. belladonna* seem to synthesize tropane alkaloids mainly in conjunction with developmental changes occurring during root formation. The developmental "switch" used is a change to a nutrient medium lacking growth regulators but containing low levels (10 mg/l) of tropic acid, a precursor of the main alkaloids (Hamilton et al. 1986). Reinvestigations by Hartmann et al. (1986) showed a total of 13 alkaloids, mostly from the roots. The alkaloid pattern of root cultures was almost identical with the pattern of roots from the intact plant. Scopolamine, which is synthesized in the roots only, was detected in the medium. Cell suspensions of this species are able to produce small amounts of norhyoscyamine and 6-hydroxyhyoscyamine, but not scopolamine when fed with hyoscyamine (Hartmann et al. 1986). Interestingly, cell suspensions produced trace amounts of 3-acetoxytropane and 3-tigloyloxytropane, which were not found in the intact plant.

In order to find callus lines of *A. belladonna* having high and stable alkaloid production and good growth rate on an inexpensive medium, the following research strategy (presented in a shorter form by Simola et al. 1988) was applied in an extensive study consisting of 156 callus cultures initiated from 500 stem or root explants:

1. Select good mother plants from different European seed samples, using as criterion the alkaloid level and composition of different plant organs (seeds, leaves, and roots).
2. Search for a possible relationship between alkaloid production of the mother plant and the respective callus lines.
3. Compare the callus growth and alkaloid levels of explants of the same plant, for the determination of somaclonal variation.
4. Search for possible influences of auxin level of the nutrient medium and the incubation temperature on alkaloid production.
5. Follow-up of the effect of callus age (number of passages) on the alkaloid production.

2.7.1 Gas Chromatographic Analysis of Hyoscyamine and Scopolamine

The alkaloids were extracted from lyophilized plant material and purified using disposable extraction columns (Baker-10 SPE). The main alkaloids (hyoscyamine and scopolamine) were analyzed using a fused silica capillary column in a gas chromatography instrument with flame ionization detector (FID) and helium as the carrier gas (Ylinen et al. 1986). Ketamine [2-(o-chlorophenyl)-2-(methylamino)-cyclo-hexanone] was used as the internal standard. Alkaloids were identified using gas chromatography-mass spectrometry.

The gas chromatographic method is relatively rapid because no derivatization is needed (Ylinen et al. 1986). It can be applied for screening the main alkaloids of *A. belladonna* having medicinal interest. For more detailed analysis of tropane alkaloids occurring at low concentration, capillary gas chromatography with a phosphorus-nitrogen-sensitive detector (cf. Hartmann et al. 1986) can be recommended because of its higher sensitivity and specificity.

2.7.2 *Alkaloid Concentration of Mother Plants*

Plants of *A. belladonna* were grown in fertilized peat in a greenhouse where day and night temperatures varied in the ranges 22–25 °C and 18–22 °C, respectively. The plants received supplementary light for 18 h daily (Simola et al. 1988).

The hyoscyamine concentration was usually much higher than that of scopolamine in leaf, root, and seed material (Tables 3, 4). Plants at the corresponding developmental stage and cultivated under the same greenhouse conditions showed considerable variation in hyoscyamine and scopolamine concentration, even when originating from the same seed sample. The wide variation in the alkaloid level of roots may reflect the difficulty in obtaining comparable samples, as there is much more variation in the morphology and anatomical structure of roots (especially the amount of xylem cells) than of leaves. Relatively

Table 3. Alkaloid contents of seed samples of *Atropa belladonna* of various origins. (Simola et al. 1988)

Sample[a] number	Hyoscyamine (g/kg)	Scopolamine (g/kg)
1	4.1	nd[b]
2	6.9	nd
3	6.5	nd
4	4.3	0.2
5	6.4	0.4
6	[c]	[c]
7	4.0	0.3
8	4.9	nd
9	1.8	0.1
10	4.8	0.3
11	5.6	0.5
12	1.2	0.1
13	3.9	nd
14	6.2	0.4
15	5.4	0.2
16	6.9	0.4

[a] 1. Italy, Reg. Marche (3), 2. Italy Reg. Marche (1), 3. Italy, Pesaro, 4. France, Mieussy, 5. France, Strasbourg, 6. France, Côte-d'Or, 7. Germany, Niedersachsen, 8. France, Nancy, 9. Germany, Bad Abbach, 10. Germany, Hessen, 11. Austria, Alpengarten Villacher Alp, 12. Germany, Hessen, 13. France, Doubs, 14. Hungary, Budakalasz, 15. Hungary (unknown), 16. Hungary, Budapest.

[b] Not detected.

[c] Small amount of seed did not permit a reliable analysis.

Table 4. Tropane alkaloid (hyoscyamine and scopolamine) content (g/kg dry wt) of leaves and roots of strains 6–16 and plants I-III of *Atropa belladonna*. (Simola et al. 1988)

Strain/plant[a]	Date[b]	Leaves		Date	Roots	
		Hyoscyamine	Scopolamine		Hyoscyamine	Scopolamine
6/I	13 Dec.	2.4	0.2	–	–	–
6/II	13 Dec.	3.2	nd[c]	–	–	–
7/I	19 Oct.	4.0	0.5	10 Jan.	2.6	0.7
7/II	10 Oct.	2.9	0.5	–	–	–
9/I	13 Dec.	4.2	0.2	19 Dec.	0.6	0.4
9/II	19 Oct.	4.9	0.2	10 Jan.	1.8	0.3
10/I	19 Oct.	2.0	0.3	19 Dec.	1.5	0.4
10/II	13 Dec.	2.2	0.1	10 Jan.	0.4	0.3
11/I	13 Dec.	2.8	0.2	10 Jan.	2.3	0.3
12/I	19 Oct.	3.4	0.3	10 Jan.	3.7	0.3
12/II	13 Dec.	1.1	0.3	21 Dec.	0.5	0.4
13/I	19 Oct.	1.7	0.3	10 Jan.	2.2	nd
15/I	13 Dec.	4.0	0.3	18 Dec.	1.5	0.9
15/II	19 Oct.	1.2	0.1	–	–	–
15/III	13 Dec.	0.5	0.2	10 Jan.	0.7	0.5
16/I	19 Oct.	1.0	nd	10 Jan.	0.5	nd
16/II	19 Oct.	1.3	0.1	–	–	–
16/III	13 Dec.	1.3	0.3	10 Jan.	1.8	nd

[a] Different mother plants are indicated by Roman numerals I, II, and III, given after the strain numbers (Arabic numerals 6–16), which correspond to the seed samples (Table 3).
[b] Time period: October 1984-January 1985.
[c] Not detected.

wide variation in alkaloid concentration was found in the seed samples as well. (Simola et al. 1988).

2.7.3 *Initiation, Cultivation, and Growth of Callus Cultures*

Young surface-sterilized stem and root pieces of the same plants from which alkaloids were analyzed (see above) were used as explants for callus cultures (Simola et al. 1988). These were transferred to a nutrient medium favorable to tissue cultures of *A. belladonna* (Simola and Sopanen 1971). This modified Wood and Braun's medium (for macronutrients, see Table 5) contained micronutrients and vitamins according to White (1943), 100 mg/l myo-inositol, 100 ml/l coconut milk, 20 g/l sucrose and, as growth regulators, 0.5 mg/l 1-naphthylacetic acid (NAA) and 0.1 mg/l kinetin. The pH was adjusted to 5.2, and the solidified (agar 6 g/l) medium was autoclaved. When this medium is used for suspension cultures it should be supplemented with $CuSO_4$ (0.1 μM) and Na_2MoO_4 (1 μM) (cf. Simola et al. 1989).

Each explant showing good callus growth was cultured separately and treated as one callus line in order to discover possible somaclonal variation (sensu lato). The callus was transferred to the same medium as in initiation but with the coconut milk omitted. Because the callus growth was not always effective, part of the lines were transferred to a medium containing a higher level of auxin (NAA 2 mg/l).

Callus lines (n = 156) could be initiated from the stems of 12 mother plants and from one young root, together representing nine European seed samples (Tables 3, 4). The growth of the lines derived from the same stem varied (Simola et al. 1988). If the induction of callus formation was frequent, poorly growing cultures were obtained as well. Although nearly all plants were able to produce one or a few very well-growing callus lines, some plants yielded several. The medium containing a relatively high auxin level (2 mg/l NAA) stimulated callus growth more than did the corresponding low-auxin (0.5 mg/l) medium.

Table 5. Macronutrient medium for callus cultures of *Atropa belladonna*. (Simola et al. 1988)

	mg/l
$NaNO_3$	1785
$Ca(NO_3)_2 \times 4\ H_2O$	235
$(NH_4)_2\ SO_4$	790
KCl	910
$MgSO_4 \times 7\ H_2O$	1740
$NaH_2PO_4 \times H_2O$	250
Na_2SO_4	190
NaFe(III)EDTA	8.3

Merck, analytical grade.

2.7.4 The Effect of Origin, Age, and Differentiation on Alkaloid Production

Variability in the alkaloid production in callus and suspension cultures of medicinal plants is a great problem. When rapid screening methods are lacking for alkaloid production at early stages of the callus growth, material must be cultured during several passages until enough is available for both subcultures and chemical analysis. In most cases the production of alkaloids is known to decline during subculturing.

Nearly all callus lines of *A. belladonna* gave a very strong positive Dragendorff's reaction in spot tests, and only a few of them could be discarded on this basis. The reaction, however, was not specific enough to be a good screening method (cf. Simola et al. 1988).

Several callus lines of *A. belladonna* having very low or no alkaloid production at the beginning of the cultivation (4th passage) produced 180–350 mg/kg dry weight hyoscyamine in the 7th to 9th passages, independent of the auxin level of the nutrient medium (Table 6). Alkaloid production declined rapidly after the 9th or 10th passage in nearly all of the lines. The highest alkaloid level (~ 0.9 g/kg dry wt.) was found in a callus line (9th passage) cultured on a medium having a low auxin level (NAA 0.5 mg/l, 25 °C) and in callus line (8th passage) originating from another plant (~ 0.8 g/kg) on a medium supplemented with a higher auxin level (NAA 2 mg/l, 25 °C) (Table 6). The mean content of hyoscyamine in callus lines of *A. belladonna* was smaller than in leaves and roots of intact plants, usually by a magnitude of 10 (Simola et al. 1988).

Table 6. Number of subculturable callus lines of *Atropa belladonna*[a], highest hyoscyamine concentrations obtained, and the corresponding growth. (Simola et al. 1988)

Strain/Plant	Organ	Number of explants	Number of lines	Passage	NAA[b] (mg/l)	Hyoscyamine (g/kg dry wt.)	Growth[c]
6/I	s	25	9	7	2.0	0.36	++
7/I	s	36	8	7	0.5	0.21	+++
7/II	s	42	6	9	0.5	0.34	++
9/I	s	60	17	4 and 9	0.5	0.30	++
9/II	s	25	3	7	2.0	0.42	++
10/I	s	42	11	4	2.0	0.30	++
10/II	s	25	8	9	0.5	0.88	++
11/I	s	25	14	8	2.0	0.80	++
12/II	s	60	12	4	2.0	0.25	+++
13/I	s	20	7	8	2.0	0.35	++
15/I	s	60	16	8	2.0	0.32	++
16/II	s	60	6	9	2.0	0.33	+++
7/I	r	20	1	4	0.5	0.30	++
Total		500	118				

[a] From stems (s) or roots (r) of 12 mother plants representing 9 strains.
[b] NAA = naphthylacetic acid.
[c] ++, 376–600 mg fresh wt. per callus piece; +++, > 600 mg fresh wt. per callus piece.

Both root and stem callus lines of *A. belladonna* could produce considerable amounts of hyoscyamine, but scopolamine was present in either undetectable or very low concentrations. Some earlier results have suggested that the synthesis of tropane alkaloids may be localized in roots and root callus in *A. belladonna* (West and Mika 1957). In electron micrographs, secondary products seem to be localized in vacuoles in the tips of roots differentiated from suspension cultures of *A. belladonna* (root origin) (Simola 1972). The results on stem callus lines instead support the idea that tropane alkaloids are produced in meristematic cells and ground tissues, not in a specific organ, although roots are the main site of synthesis in intact *A. belladonna* plants (Simola et al. 1988).

Greening suspension cultures of *A. belladonna* displaying shoot-like structures and bipolar embryoids have contained the highest tropane alkaloid concentration reported in the literature (up to 1.5 g/kg dry wt. cells) (Lindsey and Yeoman 1983). The determination of alkaloid content using the modified Dragendorff's reagent may, however, give too high a total alkaloid level.

2.7.5 *The Effect of Auxin Concentration and Temperature on Alkaloid Production*

The freeze preservation technique requires special equipment and skills. Where these are not available, slower growth rate of the calli, and thereby a lower mutation rate and greater stability of the callus lines, may be achieved by cultivating them at a lower though still effective temperature.

During the first nine passages the calli were incubated at 25 °C in the dark. Thereafter, about half of the cultures were transferred to 15 °C to determine whether time between passages could be lengthened as a result of the slower growth rate. When alkaloid production and growth of many callus lines of *A. belladonna* was followed from the 4th up to the 13th or 17th passages at 15 and 25°C on modified Wood and Braun's medium containing 0.5 or 2 mg/l of NAA (Table 7), it was found that the callus lines grew similarly on the two media but more slowly at 15°C. The time interval between passages thus was lengthened from 4 weeks (at 25°C) to 6 weeks. There was no general trend evident in the alkaloid production that could be correlated with the temperature or nutrient medium. Only a few of the lines were capable of synthesizing hyoscyamine after 13 or 14 passages (age about 1–1.5 years) (Simola et al. 1988).

Table 7. Effects of age, temperature, and auxin level on the average hyoscyamine content (g/kg dry wt.) of callus cultures of *Atropa belladonna* during subculturing. Number of lines analyzed in parentheses. (Simola et al. 1988)

Passage	NAA (0.5 mg/l, 25°C)	NAA (2 mg/l, 25°C)	NAA (0.5 mg/l, 15°C)	NAA (2 mg/l, 15°C)
4–5	0.14 (14)	0.14 (13)		
7	0.17 (5)	0.24 (4)		
8	0.13 (8)	0.27 (9)		
9	0.14 (14)	0.06 (10)		
10–11	0.04 (12)	0.02 (5)	0.03 (36)	0.03 (17)
12–13	0.02 (10)	0.04 (7)	0.04 (50)	0.03 (8)
14–15	0.04 (3)	0.01 (3)	0 (2)	0 (2)
16–17	0 (6)	0 (7)	0 (1)	0 (1)

In suspension cultures of *A. belladonna*, 1-naphthoxyacetic acid (NOA 2.5 mg/l) and tropic acid (10 mg/l), in combination with NAA (0.5 mg/l), were able to induce root formation, and alkaloid production was stimulated (Thomas and Street 1970). In a recent investigation with suspension cultures of this species, in contrast to large experiments with undifferentiated callus lines (Simola et al. 1988), alkaloid production was found only in association with rhizogenesis (Hamilton et al. 1986).

It can be concluded that the variation discovered in the alkaloid production of stem callus lines of *A. belladonna* is so great that the alkaloid level of the mother plant lacks any high value as the basis of selection. Good growth and alkaloid production were obtained in many callus lines cultured in the nutrient medium formulated for this purpose, and no negative correlation was observed between the callus growth and alkaloid production. Hyoscyamine synthesis was not affected by temperature or the NAA level of the nutrient medium, but it was slowly repressed in callus cultures of stem and root origin.

Synthesis of scopolamine was repressed in the callus cultures of *A. belladonna*, but some lines showed quite good hyoscyamine production. However, not all lines deriving from the same plant organ formed analyzable amounts of this alkaloid (Simola et al. 1988). The small number of callus lines in some earlier studies may be the reason for the contrary evidence that callus cultures deriving from shoots are unable to synthesize tropane alkaloids and that this ability is restricted to root tissue and calli originating from it (West and Mika 1957).

2.7.6 *Feeding Experiments with Alkaloid Precursors*

Limiting steps in the biosynthesis of tropane alkaloids have been studied using feeding experiments with some alkaloid precursors (hygrine and tropinone). The increased intracellular level of these compounds did not stimulate the synthesis of the subsequent intermediates or the end products (hyoscyamine and scopolamine). The results show indirectly that the enzymes evolved are neither constitutive nor easily inducible in suspension cultures with repressed alkaloid synthesis (Simola et al. 1989).

3 Summary and Conclusions

Atropa belladonna, a medicinal plant with a rich source of some tropane alkaloids has been extensively studied in vitro. Complete plants have been regenerated from callus cultures, excised anthers, and the isolated protoplasts. These plants showed somaclonal variation. Protoplast fusion has resulted in somatic hybrids with *Datura* and *Nicotiana* spp. The isolated protoplasts and pollen embryos cryopreserved in liquid nitrogen retained their morphogenetic potential. Genetic transformation of excised roots with *Agrobacterium rhizogenes* caused increased production of tropanes with hairy roots. Callus and suspension cultures of *A. belladonna* have been used as model systems of differentiation and nitrogen metabolism.

At present, tissue cultures of *A. belladonna* give much lower yields of medicinally important alkaloids than the intact plants grown in the field. Hairy root cultures are, however, a quite promising system, but the costs may be too high compared to leaf material harvested from out-of-door plantations. *A. belladonna* is able to produce the main alkaloids also under greenhouse conditions in winter (Simola et al. 1988), but extra light, heating, and watering increase the price considerably.

Better understanding of the regulation of the biosynthesis of tropane alkaloids in *A. belladonna* is needed for biotechnological applications. Unfortunately, the enzymes are poorly known and many of the alkaloid precursors are not commercially available and some of them are rather labile. Both induction and repression mechanism(s) of different steps in the biosynthetic pathway ought to be studied using different types of aseptic cultures.

References

Bajaj YPS (1974) Induction of repeated cell division in isolated pollen mother cells of *Atropa belladonna*. Plant Sci Lett 3:309–312

Bajaj YPS (1975) Formation of additional microspores in isolated tetrads of *Atropa belladonna* grown in microcultures. Z Pflanzenphysiol 75:464–466

Bajaj YPS (1976) Gene preservation through freeze storage of cell, tissue and organ cultures. Acta Hortic 63:75–84

Bajaj YPS (1977) Survival of *Atropa* and *Nicotiana* pollen embryos frozen at -196°C. Curr Sci 46:305-307

Bajaj YPS (1978) Effect of super-low temperature on excised anthers and pollen-embryos of *Atropa, Nicotiana,* and *Petunia*. Phytomorphology 28:171-176

Bajaj YPS (1988a) Cryopreservation and the retention of biosynthetic potential in cell cultures of medicinal and alkaloid-producing plants. In: Bajaj YPS (ed) Biotechnology in agriculture and forestry, vol 4. Medicinal and aromatic plants I. Springer, Berlin Heidelberg New York Tokyo, pp 169-187

Bajaj YPS (1988b) Regeneration of plants from frozen (-196°C) protoplasts of *Atropa belladonna* L., *Datura innoxia* Mill, and *Nicotiana tabacum* L. Indian J Exp Biol 26:289-292

Bajaj YPS (1989) Genetic engineering and in vitro manipulation of plant cells – technical advances. In: Bajaj YPS (ed) Biotechnology in agriculture and forestry 9. Plant protoplasts and genetic engineering II. Springer, Berlin Heidelberg New York Tokyo, pp 1-25

Bajaj YPS (1990) In vitro production of haploids and their use in cell genetics and plant breeding. In: Bajaj YPS (ed) Biotechnology in agriculture and forestry 12. Haploids in crop improvement I. Springer, Berlin Heidelberg New York Tokyo, pp 3-44

Bajaj YPS, Gosch G, Ottma M, Weber A, Gröbler A (1978) Production of polyploid and aneuploid plants from anthers and mesophyll protoplasts of *Atropa belladonna* and *Nicotiana tabacum*. Indian J Exp Biol 16:947-953

Bajaj YPS, Furmanowa M, Olszowska O (1988) Biotechnology of the micropropagation of medicinal and aromatic plants. In: Bajaj YPS (ed) Biotechnology in agriculture and forestry vol 4. Medicinal and aromatic plants I. Springer, Berlin Heidelberg New York Tokyo, pp 60-103

Benjamin BD, Roja PC, Heble MR, Chadha MS (1987) Multiple shoot cultures of *Atropa belladonna*: effect of physico-chemical factors on growth and alkaloid formation. J Plant Physiol 129:129-135

Eapen S, Rangan TS, Chadha MS, Heble MR (1978) Morphogenetic and biosynthetic studies on tissue cultures of *Atropa belladonna* L. Plant Sci Lett 13:83-89

Gleba YY, Hinnisdaeis S, Sidorov VA, Kaleda VA, Parokonny AS, Boryshuk NV (1988) Intergeneric asymmetric hybrids between *Nicotiana plumbaginifolia* and *Atropa belladonna* obtained by gamma fusion. Theor Appl Genet 76:760-766

Gosch G, Bajaj YPS, Reinert J (1975) Isolation, culture, and induction of embryogenesis in protoplasts from cell suspensions of *Atropa belladonna*. Protoplasma 86:405-410

Hamilton RM, Lang J, Pedersen H, Chin C-K (1986) Secondary metabolites from organized *Atropa belladonna* cultures. In: Scott CD (ed) 8th Symp biotechnology for fuels and chemicals. John Wiley, New York, pp 685-698

Hartmann T, Witte L, Oprach F, Toppel G (1986) Reinvestigation of the alkaloid composition of *Atropa belladonna* plants, root cultures, and cell suspension cultures. Planta Med 52:390-395

Jung G, Tepfer D (1987) Use of genetic transformation by the Ri T-DNA of *Agrobacterium rhizogenes* to stimulate biomass and tropane alkaloid production in *Atropa belladonna* and *Calystegia sepium* roots grown in vitro. Plant Sci 50:145-151

Kamada H, Okamura N, Satake M, Harada H, Shimomura K (1986) Alkaloid production by hairy root cultures in *Atropa belladonna*. Plant Cell Rep 5:239-242

Knopp E, Strauss A, Wehrli W (1988) Root induction on several Solanaceae species by *Agrobacterium rhizogenes* and the determination of root tropane alkaloid content. Plant Cell Rep 7:590-593

Konar RN, Thomas E, Street HE (1972) The diversity of morphogenesis in suspension cultures of *Atropa belladonna* L. Ann Bot 36:249-258

Krumbiegel G, Schieder O (1979) Selection of somatic hybrids after fusion of protoplasts from *Datura innoxia* Mill. and *Atropa belladonna* L. Planta 145:371-375

Krumbiegel G, Schieder O (1981) Comparison of somatic and sexual incompatibility between *Datura innoxia* and *Atropa belladonna*. Planta 153:466-470

Leete E (1990) Recent developments in the biosynthesis of the tropane alkaloids. Planta Med 339-352

Lindsey K, Yeoman MM (1983) The relationship between growth rate, differentiation and alkaloid accumulation in cell cultures. J Exp Bot 34:1055-1065

Lounasmaa M (1988) Tropane alkaloids. In: Brossi A (ed) The alkaloids, vol 33. Academic Press, New York, pp 1-81

Mathur AC, Anand M (1964) Pests of medicinal plants. In: Entomology in India. ESI, New Delhi, pp 273-277

Nag KK, Street HE (1975) Freeze preservation of cultured plant cells. II. The freezing and thawing phases. Physiol Plant 34:261-265

Narayanaswamy S, George (née Chandy) L (1972) Morphogenesis of belladonna (*Atropa belladonna* L.) plantlets from pollen in culture. Indian J Exp Biol 10:382–384
Nyman S, Simola LK (1988) The uptake of phenylalanine into suspension-cultured cells of *Atropa belladonna*. Physiol Plant 74:623–630
Petri G, Bajaj YPS (1989) *Datura* spp.: In vitro regeneration and the production of tropanes. In: Bajaj YPS (ed) Biotechnology in agriculture and forestry 7. Medicinal and aromatic plants II. Springer, Berlin Heidelberg New York Tokyo, pp 135–161
Phillipson JD, Handa SS (1976) Hyoscyamine N-oxide in *Atropa belladonna*. Phytochemistry 15:605–608
Raj Bhandary SB, Collin HA, Thomas E, Street HE (1969) Root, callus, and cell suspension cultures from *Atropa belladonna* L. and *Atropa belladonna* cultivar lutea Döll. Ann Bot 33:647–656
Rashid A, Street HE (1973) The development of haploid embryos from anther cultures of *Atropa belladonna* L. Planta 113:263–270
Rashid A, Street HE (1974) Growth, embryogenic potential, and stability of a haploid cell culture of *Atropa belladonna* L. Plant Sci Lett 2:89–94
Salonen M-L (1980) Glutamate- and aspartate-derived amino acids as nitrogen sources for the callus of *Atropa belladonna* L. Ann Bot Fennici 17:357–378
Salonen M-L (1984) Influence of ammonium and nitrate nutrition on levels of ammonium, nitrate, and total nitrogen in the callus of *Atropa belladonna* (Solanaceae). Ann Bot Fennici 21:367–381
Salonen M-L, Simola LK (1977) Dipeptides and amino acids as nitrogen sources for the callus of *Atropa belladonna*. Physiol Plant 41:55–58
Salonen M-L, Simola LK (1989) Effect of nitrate, ammonium and some amino acids on growth and nitrate reductase activity in suspension cultures of *Atropa belladonna*. Plant Cell Physiol 30:1177–1181
Simola LK (1972) Changes in the ultrastructure of cells of *Atropa belladonna* cv. *lutea* Döll during growth and differentiation in suspension culture. Z Pflanzenphysiol 68:215–227
Simola LK (1973a) Changes in the activity of several enzymes during root differentiation in cultured cells of *Atropa belladonna*. Z Pflanzenphysiol 68:373–378
Simola LK (1973b) Development of chloroplasts in intact *Atropa belladonna* and in stem callus cultures during greening and leaf differentiation. Ann Acad Sci Fenn A, IV Biologica 196:1–10
Simola LK, Sopanen T (1971) Effect of α-naphthalene- and α-naphthoxyacetic acid on the activity of certain enzymes of *Atropa belladonna* cv. *lutea* cells in suspension culture. Physiol Plant 25:8–15
Simola LK, Nieminen S, Huhtikangas A, Ylinen M, Naaranlahti T, Lounasmaa M (1988) Tropane alkaloids from *Atropa belladonna*, Part II. Interaction of origin, age, and environment in alkaloid production of callus cultures. J Nat Prod 51:234–242
Simola LK, Martinsen A, Huhtikangas A, Jokela R, Lounasmaa M (1989) Feeding experiments with precursors of tropane alkaloids using suspension cultures of *Atropa belladonna*. Acta Chem Scand 43:702–705
Thomas E, Street HE (1970) Organogenesis in cell suspension cultures of *Atropa belladonna* L. and *Atropa belladonna* cultivar *lutea* Döll. Ann Bot 34:657–669
Thomas E, Street HE (1972) Factors influencing morphogenesis in excised roots and suspension cultures of *Atropa belladonna*. Ann Bot 36:239–247
Vakkari M-L (1980) The effect of methionine on the growth, ethylene production and phenolic compounds of *Atropa belladonna* L. callus culture. Ann Bot Fennici 17:406–409
Weiner N (1985) Atropine, scopolamine, and related antimuscarinic drugs. In: Gilman AG, Goodman LS, Rall T, Murad F (eds) The pharmacological basis of therapeutics. Macmillan, New York, pp 130–144
West Jr FR, Mika ES (1957) Synthesis of atropine by isolated roots and root-callus cultures of belladonna. Bot Gaz 119:50–54
White PR (1943) A handbook of plant tissue culture. Ronald Press, New York
Wilms J, Röder E, Kating H (1977) Gaschromatographische Bestimmung von Tropan-Alkaloiden in Organen von *Atropa belladonna*. Planta Med 31:249–256
Ylinen M, Naaranlahti T, Lapinjoki S, Huhtikangas A, Salonen M-L, Simola LK, Lounasmaa M (1986) Tropane alkaloids from *Atropa belladonna*; Part I. Capillary gas chromatographic analysis. Planta Med 52:85–87
Zenkteler M (1971a) Development of new plants from leaves and roots of *Atropa belladonna* L. in the in vitro culture. Acta Soc. Bot Pol 40:305–313
Zenkteler M (1971b) In vitro production of haploid plants from pollen grains of *Atropa belladonna* L. Experientia 27:1087

II *Ageratina adenophora* (Spreng.) K. & R.: In Vitro Culture and the Production of Secondary Metabolites

P. PROKSCH[1]

1 Introduction

The family Asteraceae comprises some 20,000 species and is the second largest family of higher plants (Heywood et al. 1977). Considering the size of this family it is interesting to note that relatively few species are commercially exploited as food plants, whereas a comparatively large number is known to accumulate poisonous natural products such as sesquiterpene lactones that are cytotoxic or elicit contact allergy or liver toxic pyrrolizidine alkaloids (Wagner 1977). In evolutionary terms, the Asteraceae is still a young plant family that has nevertheless successfully spread worldwide (Turner 1977). There is now increasing evidence suggesting that the accumulation of toxic natural products forms a major ecological contribution to the successful radiation of the Asteraceae as well as of other families, enabling plants to fight off herbivors or pathogenic microorganisms (Rosenthal and Janzen 1979). In search of new sources of potentially useful natural products that may become important as key structures for designing new drugs or agrochemicals, it is appropriate to focus on those species that are the most vigorous and have successfully adapted to different habitats, since they can be expected to contain an effective arsenal of interesting biologically active compounds.

A representative of this former group of promising species from the Asteraceae is *Ageratina adenophora* (Spreng.) K. & R. (tribe Eupatorieae) (Robinson and King 1977) (Fig. 1) (common name "jediondo" or "hierba negra" in Spanish). Whereas the core of *Ageratina* is restricted to the western parts of the Americas, *A. adenophora* is widely adventive in many tropical and subtropical parts of the world (King and Robinson 1970). The weedy character, vigorous growth, and dispersal of this species cause problems, for example on several of the Canary islands, where this species had been introduced in the last century and now represents a threat to several of the autochtonous species.

Previously we could show that seedlings of *A. adenophora* accumulate several chromene derivatives (Fig. 2,*1*–*3*) but no benzofurans (Proksch et al. 1986). Chromenes as well as structurally related benzofurans (e.g., Fig. 2,*6*) are frequently encountered in species of the Asteraceae especially from the tribes Astereae, Eupatorieae, Heliantheae, Inuleae, or Senecioneae (Proksch and Rodriguez 1983; Proksch 1985). Most of the compounds known so far have been isolated from species that grow under arid or semi-arid conditions like *A. adenophora*. Encecalin (Fig. 2,*3*), as well as demethoxyencecalin (Fig. 2,*1*), is toxic to an array of insects

[1] Institut für Pharmazeutische Biologie, TU Braunschweig, Mendelssohnstr. 1, W-3300 Braunschweig, FRG

Biotechnology in Agriculture and Forestry, Vol. 15
Medicinal and Aromatic Plants III (ed. by Y.P.S. Bajaj)

Fig. 1. *Ageratina adei*

(including for example the migratory grasshopper *Melanoplus sanguinipes*, the variegated cutworm *Peridroma saucia*, or the milkweed bug *Oncopeltus fasciatus*), and they are assumed to have evolved as an effective chemical defense of plants against herbivorous insects (Isman and Proksch 1985; Proksch et al. 1987a). The toxicity of chromenes is not restricted to insects only, since encecalin (Fig. 2,*3*), for example, was also shown to be phototoxic to several bacteria and yeasts in long-wave UV light (Proksch et al. 1983). In comparison to chromenes like encecalin, the structurally related benzofurans seem devoid of insecticidal activity (Isman et al. 1987). Several benzofurans, however, have been reported to be bacteriostatic, toxic in a goldfish bioassay, phototoxic to several bacteria and yeasts, or to exhibit antitumor activity (Zalkow et al. 1962, 1979; Bonner and de Graw 1962; Wu et al. 1973; Proksch et al. 1983). Chromenes and benzofurans thus represent a group of natural products exhibiting a wide array of interesting biological activities. Very little, however, is known of the physiology and biochemistry of chromenes and benzofurans in plants. This may be partly due to problems arising in cultivating suitable plants under experimental conditions. We have now established and studied cell suspension cultures of *A. adenophora* as a model in vitro system to gain insight into the metabolism of chromenes and benzofurans and into processes leading to the induction of chromene and benzofuran accumulation.

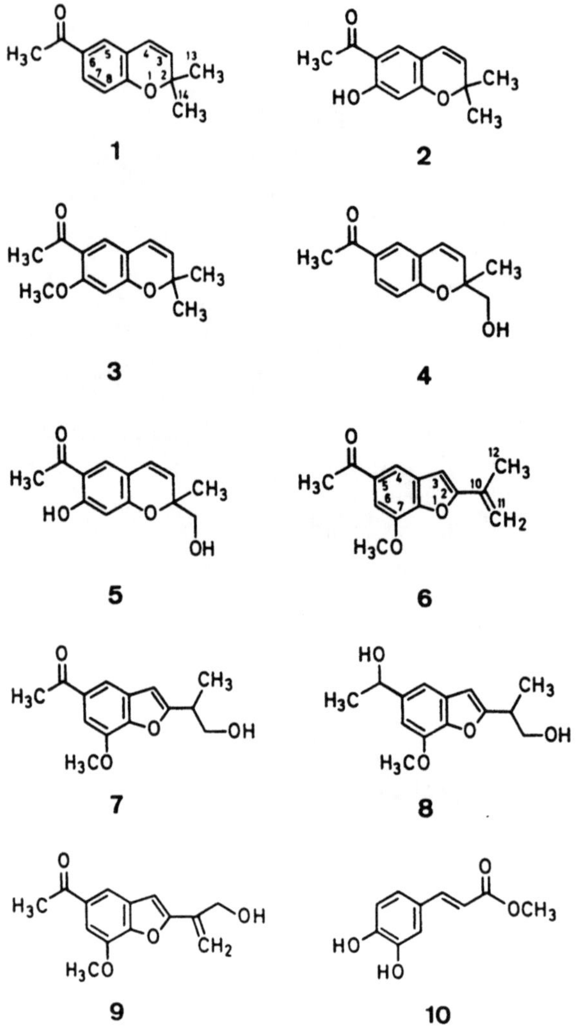

Fig. 2. Compounds from seedlings (*1–3*; *1* = demethoxyencecalin; *2* = demethylencecalin; *3* = encecalin) and cell suspension cultures (*4–10*) of *Ageratina adenophora*. (Proksch et al. 1987b)

2 In Vitro Approaches

2.1 Establishment and Growth of Cell Suspension Culture

Prior to our work no information on chromene- or benzofuran-producing cell suspension cultures either from *Ageratina* or from any other plant sources was available. The cell suspension culture of *A. adenophora* was established from young sterile seedlings grown from achenes. The cultures were cultivated according to Murashige and Skoog (1962) except for the concentrations of 2,4-D and kinetin, that were changed to 0.033 and 0.25 mg/l, respectively. The culture has been constantly kept in vitro for more than 3 years now (22 °C, 75% relative humidity, 120 rpm) Proksch et al. 1987b).

For optimum growth of the cell suspension culture, illumination (16 h daily) was found necessary (Fig. 3). Cultures grown under this light/dark rhythm reached the stationary growth phase approximately 12 days after inoculation of the cells into fresh medium. When the cultures were kept in continuous dark the growth was very slow within the first 10 days after inoculation. The stationary growth phase was reached at approximately 17 days and the total increase of biomass produced was always less than in the cultures grown under a light/dark rhythm. This growth-retarding effect also became apparent when cell cultures first grown under a light/dark rhythm were subsequently transferred to the dark (as shown in Fig. 3 for the third day after inoculation). Even though the growth curve of the latter cultures resembled that of those grown in the light, the total increase of biomass was always smaller than in those grown under a light/dark rhythm.

2.2 Accumulation of Chromenes and Benzofurans

The liquid growth media, as well as cell extracts of cell suspension cultures of *A. adenophora*, was analyzed for chromenes or benzofurans by HPLC and GC-MS during all growth stages. Whereas none of the chromenes accumulating in seedlings of *A. adenophora* could be detected as cell culture metabolites, several new compounds not accumulating in seedlings were identified. These compounds included the 2-methyl-2-hydroxymethylderivative (Fig. 2,*4*) of the seedling chromene demethoxyencecalin Fig. 2,*1* as well as several benzofurans (Fig. 2,*6*, *7*, *8*) and a caffeic acid derivative (Fig. 2,10). The benzofurans were new natural products and were unambiguously identified from their ^{1}H NMR and mass spectra (Table 1) (Proksch et al. 1987b). The concentrations of both the chromenes and the benzofurans from the cell suspension culture were small, reaching maxima at or below 0.05 μmol/g fresh weight accumulating mainly in the cells. These concentrations were approximately two orders of magnitude smaller than those of chromenes (Fig.

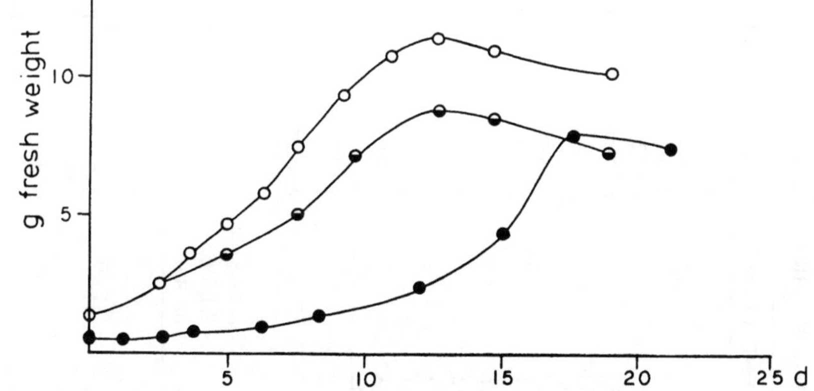

Fig. 3. Growth curve of cell suspension culture of *Ageratina adenophora*. ○ grown under light/dark rhythm; ◒ transferred to dark on day 3; ● grown in continuous dark

Table 1. ^{1}H NMR and MS data of chromenes and benzofurans. (Proksch et al. 1987b)

^{1}H NMR (Chemical shifts are relative to TMS)

	5	6	7	9
H-3	d 5.61	s 6.68	s 6.58	s 6.83
H-4	d 6.52	d 7.47	d 7.44	d 7.48
H-5	s 7.54	–	–	–
H-6	–	d 7.78	d 7.76	d 7.78
H-7	–	–	–	–
H-8	s 6.27	–	–	–
H-10	–	–	m 3.23	–
H-11	–	m 5.87, 5.23	m 3.89	m 6.02, 5.53
H-12	–	m 2.13	d 1.4	s 3.65
H-13	s 1.40	–	–	–
H-14	d 3.62, 3.57	–	–	–
CH_3CO	s 2.59	s 2.64	s 2.64	s 2.65
OMe	–	s 4.05	s 4.04	s 4.05
	$J_{3,4}$ = 10 $J_{14a,b}$ = 11	$J_{4,6}$ = 1.5	$J_{4,6}$ = 1.5 $J_{10,12}$ = 7	$J_{4,6}$ = 1.5

MS

5 m/z	R.I. 1955 rel. int.	6 m/z	R.I. 1937 rel. int.	7 m/z	R.I. 2125 rel. int.	8 m/z	R.I. 2085 rel. int.	9 m/z	R.I. 2208 rel. int.
234	6	230	65	248	33	250	27	246	75
203	100	215	100	233	6	235	4	231	100
185	17	187	49	217	100	220	11	203	34
43	6	172	3	203	4	219	100	115	6
		128	6	187	3	201	10	43	7
		115	12	174	8	175	9		
		43	7	43	11				

Numbering of compounds follows Fig. 2; R.I. = retention indices.
^{1}H NMR spectra were recorded at 300 or 400 MHz in $CDCl_3$.

2,*1*–*3*) in seedlings of *A. adenophora* (Proksch et al. 1986). The highest concentrations of the chromenes and the benzofurans in the cell suspension culture (0.05 μmol/g fresh weight) were usually found within the first 4 days after inoculation of the cells into fresh medium, followed by a gradual decline during further growth of the cell culture. This stimulation of natural product accumulation caused by inoculation of the cells into fresh medium has also been observed for other systems like cell suspension cultures of *Petroselinum* (Hahlbrock and Schröder 1975) or of *Catharanthus* (Knobloch et al. 1981). Whereas light had a pronounced effect on the growth of the cell suspension culture of *A. adenophora*, no significant differences in chromene or benzofuran concentrations were found between light- or dark-grown cultures.

2.3 Biotransformation of Chromenes

After establishing that none of the seedling chromenes was present as cell suspension culture metabolites, the biochemical capacity of the cell culture for the formation of biotransformation products from the seedling chromenes was explored. Since encecalin (Fig. 2,*3*) and demethoxyencecalin (Fig. 2,*1*) were found to give rise to unknown artifacts when exposed to light, incubation of the cell cultures with the chromenes (*1*–*3*) (1 μmol per flask, 40 μM) was carried out in the dark. Cell cultures that had grown under a light/dark rhythm were transferred to the dark on day 5 following inoculation of the cells into fresh medium. After 24 h the chromenes were applied solubilized in ethylene glycol monoethyl ether (EGME), resulting in a final concentration of EGME of 2.5% compared to the growth medium. Controls were treated only with EGME. At fixed time intervals controls and cell cultures treated with chromenes were harvested. Growth media and cell extracts were separately analyzed by HPLC for the applied chromenes and for biotransformation products.

The cell suspension cultures of *A. adenophora* were characterized by a rapid biotransformation of seedling chromenes (Fig. 2,*1* and *2*). Two to three hours after feeding of demethoxyencecalin (*1*) no substrate was detected in cells or growth media (Fig. 4), whereas a more polar biotransformation product (Fig. 2,*4*) accumulated in the cell cultures treated with demethoxyencecalin. This biotransformation product was isolated and identified as the 2-methyl-2-hydroxymethylderivative of *1*, present in the controls only as a trace constituent. The biotransformation was both rapid and effective, with a yield of roughly 80% of the applied dose of *1*. Approximately 70–80% of *4* was found in the growth media.

The biotransformation of *2* proceeded even faster than that of *1* (Fig. 5). Only 1 h after feeding no demethylencecalin (*2*) could be detected in growth media or cell extracts, whereas a more polar biotransformation product (Fig. 2,*5*) accumulated. This product was isolated and identified as the 2-methyl-2-hydroxymethylderivative of *2* based on its ^{1}H NMR and mass spectrum. The yield of this biotransformation reaction was comparable with that of the biotransformation of *1* at roughly 80% of the applied dose. Again the largest part of *5* was found in the growth media.

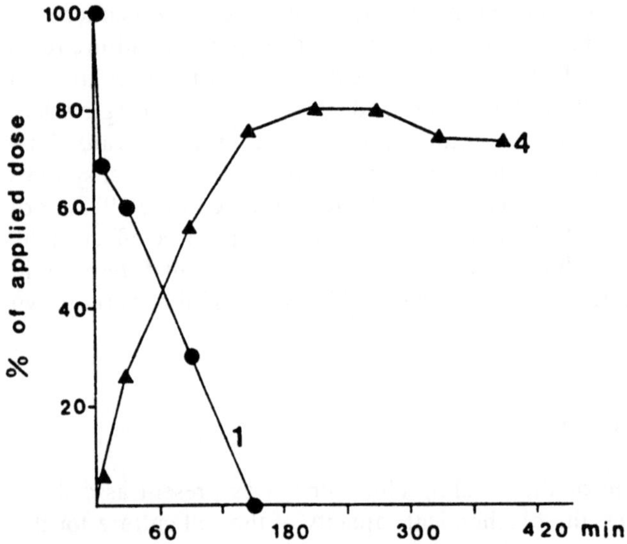

Fig. 4. Biotransformation of demethoxyencecalin (Fig. 2, 1) yielding (Fig. 2, 4). ● = compound *1*; ▲ = compound *4*. (Proksch et al. 1987b)

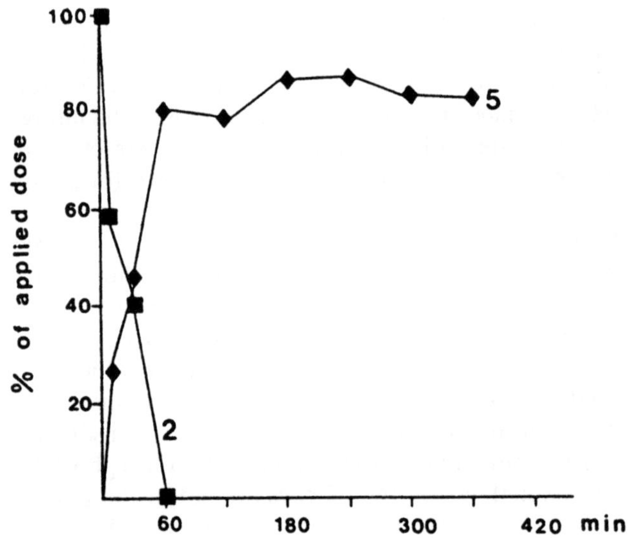

Fig. 5. Biotransformation of demethylencecalin (Fig. 2, 2) yielding (Fig. 2, 5). ■ = compound *2*; ♦ = compound *5*. (Proksch et al. 1987b)

The hydroxylations of *1* and *2* did not proceed in preconditioned growth media (media from which growing cells had been removed), indicating that the presence of cells rather than secreted enzymes was responsible for the biotransformations of the seedling chromenes. Also, the biotransformations were not observed when feeding *1* or *2* to cell suspension cultures derived from plants that do not accumulate chromenes or benzofurans such as *Atropa belladonna* or *Solanum triflorum* (both Solanaceae). This suggests that the observed hydroxylations of 2,2-dimethylchromenes are not due to unspecific enzymatic reactions, even though the hydroxylations do not reflect the major biogenetic fate of chromenes *1* and *2* in seedlings of *A. adenophora*. It was demonstrated earlier (Proksch et al. 1986) that seedlings convert *1* to *2* which, in turn, is methylated, giving rise to *3*. Recently, however, we isolated a β-glucoside of 2-methyl-2-hydroxymethyldemethoxyencecalin (*4*) from leaves of the related species *Ageratina altissima* (Proksch et al. 1988), which proves that hydroxylation of the geminal methyl groups of a chromene molecule may occur also in the bioconversion of chromenes in differentiated plants of *Ageratina*.

2.4 Induction of Benzofuran Accumulation

2.4.1 *Induction by Encecalin*

Feeding of the third seedling chromene encecalin (*3*) to cell suspension cultures of *A. adenophora* resulted in a metabolism strikingly different from that observed for feeding of demethoxyencecalin (*1*) or demethylencecalin (*2*) (Fig. 6). Whereas the concentration of encecalin added to the cell culture flasks stayed more or less stable

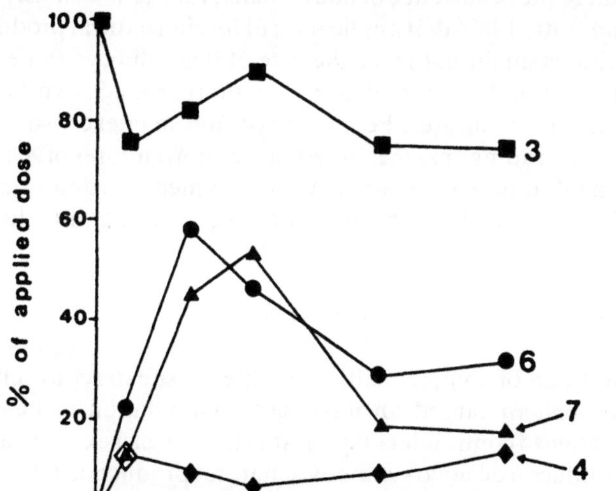

Fig. 6. Induced accumulation of benzofuran and chromene derivatives following feeding of encecalin (Fig. 2, 3). ■ = compound *3*; ● = compound *6*; ▲ = compound *7*; ♦ = compound *4*. (Proksch et al. 1987b)

at roughly 80% of the applied dose (40 μM), two prominent peaks (*6* and *7*) were detected in the growth media and cell extracts by HPLC shortly after addition of encecalin to the cell suspension cultures. Measured by peak area, the new compounds came close to the concentration of encecalin in the cell culture flasks, whereas they were barely detectable in the controls. Considering the small decrease of encecalin in the cell cultures it was obvious that the respective new compounds could not arise by biotransformations of encecalin. Compounds Fig. 2, *6* and *7* were subsequently isolated and identified by ^{1}H NMR and GC-MS (Table 1) (Proksch et al. 1987b) as novel benzofuran derivatives present also as trace components in the controls, whereas benzofurans are lacking in seedlings of this species. The accumulation maxima of the benzofurans *6* and *7* were found between 2 and 3 h after feeding of encecalin followed by a gradual decline. About 80% of the total concentration of both benzofurans was detected in the growth media, the rest residing in the cells. Two additional compounds including 2-methyl-2-hydroxymethyldemethoxyencecalin (*4*) and a further new benzofuran (Fig. 2,*8*) showed a slightly increased accumulation in presence of encecalin when compared to the controls. The structure of the latter was assigned by GC-MS and by comparison with a derivative of *7* obtained by treatment of *7* with $LiAlH_4$ in diethyl ether.

At present we do not fully understand the biochemical mechanisms leading to such a fast accumulation of benzofurans in cell suspension cultures of *A. adenophora* treated with encecalin. The time gap between application of encecalin and the increased accumulation of the benzofurans in the media is almost too short to account for a de novo enzymatic synthesis of the benzofurans. A possible liberation of compounds *6* and *7* (Fig. 2) from bound precursors (e.g., glycosides) induced by encecalin may probably also be dismissed, since acidic and alkaline hydrolysis of cell culture extracts or media from cell suspension cultures of *A. adenophora* gave no hint of the occurrence of bound benzofurans. Previously it was hypothesized (Wink and Witte 1983) that application of foreign natural products to plant cell suspension cultures might influence the rate of degrading enzymes, resulting in a short-term increase of the accumulation of natural products synthesized by the respective cell suspension cultures. Perhaps encecalin may exert such an influence on benzofuran-degrading enzymes in cell suspension cultures of *A. adenophora* (degradation of benzofurans does occur in the growth media as described in the following section). Future research will be necessary to gain insight into this interesting phenomenon.

2.4.2 Induction by Yeast Extract

Addition of commercially available yeast extract to cell suspension cultures of *A. adenophora* caused an increased accumulation of benzofurans *7* and *9* (Fig. 2) relative to controls lacking yeast extract. Yeast extract has repeatedly been reported to induce a de novo synthesis of natural products in various cell suspension cultures. Examples include an induced isoflavone and pterocarpan accumulation in cell suspension cultures of *Cicer arietinum* (Keßmann and Barz 1987) or an induced accumulation of berberine in cell suspension cultures of *Thalictrum rugosum* (Funk et al. 1987). The accumulation of benzofurans by cell suspension cultures of *Ageratina adenophora* treated with yeast extract was studied in detail (Table 2).

Table 2. Induction of benzofuran accumulation (7/9) by yeast extract. (Monir and Proksch 1989)

Time of addition of yeast extract (in days following inoculation of cells into fresh medium)		Accumulation of benzofurans 7/9 at 6 h	24 h	48 h
		following addition of yeast extract		
		μmol/g	μmol/g	μmol/g
3	+ yeast extract	0.038	0.117	0.019
	– yeast extract	0.040	0.085	0.024
6	+ yeast extract	0.047	0.120	0.040
	– yeast extract	0.053	0.038	0.024
12	+ yeast extract	0.028	0.047	0.014
	– yeast extract	0.031	0.048	0.024

The dose of yeast extract applied was 2 g/l; μmol/g are given on a fresh weight basis. Numbers of benzofurans refer to Fig. 2.

To prevent the formation of artifacts of the benzofurans in the light, all experiments employing yeast extract were carried out in the dark. Cell suspension cultures that had grown under the light/dark rhythm were transferred to the dark 24 h before addition of yeast extract. Preliminary experiments had indicated that a concentration of 2 g of yeast extract/1 growth medium was effective in inducing an increased accumulation of benzofurans in cell suspension cultures of *A. adenophora* relative to controls lacking yeast extract (Proksch et al. 1987b). This concentration of yeast extract was comparable, for example, to that used for the induction of isoflavone and pterocarpan accumulation in cell suspension cultures of *Cicer arietinum* (Keßmann and Barz 1987). Yeast extract (2 g/l) was subsequently applied to cell suspension cultures of *A. adenophora* at growth stages 3, 6 and 12 days following inoculation of the cells into fresh medium. Cells and growth media were harvested at time intervals of 6, 24, and 48 h following addition of yeast extract, analyzed by HPLC for benzofurans, and compared to controls harvested at the same time intervals (Table 2). Since the benzofurans *7* and *9* (Fig. 2) could not be resolved by HPLC, they were quantified jointly as benzofurans 7/9. The cell suspension culture showed the highest induction of benzofuran accumulation when yeast extract was added at growth stage 6 days, followed by growth stage 3 days. Growth stage 12 days was ineffective. The optimum accumulation of benzofurans at all growth stages was usually reached 24 h after addition of the yeast extract followed by a gradual decline. The decrease of benzofurans especially in the growth media 48 h after addition of yeast extract (compared to the concentration at 24 h) was also observed when growth media of cells that had received yeast extract 24 h before were separated from the cells, incubated for another 24 h, and then analyzed for benzofurans by HPLC. Therefore, the decrease in benzofuran concentration cannot be ascribed to an uptake of the benzofurans by the cells, but is assumed to be due to degradation by enzymes present in the growth media.

The stimulatory effect of yeast extract on the accumulation of benzofurans by cell suspension cultures of *A. adenophora* could be enhanced by an increase of the dose of yeast extract applied (Table 3). The highest increase of benzofurans relative to controls was found in this experiment when a concentration of 6 g yeast extract/l of growth medium was added to the cell suspension cultures at growth stage 6 days. Concentrations of yeast extract higher than 6 g/l resulted, in comparison, in a decrease of benzofuran accumulation.

When yeast extract was added at growth stage 9 days to cell suspension cultures that had already received a first dose of yeast extract at growth stage 6 days, a second increase of benzofurans relative to the controls was observed (Table 4). This second increase of benzofurans was always smaller than the first at stage 6 days, but comparable to that of cultures receiving their first dose of yeast extract at growth stage 9 days. The induction of benzofuran accumulation in cell suspension cultures of *A. adenophora* by yeast extract could thus be repeated and was dependent only upon the growth stage of the culture at the time of addition of yeast extract.

Employing root cultures of *Eupatorium cannabinum* (Asteraceae) we showed recently that the aromatic ring and acetyl function of benzofurans (C_6C_2 part of the molecule) originate biosynthetically from phenylalanine and cinnamic acid as precursors (Siebertz et al. 1989). Aminooxyacetic acid (AOA) is known as a competitive inhibitor of the phenylalanine-ammonialyase (PAL) (Amrhein 1979). Joint application of yeast extract and AOA to cell suspension cultures of *A. adenophora* resulted in a smaller accumulation of benzofurans *7/9* (Fig. 2) when compared to cell suspension cultures that had been treated with yeast extract only (Table 5). In a further experiment L-(U^{14}C)-phenylalanine (2 μCi, 0.5 mM) was applied along with yeast extract (6 g/l) to cell suspension cultures of *A. adenophora* at growth stage 6 days. Controls were treated with the ^{14}C-phenylalanine dose only. After 24 h the incorporation of radioactivity into the benzofurans was determined. Incorporation of radioactivity (in % of the applied dose) into the benzofurans (*7/9*) was calculated as 0.23% for those cultures that had been treated with yeast extract

Table 3. Benzofuran accumulation (7/9) by cell suspension cultures of *A. adenophora* in relation to increasing doses of yeast extract. (Monir and Proksch 1989)

Concentration of yeast extract (in g/l)	Relative accumulation of benzofurans 7/9 (highest accumulation = 100%)
0	26
2	71
4	70
6	100
8	60
10	50

Cells were grown under the light/dark rhythm. Yeast extract was added to the cell suspension cultures at growth stage 6 days. Accumulation of the benzofurans was determined after 24 h. Numbers of benzofurans follow Fig. 2.

Table 4. Accumulation of benzofurans *7/9* by cell suspension cultures of *A. adenophora* following single and repeated addition of yeast extract. (Monir and Proksch 1989)

Time of addition of yeast extract (in days following inoculation of cells into fresh medium)		Accumulation of benzofurans *7/9* 24 h following addition of yeast extract μmol/g
6	+ yeast extract	0.105
	− yeast extract	0.050
6+9*	+ yeast extract	0.040
	− yeast extract	0.025
9	+ yeast extract	0.042
	− yeast extract	0.025

The dose of yeast extract applied was 2 g/l; μmol/g are given on a fresh weight basis. Numbers of benzofurans refer to Fig. 2. 6+9* = first application of yeast extract at 6 days, second application at 9 days. Cell suspension cultures labeled 6, 6+9* and 9 were transferred to the dark on day 5 following inoculation of the cells into fresh medium.

Table 5. Suppression of benzofuran accumulation (*7/9*) by joint application of yeast extract and aminooxyacetic acid (AOA) to cell suspension cultures of *A. adenophora*. (Monir and Proksch 1989)

Cell suspension cultures	Benzofuran (*7/9*) accumulation 24 h following start of the experiment μmol/g
− Yeast extract − AOA	0.035
+ Yeast extract − AOA	0.156
− Yeast extract + AOA (0.125 mM)	0.052
+ Yeast extract + AOA (0.125 mM)	0.055

The dose of yeast extract was 2 g/l; μmol/g are given on a fresh weight basis. Yeast extract was applied at growth stage 6 days. Numbers of benzofurans refer to Fig. 2.

in addition to the ^{14}C-precursor, whereas no measurable incorporation into the benzofurans was found in the cultures treated with ^{14}C-phenylalanine only. The experiments with the PAL-inhibitor AOA and with ^{14}C-phenylalanine prove that the increased accumulation of benzofurans in cell suspension cultures of *A. adenophora* treated with yeast extract is due to an induced de novo synthesis of the benzofurans.

3 Conclusions and Prospects

Chromenes and benzofurans exhibit a plethora of biological activities and seem to be of importance for the chemical defense of plants. Studies on the biosynthesis and on the metabolism of chromenes and benzofurans in differentiated plants have so far been hampered by the fact that most of these compounds accumulate in species that grow under arid or semi-arid conditions, and are difficult to germinate and to grow in a greenhouse or phytotron. As shown here for *Ageratina adenophora*, cell suspension cultures offer an alternative and easily accessible experimental approach to study metabolic processes related to the biosynthesis, biotransformation, and induction of chromenes and benzofurans. However, several of the data achieved in this study seem to apply only to cell suspension culture of *A. adenophora* and do not reflect the metabolism in differentiated plants of this species. Benzofurans, for example, that are among the prominent natural products of cell suspension cultures following treatment of the cultures either with encecalin or with yeast extract are not detected in plants of *A. adenophora* and can therefore be regarded as cell culture metabolites only. The increased accumulation of benzofurans in cell suspension cultures following addition of encecalin or of yeast extract offers an interesting experimental system to study biochemical processes related to the induction of an enhanced natural product accumulation. In the case of yeast extract our data indicate an elicitation of the biosynthesis leading to an increased level of constitutively present natural products. In the case of encecalin no final explanations can be given at present. A modulation of the balance between benzofuran synthesis and degradation seems possible.

4 Protocol

For cell suspension cultures of *A. adenophora*, as well as for all experiments conducted, a modified MS medium was employed that included:

	mg/l
NH_4NO_3	1650
KNO_3	1900
$CaCl_2$ H_2O	440
$MgSO_4$ $7H_2O$	370
KH_2PO_4	170
Sucrose	30,000
NaFeEDTA	40
Kinetin	0.25
2,4-D	0.033
KI	0.75
Nicotinic acid	1.0
Thiamin-HCl	10.0
Pyridoxin-HCl	1.0
myo-Inositol	100
H_3BO_3	6.2
$MnSO_4$ $4H_2O$	22.3
$ZnSO_4$ $7H_2O$	8.6

Na_2MoO_4 $2H_2O$	0.25
$CuSO_4$ $2H_2O$	0.025
$CoCl_2$ $6H_2O$	0.025

The pH of the medium was adjusted to 5.8. The cultures were usually maintained in the light (16 h daily, 22°C, 75% relative humidity, 120 rpm).

Acknowledgments. Financial support of this project by a grant of the DFG is gratefully acknowledged.

References

Amrhein N (1979) Biosynthesis of cyanidin in buckwheat hypocotyls. Phytochemistry 18:585–589

Bonner WA, de Graw JE (1962) Ketones from white snakeroot *Eupatorium urticaefolium*. Tetrahedron 18:1295–1309

Funk C, Gügler K, Brodelius P (1987) Increased secondary product formation in plant cell suspension cultures after treatment with a yeast carbohydrate preparation (elicitor). Phytochemistry 26:401–405

Hahlbrock K, Schröder J (1975) Specific effects on enzyme activities upon dilution of *Petroselinum hortense* cell cultures into water. Arch Biochem Biophys 171:500–506

Heywood VH, Harborne JB, Turner BL (eds) (1977) The biology and chemistry of the compositae. Academic Press, New York London

Isman MB, Proksch P (1985) Deterrent and insecticidal chromenes and benzofurans from *Encelia* (Asteraceae). Phytochemistry 24:1949–1951

Isman MB, Proksch P, Yan J-Y (1987) Insecticidal chromenes from the Asteraceae: Entomol Exp Appl 43:87–93

Keßmann H, Barz W (1987) Accumulation of isoflavones and pterocarpan phytoalexines in cell suspension cultures of different cultivars of chickpea (*Cicer arietinum*). Plant Cell Rep 6:55–59

King RM, Robinson M (1970) *Ageratina*. Phytologia 19:210–229

Knobloch KH, Hansen B, Berlin J (1981) Medium-induced formation of indole alkaloids and concomitant changes of interrelated enzyme activities in cell suspension cultures of *Catharanthus roseus*. Z Naturforsch 36c:40–45

Monir H, Proksch P (1989) Yeast extract induced accumulation of benzofurans in cell suspension cultures of *Ageratina adenophora* (Asteraceae). Phytochemistry 28:2999–3002

Murashige T, Skoog F (1962) A revised medium for rapid growth and bioassays with tobacco tissue culture. Physiol Plant 15:473–497

Proksch P (1985) Vorkommen und biologische Bedeutung von Benzopyranen (Chromenen) und Benzofuranen in den Asteraceae. Plant Syst Evol 150:89–100

Proksch P, Rodriguez E (1983) Chromenes and benzofurans of the Asteraceae, their chemistry and biological significance. Phytochemistry 22:2335–2348

Proksch P, Proksch M, Towers GHN, Rodriguez E (1983) Phototoxic and insecticidal activities of chromenes and benzofurans from *Encelia*. J Nat Prod 46:331–334

Proksch P, Palmer J, Hartmann T (1986) Metabolism and bioconversion of chromene derivatives in *Ageratina adenophora* (Asteraceae). Planta 169:130–134

Proksch P, Isman MB, Witte L, Hartmann T (1987a) Metabolites of insecticidal chromenes from the migratory grasshopper *Melanoplus sanguinipes*. Phytochemistry 26:2227–2230

Proksch P, Witte L, Wray V, Rahaus I (1987b) Accumulation and biotransformation of chromenes and benzofurans in a cell suspension culture of *Ageratina adenophora*. Planta Med 53:488–492

Proksch P, Witte L, Wray V (1988) Chromene glycosides from *Ageratina altissima*. Phytochemistry 27:3690–3691

Robinson H, King RM (1977) Eupatorieae-systematic review. In: Heywood VH, Harborne JB, Turner BL (eds) The biology and chemistry of the Compositae. Academic Press, New York London, pp 437–485

Rosenthal GA, Janzen DH (eds) (1979) Herbivors, their interaction with secondary plant metabolites. Academic Press, New York London

Siebertz R, Proksch P, Wray V, Witte L (1989) Accumulation and biosynthesis of benzofurans in root cultures of *Eupatorium cannabinum*. Phytochemistry 28:789–793
Turner BL (1977) Fossil history and geography. In: Heywood VH, Harborne JB, Turner BL (eds) The biology and chemistry of the Compositae. Academic Press, New York London, pp 21–39
Wagner H (1977) Pharmaceutical and economic uses of the Compositae. In: Heywood VH, Harborne JB, Turner BL (eds) The biology and chemistry of the Compositae. Academic Press, New York, pp 411–433
Wink M, Witte L (1983) Evidence for a wide-spread occurrence of the genes of quinolizidine alkaloid biosynthesis. FEBS Lett 159:196–200
Wu CW, Lampe KF, Mende TJ (1973) Metabolic change induced in chickens by administration of tremetol. Biochem Pharmacol 22:2835–2841
Zalkow LH, Burke N, Cabat G, Grula EA (1962) Toxic constituents of rayless goldenrod. J Med Chem 5:1342–1351
Zalkow LH, Ekpo BA, Gelbaum LT, Harris RN, Keinan E, Novak JR, Ramming CT, Van Derveer D (1979) The benzofurans of *Isocoma wrightii*. Structure and stereochemistry. J Nat Prod 42:203–219

III *Ailanthus altissima* (the Tree of Heaven): In Vitro Culture and the Formation of Alkaloids and Quassinoids

M.F. ROBERTS[1]

1 Introduction

Ailanthus species (Simaroubaceae) have a history of use in traditional medicine, particularly for the treatment of dysentery (Steck 1972), *A. altissima* is particularly noted as an antibacterial, anthelmintic, amoebicide and insecticide (Polonsky 1973; Ohmoto et al. 1976; Varga et al. 1980, 1981); *A. excelsa* (Mehta and Patel 1959) is noted as a specific for respiratory problems and *A. malabarica* is noted for the treatment of dyspepsia, bronchitis, opthalmia and snake bite (Khan et al. 1982).

1.1 Distribution and Importance

A. altissima Mill. Swingle. (Syn. *A. glandulosa* Desf.) originated in China, where it has been used in traditional medicine for enteritic infections of various origins (American Herbal Pharmacology Delegation 1975). Throughout the Far East, various parts of *A. altissima* are considered to be medicinal, with the use of either the fruits or bark of either root or stem for dysentery and various other gastric and intestinal upsets. Trees harvested for medicinal purposes are usually felled in the spring or autumn, and the bark is removed and dried in the sun. It is normally used to make aqueous extracts which are bitter, astringent and cooling. Bark extract has also been used to treat anaemia and as a taeniafuge, but does not have vermifuge properties. Leaves are toxic to domestic animals, causing inflammation to the digestive tract; gardeners cutting the trees may suffer erupting pustules. The bark is reported to contain an oleoresin, a bitter essence, an aromatic essence, a resin, some mucilage, ceryl alcohol, ailanthin, calcium oxalate and isoquercetin (Perry 1980). *A. altissima*, also known as the "Tree of Heaven", has been introduced into a number of other countries, i.e., India, Japan, and northern Australia, and, more recently, has been established as an ornamental tree in cities throughout Europe and North America, where it has gained popularity due to its general compact habit, small deciduous leaves and colorful autumn fruits (Swingle 1916).

[1]Department of Pharmacognosy, The School of Pharmacy, London University, 29-39 Brunswick Square, London WC1N 1AX, United Kingdom

Biotechnology in Agriculture and Forestry, Vol. 15
Medicinal and Aromatic Plants III (ed. by Y.P.S. Bajaj)

Fig. 1A-C. *Ailanthus altissima* tree, approximately 20 years old, growing in Russell Square, London, UK. **A** Mature tree. **B** The leaf. **C** The fruits. (Photographs by Mrs. A. Cavanaugh, Department of Pharmacognosy, The School of Pharmacy, London University)

1.2 Conventional Practices in Propagation and Growing

A. altissima is a large deciduous, dioecious tree which is frequently 50–70 feet high with a trunk 2–3 feet in diameter. It has a rounded head of branches and the older bark is marked with numerous grey fissures. Leaves are pinnate, 1–1.5 feet long with 15–30 leaflets; the leaves of the male plant have a typical foetid odor. Leaflets are 3–6 inches long, ovate, with the margin almost entire. Flowers are in terminal pannicles. The fruits consist of three to five keys (similar to those of the ash tree), which form attractive orange to red clusters in August and September. *Ailanthus* has proved to be an excellent tree for planting in towns for its display of large pinnate leaves, its colorful autumn fruits and particularly for its tolerance to city pollution (Swingle 1916; Osborn 1933; Bean 1970; Dirr 1983). The plant is usually cultivated either from ripe seed (February/March) or from suckers (March/April), which thrive in any fairly good soil. Since it is preferable to produce female plants which do not have the undesirable foetid odour, propagation from root cuttings is the more popular method (Osborn 1933; Dirr 1983; MacDonald 1986). Cutting young trees back to the ground in the spring and reducing the young shoots to a single shoot helps to produce a tree with strong wood and large (4-foot) leaves (Bean 1970).

1.3 Recent Medicinal Interest in the Biological Activity of the Constituents

Phytochemical investigations into the compounds, isolated from *A. altissima*, with biological activities (Suffness and Douros 1980) have focussed on the quassinoids, which are considered to be degraded terpenes. Ailanthone is the major constituent and more than 20 other quassinoids have now been isolated from the whole plant; these are given in Fig. 2. The quassinoids isolated from *A. altissima* have demonstrated biological activity. In regard to the traditional use of aqueous extracts of bark and fruit in the treatment of dysentery, the major quassinoid constituent, ailanthone, has potent antiamoebic activity against *Entamoeba histolytica* both in vitro and in vivo (de Carneri and Casinovi 1968; Casinovi et al. 1981). More recently, other quassinoid constituents of *A. altissima* have also been shown to be active with the following IC_{50} (μg/ml) ailanthone (0.04), ailanthinone (0.06), and glaucarubinone (0.025) as compared with a standard amoebicide, metronidazole (0.35) (Gillin et al., 1982; Wright et al. 1988).

These compounds are also of interest as a source of new anticancer agents (Suffness and Douros 1980). Ailanthinone, glaucarubinone and 13(18)-dehydroglaucarubinone were found to be active in the lymphocyctic leukaemia system from mice (P388) and Eagle's carcinoma of the nasopharynx (9KB) in vitro (Ogura et al. 1977). 6-α-tigloyloxychaparrinone and 6-α-tigloyloxycharrin isolated from *A. grandis* are also found to be active against P388 leukaemia (Polonsky et al. 1980). These quassinoids are characterised by an oxygen bridge between C-11 and C-20. Structural variations related to: (a) The presence or absence of an ester substituent either at position C-6 or C-15, (b) the presence of an α-methyl or a methylene group at C-13 and (c) the oxidation state of C-1, C-2 and C-3. Compounds without a side chain, i.e., chaparrinone, are negative in the P388 leukaemia test or show weak activity at high doses. Compounds with a C-15 ester group have strong to moderate

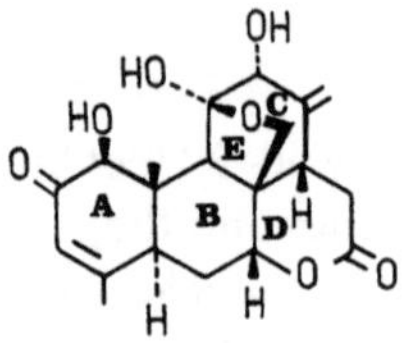

*Ailanthone [9]

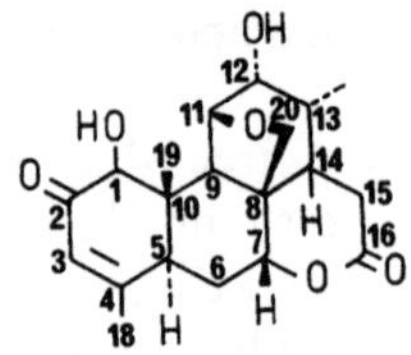

Chaparrinone [11]

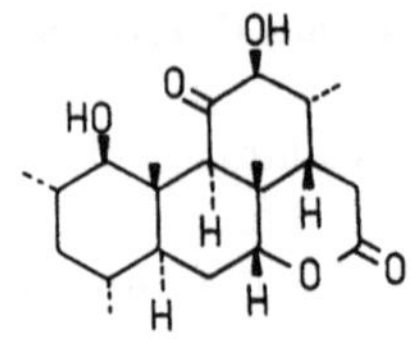

Chaparrolide [4]

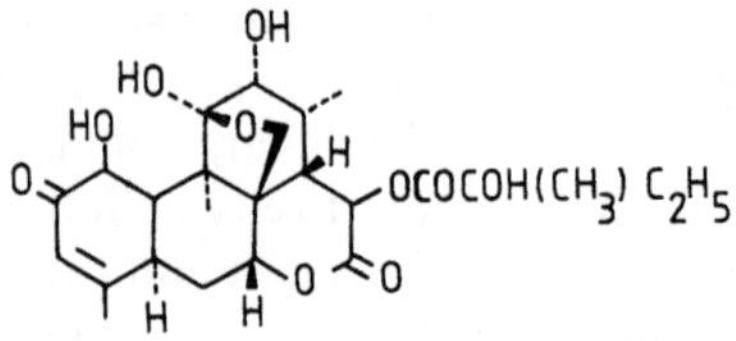

Glaucarubinone [2]

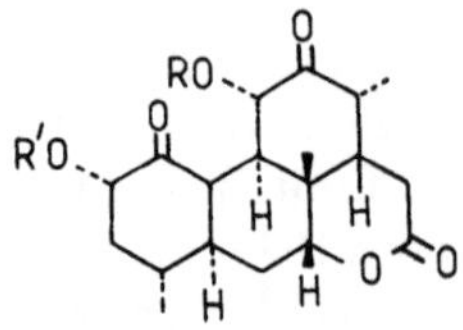

Amarolide R[1]=H, R=OH [1]

Acetylamarolide R=H R=Acetyl [1]

Shinjuglycoside C [12] R=O acetyl; R[1]=αH, β-D-glucopyranoside

Shinjuglycoside D [12] R=H; R[1]=αH,β-D-glucopyranoside

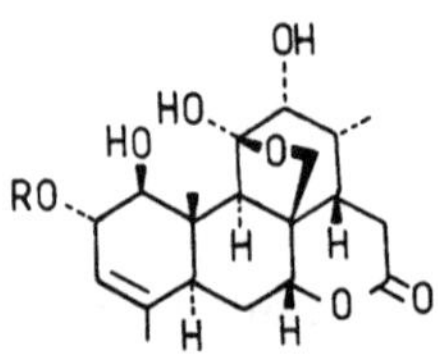

Shinjulactone A R=αOH, β-H [9]

Shinjulactone A glycoside [12]

R=2=β-D-glucopyranoside

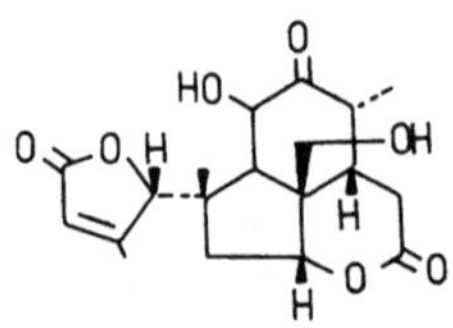

Shinjulactone B [3]

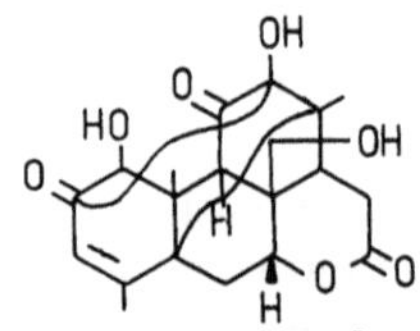

Shinjulactone C [5]

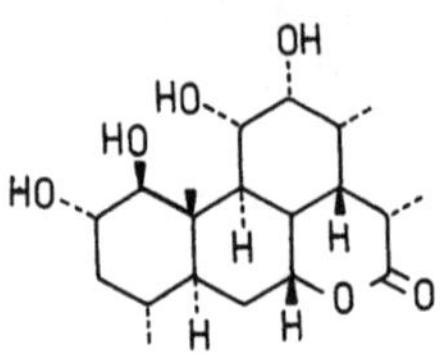

Shinjulactone D [3]

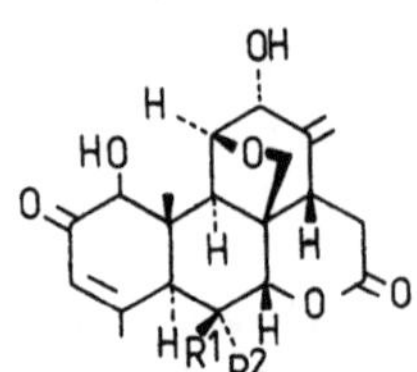

Shinjulactone E [3]

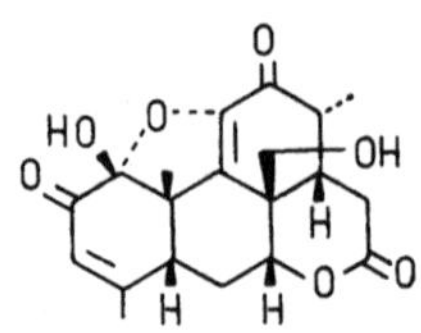

Shinjulactone F [6]

Fig. 2. Quassinoids isolated from *Ailanthus altissima*. *1* Casinovi et al. (1965); *2* Gaudemer and Polonsky (1965); *3* Furuno et al. (1984); *4* Ishibashi et al. (1981); *5* Ishibashi et al. (1983); *6* Ishibashi et al. (1984a); *7* Ishibashi et al. (1984b); *8* Ishibashi et al. (1985); *9* Naora et al. (1983); *10* Niimi et al. (1986); *11* Polonsky et al. (1975); *12* Yoshimura et al. (1984). * = also isolated from tissue cultures

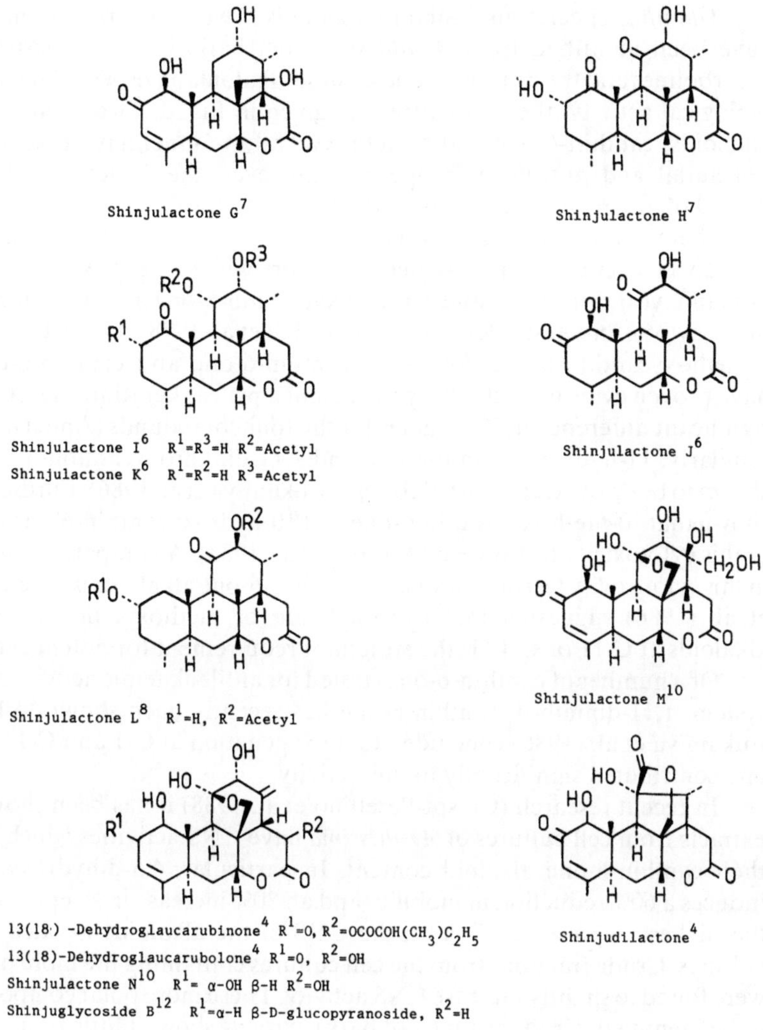

Fig. 2. cont'd.

activity, i.e., glaucarubinone, at doses in the 0.25–5 mg/kg range. Comparison with other data suggests the unsaturated ketone in ring A is also required for optimal potency and a free C-1 hydroxyl may be important for cytotoxic activity – the quassinoids from *Ailanthus* with antileukemic activity were all significantly cytotoxic (Suffness and Douros 1980).

Recently (Suffness and Douros 1980; Bray et al. 1987), crude extracts of *A. altissima* have been found to be active against *Plasmodium falciparum* in vitro and *P. berghei* in mice. It was found that the antimalarial activity was principally due to the presence of the quassinoid ailanthone (IC_{50}:0.015 μg/ml in vitro and IC_{50}:0.76 μg/ml in vivo). Glaucarubinone, whose structure also possesses an α-ketol group in ring A, has similar activity. Ailanthone and glaucarubinone, however, both showed high toxicity to mice at 9 mg/kg/day and so their potential as future antimalarials is unlikely (O'Neill et al. 1986).

Ailanthus species also contain alkaloids. Two categories of indole alkaloids have been identified from *A. altissima* root bark, i.e., the non-iridoid-derived *β*-carbolines and the canthin-6-ones. These alkaloids have been shown to have some biological activity, their structures are given in Fig. 3. Two of the canthin-6-one alkaloids, canthin-6-one and 5-methoxycanthin-6-one, have been tested for antibacterial and antifungal properties; however, the 5-methoxy derivative was devoid of any meaningful activity (Mitscher et al. 1972).

There are conflicting reports on the activity of canthin-6-one alkaloids as anticancer agents. There has been a report which suggests that canthin-6-one, 1-methoxycanthin-6-one and 5-methoxycanthin-6-one are not significantly active in the cytotoxicity test (9KB system) (Cordell et al. 1978); however, canthin-6-one, 1-methoxycanthin-6-one, 5-methoxycanthin-6-one and canthin-6-one-3N-oxide have proven cytotoxic activity against guinea pig ear keratinocytes (GPK), with no significant difference in IC_{50} values for the four compounds (Anderson et al. 1983). Similarly, 11-hydroxycanthin-6-one and 1,11-dihydroxycanthin-6-one have been shown to be cytotoxic against 9KB cells (Fukamiya et al. 1986). Further studies have shown that 10-methoxycanthin-6-one and 10-hydroxycanthin-6-one are also active in this cytotoxicity test system (Arisawa et al. 1983). A comparison of the activities in this cytotoxicity test of a number of canthin-6-one alkaloids isolated by Fukamiya et al. (1986) suggested that hydroxylation or methoxylation of canthin-6-one alkaloids at C-11 or C-10 is the structural requirement for potent cytotoxicity.

Of a number of canthin-6-ones tested for antileukaemic activity using the P388 system, 1,11-dimethoxycanthin-6-one has recently been shown to be active and Fukamiya et al. (1986) concluded that oxygenation at C-1 and C-11 of canthin-6-one contributed significantly to this activity.

In recent research (Crespi-Perellino et al. 1988) it has been shown that crude extracts from cell cultures of *A. altissima* have CNS activities which are related to their canthin-6-one alkaloid content. In particular, 4,5-dihydroxycanthin-6-one induces a 60% reduction in mobility and an 80% increase in sleeping time, showing the highest depressive effect on the CNS of the alkaloids isolated from the cell cultures. Crude fractions from the cell cultures containing the more polar alkaloids were found to slightly increase CNS activity. These more polar compounds have yet to be identified. Crude plant (root bark) extracts showed little or no depression of CNS activity, presumably due to much lower alkaloid levels and the fact that 4,5-dihydrocanthin-6-one does not accumulate.

The *β*-carboline and canthin-6-one alkaloids have strong inhibitory effects on cyclic adenosine monophosphate phosphodiesterase (Sung et al. 1984). The structure-activity relationships were studied for 31 derivatives of *β*-carboline, 2 dimeric derivatives of *β*-carboline and 12 derivatives of canthin-6-one. *β*-carboline derivatives with a methoxycarboxyl group and canthin-6-one derivatives with a hydroxymethyl group generally had a strong inhibitory effect on cAMP phos-

Fig. 3A,B. Alkaloids isolated from *Ailanthus altissima*. *1* Crespi-Perellino et al. (1986a,b); *2* Ohmoto and Koike (1984); *3* Ohmoto et al. (1976); *4* Ohmoto et al. (1981); *5* Varga et al. (1980, 1981). △ isolated from the plant; * isolated from tissue cultures

A Canthin-6-ones:

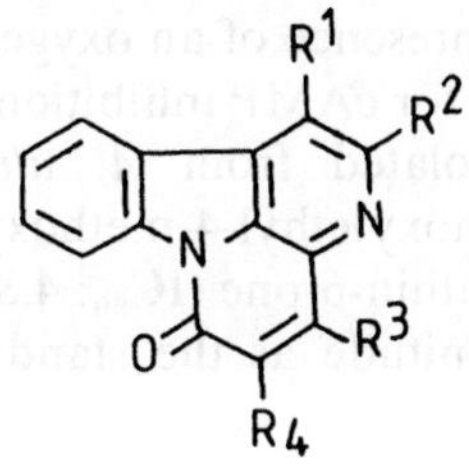

Δ*canthin-6-one[2,3,4] R^{1-4}=H

Δ*1-methoxycanthin-6-one[2,3,4] R^1=OCH_3; R^{2-4}=H

Δ*5-methoxycanthin-6-one[2,3,4] $R^{1.3}$=H R^4=OCH_3

Δ*1-hydroxycanthin-6-one[2,3,4] R^1=OH; R^{2-4}=H

Δ 2-hydroxycanthin-6-one[1] R^1=H; R^2=OH; R^{3-4}=H

Δ 4-hydroxycanthin-6-one[1] R^{1-2}=H; R^3=OH; R^4=H

Δ 5-hydroxycnathin-6-one[1] R^{1-3}=H; R^4=OH

*4,5-dihydrocanthin-6-one[1]

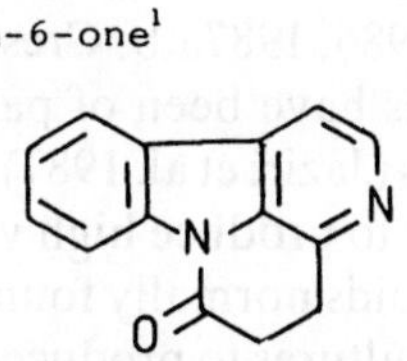

Canthin-6-one-3N-oxides:

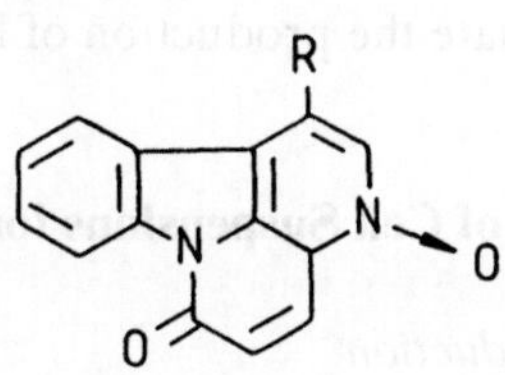

Δ*canthin-6-one3N-oxide[3,4] R=H

Δ*1-methoxycanthin-6-one-3N-oxide[3,4] R=OCH_3

B β-carbolines:

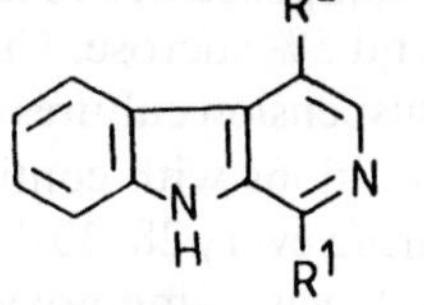

Δ*β-carboline-1-propionic acid[1] R^1=C_2H_4COOH; R^2=H

Δ 1-(1-hydroxy-2-methoxy)ethyl-4-methoxy-β-carboline[2,4,5] R^1=$CHCH_2OH$ (OCH_3); R^2=OCH_3

Δ 1-(2^1-hydroxyethyl)-4-methoxy-β-carboline[2,4,5] R^1=C_2H_4OH; R^2=OCH_3

Δ 1-(1^1,2^1-dihydroxyethyl)-4-methoxy-β-carboline[2,4,5] R^1=$C_2H_3(OH)_2$; R^2=OCH_3

Δ*1-carboxymethyl-4-methoxy-β-carboline[2,4,5] R^1=$COOCH_3$; R^2=OCH_3

phodiesterase. The presence of an oxygen atom at C-5 or at both C-4 and C-5 is apparently essential for cAMP inhibition by the canthin-6-one congeners. Of the few compounds isolated from *A. altissima* which were tested, only two, 1-(1-hydroxy-2-methoxy)ethyl-4-methoxy-β-carboline (IC_{50}: 4.6×10^{-5}M) and 5-hydroxymethylcanthin-6-one (IC_{50}: 4.8×10^{-5}M), have inhibitory activity of the same order of magnitude as the standard commonly used, papaverine (IC_{50}: 3.0×10^{-5}M).

2 In Vitro Approaches

2.1 Studies with Cell Culture

Cell cultures of *A. altissima* have been developed in at least three laboratories (Anderson et al. 1983, 1986, 1987a,b; Crespi-Perellino et al. 1986a,b; Aragozzini et al. 1988). These cultures have been of particular interest for their poor ability to produce the quassinoids (Jaziri et al. 1987), important constituents of the plant, and their remarkable ability to produce high yields (100-fold greater than the plant) of the canthin-6-one alkaloids normally found as very minor constituents. Because of the ability of these cell cultures to produce the canthin-6-one alkaloids, *A. altissima* in cell culture has been used as a model system for the study of biosynthesis and the factors which regulate the production of high yields of these alkaloids.

2.2 Establishment of Cell Suspensions for Alkaloid and Quassinoid Production

2.2.1 Alkaloid Production

Essentially three groups have been involved in the current studies of *A. altissima* in tissue culture (Table 1). Anderson et al. (1983) developed callus cultures from sterile germinated seed. Explants of hypocotyl were asceptically transferred to an agar medium composed of MS salts (Murashige and Skoog 1962), 1 mg/1 2,4-D, 0.1 mg/l kinetin and 5% sucrose. Cultures were maintained at 25°C with constant illumination. Suspension cultures of *A. altissima* were developed and maintained under the same conditions with continuous agitation (120 rpm). The cells usually required subculturing every 28–30 days. Alkaloid yields of 100 mg/l were obtained, of which 62.5 mg/l was 1-methoxycanthin-6-one, 20 mg/l was canthin-6-one and the remainder 1-hydroxycanthin-6-one. The optimum inocculum density was found to be 3 g fresh weight per 40 ml of medium. Although alteration in basal medium made some changes to alkaloid production, the conditions used above proved to be the most satisfactory over a 6-year period. The greatest variations in alkaloid production occurred as a result of seed source (Anderson et al. 1987a). Examined microscopically, these cultures were found to consist of undifferentiated cells with little pigmentation, either single or in small two- to eight-cell aggregates. Many of the aggregates showed incomplete cell division typical of many in vitro cultures (Hay 1987). Staining cells with Dragendorff's reagent demonstrated the presence of

Table 1. Summary of in vitro culture studies on *Ailanthus altissima*

Inoculum	Media mg/l	Growth response	Compounds isolated	Reference
			Alkaloids	
Hypocotyl	MS + 2,4-D(1), kin. (0.1) sucrose (5%)	Callus + Suspension	Canthin-6-one 1-Hydroxycanthin--6-one *1-methoxycanthin--6-one	Anderson et al. (1983, 1986, 1987 a,b)
Leaves Stem	MS + NAA (0.1) BA(1)	Callus	*Canthin-6-one, 1-Methoxy-	Crespi--Perillino (1986 a,b,)
Shoots	MS + 2,4-D(1)	Suspension	Canthin-6-one, Canthin-6-one--3N-oxide, 1-methoxycanthin--6-one-3N-oxide 1-hydroxycanthin-6-one, 5-hydroxycanthin-6--one, β-carboline-1-propionic acid, 4-methoxy-β-carboline-1-carboxylic acid methyl ester, 2-hydroxy-canthin-6-one 4-hydroxy-canthin-6-one 4,5-d, hydroxycanthin-6-one	
Stem	G + NAA (4) + 2,4-D(0.2)	Callus + Suspension	*Canthin-6-one	Aragozzini et al. (1988)
			Quassinoids	
Stem	MS + kin. (0.0001) IAA (0.02)	Callus	Ailanthone	Jaziri et al. (1987)
Anther	MS + 2,4D (0.0001)	Callus	Ailanthone	Jaziri et al. (1987)

*Major constituent. Basal media: MS (Murashige and Skoog 1962), G (Gamborg et al. 1968). Supplements: IAA, indoleacetic acid; NAA, naphthylacetic acid; kin, kinetin; BA, benzylaminopurine; 2,4-D, 2,4-dichlorophenoxyacetic acid.

alkaloid-accumulating cells, that is, those cells whose vacuole takes up the stain. Further inspection also revealed the presence of non-alkaloid accumulating cells (Hay 1987), i.e., those cells which do not take up the Dragendorff's reagent (Lang and Kohlenbach 1982).

Crespi-Perillino et al. (1986a,b) obtained callus cultures from sterile leaves, stem, and shoots of *A. altissima* using MS media with 1 mg/l 2,4-D. Suspension cultures were developed with the same medium and were stirred at 100 rpm at 28 °C in the dark. Cultures harvested at 25 days produced an average of 400 mg/l of alkaloid, of which 90% was canthin-6-one.

Aragozzini et al. (1988) prepared callus cultures from stems of *A. altissima* using Gamborg's agar medium with naphthalene acetic acid (4 mg/l) and 2,4-D (0.2 mg/l) as phytohormones. Suspension cultures prepared from callus cultures used similar medium and were stirred at 120 rpm at 26°C. These cultures yielded an average 50 mg/l of canthin-6-one after 30 days.

In all these variations of technique, the alkaloids remain essentially within the cells and only trace amounts have been found in the medium.

No one medium could be said to give better results, since although the yields of canthin-6-one were superior using the media of Crespi-Perillino et al. (1986a,b), these cells had poor ability to hydroxylate and methylate canthin-6-one. The extent to which the richness of alkaloid pattern in tissue culture depends on the media and supplements or on the origins of the plant material used to produce the plant cell culture has not been completely investigated. However, the more biologically active canthin-6-ones appear to require oxygenation at C-1, C-10, or C-11. To date, only the 1-, 2-, 4-, 5- and 4,5-dihydroxycanthin-6-ones- have been found in plant cell cultures of *A. altissima*.

2.2.2 *Quassinoid Production*

Jaziri et al. (1987) used sterile stem and anther explants to initiate callus cultures on MS medium supplemented with 0.1 μM kinetin, polyvinyl pyrrolidone (1%), charcoal (0.3%) and different auxins: 1 μM 2,4-D, 20 μM IAA or 20 μM IBA. Cultures were maintained at 24°C in continuous light (2000 lx). Callus from stem explants after the fifth transfer using 2,4-D were transferred to IAA or IBA and produced ailanthone. Anther explants produced ailanthone when maintained on 2,4-D; however, yields were low and averaged about 15 μg per 100 g fresh weight of callus culture.

2.3 Extraction and Stuctures of Constituents Isolated from Cell Culture

Alkaloids were extracted from cells using 80% ethanol and the residue obtained on evaporation was partitioned between aqueous NH_3 at pH 8.0 and toluene to isolate the less polar alkaloids. The aqueous phase was then brought to pH 3.0 and extracted with n-butanol. Extracts were further separated into individual compounds using either TLC or column chromatography on silica gel (Crespi-Perellino et al. 1986a,b; Aragozzini et al. 1988). Anderson et al. (1986) favoured the use of Merck Extrelut columns for analytical work with a limited range of alkaloids to be analysed.

Similar methodology, with methanol extraction followed by separation on silica gel columns, was used by Jaziri et al. (1987) for the isolation of the quassinoid ailanthone from cell cultures. Some 11 alkaloids have been isolated from *A. altissima* cell suspension cultures (Fig. 3) belonging to both the β-carboline and cathin-6-one types. The canthin-6-one alkaloids are the predominant compounds in cell suspension culture developed by Anderson et al. (1983) and Crespi-Perellino et al. (1986a). The two groups, however, report a different major alkaloid. In cultures described by Anderson et al. (1983), 1-methoxycanthin-6-one predominates with the parent compound, canthin-6-one, and 1-hydroxycanthin-6-one as minor constituents. Crespi-Perellino et al. (1986a,b) found that canthin-6-one was the major constituent of their cultures and the 1-methoxy analogue was the second most abundant alkaloid. This Italian group also reported the isolation of a further seven canthin-6-one alkaloids (Fig. 3), three of which had not been found before in nature viz. 2-hydroxycanthin-6-one, 4-hydroxycanthin-6-one and 4,5-dihydrocanthin-6-one. These three novel alkaloids represented 0.3% of the total alkaloids and their production was of the order of 2 μg/ml (Crespi-Perellino et al. 1986a,b).

In a recent study using crude enzyme preparations from *A. altissima*, ailanthone was converted to chapparinone, 13,18-dehydroglaucarubinone and 13,18-dehydroglaucarubolone. From the metabolic products produced it is apparent that the crude enzyme preparation was capable of hydroxylation at C-15, esterification at C-15 and reduction at C-18 of ailanthone. These experiments are of particular interest in view of the fact that 13,18-dehydroglaucarubinone has significant cell growth inhibition. The synthesis of this compound is not easy and it occurs in Simaroubaceae species only in low amounts (0.007% d. w. in *A. altissima* root bark).

Two factors affected the regulation of ailanthone metabolism in the plant cell cultures. Both the omission of 2,4-D from the medium [normally M & S salts with 2,4-D (1 mg/l) and kinetin (0.1 mg/l)] and the concentration of ailanthone in the culture medium increase the rate of ailanthone bioconversion. Crude enzyme preparations were made by extracting lyophilized cells with phosphate buffer (100 mM pH 7.0) containing dithioerythritol (5 mM). For metabolism of ailanthone the following cofactors were required; $MnCl_2$ (1 mM), $MgCl_2$ (5 mM), ATP (10 mM), $NADPH_2$ (1 mM) and bovine serum albumin (10 mg/ml).

Cell cultures of *A. altissima* also produce the coumarins scopoletin and isofraxidin (Anderson et al. 1987b) and there is one report of the occurrence of the quassinoid ailanthone in tissue cultures (Jaziri et al. 1987). In this instance ailanthone (Fig. 2) was isolated from callus derived from anther explants and levels of ailanthone were of the order of 0.15 μg/g fr.wt.

2.4 Biosynthesis and Regulation of 1-Methoxycanthin-6-One Production in Plant Cell Cultures

Until recently, little was shown of the biosynthesis of the canthin-6-one alkaloids. They were thought to be derived either from a β-carboline intermediate with the addition of a C-2 subunit derived from acetate or pyruvate or from tryptamine with the incorporation of a C-4 unit (Hegnauer 1973). It was suggested by Crespi-Perellino et al. 1986b) that canthin-6-one may be derived from tryptophan via β-carboline-1-propionic acid and either 4,5-dihydrocanthin-6-one or β-carbolyl-1-acrylic acid. Their large-scale experiments utilising (DL)-[methylene ^{14}C]-tryptophan and (L)-[methylene ^{14}C]-tryptophan, demonstrated the incorporation of ^{14}C from tryptophan as follows: When the canthin-6-one alkaloids were isolated, 70% of the ^{14}C-label occurred in canthin-6-one, which was 70% of the total alkaloid fraction. The rest of the ^{14}C-label was distributed among 4-hydroxycanthin-6-one, 5-hydroxycanthin-6-one, 4,5-dihydrocanthin-6-one, canthin-6-one-3N-oxide, 1-methoxycanthin-6-one and 1-methoxycanthin-6-one-3N-oxide (Fig. 4). The existence of the label in 4,5-dihydrocanthin-6-one indicates that β-carbolyl-1-acrylic acid was not involved in the formation of canthin-6-one alkaloids. This sequence was further confirmed by studies in which labelled β-carboline-1-propionic acid

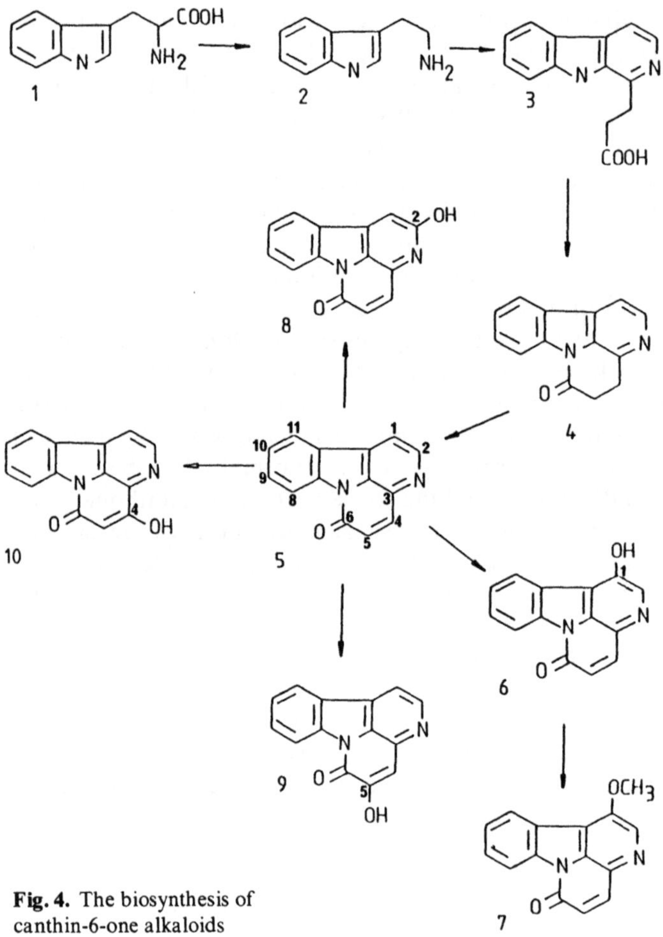

Fig. 4. The biosynthesis of canthin-6-one alkaloids

and 4,5-dihydrocanthin-6-one, fed to cell suspension cultures, yielded labelling of canthin-6-one.

In contrast to the cell cultures developed by Crespi-Perellino et al. (1986a,b), *A. altissima* cell cultures developed by Anderson et al. (1983, 1986, 1987a,b) produced 1-methoxycanthin-6-one as the major constituent with 1-hydroxycanthin-6-one and canthin-6-one as the only minor constituents.

Studies of the growth cycle of *A. altissima* cell suspensions (Fig. 5a) (Anderson et al. 1986) demonstrated that the production of canthin-6-one commences around day 16 and reaches a maximum around day 24. The level of 1-methoxycanthin-6-one, on the other hand, remains constant until day 22, when rapid production of this alkaloid occurs until day 28, at which stage maximum accumulation is reached. The yields of 1-methoxycanthin-6-one (62.5 mg/l) are always significantly greater than those of the parent alkaloid canthin-6-one (20 mg/l). The level of isolatable 1-hydroxycanthin-6-one remained low throughout the growth cycle, suggest-

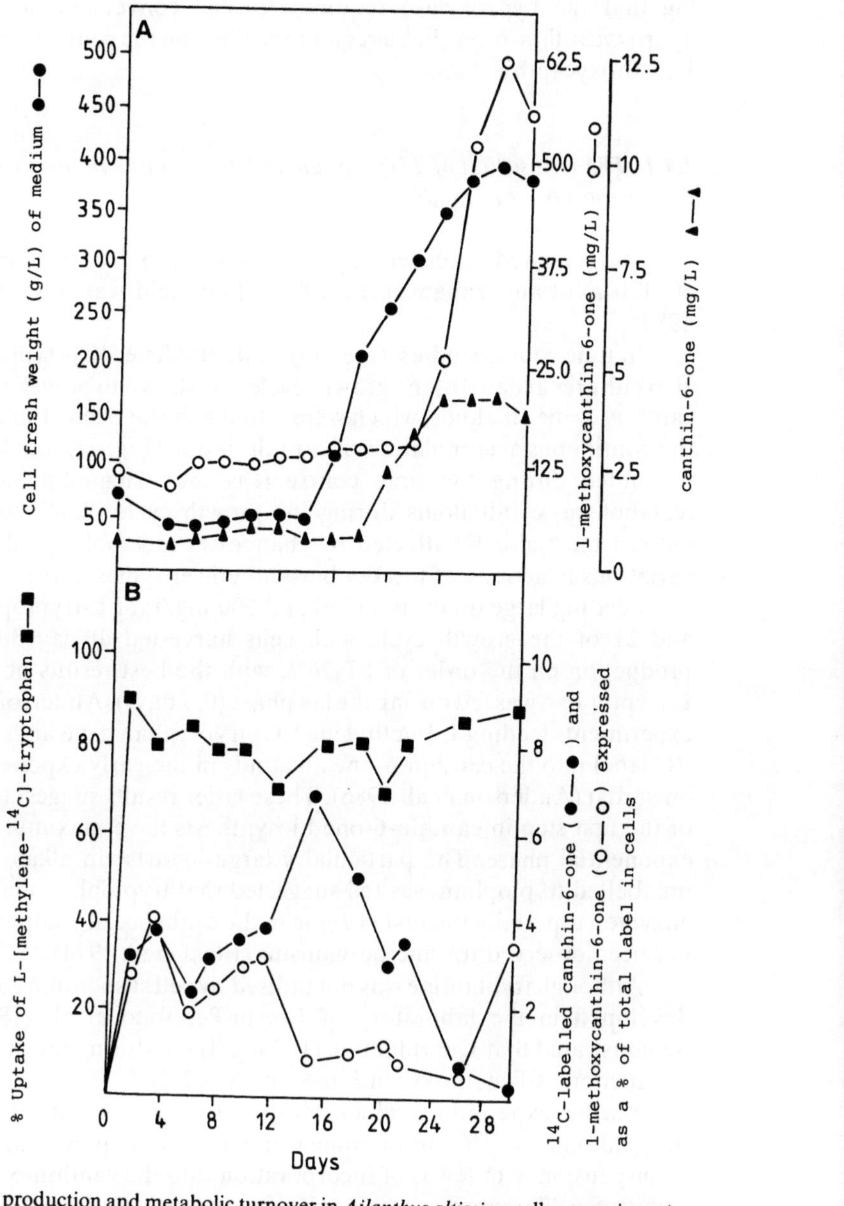

Fig. 5A,B. Canthin-6-one production and metabolic turnover in *Ailanthus altissima* cell suspensions. **A** Production of canthin-6-one alkaloids during the growth cycle. **B** Metabolic turnover of canthin-6-ones; variations in the level of incorporation of L-[methylene ^{14}C]-tryptophan. (Redrawn from data originally presented in Anderson et al. 1986)

ing that the hydroxylase required for the conversion of canthin-6-one to 1-hydroxycanthin-6-one has a regulatory effect on the conversion of canthin-6-one to 1-methoxycanthin-6-one.

2.4.1 The Utilisation of Tryptophan and Tryptamine in the Production of Canthin-6-One Alkaloids

L-tryptophan, fed to suspension cultures, was rapidly taken up by the cells (80% in 60–120 min) and enhancement of alkaloid yield was observed (Anderson et al. 1986).

In time course studies, (Fig. 5b), L-[methylene ^{14}C]-tryptophan fed to suspension cultures at day 0 of the growth cycle was shown to be well incorporated into the canthin-6-one alkaloids, which were found to be labelled after 24 h. Compared with the continuous accumulation of canthin-6-one (Fig. 5a), the label in the alkaloids fluctuated during the time course (Fig. 5b), suggesting a complex turnover (catabolism) of alkaloids during the growth cycle (Anderson et al. 1986). This pattern must also be affected by changes in metabolic pools of tryptophan and variations in activity of the key biosynthetic enzymes during the growth cycle.

Feeding large quantities (250 and 500 mg/l) of L-tryptophan at days 0, 7, 14, and 21 of the growth cycle with cells harvested at day 30 increased alkaloid production on the order of 67–76%, with the best results obtained if unlabelled L-tryptophan was fed during the lag phase (0, 7 days) (Anderson et al. 1986). Parallel experiments feeding L-[methylene ^{14}C]-tryptophan gave high incorporation of the ^{14}C label into the canthin-6-one alkaloids in the early exponential phase (14 days onwards) (Anderson et al. 1986). These latter results suggest that the key enzymes of the first step in canthin-6-one biosynthesis have maximal activity in the early exponential phase. The particularly large increase in alkaloid production when unlabelled tryptophan was fed suggested that tryptophan may well activate these enzymes, especially the first enzyme in the sequence, tryptophan decarboxylase, as has been observed for microorganisms (Floss et al. 1974).

Although tryptamine was not utilised by cell suspension cultures of *A. altissima* developed in the laboratory of Crespi-Perellino et al. (1986a,b), Hay (1987) demonstrated that L-tryptamine (100 mg/l) fed during the growth cycle increased production of 1-methoxycanthin-6-one by 17–24%.

Further experiments, which paralleled those with L-tryptophan, demonstrated that [side chain-2-^{14}C]-tryptamine is rapidly taken up by *A. altissima* cells (15 min) in suspension with levels of incorporation into the canthin-6-one alkaloids on the order of 6.2%, commensurate with incorporation observed with L-[methylene-^{14}C]-tryptophan. Unlike the feeding with L-tryptophan, the best results with [side chain-2-^{14}C]-tryptamine were obtained if feeding took place towards the end of the growth cycle although feeding unlabelled tryptamine (100 mg/l) in the lag phase, as with tryptophan, boosted alkaloid production. The results with both unlabelled tryptamine and [side chain-2-^{14}C]-tryptamine suggest that the enzymic levels required for the initial step in the conversion of tryptamine to canthin-6-one alkaloid are most active in the latter phase of growth. The difference in the patterns of incorporation of label when tryptophan, as opposed to tryptamine, was fed is attributed to a transient, developmentally induced increase in the presence of L-tryptophan decarboxylase between days 14 and 21 of the growth cycle. The activity of the decarboxylating enzyme as well as that of the tryptamine conversion enzyme apparently provides as significant a degree of regulation of alkaloid production as does the availability of these substrates (Hay 1987; Roberts et al. 1989).

2.4.2 The Role of Acetate in Canthin-6-One Synthesis

Since the biosynthetic route via L-tryptamine to the canthin-6-one alkaloids would appear to involve β-carboline-1-propionic acid as an intermediate (Crespi-Perellino et al. 1986b), Aragozzini et al. (1988) have shown a labelling pattern of canathin-6-one after feeding [1-^{13}C]-, [2-^{13}C]- and [1,2-^{13}C]-acetate that is consistent with the involvement of α-ketoglutarate as an intact precursor in the biosynthetic pathway (Fig. 4). [1-^{13}C]-acetate showed isotopic enrichment only at C-6 of canthin-6-one, while in a parallel experiment, [2-^{13}C]-acetate gave enrichment at C-3, C-4 and C-5. [1,2-^{13}C]-acetate gave enrichment at all four carbon atoms and the presence of one pair of coupled ^{13}C-satellites for the C-5 and C-6 signals gives evidence for the incorporation of only one intact acetate unit. Allowing for the fate of the acetate carbons in the TCA cycle as discussed by Turner (1971), it is possible to rule out the incorporation of a C-4 acid like succinate. The label in the C-3 to C-6 fragment of canthin-6-one is explainable if incorporation of α-ketoglutarate with loss of a carboxyl at or after the condensation step occurs. A similar route has recently been described for the harman alkaloids from *Passiflora edulus* and *Eleagnus augustifolia* (Herbert and Mann 1982).

2.4.3 The Origins of the Methoxy Group of 1-Methoxycanthin-6-One

The fact that 1-methoxycanthin-6-one was the major constituent in the *A. altissima* cultures of Anderson et al. (1983, 1986, 1987a,b) with canthin-6-one and 1-hydroxycanthin-6-one as minor alkaloids prompted that group to investigate the origins of the methyl group.

L-[methyl ^{14}C]-methionine fed to cell suspensions of *A. altissima* was rapidly taken up by the cells (20 min: 95% uptake) if fed from day 7 onwards. Incorporation into 1-methoxycanthin-6-one remained low (ca. 0.15% incorporation) except when fed to cultures in the late exponential phase (days 27 to 29) of the cycle which produced incorporation of from 2–3% into 1-methoxycanthin-6-one. Parallel experiments with L-[methyl ^{13}C]-methionine yielded 1-methoxycanthin-6-one with enhancement of the signal at 56.74 ppm (the methoxyl carbon), confirming the involvement of L-methionine in the methylation of 1-hydroxycanthin-6-one (Anderson et al. 1987b).

Roberts et al. (1989) found methyltransferase activity to remain high throughout most of the lag and exponential phases and the results, therefore, suggest that 1-methoxycanthin-6-one formation may be regulated by the onset of rapid hydroxylation of canthin-6-one in the late lag phase. Other methyltransferases are active within the cell suspensions, since cultures fed L-[methyl ^{14}C]-methionine also produced label in the methoxylated coumarin scopoletin (0.8%).

Purification of the cell-free extracts containing hydroxycanthin-6-one methyltransferase activity showed the presence of enzymes capable of the methylation of 1-, 8- and 10-hydroxycanthin-6-one. The extent to which methylated products occur therefore appears to be principally determined by the availability of the appropriate hydroxycanthin-6-one.

The ability of cell-free extracts to produce 8- and 10-methoxycanthin-6-ones, not previously reported as occurring in cell cultures, is of particular interest since 10-hydroxycanthin-6-one and its methylated form have not been reported for *Ailanthus* sp. and 8-hydroxycanthin-6-one has been recorded as occurring only in *Ailanthus excelsa* (Cordell et al. 1978). It will therefore be interesting to investigate the ability of these cultures to methylate other hydroxylated canthin-6-ones reported to occur both in the whole plant and in cell cultures.

It was concluded from the above studies (Roberts et al. 1989) that a number of factors appear to regulate the production of 1-methoxycanthin-6-one in cell suspension cultures: (1) The availability of L-tryptophan in the early stages of growth, (2) the activity of the enzymes of the early part of the biosynthetic pathway at the onset of exponential growth, (3) the activity of the hydroxylase system which controls the diversity and overall yield of alkaloids produced and (4) the activity of enzymes responsible for the further metabolism of the canthin-6-one alkaloids.

3 Conclusions

The current use of *Ailanthus altissima* in Europe and North America is principally as an ornamental tree, particularly in cities, where it withstands pollution well. In Asia, extracts of *A. altissima* bark and fruits are used as an antimicrobial, anthelmintic, and amoebicide. The biological activities of the quassinoids and canthin-6-one alkaloids isolated from this plant confirm uses in traditional medicine and interest in this plant continues as a result of these biological activities, which make some of these constituents of potential use as templates for new drugs. As a result, research continues on efforts to produce plant cell cultures of *A. altissima* high yielding in the quassinoid ailanthone and in the canthin-6-one alkaloids, since the yield of both these products from the mature plant is very low.

References

American Herbal Pharmacology Delegation (ed) (1975) Herbal Pharmacology of the Peoples' Republic of China (1975) Nat Acad Sci, Washington, DC

Anderson LA, Harris A, Phillipson JD (1983) Production of cytotoxic canthin-6-one alkaloids by *Ailanthus altissima* cell suspension cultures. J Nat Prod 46:374–378

Anderson LA, Hay CA, Roberts MF, Phillipson JD (1986) Studies on *Ailanthus altissima* cell suspension cultures. Precursor feeding of L-[methylene-^{14}C]-trytophan and L-tryptophan. Plant Cell Rep 5:387–390

Anderson LA, Roberts MF, Phillipson JD (1987a) Studies on *Ailanthus altissima* cell suspension cultures. The effect of basal media on growth and alkaloid production. Plant Cell Rep 6:239–241

Anderson LA, Hay CA, Phillipson JD, Roberts MF (1987b) Studies on *Ailanthus altissima* cell suspension cultures. Uptake of L-[methyl-^{14}C]-methionine and incorporation of label into 1-methoxycanthin-6-one. Plant Cell Rep 6:242–243

Aragozzini F, Maconi E, Gualandris R (1988) Evidence for involvement of ketoglutarate in the biosynthesis of canthin-6-one from cell cultures of *Ailanthus altissima*. Plant Cell Rep 7:213–215

Arisawa M, Kinghorn AD, Cordell GA, Farnsworth NR (1983) Plant anticancer agents. XXIV Alkaloid constituents of *Simaba multiflora*. J Nat Prod 46:222–225

Bean WJ (1970) Taylor G (ed) Trees and shrubs hardy in the British Isles, vol 1. Murrey, London, pp 265–267

Bray DH, Boardman P, O'Neill MJ, Chan KL, Phillipson JD, Warhurst DC, Suffness M (1987) Plants as sources of antimalarial drugs. 5. Activities of *Ailanthus altissima* stem constituents and of some related quassinoids. Phytother Res 1:22–24

Casinovi CG, Bellavita V, Granderling G, Ceccherelli P (1965) The occurrence of bitter substances in *Ailanthus glandulosa*. Tetrahedron Lett 27:2273–2279

Casinovi CG, Fardella G, Grandolini R, Burmato C (1981) Proprieta antiamebiche di alcuni derivati del' ailanthone e della Quassina. Farmaco 36:116–122

Cordell GA, Ogura M, Farnsworth NR (1978) Alkaloid constituents of *Ailanthus excelsa* (Simaroubaceae). Lloydia 41:166–168

Crespi-Perellino N, Guicciardi A, Malyszko G, Arlandini E, Ballabio M, Minghetti A (1986a) Occurrence of indole alkaloids in *Ailanthus altissima* cell cultures. J Nat Prod 49:1010–1014

Crespi-Perellino N, Guicciardi A, Malyszko G, Minghetti A (1986b) Biosynthetic relationship between indole alkaloids produced by cell cultures of *Ailanthus altissima*. J Nat Prod 49:814–822

Crespi-Perellino N, Guicciardi A, Minghetti A, Speroni E (1988) Comparison of biological activity induced by *Ailanthus altissima* plant or cell culture extracts. Pharmacological Res Comm 20 (suppl):45–48

de Carneri I, Casinovi CG (1986) Un potente antiamebico d'origine vegetal: L'ailanthone, prinapio altivo de *Ailanthus glandulosa*. Parasitologia 10:215–217

Dirr MA (1983) Manual of woody landscape plants; their identification, ornamental characteristics, culture, propagation and uses, 3rd edn. Stipes, Champaign, Ill, pp 82–83

Floss HG, Robbers JG, Heinstein PF (1974) Regulatory control mechanisms in alkaloid biosynthesis. In: Runeckles VC, Mabry TJ (eds) Recent advances in phytochemistry. Academic Press, New York London, pp 141–178

Fukamiya N, Okano M, Aratani T, Negoro K, McPhail AT, Ju-Ichi M, Kuo-Hsing L (1986) Antitumor agents, 79. Cytotoxic antileukemic alkaloids from *Brucea antidysenterica*. J Nat Prod 49:428–434

Furuno T, Ishibashi M, Naora H, Murae T, Hirota H, Tsuyuki T, Takahashi T, Itai A, Iitaka Y (1984) Structure determination of bitter principles of *Ailanthus altissima*. Structures of Shinjulactones B, D and E. Bull Chem Soc Jpn 57:2484–2489

Gamborg OL, Miller RA, Ojima K (1968) Nutrient requirements of suspension cultures of soybean root cells. Exp Cell Res 50:151–158

Gaudemer A, Polonsky J (1965) Structure de la guacarubinone nouveau principe amer isolé de *Simaruba glauca*. Phytochemistry 4:149–153

Gillin FD, Reiner DS, Suffness M (1982) Bruceantin, a potent amoebicide from a plant, *Brucea antidysenterica*, Antimicrob. Agents Chemother 22:342–345

Hay CA (1987) Plant tissue cultures of *Cinchona ledgeriana* and *Ailanthus altissima*. PhD Thesis, Univ London

Hegnauer R (1973) Chemotaxonomie der Pflanzen, vol 6. Birkhäuser, Basel, 387 pp

Herbert RB, Mann J (1982) The biosynthesis of the β-carboline alkaloids harman and eleagnine. J Chem Soc Perkin Trans 1:1523–1524

Ishibashi M, Murae T, Hirota H, Naora H, Tsuyuki T, Takahashi T, Itai A, Iitaka Y (1981) Shinjudilactone, a new bitter principle from *Ailanthus altissima* Swingle. Chem Lett 12:1597–1598

Ishibashi M, Tsuyuki T, Murae T, Hirota H, Takahashi T, Itai A, Iitaka Y (1983) Constituents of the rootbark of *Ailanthus altissima* Swingle. Isolation and X-ray crystal structures of shinjudilactone and shinjulactone C and the conversion of ailanthone to shinjudilactone. Bull Chem Soc Jpn 56:3683–3684

Ishibashi M, Yoshimura S, Tsuyuki T, Takahashi T, Itai A, Iitaka Y (1984a) Structure determination of bitter principles of *Ailanthus altissima*. Structures of shinjulactones F, I, J, and K. Bull Chem Soc Jpn 57:2885–2892

Ishibashi M, Yoshimura S, Tsuyuki T, Takahashi T, Matsushita K (1984b) Shinjulactones G and H. New bitter principles of *Ailanthus altissima* Swingle. Bull Chem Soc Jpn 57:2013–2014

Ishibashi M, Tsuyuki T, Takaaashi T (1985) Structure determination of a new bitter principle shinjulactone L. from *Ailanthus altissima*. Bull Chem Soc Jpn 58:2723–2724

Jaziri M, Homes J, Vanhaeien M (1987) Production of quassinoids by tissue cultures of *Ailanthus altissima*. Phytochemistry 26:999–1000

Jaziri M, Homes J, Vanhaelen M, Looze Y (1988) Biotransformation of quassinoids by a cell-free system from *Ailanthus altissima* cell suspension cultures. Med Fac Landbouww Rijksuniv Gent 53(4a): 1751–1755

Khan SA, Fozdar BI, Shamsuddin KM (1982) Quassinoids from *Ailanthus malabarica*. Indian J Chem 21:1133–1134

Lang H, Kohlenbach HW (1982) Differentiation of alkaloid cells in cultures of *Macleaya* mesophyll protoplasts. Planta Med 46:78–83

Macdonald B (1986) Plant propagation for nursery growers. Batsford, London, UK

Mehta CR, Patel CN (1959) Chemical examination of the bark of *Ailanthus excelsa* Roxb. Indian J Pharm 21:143–145

Mitscher LA, Showalter HDH, Shipchandler MT, Leu RP, Beal JL (1972) Antimicrobial agents from higher plants IV. *Zanthoxylum elephantiasis* isolation and identification of canthin-6-one. Lloydia 35:177–180

Murashige T, Skoog F (1962) A revised medium for rapid growth and bioassays with tobacco tissue cultures. Plant Physiol 15:473–497

Naora H, Ishibashi M, Furuno T, Tsuyuki T, Murae T, Hirota H, Takahashi T, Itai A, Iitaka Y (1983) Structure determination of bitter principles in *Ailanthus altissima*. Structure of shinjulactone A and a revised structure of ailanthone. Bull Chem Soc Jpn 56:3694–3698

Niimi Y, Tsuyuki T, Takahashi T, Matsushita K (1986) Structure determination of shinjulactones M and N, new bitter principles from *Ailanthus altissima* Swingle. Bull Chem Soc Jpn 59:1638–1640

Ogura M, Cordell GA, Kinghorn AD, Farnsworth NR (1977) Potential anticancer agents. VI. Constituents of *Ailanthus excelsa* (Simaroubaceae). Lloydia 40:579–584

Ohmoto T, Koike K (1984) Studies on the alkaloid constituents of *Ailanthus altissima* Swingle. III. The alkaloid constituents. Chem Pharm Bull Jpn 32:170–173

Ohmoto T, Tanaka R, Nikaido T (1976) Studies on the constituents of *Ailanthus altissima* Swingle. On the alkaloidal constituents. Chem Pharm Bull Jpn 24:1532–1536

Ohmoto T, Koike K, Sakamoto Y (1981) Studies on the constituents of *Ailanthus altissima* Swingle. II. Alkaloid constituents. Chem Pharm Bull Jpn 29:390–395

O'Neill MJ, Bray DH, Boardman P, Phillipson JD, Warhurst DC, Peters W, Suffness M (1986) Plants as sources of antimalarial drugs. Part 2 In vitro antimalarial activities of some quassinoids. Antimicrob Agents Chemother 30:101–104

Osborn A (1933) Shrubs and trees for the garden. Ward, Lock, London Melbourne, pp 245–246

Perry LM (1980) Medicinal plants of Southeast Asia. MIT, London, pp 387–388

Polonsky J (1973) Quassinoid bitter principles. In: Herz W, Grisebach H, Kirby GW (eds) Progress in the chemistry of organic natural products, vol 30. Springer, Berlin Heidelberg New York, p 102

Polonski J, Baskevitchi Z, Gottlieb HE, Hageman EW, Wenkert E (1975) Carbon-13 nuclear magnetic resonance spectral analysis of quassinoid bitter principles. J Org Chem 40:2499–2504

Polonsky J, Varon Z, Moretti C, Pettit GR, Heald CL, Rideout JA, Saha SB, Khastgir HN (1980) The antineoplastic quassinoids of *Simba cuspidata* Spruce and *Ailanthus grandis* Prain. J Nat Prod 43:503–509

Roberts MF, Hay CA, Anderson LA, Phillipson JD (1989) The formation and regulation of 1-methoxycanthin-6-one production in *Ailanthus altissima* cell suspension cultures. In: Kurz WGW (ed) Proc 2nd Int Symp Primary and secondary metabolism of plant cell cultures. Springer, Berlin Heidelberg New York, pp 175–187

Steck EA (1972) Species of *Ailanthus* used in native medicine for the treatment of dysentery. The chemotherapy of protozoan diseases, vol 1. Walter Reed Army Inst Res, US Gov Print Off, Washington, DC

Suffness M, Douros JD (1980) Miscellaneous natural products with antitumor activity. In: Cassady JM, Douros JD (eds) Anticancer agents based on natural product models. Academic Press, New York London, pp 465–476

Sung Y-I, Koike K, Nikaido T, Ohmoto T, Sankawa U (1984) Inhibitors of cyclic AMP phosphodiesterase in *Picrasma quassioides* Bennet, and inhibitory activities of related t-carboline alkaloids. Chem Pharm Bull Jpn 32:1872–1877

Swingle WT (1916) The early European history and the botanical name of the Tree of Heaven, *Ailanthus altissima*. J Washington Acad Sci 6:490–498

Turner WB (1971) Secondary metabolites derived from intermediates of the tricarboxylic acid cycle. In: Turner WB (ed) Fungal metabolites. Academic Press, New York London, pp 280–296

Varga E, Szendrei K, Reisch J, Maroti G. (1980) Indole alkaloids of *Ailanthus altissima*. Planta Med 40:337–339

Varga E, Szendrei K, Reisch J, Maroti G (1981) Indole alkaloids of *Ailanthus altissima*. II. Fitoterapia 4:183–186

Wright CW, O'Neill MJ, Phillipson JD, Warhurst DC (1988) Use of microdilution to assess in vitro antiamoebic activities of *Brucea javanica* fruits, *Simarouba amara* stem and a number of quassinoids. Antimicrob Agents Chemother 32:1725–1729

Yoshimura S, Ishibashi M, Tsuyuki T, Takahashi T, Matsushita K (1984) Constituents of seeds of *Ailanthus altissima* Swingle. Isolation and structures of shinjuglucosides A, B, C, D. Bull Chem Soc Jpn 57:2496–2501

IV *Aconitum* spp. (Monkshood): Somatic Embryogenesis, Plant Regeneration, and the Production of Aconitine and Other Alkaloids

YUKIHIRO SHOYAMA[1], ITSUO NISHIOKA[1], and KOJI HATANO[2]

1 Introduction

Aconitum spp., aconite or monkshood, is a perennial herb of the family Ranunculaceae. It is erect and grows to a height of about 1 m. The leaves are deep green and lustrous, and purple flowers bloom in the autumn. The aconite consists of both parent and daughter roots. Both are obconical in shape, dark brown in color, from 4 to 10 cm in length and from 1 to 3 cm in diameter at the crown. In northern countries, natives used the root extract of the plants as an arrow poison. In Japan this poison was also used by the Ainu for making poisoned arrows for bear hunting. It is well known that the roots contain diterpenoid alkaloids which are mainly classified into a strongly toxic group(aconitine-type) and a weakly toxic group(atisine type). The former include aconitine, mesaconitine, hypaconitine, jesaconitine, and neopelline. The latter covers anisine, kobushine, pseudokobusine, telatisine, songorine, atidine, napelline, heteratisine, ignarine, and hysognavine. The roots also contain a cardiac-activating alkaloid, higenamine (Kosuge and Yokota 1976), and coryneine (Konno et al. 1984) (Fig. 1).

The root, mainly *A. carmichaeli* Debx., has been used as one of the most important Chinese drugs (Fu-tzu in Chinese, Bushi in Japanese) prescribed, together with other herbs, as an analgesic in the treatment of rheumatism and neuralgia. The relationship between the alkaloid content and the pharmacological activities of the root, however, remains obscure. It is dangerous to use the crude drug indiscreetly since the range between a therapeutic dose and toxic dose is quite narrow. In order to decrease the toxicity of the crude drug, steam treatment (processing) has generally been carried out. During steam treatment individual aconitine-type alkaloids change to the corresponding benzoylaconine-type alkaloids by deacetylation, of lipoaconitine-type alkaloids by transesterification, or to pyraconitine-type alkaloids, whose toxicity decreases, while the pharmacological activities increase (Kitagawa et al. 1982; Mori et al. 1989) (Fig. 2).

Significant differences in the quantity and quality of alkaloids depend on the strain as well as on the place of growth(Hikino et al.1983; Kitagawa et al. 1984a). This phenomenon can be easily recognized from the chromosome variation described in Table 1 (Sakai 1933). Therefore, the breeding of a strain homogeneous with respect to alkaloid content is necessary.

[1] Faculty of Pharmaceutical Sciences, Kyushu University, Higashiku, Fukuoka 812, Japan
[2] Present address: Cooperate R and I, Plant Technology Institute, Ube Industries Ltd., 1978–5 Ogushi, Ubeshi, Yamaguchiken 755, Japan

Biotechnology in Agriculture and Forestry, Vol. 15
Medicinal and Aromatic Plants III (ed. by Y.P.S. Bajaj)

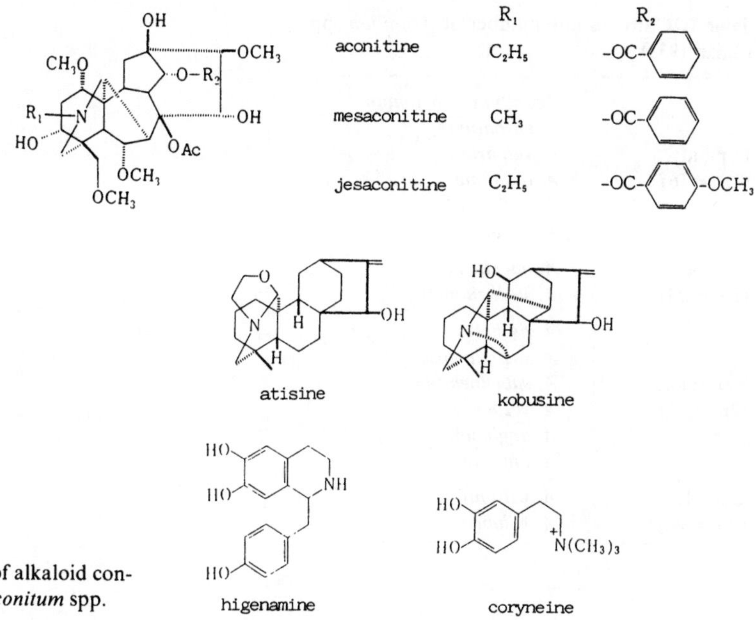

Fig. 1. Structures of alkaloid contained in root of *Aconitum* spp.

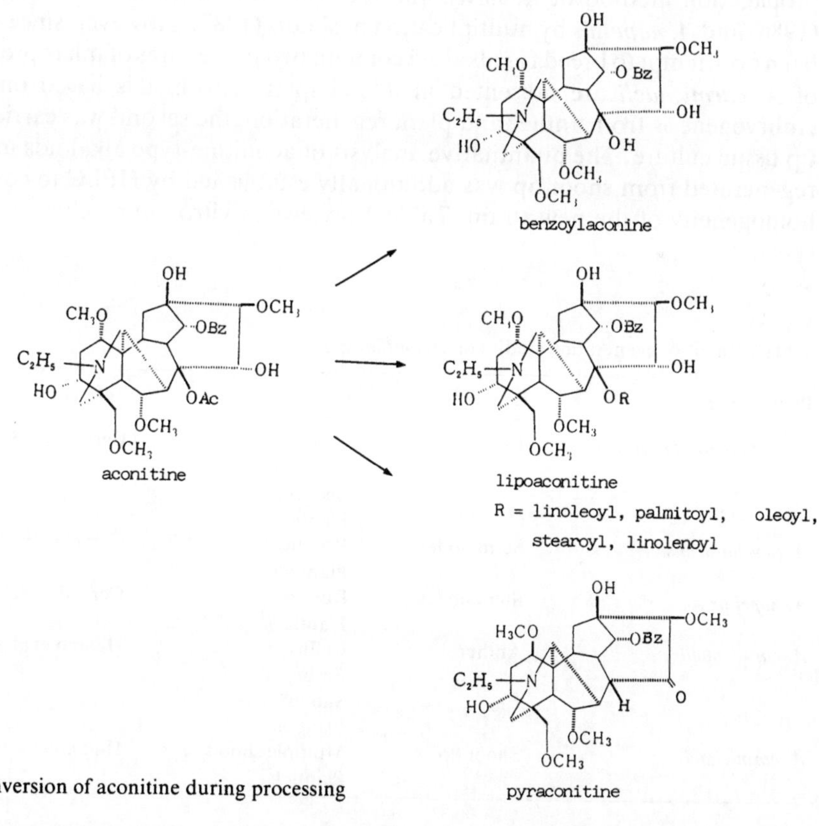

Fig. 2. Conversion of aconitine during processing

Table 1. Chromosome number of *Aconitum* spp. (Sakai 1933)

Diploid (2n = 16)	*Aconitum lycoctonum* *A. barbatum* *A. vulparia* *A. orientale* *A. ludidum* *A. yuparense*
Triploid (2n = 24)	*A. napellus* *A. stoekrianum*
Tetraploid (2n = 32)	*A. fischeri* *A. sachalinense* *A. subcuneatum* *A. yezoense* *A. anglicum* *A. anthora*
Octoploid (2n = 64)	*A. wilsonii* *A. volubile*

In the field of in vitro culture of *Aconitum* spp. Cervelli presented the micropropagation method of *A. noveboracense* from leaf segment via embryogenesis (1986) and *A. napellus* by multiplication of shoots (1987). However, since there has been no attempt to breed for alkaloid content, two procedures of micropropagation of *A. carmichaeli* are presented in this chapter. The first is based on somatic embryogenesis from anther and plant regeneration, the second was carried out by tip tissue culture. The quantitative analysis of aconitine-type alkaloids in the root regenerated from shoot tip was additionally established by HPLC to confirm the homogeneity of the new strain. Table 2 reviews in vitro approaches in *Aconitum* species.

Table 2. In vitro regeneration studies on *Aconitum* spp.

Plant species	Explant	Results	Reference
Aconitum noveboracense	Leaf	Callus Embryoids Shoots Plantlets	Cervelli (1986)
A. noveboracense	Stem nodes	Rosette Plantlets	Cervelli (1987)
A. napellus	Stem nodes	Rosette Plantlets	Cervelli (1987)
A. carmichaeli	Anther	Callus Embryos Shoots Plantlets	Hatano et al. (1987)
A. carmichaeli	Shoot tip	Multiple shoots Plantlets	Hatano et al. (1988)

2 Somatic Embryogenesis and Plant Regeneration from Anthers (Hatano et al. 1984, 1987)

2.1 Material and Methods

Young flower bud segments of *A. carmichaeli*, measuring 7–12 mm in diameter, were taken from field-grown plants. The flower buds were washed with tap water, surface sterilized with 2% sodium hypochlorite solution containing 0.1% Tween 80 for 10 min, then with 70% alcohol for 30 s, and finally thoroughly washed twice with sterilized water. The immature anthers were removed aseptically, and were then cultured individually, maintaining a consistent average of anthers in all treatments. The basal medium consisted of MS salts (Murashige and Skoog 1962). The cultures were exposed to continuous light from cool white fluorescent tubes at a temperature of 25 ± 1 °C except for callus and embryo induction treatment.

2.2 Results

In the dark, callus formation was induced by all concentrations of 2,4-D during the 15 weeks of this study, although the rate was not high (Table 3). The combination of 5 mg/l 2,4-D and 1 mg/l kinetin stimulated the callus induction rate in 36.8% of anthers. In contrast, when the culture was incubated under illumination, callus induction was completely inhibited.

Table 4 shows embryoid formation from calli induced on medium containing 0.5 to 5 mg/l 2,4-D with and without 1 mg/l kinetin. On all the media the callus segment first changed into a partly brownish color and then produced a few embryoids, whose globular forms predominated on the surface of the callus during 12 weeks of culture (Fig. 3). The best result was obtained with 1 mg/l 2,4-D; in this case all calli multiplied, and 60% of them formed embryos during the 12 weeks of culture. It was evident that in order to increase the number of embryoids and also the later stages, such as heart-shaped or cotyledonary forms, further subculture was necessary.

Table 3. Effects of hormones on callus formation through anther culture in the dark. (Hatano et al. 1987)

Hormone (mg/l)		No. of anthers examined	No. of calli formed	Fresh weight (mg)
2,4-D	Kinetin			
0	0	12	0	–
0.1	0	53	1	5
0.5	0	53	3	9
1	0	54	6	11
5	0	53	11	19
10	0	53	6	8
0.1	1	56	8	7
0.5	1	54	12	16
1	1	48	8	22
5	1	57	21	29
10	1	53	9	29

Table 4. Effects of hormone on embryogenesis. (Hatano et al. 1987)

Hormone (mg/l)			No. of calli examined	Calli propagated per culture	Embryo formed per culture	Fresh[a] weight (g)
2,4-D	IAA	Kinetin				
0	0	0	7	3	–	0.01
		1	7	3	–	0.01
0.5			10	9	5	0.04
1			10	10	6	0.04
5			10	9	2	0.04
0.5		1	7	6	–	0.04
1		1	7	5	–	0.02
5		1	7	6	2	0.02
	1		10	2	–	0.02
	5		10	4	–	0.06
	10		10	7	–	0.08
	1	1	10	10	–	0.18
	5	1	10	10	1	0.18
	10	1	10	10	–	0.04

[a] Callus weight subcultured is approximately 10 mg.

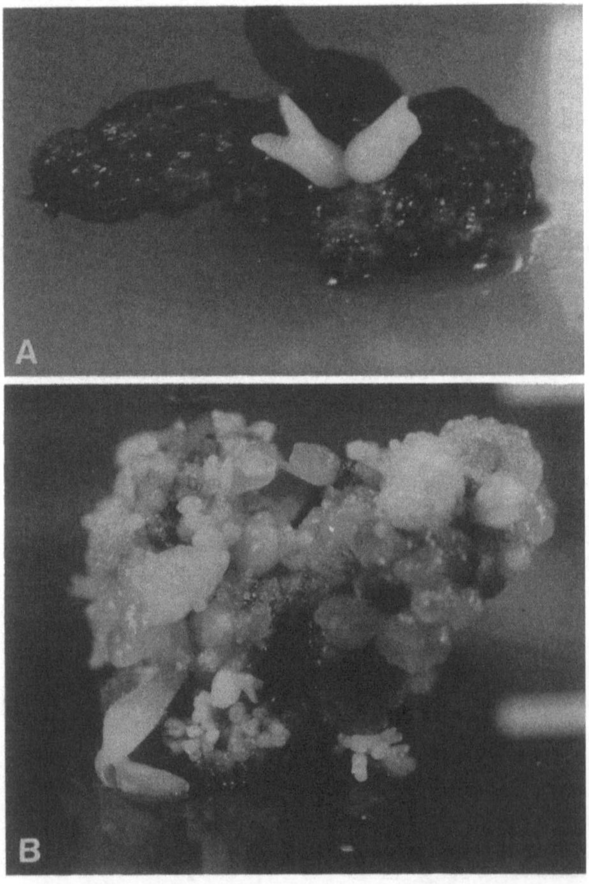

Fig. 3A,B. Embryogenesis from anthers. **A** Primary embryo induction from callus on medium supplemented with 1 mg/l 2,4-D. **B** Embryo formation on medium supplemented with 1 mg/l 2,4-D during 12 weeks of culture

Table 5 shows the effect of subculturing a callus cluster containing a single embryo onto a new medium inoculated with various concentrations of 2,4-D or of IAA. Addition of IAA did not favor the propagation of embryos in general. Lower concentrations of 2,4-D also proved ineffective, whereas higher concentrations favored embryoid propagation. In particular, a medium containing 1 mg/l 2,4-D and various concentrations of BAP was tested (Table 6). The addition of 1 mg/l 2,4-D and 1 mg/l BAP to the medium appreciably influenced shoot formation during 6 weeks of culture. A similar number of shoots was obtained during the fifth subculture. Additionally, towards the end of the culture, in an individual generation, many heart-shaped embryoids formed on the surface of the callus and the germination of these was also observed (Fig. 4).

It is well known that the addition of GA and cytokinin to the medium stimulates shoot regeneration from embryoids (Chang and Hsing 1980; Shoyama et al. 1987, 1988). Table 7 shows the effect of GA alone or with kinetin or BAP. The experiments with supplements of GA (1 mg/l) in the presence of BAP (5 mg/l) produced the largest numbers of regenerated normal plantlets. In this case 71% of the embryoids produced plantlets with phenotype identical to that of the original plant (Fig. 5) for a 6-week culturing period.

Table 5. Effects of hormone on embryo propagation. (Hatano et al. 1987)

Hormone (mg/l)		No. of calli	Embryos multiplied	Shoots formed	Fresh[a]
2,4-D	IAA	examined	per culture	per culture	weight (g) per culture
0	0	9	9	–	0.07
0.1		9	2	–	0.16
0.5		8	1	–	0.16
1		8	5	8	0.33
2		9	5	–	0.24
	1	8	2	3	0.16
	2	10	1	–	0.11
	5	10	–	–	0.11
	10	10	1	–	0.14

[a] Callus weight subcultured is approximately 20 mg.

Table 6. Effects of hormone on shoot formation in individual generation. (Hatano et al. 1987)

Generation	Hormone (mg/l)		No.	No. of	No. of shoots
	2,4-D	BAP	examined	shoots formed	per culture
	1	0	20	14	2.3
Subculture 4	1	0.5	20	20	3.1
	1	1	20	20	6.8
	1	2	20	18	3.2
	1	0	18	17	3.9
	1	0.5	17	16	5.9
Subculture 5	1	1	20	20	5.9
	1	2	20	18	5.4

Fig. 4. Embryo propagation and shoot formation on medium supplemented with 1 mg/l 2,4-D 1 mg/l BAP in 6-week-culture

Hormone (mg/l) GA	Kinetin	BAP	Germination (%)	Fresh weight(g) per culture
0	0	0	20	0.23
1	0	0	17	0.30
1	1	0	60	0.24
1	0	1	57	0.30
1	0	5	71	0.41

Fig. 5. Shoot formation from somatic embryo on medium containing 1 mg/l GA and 1 mg/5 BAP in 6-week culture

In the final stage of this study, root formation was investigated in a medium inoculated with 0.5 mg/l IAA. For 6 weeks root formation with daughter roots occurred perfectly. The average number of roots and daughter roots was 9.9 and 2.2, respectively, as shown in Fig. 6. The plantlets and daughter roots were transferred to pots filled with vermiculite at 25 °C. The transplantation was perfectly successful in all cases. The young plantlets were then transplanted in the field and matured, and flowered 5 months later (Fig. 7).

3 Clonal Propagation by Tip Tissue Culture (Hatano et al. 1985, 1988)

3.1 Materials and Methods

The plants of *A. carmichaeli* used for tissue culture and alkaloid analysis were propagated by division and cultivated in the Tochigi prefecture. Portions of the shoot top (2 cm length) and the node (2 cm length) having axillary buds were removed, sterilized with the same method as indicated in Section 2.1. The shoot tips and axillary buds were dissected from the shoot tips and nodes under a binocular microscope. Sections (0.5–1 mm in length) had two leaf primordia. The basal medium consisted of MS salts supplemented with IAA, BAP, and kinetin in the concentrations and combinations indicated in the tables. The effects of temperature on the transplantation of the plantlets in soil were investigated at 15, 20, 25 and 30 °C in a biotron and cultivated in a mixture of clay and compost in the biotron for 5 months.

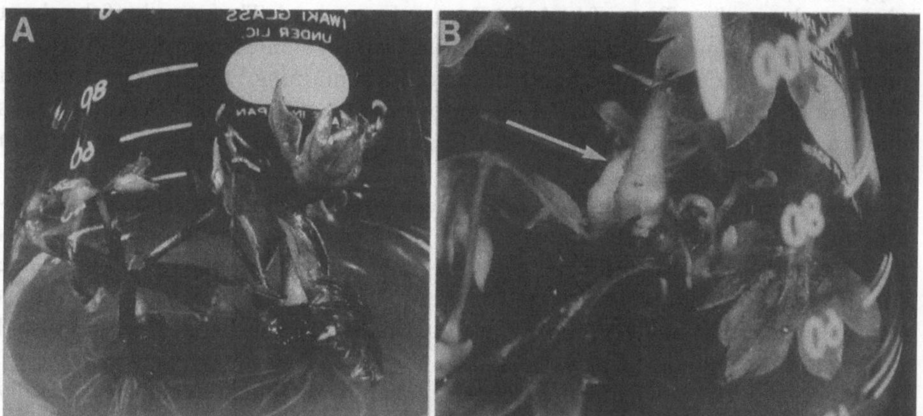

Fig. 6A,B. Root formation on medium supplemented with 0.5 mg/l IAA (**A**) and daughter root formation on medium supplemented with 0.5 mg/l IAA during 6-week-culture (**B**). *Arrow* shows daughter root

Fig. 7. Plants regenerated from somatic embryos after 5-month cultivation period in field

3.2 Results

All freshly excised shoot tips exhibited multiple shoot formation independent of the hormone concentration. The addition of IAA (0.5, 1, or 2 mg/l) and kinetin (1 mg/l) to the medium promoted multiple shoot formation. IAA (2 mg/l) kinetin (1 mg/l) supplemented medium stimulated the development of shoot tips into root-forming plantlets. Next, the effects of IAA-BAP were determined. The growth of the shoot tip was almost similar to that in the medium supplemented with IAA-kinetin as indicated above. On the other hand, the medium containing 5 mg/l BAP stimulated the most effective development of the shoot tips into multiple shoot complexes, as indicated in Table 8 (Fig. 8). Therefore, this condition became the routine medium used in the initial stage.

The culture of axillary buds was investigated on medium containing IAA-kinetin, IAA-BAP, or BAP singly (Table 9). Although the growth of axillary buds was not active as compared with shoot tips in general, approximately four or five shoots per axillary bud were obtained on the medium containing 5 or 10 mg/l. It thus becomes clear that axillary buds as well as shoot tips can be used as the clonal propagation system of *A. carmichaeli*.

Since it is well known that the regeneration ability of the callus culture generally decreases and finally disappears during long-term subculture (Shono 1965), the shoot number of the multiple shoot complex in individual generations was investigated. The shoots were transferred to MS medium containing 5 mg/l BAP and cultured under the same conditions as for the primary culture. The multiple shoot formation rate increased as compared to the primary culture (Table 10). In the fifth

Table 8. Effects of hormone on multiple shoot formation through shoot tip culture. (Hatano et al. 1988)

Hormone (mg/l) IAA	Kinetin	BAP	No. of shoot tips examined	No. of shoots formed per culture	No. of plantlets formed
0	0	0	4	4.0	–
	1		4	4.8	–
0.1	1		4	5.3	–
0.5	1		5	6.4	1
1	1		5	6.2	1
2	1		5	5.4	2
5	1		4	4.0	–
10	1		4	3.3	–
1		1	4	4.3	–
1		2.5	5	6.0	–
2		1	3	4.0	–
		1	4	5.8	–
		5	4	7.0	–
		10	4	6.8	–

Culture conditions: MS medium, 25 ± 1 °C, continuous light, 6 weeks.

Fig. 8. Multiple shoot formation cultured on medium supplemented with 5 mg/l BAP at 25 °C for 6 weeks

generation, 9.4 shoots per culture were obtained, showing that no decrease in the shoot number formed had occurred during the continuous subculture.

In order to determine the suitable basal medium and effective hormone for rooting, the MS, 1/2 MS, and 1/4 MS media supplemented with IAA were tested at 25 °C under continuous light for 8 weeks (Table 11). It was found that the MS medium supplemented with 0.5 or 1 mg/l IAA stimulated root formation. On the other hand, the addition of kinetin to the medium inhibited root formation. When the culture temperature was varied between 20 and 25 °C, a higher root formation

Table 9. Effects of hormone on multiple shoot formation through axillary bud culture. (Hatano et al. 1988)

Hormone (mg/l)			No. of buds examined	No. of shoot-forming cultures	No. of shoots formed per culture
IAA	Kinetin	BAP			
0	0	0	5	0	–
0.5	1		4	3	3.3
1	1		4	2	3.0
2	1		4	2	3.0
1		1	5	3	2.7
1		2.5	10	8	2.5
2		1	5	3	2.7
		1	10	8	2.5
		5	10	10	4.2
		10	10	10	4.8

Culture conditions: MS medium, 25 ± 1 °C, continuous light, 6 weeks.

Table 10. Number of shoots propagated in individual generation[a] (Hatano et al. 1988)

Generation	No. of shoots per culture
primary	7.0
sub. 1st	7.3
2nd	9.0
3rd	8.1
4th	8.9
5th	9.4
6th	9.0

Culture conditions: MS medium supplemented with 5 mg/l BAP, 25 ± 1 °C, continuous light, 6 weeks.
[a] Ten shoots were examined in the cultures of individual generations, except for the primary culture.

(100%) was observed at 20 °C, especially in the medium containing 0.5 or 1 mg/l IAA, which produced 14 and 12.2 roots per shoot, respectively.

Finally, the regenerated plantlets were transferred to vermiculite and cultivated in a biotron at 15, 20, 25, and 30 °C for 4 weeks. The transplantation was successful in almost all cases. The young plantlets matured and flowered 5 months later, the same as the plantlets regenerated from embryos. Figure 9 shows the 3-year cultivation of clonally propagated plantlets in the field (Fig. 9).

4 Analysis of Aconitine-Type Alkaloids

4.1 Materials and Methods

Aconitine, hypaconitine, and mesaconitine were isolated from the crude drug of *A. carmichaeli* and identified by comparisons of m.p., ^{1}H-NMR, ^{13}C-NMR, and

Medium	Hormone (mg/l)		No. of shoots examined	No. of shoots rooted	No. of roots per shoot
	IAA	Kinetin			
	0	0	9	–	–
	0.5	0	8	4	4.3
	1	0	9	5	4.2
MS	1	0	9	5	4.2
	2	0	7	–	–
	1	1	4	–	–
	2	1	4	–	–
	1	0	4	4	3.3
1/2 MS	2	0	3	–	–
	1	1	3	–	–
	2	1	3	–	–
	1	0	3	2	2.0
1/4 MS	2	0	3	–	–
	1	1	2	–	–
	2	1	2	–	–

Culture conditions: MS medium, 25 ± 1°C, continuous light, 8 weeks.

Fig. 9A,B. Cultivation of clonally propagated plant in field. **A** After 3 years **B** Harvested roots

FD-mass spectra. Quantitative analysis was carried out according to Kitagawa's method (Kitagawa et al. 1984b) with the following modification. The dried powder (500 mg) was extracted five times with MeOH (5 ml) under sonication for 15 min. The NH_3 solution (3% water, 4 ml) added to the MeOH extract was evaporated in vacuo to dryness. The residue was subjected to Al_2O_3 (Woelem, neutral, grade III; 1 g) column in chromatography eluted by $CHCl_3$ (16 ml). The solution was evaporated in vacuo to dryness. The residue (crude alkaloids) was dissolved in the mixture of CH_3CN-0.01 M dibutylamine phosphate (7:3), and subjected to HPLC. Analyses were performed on a Gasukuro Kogyo Model 576 high-performance liquid chromatograph which was connected to a Radial-PAK CN (8 mm × 100 mm, Waters) and equipped with a variable wave length detector (detection set at 235 nm) (Hikino and Konno 1981). The mobile phase was the mixture of CH_3-CN-0.01 M dibutylamine phosphate (7:3). Mesaconitine (1.60 mg), aconitine (0.79 mg), and hypaconitine (1.43 mg) were individually dissolved in the mobile phase solution, diluted stepwise, and 10 μl portions were injected into the HPLC twice. The calibration plots were obtained from the peak heights for the separate substance preparations.

4.2 Results

Table 12 shows the individual alkaloid content in the dried roots and the coefficient of variation between the regenerated and the normally grown plantlets (mother strain). Only small differences in individual alkaloid content between the two strains were observed.Since the normally grown plants were also clonally propagated by division, various characteristics were theoretically the same, but it was clearly seen that the variation among the clonally propagated plants was smaller than that of the normally grown plants.

Table 12. Aconitine-type alkaloid contents of *Aconitum carmichaeli* Debx. roots of the clonally propagated plants (1)[a] and normally grown plants (mother plant; 2). (Hatano et al. 1988)

Material	(N)	% dry weight mesaconitine	(% C.V.[b]) hypaconitine	Aconitine
1	10	0.1514 ± 0.0321 (21.2)	0.0522 ± 0.0179[c] (32.4)	0.0200 ± 0.0044 (22.0)
2	70	0.1265 ± 0.0322 (26.3)	0.0408 ± 0.0170 (41.7)	0.0202 ± 0.0087 (43.1)

[a] Cultivated in biotron at 15 ± 1°C for 5 months.
[b] Coefficient of variation.
[c] Significantly different from the normally grown strain, $P < 0.05$.

5 Conclusion

As previously described, the roots of *A. carmichaeli* contain aconitine-type alkaloids such as aconitine, hypaconitine, and mesaconitine, of which the lethal doses are 2 to 3 mg in man. In addition, the remarkable differences in the quality and quantity of aconitine-type alkaloid depend on the strain and/or the place of growth. Hikino et al. (1983) reported that the amounts of aconitine alkaloid in raw roots of *A. carmichaeli* which were cultivated in three different places in Japan and China were quantitatively and qualitatively different from each other. Therefore, it is easily expected that the variation of aconitine-type alkaloids in individual strains of *A. carmichaeli* may be larger. Therefore, the breeding of homogeneous strain for constant biological activity is necessary.

It is known that the micropropagation method by embryogenesis promotes the homogeneity of constituents, as previously reported in *Angelica acutiloba* (Miura et al. 1988) and *Bupleurum falcatum* (Hiraoka et al. 1986). Therefore, the propagation system by embryogenesis also might be used as a clonal propagation system. On the other hand, when variation is needed for alkaloid constituents, callus culture may be useful for the induction of mutants. If a favorable strain is obtained, its tip tissue can be used for clonal propagation by the second method. By this system it is theoretically possible to obtain more than approximately 1.2×10^4 clonal plants from a single shoot in 1 year. The system rapidly provides us with a strain having a more homogeneous alkaloid content.

6 Protocol

1. *Plantlet regeneration by embryogenesis from anthers*
   1. Culture anthers on MS medium containing 2,4-D (5 mg/l) and kinetin (1 mg/l) under dark at 25°C for 15 weeks to form callus.
   2. Transfer callus on MS medium supplemented with 2,4-D (1 mg/l) under dark at 25°C for 12 weeks to give embryoids.
   3. Culture embryos on MS medium supplemented with GA (1 mg/l) and BAP (5 mg/l) under continuous light at 25°C for 6 weeks to regenerate shoots.
   4. Culture regenerated shoots on MS medium supplemented with IAA (0.5–1 mg/l) at 25°C for 6 weeks to give plantlet.
2. *Propagation by tip culture*
   1. Culture tip tissue on MS medium containing BAP (5 mg/l) under continuous light at 25°C for 6 weeks to form multiple shoot complex.
   2. Transfer shoot on MS medium supplemented with IAA (1 mg/l) under continuous light at 20°C for 8 weeks to give plantlet.
3. *Quantitative analysis of aconitine-type alkaloids*
   Analyze the contents of aconitine-type alkaloid contained in the regenerated plants by HPLC to confirm the coefficient of variation.

References

Cervelli R (1986) Somatic embryogenesis and shoot multiplication of the threatened perennial: northern monkshood *Aconitum noveboracens*. In: 6th Int Congr Plant tissue and cell cultures, Minneapolis, MN, Abstr, p 189

Cervelli R (1987) In vitro propagation of *Aconitum noveboracense* and *Aconitum napellus*. HortSci 22:304–305

Chang W, Hsing Y (1980) In vitro flowering of embryoid derived from mature root culture of ginseng (*Panax ginseng*). Nature 284:340–342

Hatano K, Shoyama Y, Nishioka I (1984) Study on *Aconitum* spp. (2) Propagation of *Aconitum carmichaeli* by anther culture. In: 31st Annu Meet Jpn Soc Pharmacognosy, Abstr, p 69

Hatano K, Shoyama Y, Nishioka I (1985) Study on *Aconitum* spp. (3) Clonal propagation of *Aconitum carmichaeli* by tip tissue culture. In: 105th Annu Meet Pharmaceut Soc Jpn, Abstr, p 429

Hatano K, Shoyama Y, Nishioka I (1987) Somatic embryogenesis and plant regeneration from the anther of *Aconitum carmichaeli* Debx. Plant Cell Rep 6:446–448

Hatano K, Kamura K, Shoyama Y. Nishoka I (1988) Clonal multiplication of *Aconitum carmichaeli* by tip tissue culture and alkaloid contents of clonally propagated plant. Planta Med 54:152–155

Hikino H, Konno C (1981) Determination of aconitine alkaloids by high-performance liquid chromatography. J Chromatogr 211:123–128

Hikino H, Murakami M, Konno C, Watanabe H (1983) Determination of aconitine alkaloids in *Aconitum* root. Planta Med 48:67–71

Hiraoka N, Kodama T, Oyanagi M, Nakano S, Tomita Y, Yamada N, Iida O, Satake M (1986) Characteristics of *Bupleurum falcatum* plants propagated through somatic embryogenesis of callus culture. Plant Cell Rep 5:319–321

Kitagawa I, Yoshikawa M, Chen ZL, Kobayashi K (1982) Four new lipo-alkaloids from aconiti tuber. Chem Pharm Bull 30:758–761

Kitagawa I, Chen ZL, Yoshihara M, Yoshikawa M (1984a) Chemical studies on crude drug processing II. Aconiti tuber (1). On the constituents of "Chuon-wu", the dried tuber of *Aconitum carmichaeli* Debx. Yakugaku Zasshi 104:848–857

Kitagawa I, Chen ZL, Yoshihara M, Yoshikawa M (1984b) Chemical studies on crude drug processing IV. Aconiti tuber (3). Quantitative determination of aconitine alkaloids in aconiti tuber by means of high performance liquid chromatography. Yakugaku Zasshi 104:867–872

Konno C, Shirasaka M, Hikino H (1984) Cardioactive principle of *Aconitum carmichaeli* roots. Planta Med 35:150–155

Kosuge T, Yokota M (1976) Studies on cardiac principle of aconiti root. Chem Pharm Bull 24:176–178

Miura Y, Fukui H, Tabata M (1988) Reduced inhomogeneity of *Angelica acutiloba* plants propagated clonally through somatic embryoids. Planta Med 54:79–81

Mori T, Ohsawa T, Murayama M, Bando H, Wada K, Amiya T (1989) Studies on *Aconitum* species. VIII. Components of "Kako-bushimatsu". Heterocycles 29:873–885

Murashige T, Skoog F (1962) A revised medium for rapid growth and bioassays with tobacco tissue culture. Physiol Plant 15:473–497

Sakai K (1933) A brief note on the chromosomes in some *Aconitum* species. Trans Sapporo Nat Hist Soc 13:74–77

Shono K (1965) Changes in organ forming capacity of carrot root calluses during subcultures. Plant Cell Physiol 6:403

Shoyama Y, Nishioka I, Fujioka N, Kohda H, Yamasaki K (1987) Clonal multiplication of *Panax japonicus* by tissue culture. Shoyakugaku Zasshi 41:333–337

Shoyama Y, Kamura K, Nishioka I (1988) Somatic embryogenesis and clonal multiplication of *Panax ginseng*. Planta Med 54:155–156

V *Apium graveolens* L. (Celery): In Vitro Culture and the Production of Flavors

H.A. COLLIN and S. ISAAC[1]

1 Occurrence and Importance of Celery

Celery (*Apium graveolens* L.) is an umbellifer and is therefore a close relative of parsley, parsnip, and carrot. Like many damp-loving umbellifers, wild celery has a wide distribution, extending from Sweden to Algeria, Egypt, Ethiopia, and in Asia from the Caucasus to Baluchistan and to the mountains of India (Atal and Kapur 1977). The wild plant is bitter and was probably used for medicinal purposes before gaining popularity as a herb and a vegetable (Small 1948). The curative powers attributed to celery are many and varied. Both the roots and seeds were used medicinally, especially in obstructions of the liver and spleen and in the treatment of fevers, jaundice and diaorrhea, pains in the chest, windy cholic and as a diuretic (Phillips 1827). Celery seed (or its oil) is also apparently a cure for rheumatism, gout, bronchitis and asthma (Karim and Bhatty 1976), flatulence and colic (Bjeldanes and Kim 1978) and is ascribed the properties of being abortifacient, antiseptic (Bjeldanes and Kim 1978), deobstruent, anti-inflammatory, a cardiac tonic, a sedative (Guenther 1950) and finally and in contradiction to the last, a stimulant (Karim and Bhatty 1976). Despite this wealth of medicinal qualities ascribed to celery and its seed, no pharmaceutical value is at present attributed to it, being regarded as one of many folk remedies (Crosby and Anderson 1963). The exceptions to this view are publications by Kohli et al. (1967) and Bjeldanes and Kim (1978) on the sedative activity of celery seed constituents.

The main present day use of celery is in the food industry, largely in the form of seed oil or oleoresin. Seed oil is used in the flavouring of many kinds of convenience foods; soups, sauces, meats, pickles and vegetable juices. The main demand comes from the canned soup industry, however, where celery is important as a flavour in its own right. The other outlet for celery seed is to the vegetable seed trade, as the herb itself is a very popular vegetable, either as the strongly flavoured green varieties, or as the increasingly popular self-blanching varieties. This market takes the high quality fertile seed, leaving the cheaper seed for oil extraction. Celery herb oil can also be distilled from the aerial plant parts, but according to Robbins and Greenhalgh (1979), this source is of negligible significance since the oil content of the shoot is so small as to make production costs prohibitive. American workers (Gold and Wilson 1961; Wilson et al. 1967, 1969; Wilson 1970) however, have made a series of investigations and improvements on the efficiency of flavour extractions

[1]Department of Genetics and Microbiology, University of Liverpool, Liverpool L69 3BX, United Kingdom

Biotechnology in Agriculture and Forestry, Vol. 15
Medicinal and Aromatic Plants III (ed. by Y.P.S. Bajaj)

from the whole plant. They point out (Gold and Wilson 1961) that since fresh celery is subject to high variations in the market, flavour extraction from such surplus could provide at least some return. Lund et al. (1974) and Buttkus (1978) have also drawn attention to the possibilities of flavour extraction from celery packing-house waste. If a rapid, large-scale production method for growth of fresh celery tissue (such as tissue culture) was available, prospects for herb oil could be further improved.

The main celery seed oil distilling countries are France, India, the USA and the UK. Much of the seed which enters international trade markets for eventual distillation originates in India (Robbins and Greenhalgh 1979). Currently, India produces about 1500 tonnes of celery seed and with a yield of 2% oil, the total production of oil would be around 30 tonnes. Of the other major distillers, the USA production is mainly for its own internal requirements, as is the high quality product distilled in the UK, although this oil is also exported in small quantities to a number of other countries.

The other major product from celery seed used by the flavour industry is that of oleoresin, which is produced by extraction of the seed with a hydrocarbon solvent or ethanol. In 1970, the report by Adamson stated that this product did not compete on a large scale with seed oil, being largely sold as a dispersed spice extract. However, by 1979 (Robbins and Greenhalgh) the rapid growth of celery seed oleoresin market had limited the expansion of seed oil trade. Usually oleoresins dispersed as above compete more directly with ground spice or seed products, avoiding the high bacterial loads associated with these commodities.

2 Composition of Celery Essential Oil

Since celery has a long history of use as a vegetable and a drug, much work has been carried out on the constituents of celery and celery seed oil. Fatty acids, for example, include oleic, linoleic, palmitic, palitoleic, petroselinic, petroselaidic, stearic, myristic and myristoleic (Gold and Wilson 1961; Balbaa et al. 1976) and several shorter chain fatty acids (Gold and Wilson 1961). The presence of acids of the tricarboxylic acid cycle has been demonstrated in celery tissue, e.g., citric, isocitric, fumaric, malic and tartaric acids were reported by Schramm (1961) in celery root, while malic and citric acids were detected in the edible portions in some early work by Hartman and Hillig (1934) and pyruvic (Gold and Wilson 1961), malonic (Bentley 1952) and glycolic (Andrews and Viser 1951) acids have also been found. Many macromolecular compounds have also been isolated from celery herb and seed oils. For compounds such as proteins, carbohydrates and sugars, see the review published by Crosby and Anderson (1963). However, the importance of celery as a flavour is due more to the components of the essential oil.

Terpenes of celery oil have been thoroughly investigated since the early work by Ciamician and Silber (1897). Simonsen and Barton isolated δ- and ε-selinene as well as α- and β- in 1952. Other terpenes include α- and β-pinene, myrcene (Wilson 1969; Fehr 1974; Karim and Bhatty 1976), γ-terpenine, β-elemene, α- and β-humulene (Wilson 1969; Fehr 1974), p-cymene, β-caryophyllene (Karim and Bhatty 1976; Wilson 1970), p-cymol (Fehr 1974), γ-terpineol (Karim and Bhatty 1976) and α-ionone (Wilson 1970).

Two other interesting aromatic compounds which occur in celery are apiole (1-allyl-2,5-dimethoxy-3,4-methylene dioxybenzene) and myristicin (1-allyl-3-methoxy-4,5-methylene dioxybenzene). These compounds are closely related to safrole (1-allyl-3,4-methylene dioxybenzene), which has now fallen into disuse as a flavouring agent because of its alleged carcinogenicity (Crosby and Anderson 1963). Other aromatic compounds which have been demonstrated in celery are coumaric, caffeic, ferulic, chlorogenic and isochlorogenic acids (Herrmann 1978). One other compound found in celery of general interest is the photodynamic furanocouramin bergapten (furanocoumarin-5-methoxysporalen), an oxygen-containing heterocyclic compound (Opdyke 1974), and more recently two other furanocoumarins have been identified in celery (Innocenti et al. 1976).

A comprehensive list of compounds isolated from celery has been collected together by van Straten et al. 1977) and is reproduced in Table 1. The major analysis of celery has been carried out by Wilson and Gold, and further details of alcohols, esters, aldehydes, phthalides, hydrocarbons, and carbonyls can be found in their publications (Gold and Wilson 1961, 1963a,b; Wilson 1969, 1970; Wilson et al. 1969). Recently Macleod et al. (1988) detected nine additional compounds in the volatiles of fresh celery (Table 2). The major aroma constituents were apiole, myristicin and the two phthalides, 3-butylphthalide and 3-butyltetrahydrophthalide (sedanenolide), although the relative proportions of each differed from the earlier finding of Gold and Wilson.

Although the chief flavour constituents of celery are the C_{12} lactones known as phthalides, many of the diverse compounds listed by van Stratten et al. (1977) and Macleod et al. (1988) may contribute to the finer nuances of odour and flavour, giving it depth and variation. The first qualitative attempt at evaluating flavour in celery was that of Hall (1957, 1959), where the position of the petiole was given an organoleptic rating; the outer greener petioles were much more bitter and less palatable than the inner ones. Pan (1960) reported a burning-numbing taste associated with the outer green layer of the petiole, as distinct from a bitter principle (Pan 1961) which was a non-flavonoid D-glycoside. The burning-numbing factor has since been attributed to dihydrophthalides, the isobutylidene being more intense than the isovalidene (Gold and Wilson 1963b). Although the terpenes comprise some 80% of the essential oil, only β selinene is attributed any role in celery flavour (Wilson 1969; Lund 1978). Of the phthalides present in the essential oil sedanenolide, sedanolide, 3-isobutylidene-3a,4-dihydrophthalide, 3 isovalidene-3a,4-dihydrophthalide, 3-butyl phthalide, 3-isobutylidene phthalide, 3-isovalidene phthalide (Fig. 1), greatest importance is given to the last five, particularly the dihydrophthalides. In addition to the phthalides, cis-3-hexenyl-1-yl pyruvate and 2,3-butanedione are also said to have a minor role as flavour components (Gold and Wilson 1963a). Work previously carried out in this laboratory has isolated 3-isobutylidene-3a,4,5,6-tetrahydrophthalide and 3-isobutylidene-3a,4-dihydrophthalide from celery tissue cultures (Al-Abta et al. 1979) but only from morphologically differentiated cultures; undifferentiated callus so far appears unable to synthesise celery flavours.

Lund (1975) reported that the major phthalides are obtained in higher yields from celery waste with a higher proportion of leaves to stalks. Subsequent work by Fehr (1979) showed that the chief phthalide found is strongly dependent on the plant organ from which the essential oil is extracted. Some of the phthalides are

Table 1. Compounds isolated from celery tissue. (van Straten et al. 1977)

Hydrocarbons	Acids	Esters
Ocimene	Acetic	cis-3-Hexanyl acetate
Myrcene	Glyoxylic	2-Methylhexyl acetate
a-Terpinene	2-Methylpropanoic	Decyl acetate
g-Terpinene	2-Oxopropanoic	Citronellyl acetate
Terpinolene	cis-2-Methylbut-2-enoic	Geranyl acetate
Limonene	trans-2-Methylbut-2-enoic	Neryl acetate
b-Elemene	Pentanoic	Linalyl acetate
a-Humulene	Hexadecanoic	Terpinyl acetate
b-Humulene	9-Octadecanoic	Dihydrocarvyl acetate
a-Pinene	9,12-Octadecadienoic	cis-p-Metha-1(7),8-dien-2-yl acetate
b-Pinene	4-Hydroxycinnamic	cis-Carvyl acetate
Sabinene	3,4-Hydroxycinnamic	trans-Carvyl acetate
b-Selinene	4-Hydroxy-3-methoxy cinnamic	Pinocarvyl acetate
g-Selinene		Terpinyl propanoate
d-Selinene		cis-3-Hexenyl 2-oxopropanoate
e-Selinene		Geranyl butanoate
b-Caryophyllene		Ethyl 3-methylbutanoate
4-Isopropyl-1-methylbenzene		Benzyl benzoate

Alcohols	Carbonyls	Miscellaneous
Methanol	Formaldehyde	2-Methoxyphenol
1-Propanol	Acetaldehyde	1-Allyl-5-methoxy-3,4-methylenedioxybenzene
1-Butanol	Propanal	1-Allyl-2, 5-methoxy-3, 4-methylenedioxybenzene
3-Methylbutan-1-ol	2-Methylpropanal	Furfural
1-Hexanol	2-Oxopropanal	cis-Limonene oxide
trans-2-Hexen-1-ol	2-Butenal	trans-Limonene oxide
3-Hexen-1-ol (cis and trans)	3-Methylbutanal	5-Methoxypsoralen
1-Heptanol	Pentanal	7-Methoxypsoralen (diseased)
1-Octanol	Hexanal	4-5',8-Trimethylpsoralen (diseased)
1-Nonanol	Heptanal	3-Butylhexahydrophthalide
1-Decanol	Octanal	Sedanolide
1-Undecanol	Nonanal	Sedanoic anhydride
1-Dodecanol	Decanal	3a,4-Dihydro-3-isobutylidene-phthalide
a-Terpineol	Undecanal	3a,4-Dihydro-3-isovalidene-phthalide
Dihydrocarveol	Dodecanal	3-Butylphthalide
cis-p-Metha-2, 8-dien-1-ol	Citronellal	3-Isobutylidenephthalide
trans-p-Metha-2, 8-dien-1-ol	Geranial	3-Isovalidenephthalide
cis-p-Metha-1(7), 8-dien-2-ol	Neral	
trans-p-Metha-1(7), 8-dien-2-ol	2-Propanone	
cis-Carveol	2-Butanone	
trans-Carveol	2,3-Butanedione	
p-Menth-8-ene-1, 2-diol	2-Pentanone	
	a-Ionone	
	1-Phenylpentan-1-one	
	Dihydrocarvone	
	Carvone	

Table 2. Volatile components of celery. (Macleod et al. 1988)

Component	% relative abundance
Ethanol	2.2
Pentane	1.5
2-Pentylfuran	0.6
4-Isopropenyl-1-1-methylbenzene	0.7
4-Isopropylcyclohex-2-enone (cryptone)	1.0
p-Mentha-1,3,8-triene	2.3
p-Methylacetophenone	1.2
Valencene	0.5
α-Selinene	0.3

reported to be unstable, for example, ligustilide (from celery and from the umbellifers, *Ligusticum acutilobum*, *Levisticum officinale* and *Cnidium officinale*) decomposes rapidly on exposure to air (Mitsuhashi et al. 1960). Several other closely related phthalides have been reported in celery and also in other umbelliferae species, particularly those used as drugs in Japan; i.e., in *Cnidium officinale* Makino, *Levisticum officinale* Koch., and *Ligusticum acutilobum* Sieb. et Zucc. The chemical relationship between phthalides and celery aroma has been investigated by Berlingozzi (1927) working with dihydro-, tetrahydro- and hexahydrophthalides. He reported that when one of the γ carbon hydrogens was replaced by an alkyl group, a celery odour was less intense. When the γ carbon hydrogens were replaced by an alkylidene group, aroma intensity was found to increase as the number of carbons was increased from one to four. This last finding has been confirmed by Kariyone and Shimuzu (1953).

Little work has been carried out on the biogenesis of these compounds. Barton and De Vries (cited by Mitsuhashi and Nagai 1963) suggested that they may be constructed by head-to-tail linkages of acetate or malonate units. This hypothesis has subsequently been tested in *Ligusticum acutilobum* (Mitsuhashi and Nomura 1966) by feeding of labelled acetic acid to the plant. The resulting alternate carbon-labelling pattern found in 3-butyl phthalide supported the above hypothesis fully. The acetic-malonate pathway has long been accepted as a means of fatty acid synthesis, but the synthesis of aromatic polyketides as an extension of this hypothesis was first put forward by Birch and Donovan in 1953. The actual assembly process is common to both straight-chain polyketides, such as fatty acids, and to aromatics. A starter unit of acyl coenzyme-A, usually acetyl, is extended by malonyl Co-A units, usually previously formed by carboxylation of acetyl Co-A. The intermediate carbon dioxide donor in this reaction is an enzyme bound biotin derivative. In the assembly sequence, both the starter acetyl Co-A and malonyl Co-A are converted to thiolesters of a carrier protein, which has a reactive centre similar to that of coenzyme-A. The acetyl- and malonyl-protein then react together, releasing one molecule of protein and one of carbon dioxide, the protein then being recycled to malonyl Co-A until the required chain length is reached.

MAJOR TETRAHYDROPHTHALIDES AND THEIR ACIDS

Sedanonic acid (1- (1-carboxycyclohex-1-en-6-yl) pentan-1-one)

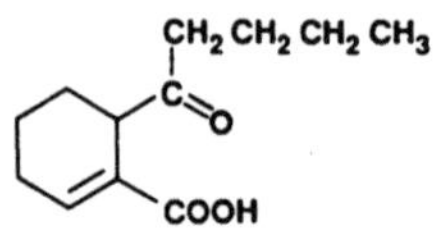

Sedanolic acid (ii) (1-(3-carboxycyclohex-1-en-2-yl) pentan-1-ol)

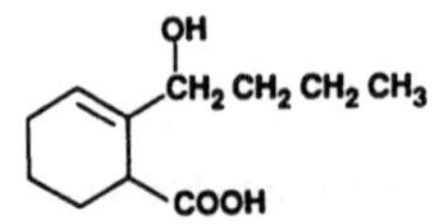

Sedanonic anhydride (3-butylidene-3a, 4, 5, 6-tetrahydrophthalide)

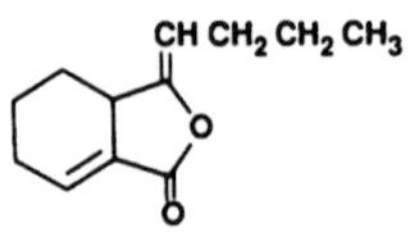

Sedanolide (i) neocnidilide (3-butyl-3a, 4, 5, 6-tetrahydrophthalide)

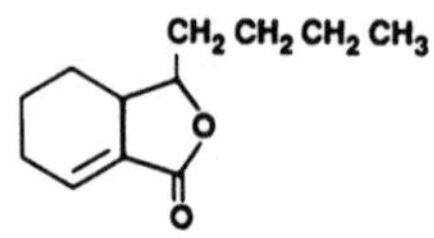

Sedanolic acid (i) (1-(3-carboxycyclohex-2-en-4-yl) pentan-1-ol)

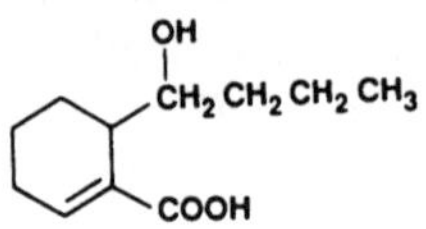

Sedanolide (ii) (3-butyl-5, 6, 7, 7a-tetrahydrophthalide)

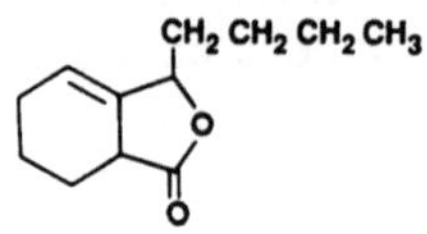

OTHER TETRA- AND HEXAHYDROPHTHALIDES

3-isobutylidene-3a, 4, 5, 6-tetrahydrophthalide

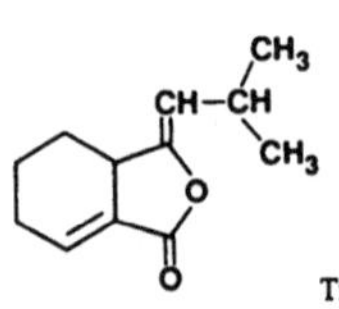

Tissue culture

3-butyl-hexahydrophthalide

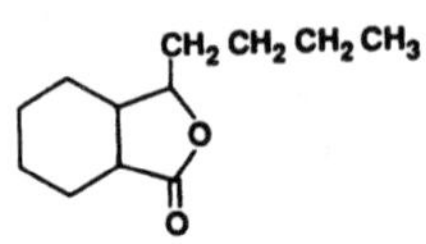

Fig. 1. Structure of phthalides found in celery (*Apium graveolens* L.). (Van Straten et al. 1977)

DIHYDROPHTHALIDES

Sedanenolide, senkyunolide (3-butyl-4, 5-dihydrophthalide)

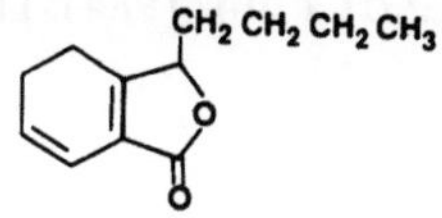

Lingustilide (3-butylidene-4, 5-dihyrophthalide)

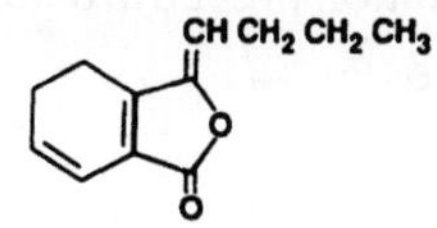

3-isobutylidene-3a, 4-dihydrophthalide

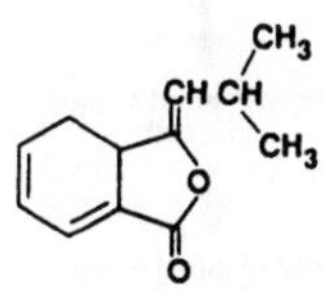

Tissue culture

3-isovalidene-3a, 4-dihydrophthalide

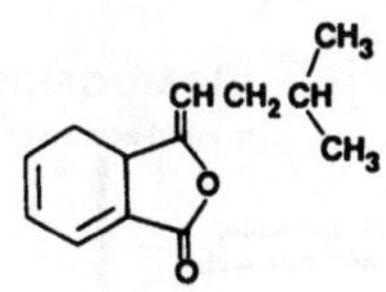

PHTHALIDES

3-butyl phthalide

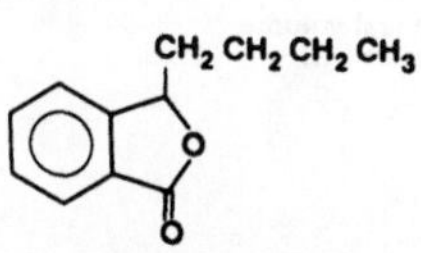

3-butylidene phthalide

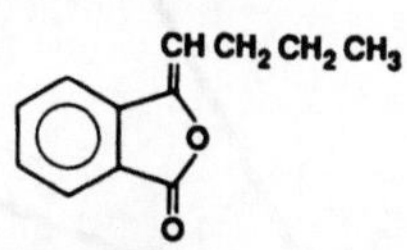

3-isobutylidene phthalide

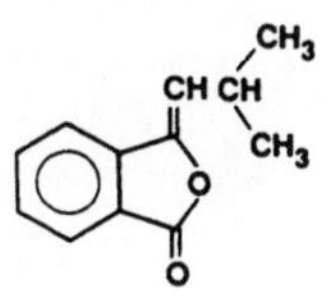

3-isovalidene phthalide

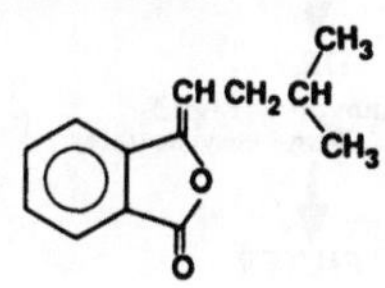

Fig. 1. cont'd.

3 Extraction and Separation of Celery Essential Oil

As previous extractions from celery herb were carried out on a very large scale (Gold and Wilson 1961; Wilson et al. 1967, 1969; Wilson 1970) a small-scale, reliable extraction procedure was developed by Watts and Collin (1984). This is summarised in Fig. 2.

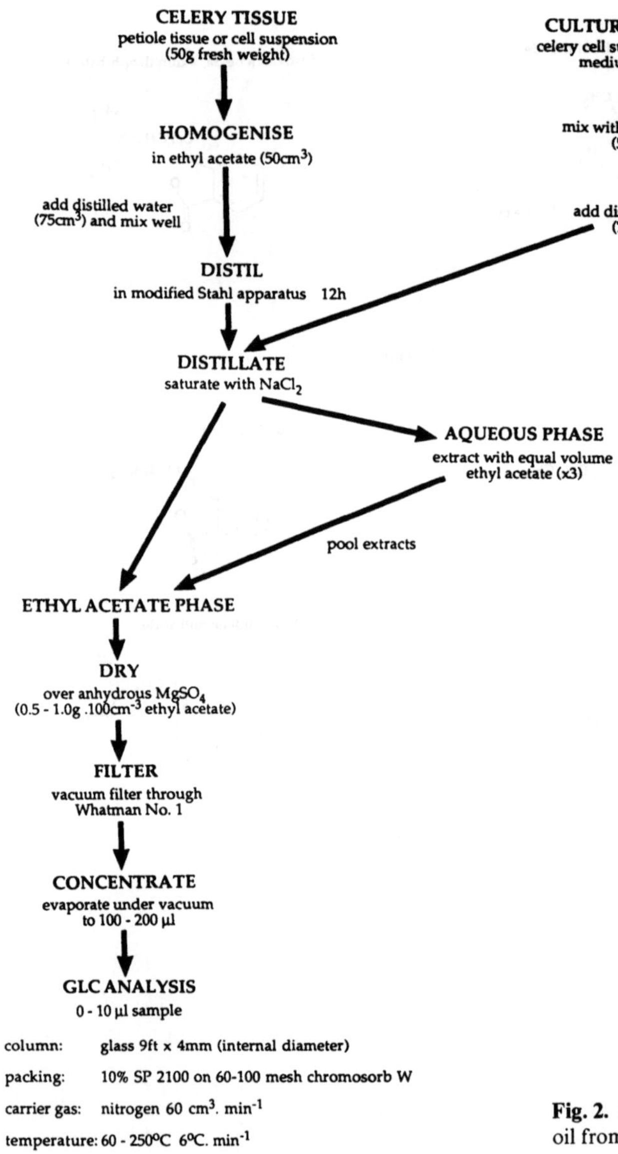

Fig. 2. Protocol for extraction of essential oil from celery. (Collin and Watt 1981)

4 Tissue Culture of Celery

Celery cultures have been initiated, and the effects of various cultural parameters on growth and differentiation investigated by Williams and Collin (1976); Chen (1976) and Rappaport et al. (1980). A comparison of tissue culture methods for celery has been made by Orton (1984). The procedures used by Williams and Collin (1976); Al-Abta and Collin (1978a) to initiate and to regenerate celery tissue cultures are summarised in Fig. 3, see also Browers and Orton (1986).

Regenerating callus on solid medium and in liquid medium showed globular, heart-shaped and torpedo embryos attached to the surface of the callus, or cell aggregates. Alternatively, the embryos were free-floating in the liquid medium (Al-Abta and Collin 1978a) (Fig. 4). Examination of the surface of the callus showed that embryos arose from single cells, or small groups of cells. The embryonic cells were characterised by having a dense cytoplasm and large nuclei (Dunstan et al. 1982) (Fig. 5). The pattern of development of the newly initiated embryos could be controlled by modifying the concentration of 2,4-dichlorophenoxyacetic acid in the medium (Al-Abta and Collin 1978a). Thus at high concentrations of 2,4-D (5 mg/l) embryo development was restricted to the globular stage, and after two subcultures on solid medium it was totally repressed, while after ten subcultures the potential for embryogenesis was completely lost. Where the callus was maintained on a high 2,4-D medium (5 mg/l) for two subcultures, then tranferred to a liquid medium without growth regulators, the numbers of embryos initiated was very large, and the embryos also showed a sequence starting from the globular form and developing to largely torpedo forms after 3 weeks, and plantlets after 6 weeks (Al-Abta and Collin 1978a; Al-Abta et al. 1979) (Fig. 6). Examination of the structure of the embryos revealed that the globular and heart-shaped stage showed no organisation of the cells into tissues (Dunstan et al. 1982) (Fig. 7), but torpedo-stage embryos contained easily recognisable xylem strands and showed a polarised orientation of cells (Al-Abta and Collin 1978b) (Fig. 8). The sequence of embryo development in the liquid medium was partially synchronised and provided therefore a means of relating morphological and structural changes in the embryo to biosynthetic capacity.

5 Production of Flavor Compounds in Celery Tissue Cultures

Flavours have had a long history of use in food and the food industry. Some flavours are determined by single compounds, and hence are suitable for production on a large scale by tissue culture. The role of tissue culture in flavour production has been described previously (Collin and Watts 1981; Collin 1987, 1988). Flavour production in celery is determined largely by one group of compounds, the phthalides, which are found in celery oil.

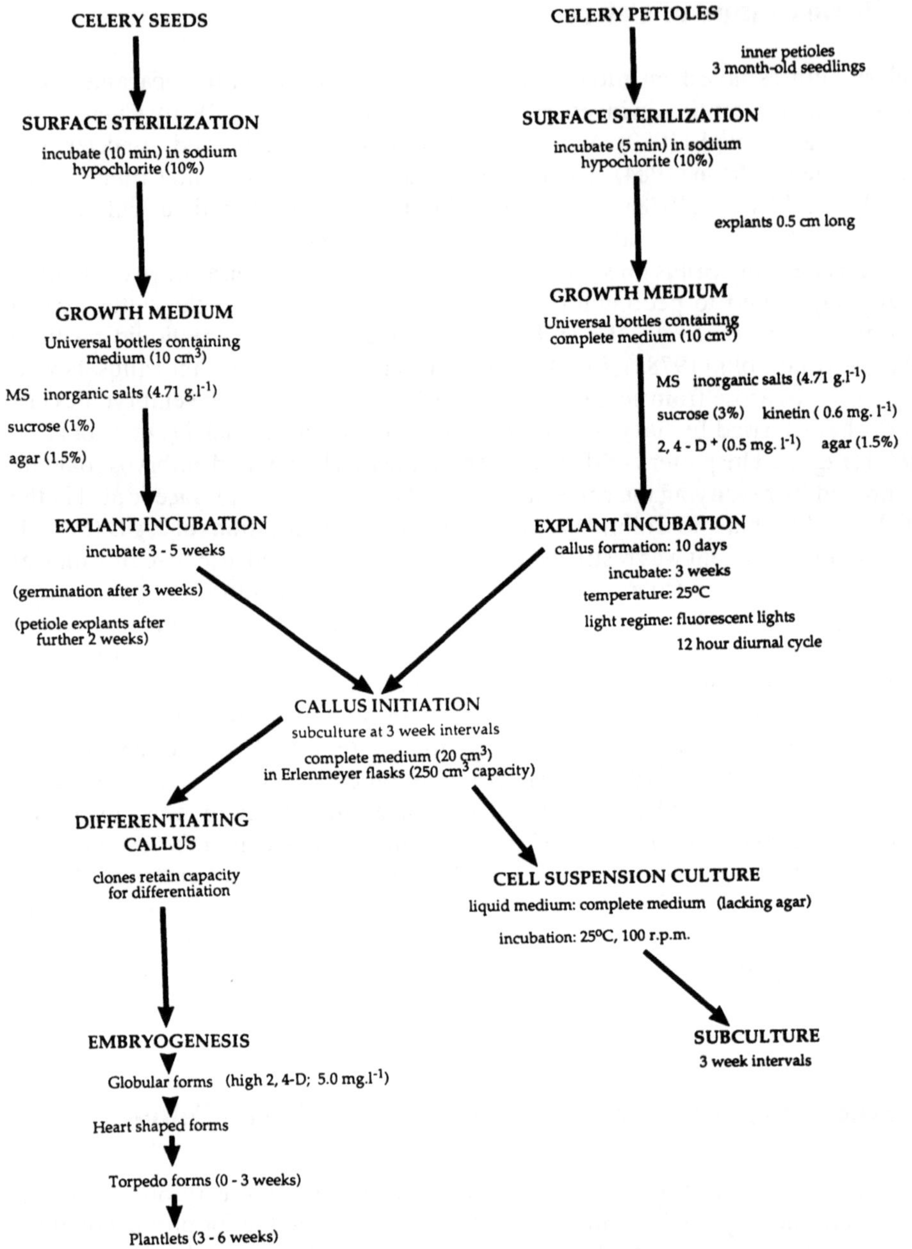

Fig. 3. Celery tissue culture – initiation and embryogenesis (modified method of Williams and Collin 1976)

Fig. 5A,B. Embryonic cells on surface tissue culture aggregate (**A** ×450; **B** ×4050). Dunstan et al. 1982)

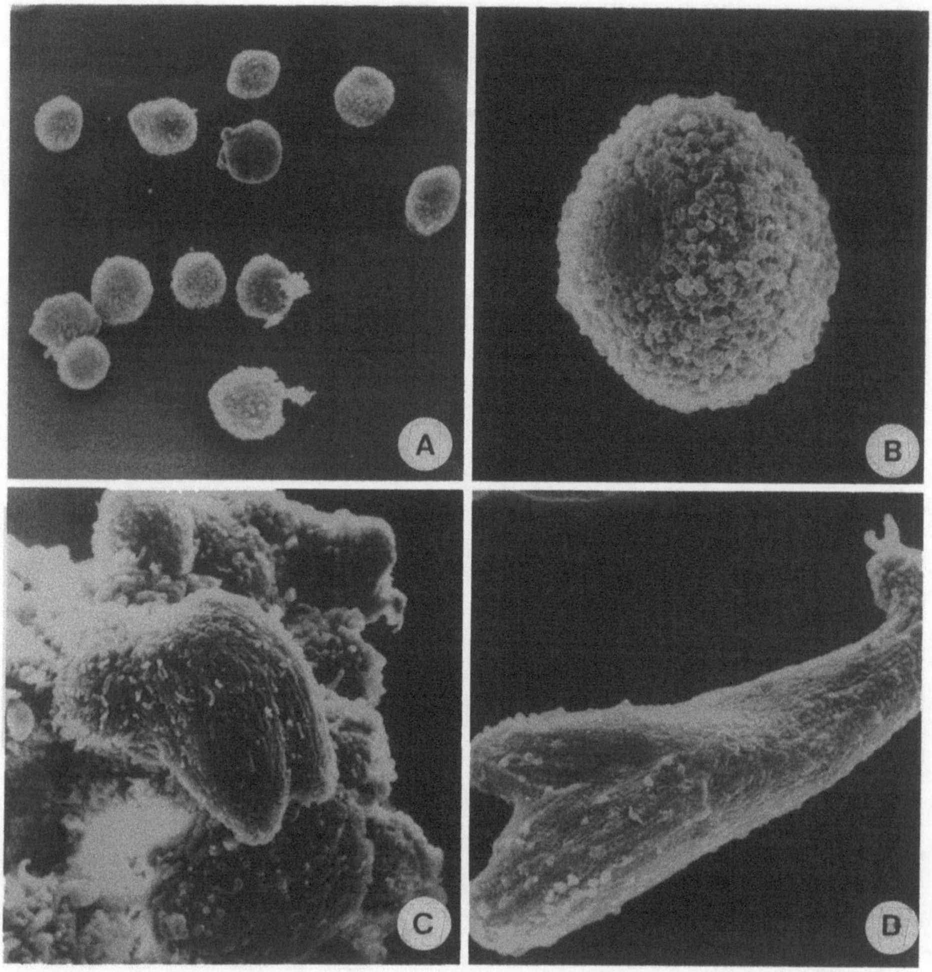

Fig. 4A-D. Formation of embryos on celery callus. **A,B** Globular embryos (× 18, × 85.5 respectively). **C** Heart-shaped embryos (× 17). **D** Torpedo embryos (× 18). (Al-Abta and Collin 1978a)

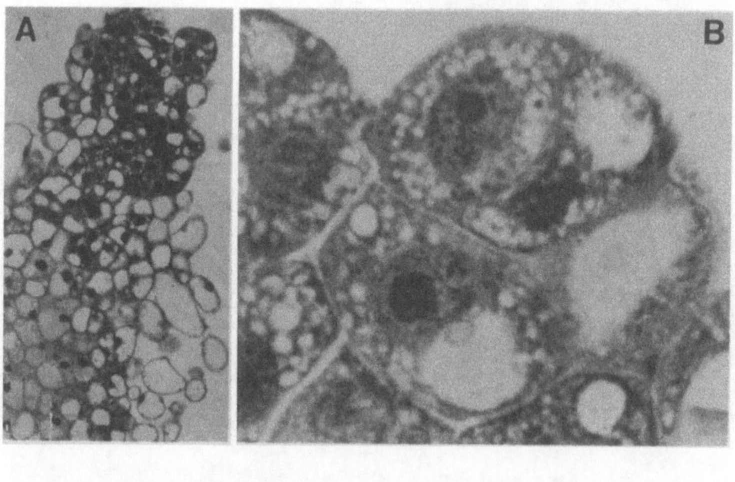

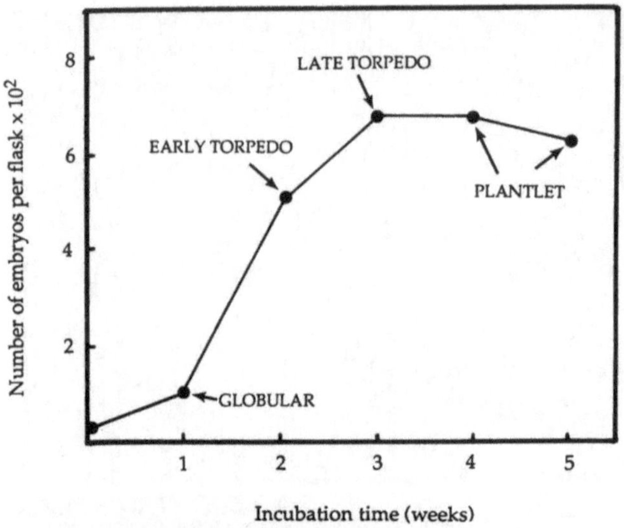

Fig. 6. Changes in embryo number and stage of development during incubation of suspension culture. (Al-Abta et al. 1979)

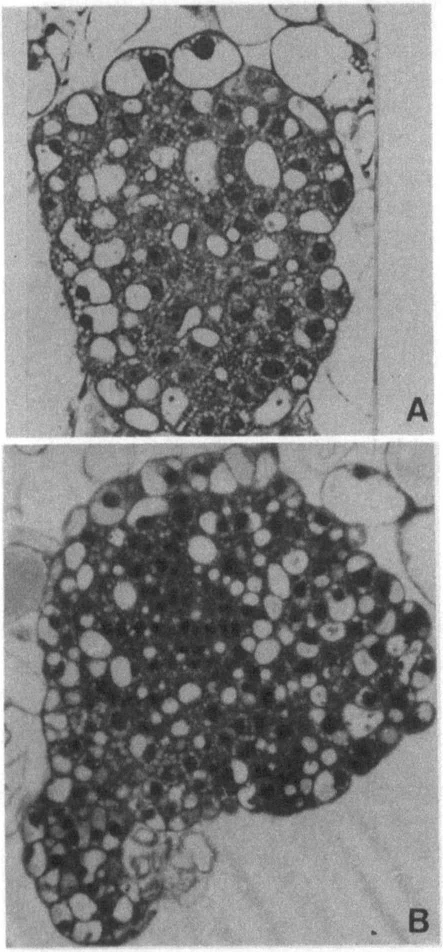

Fig. 7A,B. Internal structure of embryos showing no cell differentiation. **A** Globular (×450). **B** Heart-shaped (×450). (Dunstan et al. 1982)

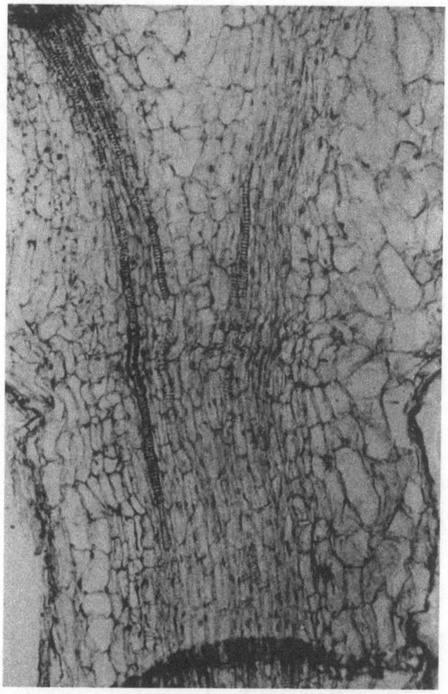

Fig. 8. Section of torpedo stage embryo showing xylem development (× 160). (Al-Abta and Collin 1978b)

5.1 Effect of Differentiation

Celery essential oil is accumulated in intact plants in specialised oil ducts, which are located near each vascular bundle of both the lamina and petiole. In section the oil duct appears as a space surrounded by smaller cells which probably have a secretory function (Fig. 9). Examination of the undifferentiated callus and differentiated embryos up to the late torpedo stage did not show any oil ducts, whereas these structures were detected in the early plantlet stage (Al-Abta and Collin 1978b). A comparison was made between the amount and composition of the essential oil from tissue cultures at different levels of organisation and the intact plant. (Al-Abta et al. 1979). Analysis by GLC showed that the undifferentiated tissue culture contained traces of flavour compounds, whereas differentiating callus which was covered with green embryos from the globular to the torpedo stage had a more complex composition. The peaks present on the trace were identified as the phthalides 3, isobutylidene-3a, 4,5,6-tetrahydrophthalide; 3-isobutylidene-3a, 4-dihydrophthalide and propylidene phthalide. Embryos at a more advanced stage of differentiation were obtained by inoculating medium with differentiating callus then sampling the culture at 3 weeks, when the majority of the culture was at the torpedo stage (Figs 6 and 10). The extract showed a much more complex composition and a higher level of the same phthalides. The number of terpenes and amounts of phthalides in the extract from the intact plants were significantly greater than the differentiated callus. This early investigation showed that the production of the flavour compounds was at a low level in the undifferentiated tissue but, with

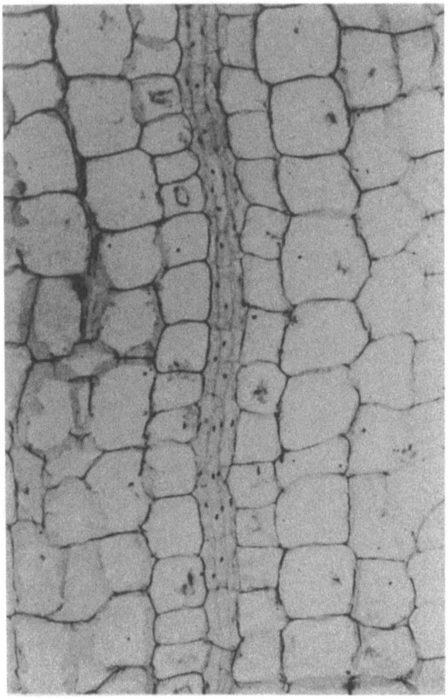

Fig. 9. Longitudinal section of oil duct in celery petiole showing small surrounding cells (× 100). (Al-Abta and Collin 1978b)

increasing tissue organisation and structure, the amounts and number of compounds in the essential oil increased.

Examination of the embryos showed no specialised oil ducts although the torpedo embryos contained significant amounts of flavour compounds. This implied that the oil ducts were not necessary for synthesis of the oil and only functioned as sites of accumulation. Although the synthesis of the flavour compounds did not appear to be dependent on the presence of specialised cells, it was still related to levels of differentiation. Thus the torpedo embryos showed increasing differentiation of form and polarity, and internal tissue differentiation into vascular strands. In order to determine at which level of organisation biosynthesis was stimulated, a cell suspension was separated into aggregates and single cells, then the two types of tissue subcultured three times (Watts et al. 1984). Analysis of the tissue showed that only the aggregates contained phthalides (butyl phthalide and sedanenolide) and the terpene, limonene (Table 3). It has been suggested that cells in aggregates as in cell suspensions, callus, embryo or plant, are in close cell-cell contact, which reduces loss of compounds from leaching and facilitates exchange of the same compounds between cells (Lindsey and Yeoman 1983). One result of this increased cell-cell contact is increased synthesis and accumulation of secondary products.

To separate biosynthesis from stimulatory effects of differentiation, undifferentiated cultures were exposed to modified culture conditions. Thus the standard auxin source, 2,4-D, was replaced by a number of auxins and anti-auxins

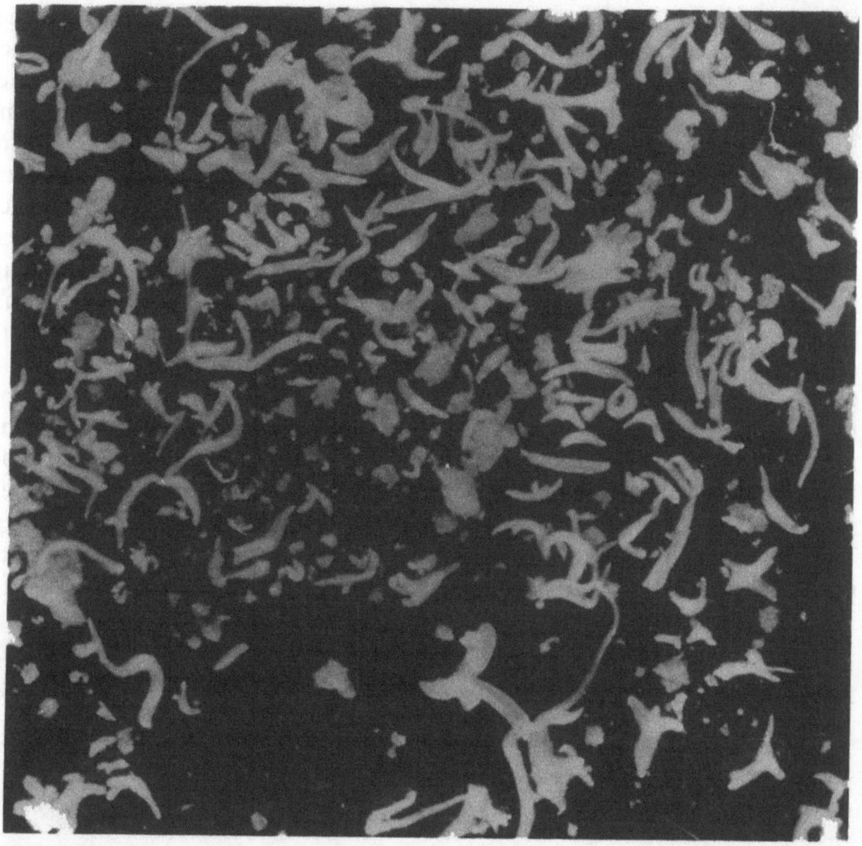

Fig. 10. Differentiated cell suspension showing mainly torpedo embryos after 3 weeks incubation. (Al-Abta et al. 1979)

Table 3. Fresh weight and essential oil components of aggregated and finely divided cell suspensions after selecting for aggregation or fine suspensions during three subcultures. (Watts et al. 1984)

Form of suspension	Fresh weight (g/l)	Components of essential oil	
		Compound	Amount (μg/g fresh weight)
Aggregated (< 0.7 mm)	71.38 ± 4.95	Limonene	3.1
		3-Butylphthalide	6.7
		Sedanenolide	4.5
Fine (< 0.7 mm)	23.11 ± 2.97	None	
Control	80.0 ± 0.20	None	

(Watts et al. 1984). These were naphthalene acetic acid (NAA), indoleacetic acid (IAA), 2,3-dichlorophenoxyacetic acid (2,3-D), 2-chlorophenoxyisobutyric acid (CIBA), 3–5-dichlorophenoxy acetic acid (3,5-D). All compounds were tested separately at 0.5, 1 and 5 mg/l concentrations in combination with 0.6 mg/l kinetin. The only treatment which showed more than trace levels of components of the essential oil (butyl phthalide, sedanenolide and limonene) was the 3,5-D treatment. Since this treatment also caused the cells to become green, it seemed that the increase in biosynthetic capacity of green cells may have contributed to the increase in flavour compounds. The effect of greening on flavour production was investigated by comparing green and non-green tissue from both intact plants and tissue cultures (Watts et al. 1985). In the plant material (Table 4) the level of phthalides (butyl phthalide and sedanenolide) and terpenes (limonene, β pinene, selinene and caryophyllene) increased with greening. In the cultures where 2,4-D was replaced by 3,5-D in order to stimulate greening, phthalides were produced during the growth phase of the first subculture, but were present in trace amounts only during the second and third subculture. Limonene, however, increased during the later subcultures to a significant level, comparable to the concentration in a young plant. The rise and fall of the different compounds in the essential oil suggested that phthalide production was stimulated initially by the transfer from a 2,4-D- to a

Table 4. Chlorophyll and essential oil component in the leaves and petioles of New Dwarf white celery plants at 10 and 20 weeks after germination. (Watts et al. 1985)

	Plant age					
	10 weeks			20 weeks		
Plant components	Leaves	Green petioles	Blanched petioles	Leaves	Green petioles	Blanched petioles
Chlorophyll (mg/g fr. wt.)	0.065 ± 0.100	0.041 ± 0.001	0.003 ± 0.001	1.59 ± 0.070	0.13 ± 0.010	0.03 ± 0.004
Limonene (mg/g fr. wt.)	0.253 ± 0.039	0.141 ± 0.011	0.013 ± 0.009	0.136 ± 0.020	0.064 ± 0.002	0.045 ± 0.004
β-pinene (mm^2/g fr. wt.)	Trace	Trace	Trace	158 ± 19	76 ± 8	98 ± 5
γ-terpenine (mm^2/g fr. wt.)	0	438 ± 24	Trace	276 ± 36	277 ± 43	350 ± 5
Caryophyllene (mm^2/g fr. wt.)	746 ± 256	305 ± 55	135 ± 20	630 ± 39	277 ± 59	273 ± 12
Selinene (mm^2/g fr. wt.)	651 ± 200	247 ± 120	Trace	443 ± 39	255 ± 55	289 ± 6
3-*n*-butylphthalide (μg L.E./g fr. wt.)	0.083 ± 0.023	Trace	Trace	Trace	Trace	Trace
Sedanenolide (μg L.E./g fr. wt.)	0.92 ± 0.33	0.091 ± 0.02	0.085 ± 0.01	0.05 ± 0.005	0.016 ± 0.002	0.017 ± 0.002

Values are the means of three replicates ± S.E.
mm^2/g fr. wt = peak area per gram fresh weight.
L.E. = ligustilide equivalents.

3,5-D-containing medium, but as the level of aggregation and greening increased in the culture, phthalide production was reduced and limonene production increased. Examinations of the cultures showed that despite greening and aggregation, there was very limited thylakoid formation in the tissue cultures incubated in 3,5-D.

A technique which enables aggregated or partially differentiated green cultures or embryos to be maintained in a protected environment is the technique of cell immobilisation. Cells are in an enclosed micro-environment which is thought to increase cell-cell interaction and reduce cell leaching. Consequently, synthesis and accumulation of flavour compounds can be increased (Lindsey and Yeoman 1983). Originally, cell aggregates were enclosed in alginate beads but because these were unstable, the alginate has been replaced by polyurothane foam blocks (Lindsey et al. 1983). Since slow growth and greening enhanced the accumulation of flavour compounds in celery cultures, the ability to immobilise such cultures was investigated. Cultures previously grown in 2,4-D were enclosed in alginate beads, as described by Brodelius et al. (1979) then incubated in a 2,4-D- and a 3,5-D-containing medium (Watts et al. 1985). Growth and metabolism were much more active in the 2,4-D medium and eventually the beads split (Fig. 11). In the 3,5-D medium the cultures showed no growth and became green. The results showed that it was possible to immobilise celery cells in a growth-limiting medium, but no analyses were made of the cells or medium to see whether the flavour compounds had increased.

5.2 Effect of Culture Environment

The effect of changes in the culture environment on flavour accumulation have also been investigated. Cell suspension cultures were grown in periods of light or dark during the different stages of the growth cycle, or in continuous light or dark (Watts et al. 1984). None of these light treatments had any effect on phthalide or terpene accumulation. Altering the temperature during all or part of the growth cycle, however, was more effective. A low temperature (4 °C) for the first 5 days of the culture period stimulated production of the terpenes, limonene, myrcene, terpinene, selinene and the phthalide, butylphthalide. The limonene level was particularly high, equivalent to the values in a celery plant, but the level of phthalides was still low. It was interesting to note that abiotic stimuli, such as low temperature, heavy metals and UV light, stimulated synthesis of a group of secondary products in celery, the linear furanocoumarins (Beier and Oertli 1983). It has also been shown that celery infected with *Sclerotina sclerotiorum* contained enhanced levels of these compounds (Scheel et al. 1963; Floss et al. 1969; Wu et al. 1972).

Exposure of tissue culture cells to biotic stimuli such as extracts of fungal mycelium, spores, or culture filtrates will often elicit phytoalexin production by the host cells, which may include the production of valuable secondary products (Heinstein 1985). It was interesting, therefore, to find that in a preliminary investigation where the undifferentiated celery callus cultures were infected with fungal spores of the celery pathogen, *Septoria apiicola*, the concentration of individual components of the oil were altered. The terpene, β-eudesmol, was 20-fold higher in infected callus than in uninfected callus. Other compounds such as the phthalides,

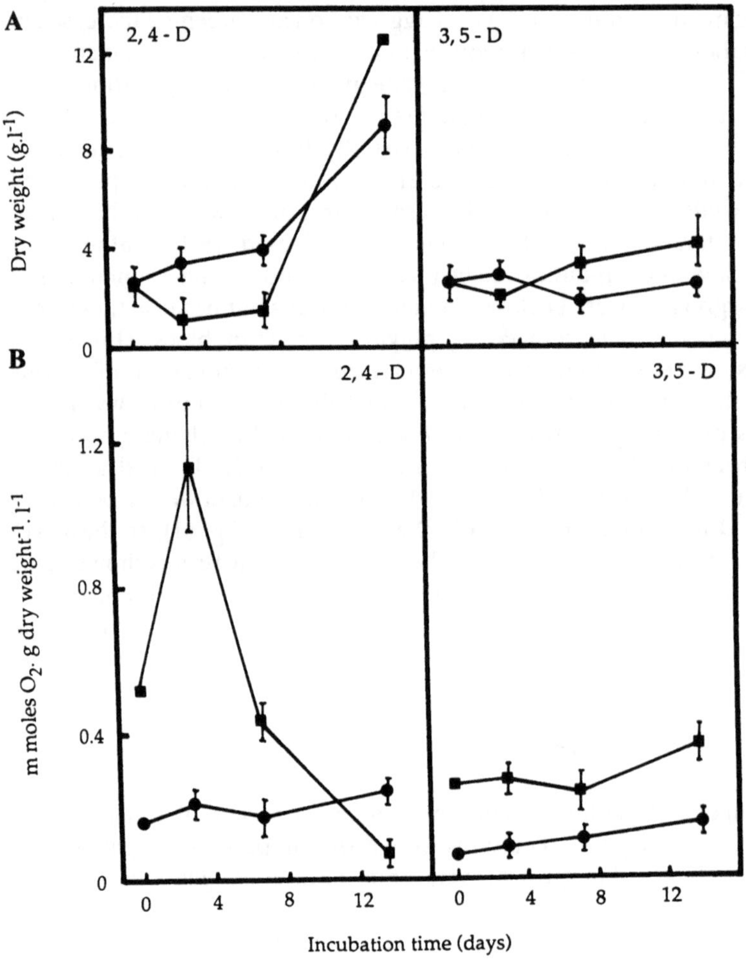

Fig. 11A,B. Dry weight (**A**) and respiration rate (**B**) of immobilised (-●-) and free (-■-) cell suspensions of celery grown in medium containing 2,4-D or 3,5-D. *Bar markers* represent S.E. of three replicates. (Watts and Collin 1985)

3-butylphthalide and sedanenolide, normally detected at trace levels, were present at enhanced concentrations (Donovan et al. 1989). Biotic factors stimulated the terpenoid metabolism of celery but it is not clear at this early stage in the investigation whether the altered metabolism is a sustained or a transient response. However, it is encouraging to see that the undifferentiated callus is able to respond to elicitation. Flavour compounds of celery do have a role in the resistance of intact celery to *Septoria apiicola* (Donovan et al. 1989) so it is not surprising that elicitation of the flavour compounds can occur in celery cultures.

An attempt was made to select for high-yielding clones using the more strongly flavoured variety, French Celery Dimant (Watts pers. communi.). A total of over

5000 hypocotyl explants derived from seedlings were placed on a standard callus initiation medium then after one passage transferred to liquid medium. After 2 weeks, the cultures were assessed for celery aroma by smelling the open flask, then subcultured if an aroma was detected. Most of the cultures possessed a celery aroma initially but this declined to 50% of the cultures after five subcultures. Selected cultures showed considerable variation in structural organisation from undifferentiated cells to embryos, roots and shoots. However, those cultures showing organogenesis retained the capacity to produce a celery aroma, whereas in an undifferentiated cell suspension, the level of terpenes and phthalides declined with subculture. At the tenth subculture, the composition of the essential oil of one undifferentiated cell line was assessed every 2–3 days. The accumulation of limonene and selinene occurred during the exponential phase then declined, whereas the phthalide, methyl phthalide, accumulated only during the stationary phase, as though it was a product of cell senescence. The medium contained measurable levels of limonene and selinene only during the exponential phase, and subsequently these compounds declined. This work with the stronger flavoured French Celery Dimant confirmed the earlier reports that the strength of the celery flavour was related to the level of differentiation.

6 Conclusions

The yield of phthalides and other components of the essential oil in celery tissue cultures was too low to be of commercial value. Normally, the capacity to synthesise the essential oil in tissue cultures is dependent on redifferentiation of the tissue culture, and celery was no exception. Redifferentiation of the cultures from the stage of late torpedo formation to shoot formation led to a renewed synthesis of components of the essential oil. It was possible to select high-yielding undifferentiated cultures but yield declined on subculture. Undifferentiated cultures showed increased yield after abiotic and biotic elicitation, but the stimulatory effect on production applied to specific compounds rather than the complete oil. The yield, however, was not increased sufficiently to be of immediate commercial interest. Although celery is not a major flavour, the attraction of the species is that it produces a range of terpenoids in the intact plant and culture. The tissue can be easily grown as a cell suspension, as immobilised cells and as partially differentiated units. It provides therefore a useful species to study the factors, both internal and external, which control terpene production in plants.

The most serious problem for secondary product formation in tissue cultures is the low yield of valuable secondary products. In addition, cultures are often unstable and average yields do decline with subculture. Preliminary experiments with celery have shown that it is possible to elicit a significant response by a reduced temperature and by exposure to a fungal pathogen. These two encouraging responses are being explored more fully by attempting to specify the conditions that give maximum response and also to examine the metabolic controls that are affected by the elicitation.

7 Protocol for Extraction of Essential Oil from Celery

A total of 50 g fresh weight of celery (plants or tissue cultures) was homogenised in 50 cm^3 ethyl acetate, then 75 cm^3 distilled water added, mixed and transferred to a 500-cm^3 spherical flask with a few antibumping granules. To extract nutrient medium, 120 cm^3 of medium was shaken with 50 cm^3 ethyl acetate. The homogenate was distilled in a modified Stahl apparatus (1969) for 12 h, running off the distillate every 2 h into an aluminium foil-capped flask. The aqueous phase of the distillate was saturated with sodium chloride, then the ethyl acetate fraction partitioned off. The remaining aqueous fraction was partitioned three times with an equal volume of ethyl acetate. All the ethyl acetate fractions were pooled, dried over anhydrous magnesium sulphate (0.5–1.0 g $MgSO_4$ 100 cm^{-3} ethyl acetate) then filtered and evaporated under vacuum to 100–200 μl. Up to 10μl of this extract was analysed by gas liquid chromatography (9 ft × 4 mm I.D. glass column, packed with 10% SP 2100 on a 60–100 mesh chromosorb using nitrogen at a flow rate of 60 cm^3/min and a temperature programme of 60–250°C at 6°C/min). Phthalide levels were calculated from the peak areas and expressed as equivalents to ligustilide (3-butylidene-4,5-dihydrophthalide), kindly provided by Professor H. Mitsuhashi, Faculty of Pharmaceutical Science, Hokkaido University, Sapporo, Japan.

References

Adamson AD (1971) The market for certain herbaceous essential oils. Trop Prod Inst Rep G54:3–7

Al-Abta S, Collin HA (1978a) Control of embryoid development in tissue cultures of celery. Ann Bot (London) 42:773–782

Al-Abta S, Collin HA (1978b) Cell differentiation in embryoids and plantlets of celery tissue cultures. New Phytol 80:517–521

Al-Abta S, Galpin IJ, Collin HA (1979) Flavour production in tissue cultures of celery. Plant Sci Lett 16:129–134

Andrews JC, Viser ET (1951) The oxalic acid content of some common foods. Food Res 16:306–312

Atal CK, Kapur BM (eds) (1977) Cultivation and utilisation of medicinal and aromatic plants. Leipzig Press, New Delhi, pp 330–334

Balbaa SI, Hilal SH, Haggae MY (1976) A study of the fixed oils of the fruits of *Carum copticum* Benth. and Hook, *Apium graveolens* L. and *Petroselinum sativum* Hoffm. growing in Egypt. J Pharm Sci 16:383-390

Barton DHR, De Vries JX (1963) The constitution of sedanolide. J Chem Soc 2:1916–1919

Beier RC, Oertli EH (1983) Psoralen and other linear furocoumarins as phytoalexins in celery. Phytochemistry 22:2595–2597

Bentley LE (1952) Occurrence of malonic acids in plants. Nature (London) 170:847–848

Berlingozzi S (1927) Hydrophthalides. I. Contribution to the study of the relationship between constituent and odour. Gazz Chim Ital 57:264–268

Birch AJ, Donovan FW (1953) Studies in relation to biosynthesis. I. Some possible routes to derivaties of orcinol and phloroglucinol. Aust J Chem 6:360–368

Bjeldanes LF, Kim IS (1978) A sedative activity of celery oil constituents. J Food Sci 43:143–144

Brodelius P, Nilsson (1980) Entrapment of plant cells in different matrices. FEBS Lett 122:312–316

Brodelius P, Deus B, Mosbach K, Zenk MH (1979) Immobilised plant cells for the production and transformation of natural products. FEBS Lett 103:93–97

Browers MA, Orton TJ (1986) Celery (*Apium graveolens* L.). In: Bajaj YPS (ed) Biotechnology in agriculture and forestry, vol 2: Crops I. Springer, Berlin Heidelberg New York Tokyo, pp 405–420

Buttkus HA (1978) Celery leaf juice – evaluation and utilisation of a product from harvest debris. J Agric Food Chem 26:827-830

Chen CH (1976) Vegetative propagation of the celery plant by tissue culture. Proc S Dakota Acad Sci 55:44-48

Ciamician G, Silber P (1897) Über die hochsiedenden Bestandteile des Sellerieöls. Ber Dtsch Chem Ges 30:492–501

Collin HA (1987) Determinants of yield of secondary products in plant tissue cultures. In: Callow JA (ed) Advances in botanical research, vol 13. Academic Press, New York London, pp 145–199

Collin HA (1988) Flavours. In: Constabel F, Vasil IK (eds) Cell cultures and somatic cell genetics of plants, vol 5. Academic Press, New York London, pp 569–585

Collin HA, Watts M (1981) Flavour production in culture. In: Evans DA, Sharp WR, Ammirato PV, Yamada Y (eds) Handbook of plant cell culture, vol 1. Techniques for propagation and breeding. Macmillan, New York, pp 729–747

Collin HA, Donovan A, Isaac S (1987) Selection for disease resistance using celery tissue cultures. In: Boccon-Giboud J, Benbadis A, Short KC (eds) Cell culture techniques applied to plant production and breeding. Proc IAPTC, Angers, ENITHP, pp 21–33

Crosby DG, Anderson LJ (1963) The organic constituents of food. II. Celery J Food Sci 28:640–648

Donovan A, Isaac S, Collin HA (1989) Dual fungal and plant cell culture. In: Walker J (ed) Methods in molecular biology. Humana, NJ

Dunstan DI, Short KC, Merrick MA, Collin HA (1982) Origin and early growth of celery embryoids. New Phytol 91:121–128

Fehr D (1974) On the essential oil of celery (*Apium graveolens* L.) Pharmazie 29:349–351

Fehr D (1979) Study on the aroma substances of celery (*Apium graveolens* L.), pt 1. Pharmazie 34:658–662

Floss HG, Guenther H, Hardwiger LA (1969) Biosynthesis of furanocoumarins in diseased celery. Phytochemistry 8:585–588

Gold HJ, Wilson CW (1961) Techniques in the isolation of volatile materials from celery and the identification of some compounds with acidic properties. Proc Fla State Hortic Soc 74:291–296

Gold HJ, Wilson CW (1963a) The volatile flavour substances of celery. J Food Sci 28:484–488

Gold HJ, Wilson CW (1963b) Alkylidene phthalides and dihydropthalides from celery. J Org Chem 28:985–987

Guenther E (1950) The essential oils, vol 4. D Van Nostrand Reinhold, New York, pp 591–602

Hall CB (1957) Composition and organoleptic evaluation of portions of celery stalks. Proc Fla State Hortic Soc 70:204–208

Hall CB (1959) Observations on celery flavours. Proc Fla State Hortic Soc 72:284–285

Hartman BG, Hillig F (1934) Acid constituents of food products with special reference to citric, malic and tartaric acids. J Assoc Off Agric Chem 17:522–531

Heinstein PF (1985) Future approaches to the formation of secondary natural products in cell suspension cultures. J Nat Prod 48:1–9

Herrmann K (1978) Review on non-essential constituents of vegetables. III. Carrots, celery, parsnips, beets, spinach, lettuce, endives, chicory, rhubarb and artichokes. Z Lebensmittel Unters Forsch 167:262–273

Innocenti G, Dall'Acqua F, Caporale G (1976) Investigations of the content of furocoumarins in *Apium graveolens* and *Petroselinum sativum*. Planta Med 29:165–170

Karim A, Bhatty MK (1976) Studies on the essential oils of the Pakistani species of the family Umbelliferae. IV. *Apium graveolens* Linn. (celery, ajmodn) seed oil. Pak J Sci Ind Res 19:243–246

Kariyone T, Shimizu S (1953) Synthesis of alkylidene phthalides and their odour. J Pharm Soc Jpn 73:336–338

Kohli RP, Dua PR, Shanker K, Saxena RC (1967) Some central effects of an essential oil of *Apium graveolens* Linn. Indian J Med Res 55:1099–1102

Lindsey K, Yeoman MM (1983) The relationship between growth rate, differentiation and alkaloid accumulation in cell cultures. J Exp Bot 34:1055–1065

Lindsey K, Yeoman MM, Black GM, Mavituna F (1983) A novel method for the immobilisation and culture of plant cells. FEBS Lett 155:143–149

Lund ED (1978) Thin layer and high pressure liquid chromatographic analysis of celery seed oil. J Assoc Off Anal Chem 61:1083–1088

Lund ED, Wagner CJ, Bryan WL (1974) Oils recovered from celery packing house waste. Proc Fl State Hortic Soc 86:255–259

Macleod AJ, Macleod G, Subramanian G (1988) Volatile aroma constituents of celery. Phytochemistry 27:373–375

Mitsuhashi H, Nagai U (1963) Studies on the constituents of Umbelliferae plants. VII. Structure of Ligustilide 2. Tetrahedron 19:1277–1283

Mitsuhashi H, Nomura M (1966) Studies on the constituents of Umbelliferae plants. XII. Biogenesis of 3-butylphthalide. Chem Pharm Bull 14:777–778

Mitsuhashi H, Nagai U, Muramatsu T, Tashiro H (1960) Studies on the constituents of Umbelliferae plants. II. Isolation of the active principles of *Ligusticum* root. Chem Pharm Bull 8:243–245

Murashige T, Skoog F (1962) A revised medium for rapid growth and bioassays with tobacco tissue cultures. Physiol Plant 15:473–497

Opdyke DHJ (1974) Monographs on fragrance raw materials. Celery seed oil. Food Cosmet Toxicol 12(Suppl):849–850

Orton TJ (1984) Celery. In: Sharp WR, Evans DA, Ammirato PV, Yamada Y (eds) Handbook of plant cell culture, vol 2: Crop species. Macmillan, New York, pp 243–267

Pan H (1960) Burning-numbing taste in celery. Proc Fl State Hortic Soc 73:223–227

Pan H (1961) Bitterness in celery. J Food Sci 26:337–341

Phillips H (1827) History of cultivated vegetables. Coburn, London, pp 38–40

Rappaport L, Fujii DS, Thompson RH (1980) From cells to celery (*Apium graveolens*): callus formation and plant regeneration in tissue and liquid suspension cultures. Hort Sci 15:416

Robbins SRJ, Greenhalgh P (1979) The markets for selected herbaceous essential oils. Trop Prod Inst Rep G120:33–36

Scheel LD, Perone VB, Larkin RL, Kupel RE (1963) The isolation and characterisation of two phytotoxic furanocoumarins (Psoralens) from diseased celery. Biochemistry 2:1127–1131

Schramm R (1961) A comparative paper chromatography of organic acids in storage roots of some cultivable Umbelliferae plants. Acta Soc Bot Pol 30:285–292

Simonsen J, Barton DHR (1952) The terpenes, vol 3. Univ Press, Cambridge

Small J (1948) Celery seed. Food 17:181–183

Stahl E (1969) Thin layer chromatography. A laboratory handbook, 2nd edn. Allen & Unwin, London

van Straten S, de Vrijer F, Beauveser JC (eds) (1977) Volatile compounds in food. Cent Inst Nutrit Food Res. TNO Krips Repro BV, Meppel, Neth

Watts MJ, Collin HA (1985) Growth and nutrient uptake by immobilised tissue culture cells of celery (*Apium graveolens*). Plant Science 42:67–72

Watts MJ, Galpin IJ, Collin HA (1984) The effect of growth regulators, light and temperature on flavour production in celery tissue cultures. New Phytol 98:583–591

Watts MJ, Galpin IJ, Collin HA (1985) The effect of greening on flavour production in celery tissue cultures. New Phytol 100:45–56

Williams L, Collin HA (1976) Embryogenesis and plantlet formation in tissue cultures of celery. Ann Bot (London) 40:325–332

Wilson CW III (1969) Terpene and sesquiterpene hydrocarbons in the essential oil from fresh celery. J Food Sci 34:521–524

Wilson CW III (1970) Relative recovery of carbonyl compounds of celery essential oil. J Food Sci 35:766–768

Wilson CW III, Neubert AM, Veldhuis MK, Miller WH (1967) Recovery of volatile flavours from celery. Proc Fl State Hortic Soc 80:292–296

Wilson CW III, Wagner CJ, Berry RE, Veldhuis MK (1969) Relationship of method for recovering essential oil from fresh celery on the chemical composition and flavour. Proc Fl State Hortic Sci 82:187–191

Wu CM, Koehler PE, Ayres JC (1972) Isolation and identification of xanthotoxin (8-methoxypsoralen) and bergapten (5-methoxypsoralen) from celery infected with *Sclerotina sclerotiorum*. Appl Microbiol 23:852–856

VI *Aloe barbadensis* Mill. (= *A. vera* L.)

A. Cavallini, L. Natali, and I. Castorena Sanchez[2]

1 General Account

Aloe belongs to the Liliaceae family. The name is derived from the Arabic "alloeh", which is the name used by the natives of eastern Africa, where the plant was first cultivated. The genus comprises about 300 perennial species (Reynolds 1985), native to both eastern and southern Africa. The juice from the leaves of certain species yields a medical substance, the "aloe" drug. The drug was found cited in Ebers' papyrus and in the works of Dioscorides. Various species were introduced into and cultivated in the dry area of Africa, in the mountains of tropical Africa, in the Antilles, in India, and in the Mediterranean basin.

1.1 Botany, Distribution, and Importance

Aloe plants may be either stemless, bearing a rosette of large, thick, succulent leaves, or have a stem (up to 1.5 m in length) along with or at the end of which the leaves are borne. Fleshy leaves are usually lance-shaped, up to 50 cm in length, with a sharp apex and a spiny margin, differently colored, from grey to bright green, sometimes striped or mottled; the upper leaf surface is flat or lightly concave, the lower surface strongly convex. The inflorescence (a raceme), is borne on a simple or branched scape, originating from the rosette; flowers are small, tubular, red, yellow, or white. The fruit is a capsule.

The majority of the species belonging to the genus *Aloe* are diploids (Brandham 1971) with the chromosome number 2n = 14, and with a bimodal karyotype complement including eight long and six short acrocentric chromosomes (Sapre 1978).

Examination of *Aloe* leaf sections reveals the presence of three types of cells at the phloem pole of the vascular bundles: aloin cells, outer bundle sheath cells, and fibers (Cutler et al. 1980). The majority of the species have aloin cells of various sizes which produce a copious exudate containing the active principles of the aloe drug (Cutler 1972). The theory that aloin cells are storage cells rather than secretory cells remains untested. It was suggested that the aloin cells act as storage tissue and that compounds are synthesized in the surrounding layer of cells of smaller diameter, many of which can be seen to contain globules of unknown constitution (Beaumont

[1]Department of Agricultural Plant Biology, Genetics Section, University of Pisa, Via Matteotti 1/B, 56100 Pisa, Italy

Biotechnology in Agriculture and Forestry, Vol. 15
Medicinal and Aromatic Plants III (ed. by Y.P.S. Bajaj)

et al. 1986). In many species, cells in this region form a more or less distinct layer, usually one cell in thickness (Beaumont et al. 1986).

The most frequently employed species of the genus are *Aloe barbadensis*, *A. ferox*, and *A. perryi*, in addition to the hybrids *A. ferox* × *africana* and *A. ferox* × *spicata* (Capasso and Donatelli 1982); other medicinal aloes are *A. chinensis* and *A. ciliaris* (Byra Graf 1978). Aloes are generally much cultivated in the warm areas of the world as ornamentals, especially in public gardens, owing to their stiff habit.

Aloe ferox Mill. ("Cape Aloe") is commonly cultivated in South Africa for medicinal and cosmetic uses; its cultivation gives the highest production of the drug named Cape aloe. The second-ranking species in drug production is *A. barbadensis* Mill. (= *A. vera* L., "Bitter Aloe"), which originated in northern Africa and was then introduced in the Antilles (Barbados, Jamaica, Curacao islands) and Central America (mainly Mexico and Venezuela), where it is commonly cultivated (Fassina 1974); the drug is called either Barbados- or Curacao aloe and shows a higher content of active principles (more than double) than Cape aloe (Lister and Pride 1959). Finally, a lesser quantity of the drug (the so-called Socotrine aloe) is produced in eastern Africa, where the cultivated species is *A. perryi* (= *A. succotrina*, "Socotrina Aloe").

The last available data on world drug production date back many years: Fassina (1974) reported a production of 500 tons per year for Cape aloe and 350 tons per year for Barbados aloe. Drug production should have remained practically the same since then.

The drug shows very different morphological characteristics (color, brightness, consistency) according to the species from which it is extracted and preparation methods followed; it has a purgative, eupeptic, and cholagogue effect and is also largely used in the food industry as a bittering substance.

1.2 Conventional Propagation and Production of the Drug

Aloe barbadensis plants are usually vegetatively propagated. New suckers arise from the base of the stem and may be easily cultivated as cuttings, the adventitious rooting of cuttings being rapid.

Vegetative propagation is, in many cases, a forced practice owing to the widespread male sterility affecting this species (see Keijzer and Cresti 1987). Sapre (1975) described irregularities during pollen meiosis in *A. barbadensis*, which gave a substantial clue to the high percentage of pollen sterility (58.3%); nevertheless, even when pollen grains seemed normal, failure in fruit set was observed (Sapre 1975).

As a general rule, even when seed production would be possible, vegetative propagation gives more rapid results and is preferred. This method of reproduction has been instrumental in determining a strong reduction in genetic variability in cultivated *Aloe* varieties, which are expected to face weak environmental modifications, such as particular climatic conditions and/or pests and diseases.

Aloe barbadensis plants are cultivated in spaced rows, planted out immediately at the end of the rainy season, and outer leaves are harvested from the second year.

Aloe juice is collected from August to October, by cutting leaves from which the juice flows out.

1.3 Extraction, Structure, and Properties of Active Principles

The aloe drug is obtained by different procedures, depending on the country where the plants are cultivated. Generally, the juice is obtained by simply draining or squeezing the leaves, or by warm or cold water retting of leaf pieces. Barbados aloe juice is then vacuum evaporated and concentrated to make the juice components crystallize, and after cooling, produces an opaque resinous mass, variable in color from dark brown to red brown. The opaque fragments of the drug resemble liver tissue (the so-called hepatic aloe, Fassina 1974) owing to the presence, in the solidified mass, of aloin crystals interspersed with a brown-yellow resin. The drug is both water- and 80% alcohol-soluble.

The drug contains anthracene derivatives occurring either free (anthraquinones: 0.05–0.5%) or in the form of glycosides, usually containing glucose (5–25%; mean 18%) (Fairbairn 1949; Capasso and Donatelli 1982). Free anthraquinone compounds are aloe-emodin (Fig. 1A) and chrysophanic acid (Fig. 1B) (Fassina 1974).

The main active principle has long been recognized to be a C-glycoside of aloe-emodin (Birch and Donovan 1955), known as barbaloin (Fig. 1C) and finally characterized as 10-D-glucopyranosyl-1,8-dihydroxy-3-hydroxymethyl-9(10H)-

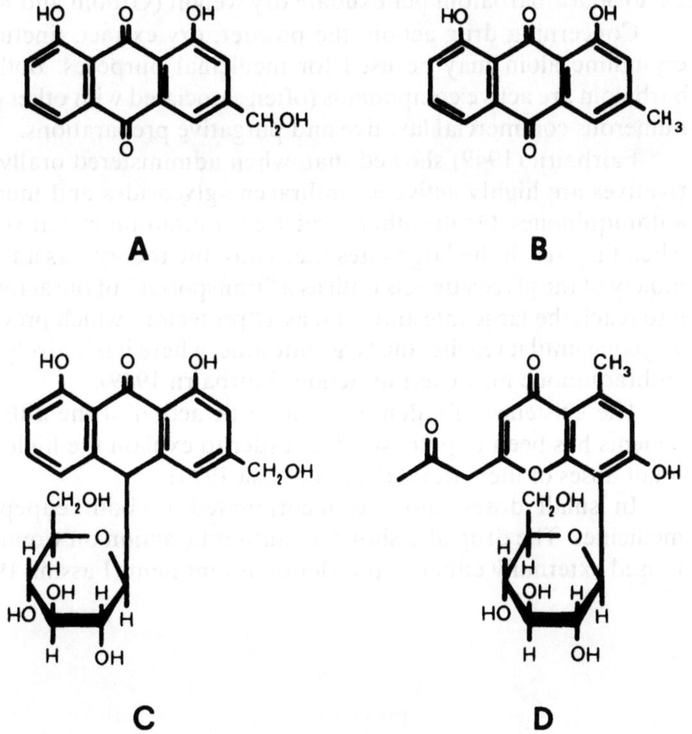

Fig. 1A-D. Chemical structures of four components of the aloe drug. **A** Aloe emodin. **B** Chrysophanic acid. **C** Barbaloin. (After Fassina 1974). **D** Aloesin. (After Gramatica et al. 1982)

anthracenone (Hay and Haynes 1956; see also Reynolds 1985). It is obtained by extraction of the resin with hot water, then insoluble calcium salts are precipitated with lime water and ammonia. Salts are then collected and treated with strong hydrochloric acid to give a yellow powder from which barbaloin, the major component, is obtained by careful recrystallization (Hay and Haynes 1956).

The name aloin was given to the crude material, amorphous or crystalline according to the degree of purification, from which barbaloin could be isolated (Lister and Pride 1959); in recent literature, the two names have become interchangeable (Groom and Reynolds 1987).

Other major components of the drug are two 5-methylchromone-C-glycosides, aloesin (also called aloeresin B, Haynes et al. 1970) (Fig. 1D) and aloeresin A (Gramatica et al. 1982), and the aloesin aglycone, aloesone (Holdsworth 1972).

Anthracene derivatives respond to the Bornträger test or suitable modifications, by means of which they are all converted into free anthraquinone compounds which give pink to red colors in alkaline solution (Fairbairn 1949; Fassina 1974).

It was shown by Groom and Reynolds (1987) that large variations in the content of the main active principle (i.e., barbaloin) within one and the same plant do occur: the highest concentration of anthracene-C-glycosides in leaf exudates was found in younger mature leaves and concentration decreased with age by 20–40% between the apex and the base (Beaumont et al. 1985; Groom and Reynolds 1987). A great variability in barbaloin content was also shown within the genus *Aloe*, ranging from 2.5 to 30.8% barbaloin per exudate dry weight (Groom and Reynolds 1987).

Concerning drug action, the powder, dry extract, tincture, or even the pure crystalline aloin may be used for medicinal purposes. Both the aloe drug and barbaloin are active components (often associated with other coadjuvant drugs) of numerous commercial laxative and purgative preparations.

Fairbairn (1949) showed that, when administered orally, the anthracene derivatives are highly active as anthracene-glycosides and much less active as free anthraquinones. On the other hand, free anthraquinones are the active compounds when they reach the large intestine. Thus, the theory was advanced that the sugar moiety of the glycoside acts both as a "transporter" of the active aglycone, enabling it to reach the large intestine, and as a "protector" which prevents oxidation of the aglycone until it reaches the large intestine, where it is hydrolyzed and the liberated anthraquinone may exert its action (Fairbairn 1949).

The existence of a definite synergistic action of the different aloe drug components has been hypothesized, in order to explain the high biological activity of small doses of the natural drug (Fassina 1974).

In small doses, aloin is incorporated in both eupeptic and cholagogue medicines. The drug also shows an antiseptic action on wounds and eczemas, and is used externally either as powder or as ointment (Fassina 1974).

2 In Vitro Approaches

2.1 Review of Tissue Culture Studies in the Genus *Aloe*

In spite of the importance of the aloe drug in officinal and industrial fields, in vitro culture of species of *Aloe* has received little attention (see Table 1).

Yagi et al. (1983) cultivated various organs of *Aloe saponaria* (Hill.) Haw. to induce callus formation on Murashige and Skoog (1962) basal medium (MS) supplemented with various combinations of different auxins and kinetin in the dark. The best results were obtained using root tissues and with a combination of 1 ppm indoleacetic acid (IAA) and 0.5 ppm kinetin or 0.5 ppm 2,4-dichlorophenoxyacetic acid (2,4-D) and 2 ppm kinetin; callus tissue was pale white and friable. The occurrence of plant regeneration from these calli was not reported (Yagi et al. 1983).

Groenewald et al. (1975) reported the successful establishment of calli and subsequent plantlet regeneration in *Aloe pretoriensis* Pole Evans. They induced callus formation from seed tissues (regardless of whether callus developed from either embryo or endosperm tissues) in 3–4 weeks, and observed shoot and root regeneration after 8–10 weeks. Explants were cultivated on Linsmaier and Skoog (1965) basal medium, supplemented with 2,4-D 0.2 mg/l and kinetin 1 mg/l, under continuous light (3450 lx) in a growth chamber at 27°C. Percentages of callus formation and plant regeneration were not reported.

This method was tested by Racchi (1987), who used seed fragments of *Aloe ferox* Mill.: initial callus grew poorly and showed no regeneration tendency, even after four to five subcultures. Racchi (1987) also tested callus formation and regeneration ability of different explants of *A. ferox*: roots and the basal regions (containing the meristem) of young leaves were aseptically cultivated on MS medium supplemented with 0.5 mg/l 2,4-D and 1 mg/l kinetin for root explants, and 0.2 mg/l 2,4-D and 1 mg/l kinetin for leaf meristems, under continuous light (3450 lx) at 27°C. From the basal leaf region, only occasional callus formation was observed; pale yellow and soft white callus was achieved rapidly from root explants, but no embryo, shoot, or root formation was obtained during 9 subsequent months of culture. A good frequency of plant regeneration was obtained from enlarged cotyledons and embryos expanded after soaking the seeds for 5 days in water: segments of both cotyledon and embryo, cultured under continuous light on N6 medium (Chu 1978) supplemented with 0.5 mg/l 2,4-D, 1 mg/l kinetin and 0.5 mg/l 6-benzylaminopurine (BAP), gave a good frequency of callus formation, within 1 month; callus subcultured monthly on the same medium produced nodular structures which then developed clusters of somatic embryos. Most embryos grew to plantlets, and the callus maintained its embryogenic potential for up to 1 year (Racchi 1987).

2.2 *Aloe barbadensis* Mill. Tissue Culture

Tissue cultures of *Aloe barbadensis* were established by Castorena Sanchez et al. (1988) using vegetative meristems and leaf explants from not yet flowering healthy

Table 1. Tissue culture and growth response of different species of *Aloe*

Plant species	Explant	Medium (mg/l)	Growth response	Reference
Aloe barbadensis Mill.	Shoot meristem	MS + 2,4-D(0.25) + kin(1)	Shoot micropropagation	Castorena Sanchez et al. (1988)
	Leaf meristem	MS + 2,4-D(0.25) + kin(1) then MS + 2,4-D(0.025) + BAP(1)	Callusing and rare plant regeneration	Castorena Sanchez et al. (1988)
A. ferox Mill.	Root tissues	MS + 2,4-D(0.5) + kin(1)	Callusing	Racchi (1987)
	Leaf meristem	MS + 2,4-D(0.2) + kin(1)	Rare callusing	Racchi (1987)
	Seed tissues	N6 + 2,4-D(0.5) + kin(1) + BAP(0.5)	Callus, embryos, plants	Racchi (1987)
A. pretoriensis Pole Ev.	Seed tissues	LS + 2,4-D(0.2) + kin(1)	Callusing and plants	Groenewald et al. (1975)
A. saponaria (Hill.) Haw.	Root tissues	MS + 2,4-D(0.5) + kin(2)	Callusing	Yagi et al. (1983)
		MS + IAA(1) + kin(0.5)	Callusing	Yagi et al. (1983)

Abbreviations: *Basal media*: MS: Murashige and Skoog (1962); LS: Linsmaier and Skoog (1965); N6: Chu (1978).
Supplements: 2,4-D: 2,4-dichlorophenoxyacetic acid; kin: kinetin; IAA: indoleacetic acid; BAP: 6-benzylaminopurine.

plants cultivated either in the garden or the greenhouse, and native to both Yucatan (Mexico) and Italy.

In order to achieve rapid shoot micropropagation, cultures of vegetative meristems were established. Apical meristem tips were deprived (under a dissection microscope) of accessory tissues and then surface sterilized by immersion into a 70% ethanol solution for 1 min, followed by stirring in a 9% sodium hypochlorite solution for 5 min and rinsing in sterile distilled water. Explants were then transplanted in Erlenmeyer flasks (two to three explants per flask; Fig. 2A), containing 50 ml of agar-solidified Murashige and Skoog (1962) medium supplemented with 2,4-D

Fig. 2A-D. Different steps of clonal micropropagation in *Aloe barbadensis*. **A** Meristem explant. **B** Cluster of micropropagated plantlets after 2 months of culture on MS + 2,4-D (0.25) + kin (1). **C** Rooted plantlet after 1 month of culture on hormone-free MS. **D** Mature plant after 1 year from transplanting in soil (Cavallini, original illustrations)

(0.25 mg/l), kinetin (1 mg/l), p-aminobenzoic acid (0.1 mg/l) and tyrosine (100 mg/l). The pH of the medium was adjusted to 5.7 prior to autoclaving. The cultures were kept in a temperature-controlled room (27 + 1 °C) under continuous white fluorescent light at 3400 lx.

No callus induction was observed, but after 45 days, micropropagation of normal shoots was achieved (Fig. 2B). Further production of shoots was obtained by subculturing the explants monthly on MS medium containing a reduced concentration of 2,4-D (0.025 mg/l) plus 1 mg/l BAP. On this medium, propagation was continuous for more than 1 year.

To induce callus formation and subsequent plant regeneration, leaves approximately 20 cm long were surface sterilized as described above, cut aseptically into fragments ca. 1 cm in length, and then placed on the same initiating medium described for meristem culture (MS medium containing 0.25 mg/l 2,4-D plus 1 mg/l kinetin).

Explants from leaf distal portions (tips) never originated callus and/or plants, and died after 2–3 weeks. When excised from leaf base, in 95% explants an injury callus protruded at the cut edges. This callus stopped growing as soon as the cut surface was covered after 8 days of culture. Some 5% of leaf base explants formed soft callus, which originated from the epidermal tissues and soon became yellowish and developed (in 3 weeks) persistent meristematic centers (i.e., mitotically active centers; Castorena Sanchez et al. 1988). These calli, transplanted on MS medium supplemented with 4 mg/l naphtaleneacetic acid plus 1 mg/l kinetin, regenerated roots in less than 1 week, but no shoot regeneration was observed (Castorena Sanchez unpubl.) Occasionally, some leaf base explants formed small quantities of compact callus which, subcultured for 20 days on MS medium supplemented with 0.025 mg/l 2,4-D plus 1 mg/l BAP, regenerated sterile flowers (lacking carpels) and, later, shoots (Fig. 3). Regeneration ability was maintained for a long time (Castorena Sanchez et al. 1988).

In subsequent plant development, 100% root formation was achieved in both micropropagated and regenerated shoots, after 8-day culture on solid hormone-free MS medium (Fig. 2C). Rooted plantlets were then transferred to earthen pots (containing a mixture of soil, sand, and pumice); they grew easily to maturity (Fig. 2D).

Fig. 3. Regenerated shoots on callus from leaf base of *Aloe barbadensis*, cultured for 1 month on MS + 2,4-D (0.025) + BAP (1) (Cavallini, original illustration)

Root tips from both regenerated and micropropagated plantlets were collected, pretreated in saturate solution of α-bromonaphtalene, and then fixed. Apices were stained by the Feulgen technique and squashed. No deviations from the normal karyotype (2n = 14) were observed in either propagated or regenerated plants (Castorena Sanchez et al. 1988).

Microdensitometric analysis of cell nuclei of *A. barbadensis* clearly shows that nuclear DNA variations occur in vivo with cell differentiation: an increasing frequency of 4C and endopolyploid nuclei was found in differentiated portions of both shoot and root in respect of meristematic portions. Leaf cytodifferentiation occurs in a diploid state, but a significant reduction of nuclear Feulgen/DNA content was observed in the leaf tip (Castorena Sanchez et al. 1988).

A relationship was shown between morphogenetic ability and the cytological status of explant and calli, as determined by cytophotometric analyses: meristems cultivated in vitro maintained the diploid state (without polyploidization) and originated hundreds of shoots, via micropropagation. As to more differentiated tissue explants, leaf tips (with nuclear DNA content lower than 2C) were cultured in vitro but gave neither calli nor shoots; leaf base explants showed low frequencies of callus formation and plant regeneration: at the beginning of culture, all leaf base explants showed a resumption of mitotic activity, the appearance of some endopolyploid nuclei and nuclei with DNA content lower than 2C; when callus was established, most calli showed a complete cell polyploidization and no regeneration ability, while, when the amount of DNA per nucleus fell strictly within the diploid limits (2C and 4C) of the species, calli regenerated plantlets (Castorena Sanchez et al. 1988; Fig. 4). However, since the frequency of callus formation and plant

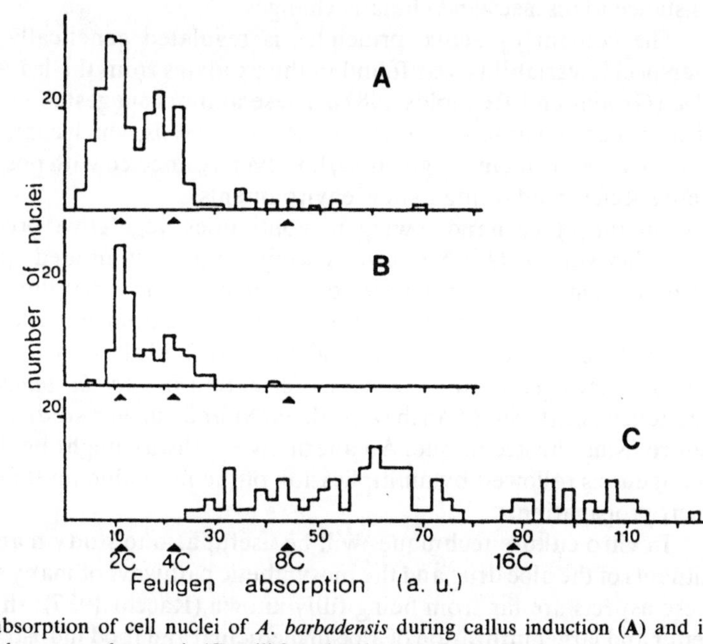

Fig. 4A-C. Feulgen absorption of cell nuclei of *A. barbadensis* during callus induction (**A**) and in morphogenetic (**B**) and non-morphogenetic (**C**) calli from leaf base explants. C-values were calculated on measurements of early prophases in the shoot meristem. (Castorena Sanchez et al. 1988).

regeneration was very low, other important factors (genetic, nutritional, phytohormonal, and environmental factors) must be taken into consideration to explain the difficulties in achieving regenerants of this species.

It is worth noting that calli originated from the same explant and cultured on the same medium showed two different cytological conditions (diploidy or polyploidy): this suggests that polyploidization in *Aloe* tissue culture is not induced by the culture conditions. We may suppose that cells of the initial explant may undergo polyploidization or remain diploid because of local physiological factors in the explant and/or in the callus (Castorena Sanchez et al. 1988).

3 Conclusions and Prospects

At present, even though derived from scanty and conventional culture practices, *Aloe barbadensis* production seems to be enough to meet the demand of the market. In the near future, if the demand for Barbados aloe increases, which may be assumed from the increasing utilization of natural medicinal products, better planning of the cultures and breeding strategy will become necessary, and the utilization of in vitro culture might play a central role. In particular, the clonal micropropagation method set up for *Aloe barbadensis* by Castorena Sanchez et al. (1988) is very efficient in terms of both the number of propagated shoots and the rapidity with which new rooted shoots are achieved, even in comparison with normal vegetative propagation. Micropropagation might be used to achieve a rapid multiplication of very favorable genotypes, i.e., showing higher content in active principles and/or resistance to diseases and climatic changes.

The content of active principles is regulated genetically: in fact, a great interspecific variability was found in the exudates from the leaves of 68 species of *Aloe* (Groom and Reynolds 1987). These authors suggested cultivating different African species, namely *A. classenii* or *A. turkanensis*, which contain very high levels of barbaloin. Problems might arise, however, connected with possible difficulties of these species in adapting to new environments.

On the other hand, owing to continuous vegetative propagation, genetic variability within *Aloe barbadensis* seems to be very limited; plant regeneration from callus might be useful to introduce new genetic variability and to select plant variants. Indeed, it has been shown that callus formation may allow the regeneration of somaclonal variants (Larkin and Scowcroft 1981; D'Amato 1985). Unfortunately, plant regeneration via callus formation in *Aloe barbadensis* occurs at low frequency (Castorena Sanchez et al. 1988) and further studies will be necessary before using this technique. An alternative pathway might be the use of physical mutagenesis followed by multiplication of the favorable mutants through in vitro micropropagation.

In vitro culture techniques will be useful also to study many secondary constituents of the aloe drug and the biosynthetic pathways of many of its components; these aspects are far from being fully known (Racchi 1987); this is necessary for better and larger utilization of aloe in medicinal and food industry. Studies on these aspects have been undertaken in callus from root tissues of a kindred species, *Aloe saponaria* (Yagi et al. 1983), and in calli and regenerated plants of *Aloe ferox* (Racchi et al. 1988).

Acknowledgments. We wish to thank Prof. F. D'Amato for his critical reading of the manuscript. This study was supported by CNR, Italy, Contract no. 89.04129.04.

References

Beaumont J, Cutler DF, Reynolds T, Vaughan JG (1985) The secretory tissues of aloes and their allies. Isr J Bot 34:265–282

Beaumont J, Cutler DF, Reynolds T, Vaughan JG (1986) Secretory tissues in the East African shrubby aloes. Bot J Linn Soc 92:399–403

Birch AJ, Donovan FW (1955) Barbaloin. I. Some observations on its structure. Aust J Chem 8:523–528

Brandham PE (1971) The chromosomes of the Liliaceae: II. Polyploidy and karyotype variation in the Aloineae. Kew Bull 25:381–399

Byra Graf A (1978) Exotic plant manual, 5th edn. Rohrs, Jersey City, NJ

Capasso F, Donatelli L (1982) Farmacognosia. Le droghe della FUI. Piccin, Padova, Italy

Castorena Sanchez I, Natali L, Cavallini A (1988) In vitro culture of *Aloe barbadensis* Mill.: morphogenetic ability and nuclear DNA content. Plant Sci 55:53–59

Chu CC (1978) The N6 medium and its applications to anther culture of cereal crops. In: Proc Symp Plant tissue culture. Science Press, Peking, pp 43–50

Cutler DF (1972) Leaf anatomy of certain *Aloe* and *Gasteria* species and their hybrids. In: Ghouse AKM, Yunus M (eds) Research trends in plant anatomy. Tata McGraw Hill, New Delhi, pp 103–122

Cutler DF, Brandham PE, Carter S, Harris SJ (1980) Morphological, anatomical, cytological and biochemical aspects of evolution in East African shrubby species of *Aloe* L. (Liliaceae). Bot J Linn Soc 80:293–317

D'Amato F (1985) Cytogenetics of plant cell and tissue cultures and their regenerates. CRC Crit Rev Plant Sci 3:73–112

Fairbairn JW (1949) The active constituents of the vegetable purgatives containing anthracene derivatives. Part I. Glycosides and aglycones. J Pharm Pharmacol 1:683–694

Fassina G (1974) Lezioni di Farmacognosia – Droghe vegetali. CEDAM, Padova, Italy

Gramatica P, Monti D, Speranza G, Manitto P (1982) Aloe revisited. The structure of aloeresin A. Tetrahedron Lett 23:2423–2424

Groenewald EG, Koeleman A, Wessels DCJ (1975) Callus formation and plant regeneration from seed tissue of *Aloe pretoriensis* Pole Evans. Z Pflanzenphysiol 75:270–272

Groom QJ, Reynolds T (1987) Barbaloin in *Aloe* species. Planta Med 35:345–348

Hay JE, Haynes LJ (1956) The aloins, pt 1. The structure of barbaloin. J Chem Soc 3141–3147

Haynes LJ, Holdsworth DK, Russell R (1970) C-glycosyl compounds, pt 4. Aloesin, a C-glycosylchromone from *Aloe* sp. J Chem Soc (C):2581–2586

Holdsworth DK (1972) Chromones in *Aloe* species. Part II. Aloesone. Planta Med 22:54–58

Keijzer CJ, Cresti M (1987) A comparison of anther tissue development in male sterile *Aloe vera*, and male fertile *Aloe ciliaris*. Ann Bot (London) 59:533–542

Larkin PJ, Scowcroft WR (1981) Somaclonal variation: a novel source of variability from cell cultures for plant improvement. Theor Appl Genet 60:197–214

Lister RE, Pride RRA (1959) The characterization of crystalline and amorphous aloin. J Pharm Pharmacol 11:278T–282T

Linsmaier EM, Skoog F (1965) Organic growth factor requirements of tobacco tissue cultures. Physiol Plant 18:100–127

Murashige T, Skoog F (1962) A revised medium for rapid growth and bioassay with tobacco tissue cultures. Physiol Plant 15:473–497

Racchi ML (1987) Plant regeneration from callus cultures of *Aloe ferox*. In: Proc Int Congr Plant tissue culture, Bogota (in press)
Racchi ML, Speranza G, Martignoni A, Monti D (1988) Analisi di C-glucosidi in culture in vitro di *Aloe ferox* Mill. In: Atti Conv Ann Soc Fitochim, Como (in press)
Reynolds T (1985) Observations on the phytochemistry of the *Aloe* leaf exudate compounds. Bot J Linn Soc 90:179–199
Sapre AB (1975) Meiosis and pollen mitosis in *Aloe barbadensis* Mill. (*A. perfoliata* var. *vera* L., *A. vera* Auth. non Mill.). Cytologia 40:525–533
Sapre AB (1978) Karyotype of *Aloe barbadensis* Mill.: a reinvestigation. Cytologia 43:237–241
Yagi A, Shoyama Y, Nishioka I (1983) Formation of tetrahydroanthracene glucosides by callus tissue of *Aloe saponaria*. Phytochemistry 22:1483–1484

VII *Akebia quinata* Decne (Akebi): In Vitro Culture and the Formation of Secondary Metabolites

A. Ikuta[1]

1 Introduction

The lardizabalaceous family occurs in central (the Himalayas) to eastern Asia (Japan), and in Chile, and there are eight genera [*Decaisnea* (India, China), *Sinofranchetia* (China), *Holboellia* (E. Asia), *Akebia* (E. Asia), *Parvatia* (E. Himal.), *Boquila* (Chile), *Stauntonia* (E. Asia), and *Lardizabala* (Chile)]. About 38 species are recorded (Huchinson 1973). The plants are woody vines or sometimes shrubs. Leaves are palmate or rarely pinnate alternate. Flowers are usually unisexual. Ovules are usually many, and fruit berrylike, dehiscing lengthwise. Some genera are cultivated in the United States, where *Akebia* is more common (*Decaisnea*, *Lardizabala*, *Stauntonia*, and rarely *Sargentadoxa* along the southern border of the United States) (Benson 1979). Futhermore, some few species are found in East Asia and three species in Japan, i.e., *Akebia quinata* Decne (Akebi in Japanese), *A. triforiata* Koidz (Mitubaakebi in Japanese), and *Stauntonia hexaphylla* Decne (Mube in Japanese). *A. quinata* (Fig. 1) is widely distributed in thickets in hills and mountains in Japan, Korea, and China. It is a glabrous climber with woody vines or sometimes shrubs, with plants reaching to more than 3 m high, whose flowers, usually unisexual, bloom pale purple in April-May. The ovules are usually many and the fruit is 5–8 cm long, oblong and is a fleshy follicle, bacciform, dehiscing lengthwise. It becomes dark purplish, and when mature in September-October, the seeds are black (Ohwi 1984). Its air-dried stems are used as a crude drug (Mokutsu in Japanese) in Japan and China, for example as an antiphlogistic, diuretic, and analgesic. The drug is also prescribed in China as a diuretic and antiphlogistic to treat rheumatism, lumbago, hernia, dropsy, diabetes, amenorrhea, inflammation of the stomach and kidneys, breast trouble, headache, cold, and general debility. Another species, *A. trifoliata*, is occasionally used in China; further, a decoction of the stem and root of *Stauntonia hexaphylla* or the pericarp of the fruit is used as a diuretic in Japan and China (Keys 1976; Perry and Metzger 1980). A large number of triterpenoid saponins, which on acid hydrolysis yield oleanolic acid (**1**), hederagenin (**10**), arjunolic acid (**11**), and norarjunolic acid (**12**), respectively, have been reported as sapogenins from the stems (Fujita et al. 1974; Kumekawa et al. 1974), seed (Higuchi et al. 1972; Higuchi and Kawasaki 1972), and pericarp (Higuchi and Kawasaki 1976a,b) (Table 1, Fig. 2). However, free triterpenoids from the plant

[1]The Research Institute for Biosciences, Science University of Tokyo, 2669 Yamazaki, Noda City, Chiba 278, Japan

Biotechnology in Agriculture and Forestry, Vol. 15
Medicinal and Aromatic Plants III (ed. by Y.P.S. Bajaj)

Fig. 1. Plant of *Akebia quinata* growing in the fence of a garden

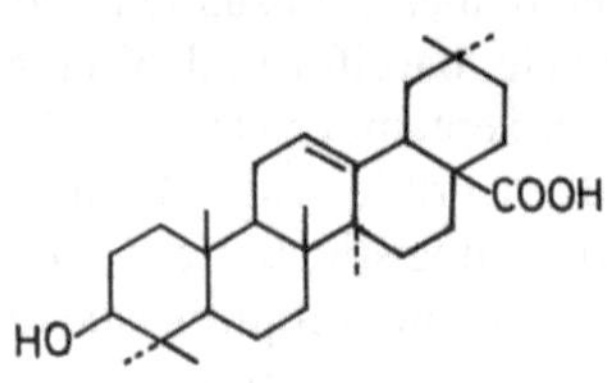

Oleanolic acid (1)
Stem
Pericarp

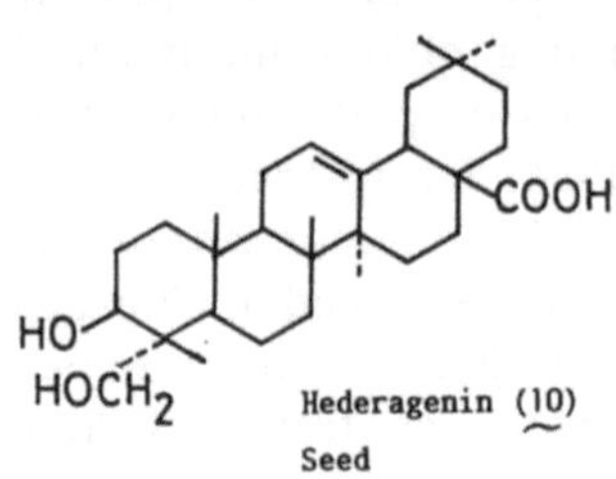

Hederagenin (10)
Seed
Stem
Pericarp

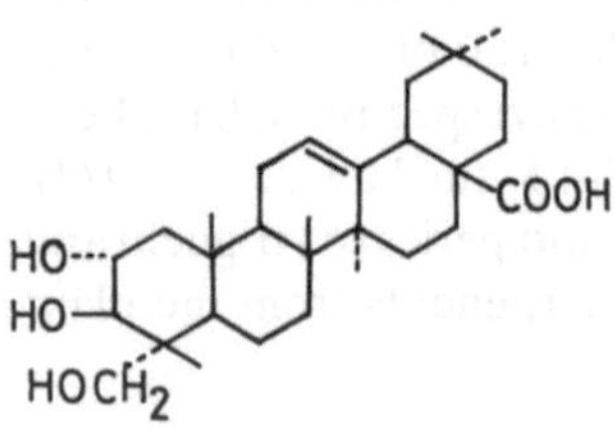

Arjunolic acid (11)
Pericarp

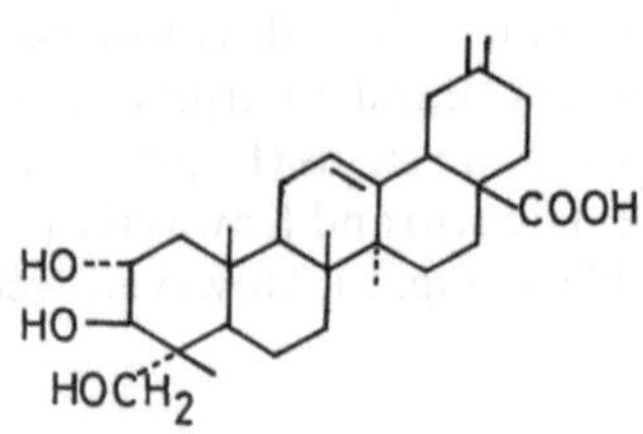

Norarjunolic acid (12)
Pericarp

Fig. 2. Structure of sapogenins from *Akebia quinata*

Table 1. Compounds isolated from plant and callus tissue of *Akebia quinata*

Part	Compound	Genin	Sugar moiety
Seed[a]	Saponin A (*1)	Hederagenin	R_1 = Ara- R_2 = H
	Saponin B	Hederagenin	R_1 = Xyl-^{2}Ara- R_2 = H
	Saponin C (*3)	Hederagenin	R_1 = Glc-^{2}Ara- R_2 = H
	Saponin D	Hederagenin	R_1 = Ara- R_2 = Glc-^{6}Glc-
	Saponin E	Hederagenin	R_1 = Xyl-^{2}Ara- R_2 = Glc-^{6}Glc-
	Saponin F	Hederagenin	R_1 = Glc-^{2}Ara- R_2 = Glc-^{6}Glc-
	Saponin G	Hederagenin	R_1 = Rha-$^4_2$Ara- Glc R_2 = Rha-^{4}Glc-^{6}Glc-
Stem[b]	Akeboside Stb (*1)	Hederagenin	R_1 = Ara- R_2 = H
	Akeboside Stc (*2)	Hederagenin	R_1 = Rha-^{2}Ara- R_2 = H
	Akeboside Std (*3)	Hederagenin	R_1 = Glc-^{2}Ara- R_2 = H
	Akeboside Stf	Hederagenin	R_1 = Rha-^{6}Glc-^{2}Ara- R_2 = H
	Akeboside Sth (*4)	Hederagenin	R_1 = Rha-^{2}Ara- R_2 = Rha-^{4}Glc-^{6}Glc-
	Akeboside Stk	Hederagenin	R_1 = Rha-^{6}Glc-^{2}Ara- R_2 = Rha-^{4}Glc-^{6}Glc-
	Akeboside Ste	Oleanolic acid	R_1 = Rha-^{6}Glc-^{2}Ara- R_2 = H
	Akeboside Stj	Oleanolic acid	R_1 = Rha-^{6}Glc-^{2}Ara- R_2 = Rha-^{4}Glc-^{6}Glc-
Pericarp[c]	Saponin Pa (*1)	Hederagenin	R_1 = Ara- R_2 = H
	Saponin Pc	Hederagenin	R_1 = Xyl-^{3}Ara- R_2 = H
	Saponin Pd (*2)	Hederagenin	R_1 = Rha-^{2}Ara- R_2 = H
	Saponin Pf (*3)	Hederagenin	R_1 = Glc-^{2}Ara- R_2 = H
	Saponin Pg	Hederagenin	R_1 = Xyl-^{3}Rha-^{2}Ara- R_2 = H
	Saponin Pk (*4)	Hederagenin	R_1 = Rha-^{2}Ara- R_2 = Rha-^{4}Glc-^{6}Glc-
	Saponin Pj$_2$	Hederagenin	R_1 = Ara- R_2 = Rha-^{4}Glc-^{6}Glc-
	Saponin Pb	Oleanolic acid	R_1 = Rha-^{2}Ara- R_2 = H
	Saponin Pe	Oleanolic acid	R_1 = Glc-^{2}Ara- R_2 = H
	Saponin Pj$_3$	Oleanolic acid	R_1 = Rha-^{2}Ara- R_2 = Rha-^{4}Glc-^{6}Glc-

(to be continued)

Table 1. (*Continued*)

Part	Compound	Genin	Sugar moiety
	Norarjunolic acid		
	Saponin Ph	Norarjunolic acid	R = Rha-[4]Glc-[6]Glc-
	Arjunolic acid		
	Saponin Pj_1	Arjunolic acid	R = Xyl-[3]Rha-[4]Glc-[6]Glc-
Callus Tissue			
Free triterpenoids[d]			
Oleanolic acid			
30-Noroleanolic acid			
3-Epi-30-noroleanolic acid			
Mesembryanthemoidigenic acid			
3β-Hydroxy-29-al-olean-12-en-28-oic acid			
3α,24-Dihydroxy-30-norolean-12,20(29)-dien-28-oic acid			
Triterpenoid saponins[e]			
	Quinatoside A	30-Nor-hederagenin	R = Ara-
	Quinatoside B	30-Nor-hederagenin	R = Glc-[3]Ara-
	Quinatoside C	30-Nor-hederagenin	R = Xyl-[2]Ara-
	Quinatoside D	30-Nor-oleanolic acid	R = Xyl-[2]Ara-

(*1–4) are the same compounds.
[a] Higuchi and Kawasaki (1972); Higuchi et al. (1972).
[b] Kumekawa et al. (1974).
[c] Higuchi and Kawasaki (1976a,b).
[d] Ikuta and Itokawa (1986a).
[e] Ikuta and Itokawa (1989a).

have not been reported except for **11** and **12** obtained from the pericarp. Anthocyanins have been reported from the flowers (Ishikura and Nagamino 1976).

The antiulcerogenic effect of the saponins of *A. quinata* has been examined by oral administration in rats. Its saponins showed an activity at 2000 mg/kg, p.o. (crude methanol extracts) and 500 mg/kg, p.o. (crude saponins) (Yamahara et al. 1975). In screening tests using a model system of congestive edema in rats, the saponins showed antiedomatous and diuretic effects at 2000 mg/kg, p.o. (methanol extracts) on oral administration (Yamahara et al. 1979). Aqueous extract of stems of *A. quinata* is administered to rats fed a high fat diet containing 1.5% cholesterol and 1% cholelic acid in order to screen the hypocholesterolic activity of these crude drugs. Significant decreases in the serum TC, FC, and PL were observed in the high-fat groups on being given these crude drugs. TC in the liver showed no reduction in the groups administered Moktsu as compared with the fat-loaded control group (Onishi et al. 1984). (TC = total cholesterol; FC = free cholesterol; CE = cholesterol ester). In this respect, the *A. quinata* plant is a very important crude drug in Japan.

2 In Vitro Approaches

2.1 Derivation and Culture of Callus Tissue

The plant material of *A. quinata* was collected in October 1981 at the Medicinal Plant Garden of Tokyo College of Pharmacy. Callus tissues from the stalk were established using MS medium (minus glycine) (Murashige and Skoog 1962) containing 2,4-D (3 mg/l, 1 mg/l) and kin (0.1 mg/l) as plant growth regulators. They were subcultured every 5–6 weeks onto fresh MS medium containing 2,4-D (0.1 and 1 mg/l) with kin (0.1 mg/l) at 26 ± 1 °C in the dark (Ikuta and Itokawa 1986a).

2.2 Isolation and Identification of the Compounds (Figs. 3, 4)

2.2.1 *Isolation of Triterpenes*

The fresh callus tissues were extracted successively with cold methanol in a Waring blender and then with hot ethylacetate. Both the extracts were combined and concentrated under reduced pressure to yield the extracts, which were chromatographed over a column of silica gel using hexane containing increasing proportions of EtOAc and MeOH as the solvent system of the eluent. Compounds **1** and **2** were eluted with hexane-EtOAc (5:1), and compounds **2**, **3**, **4**, **5**, and **6** were eluted with hexane-EtOAc (2:1). The compounds **1–5** were finally purified by methylation with freshly prepared diazomethane and then by HPLC (silica gel) using the solvent system of hexane-EtOAc-CH_3CN (2:1:0.5) to give **1a**, **2a**, **3a**, **4a**, and **5a**, respectively.

2.2.2 *Identification of Triterpenes*

Compound **1** is oleanolic acid and compounds **2**, **3**, **5**, and **6** are new triterpenoids, also compound **4** is mesenbryanthemoidigenic acid, and compounds **7** and **8** are a mixture of sitosterol and stigmasterol (Ikuta and Itokawa 1986a,b, 1988c).

The structure of **1** was established by spectroscopic means and identified as oleanolic acid by comparison with an authentic sample. Compound **2**, on treatment with diazomethane, yielded a methylester (**2a**). The mass spectrum of **2a** showed molecular ion $[M]^+$ at m/z 454 and exhibited the typical retro-Diels-Alder (r.D.A.) fragmentation of the C-ring of the β-amyrin Δ^{12}-skeleton at m/z 246 (a) and 187 [a − COOMe] (Fig. 4) (Budzikiewicz et al. 1963). The mass number of the peak(a) of **2a** was 16 mass units less than that of the peak due to the r.D.A. type of fragmentation of oleanolic acid methyl ester (**1a**). The ^1H NMR spectrum ($CDCl_3$)

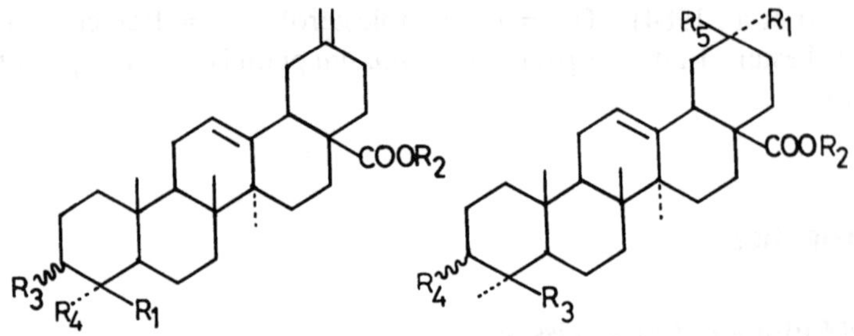

	R_1	R_2	R_3	R_4
2	CH_3	H	▶ OH	CH_3
2a	CH_3	CH_3	▶ OH	CH_3
3	CH_3	H	··· OH	CH_3
3a	CH_3	CH_3	··· OH	CH_3
6	CH_2OH	H	··· OH	CH_3
6a	CH_2OH	CH_3	··· OH	CH_3
6b	CH_2OAc	H	··· OAc	CH_3
14	CH_3	H	▶ OH	CH_2OH

	R_1	R_2	R_3	R_4	R_5
1	CH_3	H	CH_3	▶ OH	CH_3
1a	CH_3	CH_3	CH_3	▶ OH	CH_3
4	CH_2OH	H	CH_3	▶ OH	CH_3
4a	CH_2OH	CH_3	CH_3	▶ OH	CH_3
5	CHO	H	CH_3	▶ OH	CH_3
5a	CHO	CH_3	CH_3	▶ OH	CH_3
9	CH_3	H	CH_2OH	··· OH	CH_3
13a	CH_3	CH_3	CH_3	OH	CH_2OH

Fig. 3. Structure of triterpenes isolated from *Akebia quinata* callus tissues

of **2a** shows five tertiary methyl signals, two protons at δ 4.61 and 4.62 ascribable to the exomethylene protons, and the H-18 proton at 2.79 (dd). Therefore, compound **2a** is presumed to be a 30-norolean-12-ene derivative carrying an exomethylene group in ring D or E. Furthermore, the ^{13}C NMR chemical shifts (Table 2) of **2a** were found to be similar to the authentic sample of **1a**, but with predictable differences and the low field carbon signals at δ 148.0(s) and 106.8(t) were assigned to the exomethylene C-20 and C-29, respectively. The hydroxyl group at C-3 must have a β-equatorial configuration according to the signal at δ 78.7(d) (Tori et al. 1974) and also from the value of the ^{1}H NMR chemical shift and the coupling pattern of the proton at C-3. The other assignments were supported by comparison of the ^{13}C NMR spectrum of **2a** with that of **1a**. Thus, the structure of compound **2** was established as 3β-hydroxy-30-norolean-12,20(29)-dien-28-oic acid (akebonoic acid). Compound **3**, on treatment with diazomethane, yielded a methylester (**3a**), which was found to be similar to **2a** but with predictable differences. The ^{13}C NMR signal of the low field at δ 76.2(d) was assigned to C-3 with the α-configuration of the hydroxyl group. Other signals at δ 33.0(t), 25.2(t), 37.4(s), and 49.0(d) were due to C-1, C-2, C-4, and C-5, respectively (Table 2). These values for the chemical shifts also agreed with the α-configuration of the C-3 hydroxyl group (Asakawa et al. 1977), and this was further supported by the ^{1}H NMR chemical shift and the coupling pattern of the C-3 proton.

On the other hand, the values of the chemical shifts of the ^{13}C NMR for other carbons were similar to those of **2a**. The mass spectrum of **3a** exhibited the typical r.D.A. fragmentation and its fragments had the same m/z values to those of **2a**

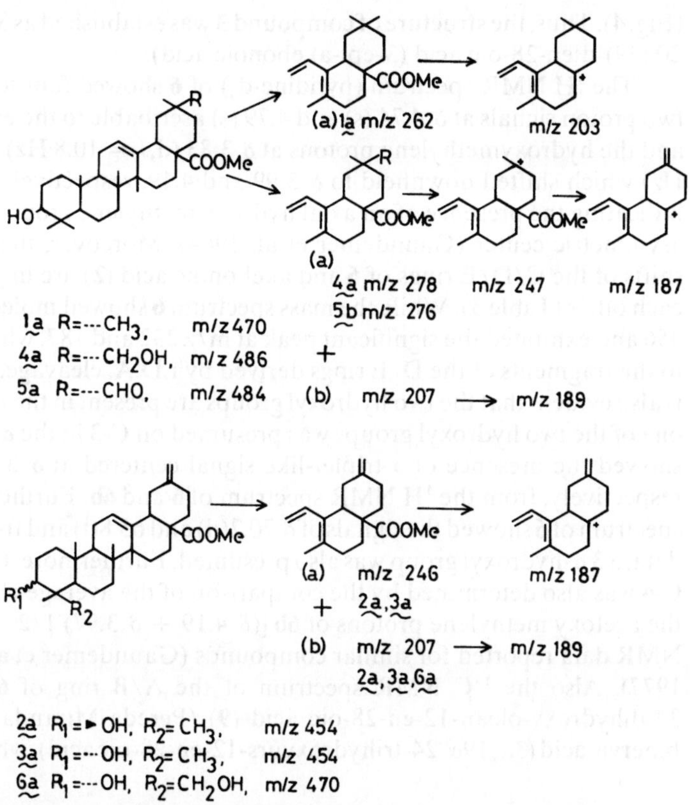

Fig. 4. Fragmentations of compounds **1–6**. (Ikuta and Itokawa 1986a)

Table 2. ^{13}C NMR chemical shifts of compounds **1a**, **2a**, **3a**, **4a**, and **13a** ($CDCl_3$, δ ppm). (Ikuta and Itokawa 1986a)

Carbon No.	Chemical shift					Carbon No.	Chemical shift				
	1a	**2a**	**3a**	**4a**	**13a**		**1a**	**2a**	**3a**	**4a**	**13a**
1	38.5	38.5	33.0	38.5	38.7	16	23.4	23.5	23.4	23.5	23.4
2	27.1	27.2	25.2	27.2	27.0	17	46.6	47.0	47.1	47.0	46.8
3	78.7	79.2	76.2	79.1	78.9	18	41.3	41.4	41.4	40.5	40.3
4	38.7	38.8	37.4	38.8	38.7	19	45.8	47.2	47.2	40.1	40.3
5	55.2	55.2	49.0	55.3	55.3	20	30.6	148.0	148.2	35.8	35.2
6	18.3	18.3	18.3	18.4	18.5	21	33.8	30.0	30.0	31.6	32.8
7	32.6	32.7	32.7	32.7	32.8	22	32.3	37.5	37.5	32.7	32.1
8	39.4	39.4	39.5	39.3	39.3	23	28.1	28.1	28.3	28.1	28.1
9	47.6	47.7	47.5	47.7	47.7	24	15.6	15.6	22.3	15.6	15.7
10	37.0	37.1	37.1	37.1	37.2	25	15.3	15.3	15.3	15.3	15.4
11	23.1	23.3	23.3	23.1	23.4	26	16.8	16.8	16.9	16.9	16.9
12	122.1	123.2	123.0	122.8	122.8	27	26.0	26.0	26.2	25.9	26.0
13	143.4	143.1	143.2	143.2	143.2	28	177.9	177.5	177.5	178.1	178.5
14	41.6	41.6	41.8	41.6	41.7	29	33.1	106.8	107.0	74.4	28.9
15	27.7	27.7	27.7	27.7	27.7	30	23.6	–	–	19.0	65.8
						COOMe	51.3	51.6	51.6	51.6	50.6

(Fig. 4). Thus, the structure of compound **3** was established as 3-epi-30-norolean-12, 20 (29)-dien-28-oic acid (3-epi-akebonoic acid).

The ^{1}H NMR spectrum (pyridine-d_5) of **6** showed four tertiary methyl signals, two proton signals at δ 4.74 (s) and 4.79 (s) ascribable to the exomethylene protons and the hydroxymethylene protons at δ 3.83 (d,J = 10.8 Hz) and 4.07 (d, J = 10.8 Hz) which shifted downfield to δ 3.99 and 4.19, respectively, on acetylation (**6b**), indicating the presence of an axial hydroxy methylene group attached to C-4 of an asymmetric center (Gaundemer et al. 1964). Moreover, the ^{13}C NMR chemical shifts of the C/D/E rings of **6** and akebonoic acid (**2**) are in good agreement with each other (Table 3). While the mass spectrum **6** showed molecular ion $[M]^+$ at m/z 456 and exhibited the significant peak at m/z 232 and 187, which could be assigned to the fragments of the D/E rings derived by r.D.A. cleavage. From these results it is also evident that the two hydroxyl groups are present in the A/B ring portion, and one of the two hydroxyl groups was presumed on C-3 in the α-configuration which showed the presence of a triplet-like signal centered at δ 3.70 (**6**) and 4.92 (**6b**), respectively, from the ^{1}H NMR spectrum of **6** and **6b**. Furthermore, the ^{13}C NMR spectrum of **6** showed the signals of δ 70.2(d) and 65.8(t) and from the above spectral data, a 3 α-hydroxyl group was also presumed. Furthermore, the stereochemistry of C-4 was also determined by the comparison of the average chemical shift value of the acetoxy methylene protons of **6b** [(δ 4.19 + δ 3.97) 1/2 = δ 4.08] with the ^{1}H NMR data reported for similar compounds (Gaundemer et al. 1964; Takani et al. 1977). Also the ^{13}C NMR spectrum of the A/B ring of **6** coincided with 3α, 24-dihydroxy-olean-12-en-28-oic acid (**9**) (Pereda-Miranda et al. 1986) or barbinervic acid (3α, 19α, 24-trihydroxyurs-12-en-28-oic acid), which was isolated from

Table 3. ^{13}C NMR chemical shifts of compounds **2**, **6**, and **9**. (C_5D_5N, δ ppm). (Ikuta and Itokawa 1988c)

Chemical shift				Chemical shift			
Carbon No.	**6**	**2**	**9**	Carbon No.	**6**	**2**	**9**
1	33.7	39.0	33.9	16	24.0	23.8	23.5
2	26.4	28.1	26.4	17	47.1	47.1	47.0
3	70.2	78.2	70.0	18	47.9	48.0	41.9
4	43.9	39.4	43.9	19	42.0	42.0	46.2
5	50.1	55.9	50.1	20	149.2	148.5	30.8
6	19.1	18.8	19.1	21	38.4	38.4	34.0
7	33.7	33.3	33.6	22	30.4	30.4	32.8
8	40.0	39.8	39.9	23	23.6	28.2	23.4
9	48.1	48.1	48.1	24	65.8	16.5	65.8
10	37.5	37.4	37.5	25	15.9	15.6	15.9
11	23.8	23.8	24.0	26	17.3	17.4	17.1
12	123.0	123.3	120.0	27	26.1	26.2	26.1
13	144.2	143.5	144.1	28	179.4	177.3	177.9
14	42.2	42.1	42.0	29	107.1	107.1	33.1
15	28.3	28.3	28.1	30	–	–	23.7
				COOMe			51.5

All signals were corroborated from DEPT techniques. The measurements were made on a Brucker AM-400 spectrometer in pyridine-d_5 with TMS as an internal standard.

leaves of *Clethra barbinervis* (Table 3) (Takahashi and Takani 1978). The finding was further confirmed by the 2D-NOE spectrum of compound **6a**. Therefore, **6** was established as 3α, 24-dihydroxy-30-norolean-12,20 (29)-dien-28-oic acid (quinatic acid). 24-Epi-hederagenin-type triterpene has not been reported from the original plant. The compounds **2**, **3**, and **6** are reported for the first time from natural sources and these 30-noroleanane-type triterpenoids are rare as natural products except for norhederagenin (**14**), norarjunolic acid (**12**), and eupteleogenins isolated from the callus tissues of *Paeonia japonica* (Ikuta and Itokawa 1988b), from the pericarp of *A. quinata* (Higuchi and Kawasaki 1972), and from the leaves of *Euptelea polandra* (Eupteleaceae) (Murata et al. 1965), respectively. Furthermore, the mass spectrum of **4a** exhibited significant peaks at m/z 278(a) and 247 [a-CH_2OH], which could be assigned to fragments of the D and E rings derived by r.D.A. cleavage and at m/z 207 (b) and 189 which could be unequivocally assigned to fragments of the A and B rings by a cleavage of the C ring and elimination of the hydroxyl groups (Fig. 4). The ^{1}H NMR spectrum of **4b** showed six singlet methyl protons between δ 0.70 and 1.12, hydroxymethylene protons at 3.30 (s, 2H), carbomethoxy protons at 3.68(s), and an olefinic proton at 5.33 (t-like). The ^{13}C NMR spectrum (Table 2) of **4a** was very similar to that of queretaroic acid (**13**) (Tori et al. 1974) except that the former exhibited signals δ 74.4(C-29) and 19.0(C-30), the latter exhibited signals 28.9(C-29) and 65.8(C-30). This evidence supports the α (equatorial, i.e., C-29) rather than β (axial, i.e., C-30) orientation of the hydroxymethylene group at C-20 in the compound **4a**. Thus **4a** is represented as 3β,29-dihydroxy-olean-12-en-28-oic acid (mesembryanthemoidigenic acid). This is the second report of compound **4** from a natural source. Compound **5**, on treatment with diazomethane, yielded a methyl ester (**5a**), which showed in the ^{1}H NMR spectrum signals attributable to an aldehyde group at δ 9.39, a vinyl proton at 5.35 and six tertiary C-methyl groups at δ 0.7, 0.75, 0.77, 0.91, 0.98, 1.13, and 1.16, suggesting it to have an aldehyde group at C-20. The mass spectrum of **5a** exhibited the molecular ion $[M]^+$ at m/z 484 and its fragment ions (a) and (b) originating from the r.D.A. showed an analogous cleavage pattern (Fig. 4), comparable with the ions of compounds **1-4** at m/z 247 (a) [276-CHO], 187 and 207 (b), and 189. Therefore, compound **5** was determined as 3β-hydroxy-29-al-olean-12-en-28-oic acid. The compounds isolated were present in small amounts except for **1**, **2**, and **6**, but it is very interesting from the biogenetic point of view that callus tissues of *A. quinata* produced the compounds **1-5** showing the different degrees of oxidation at C-20. Further, these free triterpenoids **1**, **2**, **3**, **4**, **5**, and **6** have not been reported from the original plant material. The ^{1}H NMR, ^{13}C NMR, and mass spectra of compounds **7** and **8** demonstrated the presence of two sterol mixtures, whose compositions were confirmed as equal amounts of sitosterol (**7**) and stigmasterol (**8**) by comparison with authentic samples on GLC (Ikuta and Itokawa 1986a, 1988c). Furthermore, four new saponins were recently isolated from the callus tissues of *A. quinata* (Table 1) (Ikuta and Itokawa 1989a).

2.3 Chemotaxonomic Studies

From the chemotaxonomic point of view, lardizabalaceous plants *A. trifoliata* and *S. hexaphylla* callus tissues were further established from the stem and their methanolic extract was investigated for triterpenoid content. Each callus tissue produced the above-mentioned 30-noroleanan-type triterpenes along with the oleanan type, and these callus tissues all produced the same triterpenoids, as shown in Table 4 (Ikuta and Itokawa 1986b; Ikuta 1987). Similar results have also been reported that the callus tissues of the papaveraceous, ranunculaceous, berberidaceous and menispermaceous plants all produced the isoquinoline type alkaloids, benzophenanthridine, protoberberine, and aporphine as main alkaloid compounds (Ikuta et al. 1974; Ikuta 1987; Ikuta and Itokawa 1988a). The different plant species show generally different constituent patterns, even among the same family. However, the metabolic sequence of callus tissues derived from the different species belonging to the same genera or family, or in the vicinity of the families, tended to show simple biogenetic pathways of the secondary compounds that produced a similar type of compounds, such as triterpenoids or alkaloids. Therefore, this method of comparison of the constituents of callus tissues may open the possibility of chemotaxonomical studies, and is interesting from the phylogenetic point of view.

Table 4. Comparison of triterpenes from lardizabalaceous plant callus tissues. (Ikuta 1987)

Lardizabalaceae	1	2	3	4	5	6
Akebia						
Akebia quinata	+	+	+	+	+	+
Akebia trifoliata	+	+	+			
Stauntonia						
Stauntonia hexaphylla	+	+	+	+	+	+

3 Conclusion

These new 30-norolean type triterpenes from *A. quinata* callus tissue were not reported from the original plant source, which suggests biochemical differences between intact plants and the callus tissues derived from them. The result may make it possible to obtain new biologically active substances from plant tissue cultures as new sources of natural products.

For this purpose, plant tissue cultures must be developed for new phytochemical studies. Furthermore, the constituents of the callus tissues of *A. quinata* were deduced based on the degrees of oxidation that the speculative biogenesis of 30-norolean type triterpenes (ex. akebonoic acid) may proceed via the route of oxidation [$-CH_3 \rightarrow -CH_2OH \rightarrow -CHO \rightarrow (-> = O) \rightarrow > =$]. In this respect, *A. quinata* tissue culture will provide one of the most useful experimental systems to study the biosynthesis of 30-noroleanan-type triterpenes. Studies of a more detailed speculative biogenetic pathway using callus tissues of *S. hexaphylla* was reported (Ikuta and Itokawa 1989b).

4 Protocol

Callus cultures of *A. quinata* were established by growing surface sterilized stem segments (0.5–1.5 cm) under aseptic conditions on solid MS medium (minus glycine) containing sucrose (3%, w/v), agar (0.9%, w/v), 2,4-D(3 mg/l, 1 mg/l) and kin (0.1 mg/l). Subcultures were made every 4–5 weeks on MS containing sucrose (3%, w/v), agar (0.9%, w/v), 2,4-D(1 mg/l) and kin (0.1 mg/l) and were maintained in the dark at 26 ± 1 °C. Examination of the optimum conditions for triterpenoid production has not been carried out, except for those above. However, the cells produced oleanolic acid (**1**), akebonoic acid (**2**), and quinatic acid (**6**) as main triterpenoid compounds from the triterpenoid fraction excluding the other secondary metabolites. The methanolic extract was further carefully investigated by TLC, using several solvent systems and color-developing reagents, but other secondary compounds (e.g., alkaloids, flavonoids, etc.), except for large amounts of triterpenoids and phytosterols, could not be detected.

Acknowledgement. The author thanks Prof. H. Itokawa, Tokyo College of Pharmacy for helpful discussions.

References

Asakawa J, Kasai R, Yamasaki K, Tanaka O (1977) ^{13}C NMR study of ginseng sapogenins and their related dammarane type triterpenes. Tetrahedron 33:1935–1939

Benson L (1979) Plant classification, 2nd edn. Heath, Lexington Toronto, 146 pp

Budzikiewicz H, Wilson JM, Djerassi C (1963) Mass spectrometry in structural and stereochemistry problems. XXXII. Pentacyclic triterpenes. J Am Chem Soc 85:3688–3699

Fujita M, Itokawa H, Kumekawa Y (1974) The study on the constituents of *Clematis* and *Akebia* spp. II. On the saponins isolation from the stem of *Akebia quinata* Decne. Yakugaku Zasshi 94:194–198

Gaundemer A, Polonsky J, Wenkert E (1964) Determination of the conformation of the CH_2OH and CH_2OAc groups of some triterpenes by nuclear magnetic resonance. Bull Soc Chem Fr 407–411

Higuchi R, Kawasaki T (1972) Seed saponins of *Akebia quinata* Decne. II. Hederagenin 3,28-O-bisglycosides. Chem Pharm Bull 20:2143–2149

Higuchi R, Kawasaki T (1976a) Pericarp saponins of *Akebia quinata* Decne. I. Glycosides of hederagenin and oleanolic acid. Chem Pham Bull 24:1021–1032

Higuchi R, Kawasaki T (1976b) Pericarp saponins of *Akebia quinata* Decne. II. Arjunolic and norarjunolic acids, and their glycosides. Chem Pharm Bull 24:1314–1323

Higuchi R, Miyahara K, Kawasaki T (1972) Seed saponins of *Akebia quinata* Decne I. Hederagenin 3-O-glycosides. Chem Pharm Bull 20:1935–1939

Hutchinson J (1973) The families of flowering plants (arranged according to a new system based on their probable phylogeny), 3rd edn.Clarendon, Oxford, pp 503–505

Ikuta A (1987) Callus constitutions and chemotaxonomy. Plant Tiss Cult Lett 4:1–7

Ikuta A (1988) Isoquinoline. In: Constabel F, Vasil IK (eds) Cell culture and somatic cell genetics of plants, vol 5. Academic Press, New York London, pp 289–314

Ikuta A, Itokawa H (1986a) Triterpenoids *Akebia quinata* callus tissue. Phytochemistry 25:1625–1628

Ikuta A, Itokawa H (1986b) The isolation of new 30-noroleanan type triterpenes from callus of lardizabalaceous plants. In: Somers DA, Gengenbach BG, Biesboer DD, Hackett WP, Green CE (eds) 6th Int Congr Plant tissue culture, Minneapolis, Abstr, 351 pp

Ikuta A, Itokawa H (1988a) Alkaloids of tissue cultures of *Nandina domestica*. Phytochemistry 27:2143–2145

Ikuta A, Itokawa H (1988b) Triterpenoids of *Paeonia japonica* callus tissue. Phytochemistry 27:2813–2815

Ikuta A, Itokawa H (1988c) A triterpene from *Akebia quinata* callus tissue. Phytochemistry 27:3809–3810

Ikuta A, Itokawa H (1989a) 30-Noroleanane saponins from callus tissue of *Akebia quinata*. Phytochemistry 28:2663–2665

Ikuta A, Itokawa H (1989b) The triterpenes from *Stauntonia hexaphylla* callus tissue and their biosynthetic significance. J Nat Prod 52:623–628

Ikuta A, Syono K, Furuya T (1974) Alkaloids of callus tissues and redifferentiated plantlets in the Papaveraceae. Phytochemistry 13:2175–2179

Ishikura N, Nagamino N (1976) Anthocyanins *Akebia* and *Stauntonia*. Phytochemistry 15:442–443

Keys JD (1976) Chinese herbs. Their botany, chemistry, and pharmacodynamics. Tuttle, Rutland Tokyo, 110 pp

Kumekawa Y, Itokawa H, Fujita M (1974) The study on the constituents of *Clematis* and *Akebia* sp. III. The study of the structures of akebosides isolated from the stem of *Akebia quinata* Decne. Chem Pharm Bull 22:2294–2300

Murashige T, Skoog F (1962) A revised medium for rapid growth and bioassays with tobacco tissue cultures. Physiol Plant 15:473–497

Murata T, Imai S, Imanishi M, Goto M, Morita K (1965) The structure of eupteleogenin, the aglycone of anti-fungal glycosides, eupteleoside A and eupteleoside B. Tetrahedron Lett 3215–3222

Ohwi J (1984) A combined, much revised, and extended translation by the author of his Flora of Japan (1953) and Flora of Japan-Pteridophyta (1957). In: Meyer FG, Walker EH (eds) Flora of Japan. Smithsonian Inst, Washington, 461 pp

Onishi E, Yamada K, Yamada T, Kaji K, Inoue H, Seyama Y, Yamashita S (1984) Comparative effects of crude drugs on serum lipids. Chem Pharm Bull 32:646–650

Pereda-Miranda R, Delgado G, Romo de Vivar A (1986) New triterpenoids from *Salvia nicoesoniana*. J Nat Prod 49:225–230

Perry LM, Metzger J (1980) Medicinal plants of East and Southeast Asia: Attributed properties and uses. MIT, Cambridge, Mass London, pp 195–196

Takahashi K, Takani M (1978) Studies on the constituents of the medicinal plants. XXI. Constituents of the leaves of *Clethra barbinervis* Sneb et Zucc (2) and ^{13}C-Nuclear Magnetic Resonance spectra of 19α-hydroxy-12-en-28-oic acid type of triterpenoids. Chem Pharm Bull 26:2689–2693

Takani M, Kubota K, Nozawa M, Ushiki T, Takahashi K (1977) Studies on constituents of medicinal plants. XVIII. Constituents of the leaves of *Clethra barbineruis* Sieb et Zucc (1). Chem Pharm Bull 25:981–985

Tori K, Seo S, Shimaoka A, Tomita Y (1974) Carbon-13 NMR spectra of olean-12-enes. Full signal assignments including quaternary carbon signals assigned by use of indirect ^{13}C, ^{1}H spin couplings. Tetrahedron Lett 4227–4230

Yamahara J, Shintani Y, Konoshima T, Sawada T, Fujimura H (1975) Biological active principles of the crude drugs. II. Antiulcerogenic and anti-inflammatory actions of the crude drugs containing saponin. Yakugaku Zasshi 95:1179–1975

Yamahara J, Takagi Y, Sawada T, Fujimura H, Shirakawa K, Kitagawa I, Yoshikawa M (1979) Effects of crude drugs on congestive edema. Chem Pharm Bull 27:1464–1468

VIII *Bidens alba* (Smooth Beggar-Tick) and *Bidens pilosa* (Hairy Beggar-Tick)

R.A.Norton[1]

1 Introduction

The genus *Bidens* (Compositae) is composed of approximately 230 species having a worldwide distribution in tropical and temperate regions. It is primarily a continental group, which has become established on some islands, notably the Hawaiian islands. The centers of diversity are Africa and the New World, with each center having about 100 species (Gillett 1975). Several species are so abundant that they are considered serious weeds. Two will be of particular concern here: *Bidens alba* var. *radiata* (Schultz-Bip.) Ballard (Ballard 1986) (Fig. 1) and *B. pilosa* var. *minor* (Blume) Sherff, another member of the complex. *B. alba* var. *radiata* (smooth beggar-tick) occurs in south eastern Mexico into Central America and in Florida, U.S.A.; *B. pilosa* var. *minor* (hairy beggar-tick) is primarily restricted to Central America (Ballard 1986).

B. pilosa var. *minor* and *B. alba* var. *radiata* are erect annual herbs with opposite pinnate leaves (Fig. 1). Flowers are organized into a capitulum with yellow disc flowers and five or six white (occasionally purple) ray flowers which are 5–7 mm long and have a nonfunctional style in the former species and ray flowers 15–18 mm long with no style in the latter. Both plants, in common with most species of *Bidens*, are found in moist, disturbed areas. Mature plants are easily recognized by the spherical heads radiating black, barbed seeds in all directions from the receptacle. The two species are cross-fertile, although *B. alba* var. *radiata* is a tetraploid (x = 24) and *B. pilosa* var. *minor* is a hexaploid (x = 36) (Ballard 1986; Norton unpubl.). Hybrids are fertile pentaploids (x = 30) which are intermediate for flower size, hairiness, time of flowering, and leaf polyacetylene level (unpubl.). The base chromosome number for most species of *Bidens* appears to be x = 12, although many species are tetra- and hexaploid.

Although normally not thought of as a significant biomedical resource, several species of *Bidens* have attained widespread use in folk medicine, or in one case agriculturally, and have been investigated by modern chemical methods. Biologically active constituents have been isolated, characterized, and tested as single compounds against a spectrum of organisms. In this review some of the ethnopharmacology of *Bidens* will be discussed along with the results from modern methodology assessing biological activity. Results for two tissue culture systems for producing some of the compounds shown to be of interest will also be presented.

[1]Department of Botany, University of British Columbia, Vancouver, British Columbia V6T 2B1, Canada. Present address: USDA, Agricultural Research Service, Plant Physiology Research Unit, P.O. Box 5677, Athens, GA 30613, USA

Biotechnology in Agriculture and Forestry, Vol. 15
Medicinal and Aromatic Plants III (ed. by Y.P.S. Bajaj)

Fig. 1. Mature flowering plant of *Bidens alba* var. *radiata*

1.1 Traditional Uses of *Bidens*

Bidens pilosa is a major weed in Africa and Latin America, seriously affecting yields of maize, dryland rice, cereals, coffee, and tea (Akobundu 1987). Holm et al. (1979) list 20 species of *Bidens* that are considered important weeds. *Bidens pilosa* is the most widespread; with documented occurrence in 59 countries around the world. It is especially a problem in agriculture in the humid tropics in the Central American lowland (Campbell et al. 1982). However, in the tropics, *B. pilosa* and the segregate species *B. alba* are also widely used as medicinal herbs. This is not without its risks, however, Rose and Guillarmod (1974) note that *B. pilosa* leaves are boiled and eaten as a vegetable with maize and spinach as the staple diet in parts of the Transkei region of South Africa. However, this region is known to have a high incidence of esophageal cancer. Studies by Mirvish et al. (1979, 1985) show that ground-dried leaves of *B. pilosa* acted as a weak cocarcinogen when added to the diet of rats along with methyl-n-amylnitrosamine (MNAN), a known cocarcinogen for papilloma and carcinoma of the esophagus. *B. pilosa* is used in parts of Central America in traditional medicine for treatment of wounds and such (see Table 1) but it is also sometimes mixed with half-boiled rice and fermented to make a kind of saki. (Duke

Table 1. Uses and biological activities of *Bidens* spp.

Species (locale)	Use/sensitive organism	Plant part or compounds tested	Reference
B. amplissima			
	Candida albicans	Root	Camm et al. (1975)
	" "	Leaf, seed	Towers et al. (1977)
B. bipinnata			
(Brazil)	Diuretic	–	References in Craveiro et al. (1986)
(P.R. China)	Rheumatism, weakness, furunculosis, anticancer	–	References in Cräveiro et al. (1986)
	Allelopathic	Whole plant	Meissner et al. (1986)
	Antibacterial	Plant extract	Pates and Madsen (1955)
B. campylotheca			
Hawaii	Tea, blood purifier treatment for thrush	Leaf	Degener (1975)
B. cernua			
	Mosquito larvae	Defatted etoh extract	Farnsworth et al. 1966
	Anti-insect	Whole plant	Farnsworth et al. (1966)
	bacteria, yeasts, dermatophytic fungi	PHT	Bondarenko et al. (Reference no. 121, cited in Towers 1980)
	Candida albicans	All parts[b]	Towers et al. (1977)
B. frondosa			
	Escherichia coli, Staphylococcus aurea	Leaf: ether extract saline extract	Carlson et al. (1948); Carlson and Douglas (1948)
	Periplanata americanus (roach)	Stem, leaves, flowers	Heal et al. (1950)
	Candida albicans	Root, seed	Towers et al. (1977)
B. laevis			
	Seed germination/ root growth (lettuce, radish, cucumber)	Soil, plant extracts	Bonasera et al. (1979)
B. pilosa			
(P.R. China)	Wounds and ulcers[a]	Leaf	cited in Wat et al. (1980)
(Mexico)	Food, agricultural system	Leaf, plants	Kahl (1987)
(Africa)	Food	Leaf	Rose and Guillarmod (1974)
(U.S.A.)	Food	Leaf	Morton (1962)
(Central Am.)	Wounds, rheumatism, colds, hepatic, infections, saki	Various parts	Duke (1986)
	Various bacteria	Whole plant extract	Bushnell et al. (1950)
	Insect antifeedant	Various parts	Heal et al. (1950)
	Cocarcinogen	Leaf	Mirvish et al. (1979, 1985)
	"	Leaf	Parry et al. (1986)
	Food, high niacin level	Leaf	Oliveira and De Carvalho (1975)

(*to be continued*)

Table 1. (*Continued*)

Species (locale)	Use/sensitive organism	Plant part or compounds tested	Reference
(Uganda)	Food, flavoring herb	Leaf, stem	Goode (1989)
	Algae	PHT	Arnason et al. (1981a)
B. pilosa	Allelopathic	PHT	Campbell et al. (1982)
	Analgesic, anti-pyretic, anti-inflammatory	Leaf	Solevilla (1984)
	Lung cancer	PHT	Fleischer (1981)
	Allelopathic	Root exudate	Stevens and Tang (1985)
	Bacteria, yeasts, filamentous fungi, Skin fibroblasts	PHT	Wat et al. (1979)
	Trematode Cercariae	PHT	Graham et al. (1980)
	Cryptococcus laurentii yeast	PHT, PDE-OAc, PDE, ETE[b]	Marchant and Towers (1986)
	Candida albicans	PDE, ETE and and others PA's	Towers et al. (1977)
	Fusarium culmorum	PHT	Bourque and Arnason (1985); Arnason et al. (1986)
	Drosophila melonogaster eggs	PHT	Kagan and Chan (1983)
	" " larvae, Eggs	PHT	Nakajima and Kawazu (1980)
	Yeasts, filamentous fungi, various bacteria	Various PA's	Towers (1980)
	Antiviral	PHT	Hudson et al. (1986a,b)
	Bacteriophages	PHT	Warren et al. (1980)
	Mosquito larvae	PHT, PDE	Arnason et al. (1981b)
	E. coli, S. ceriviseae	PHT	Arnason et al. (1980)
B. polylepis	Antitumor	Defatted Etoh extract[b]	Fong et al. (1972)
	Candida albicans	Seed	Towers et al. (1977)
B. reptans	*Candida albicans*	Seed	Towers et al. (1977)
B. trinervia	*Candida albicans*	All parts[b]	Towers et al. (1977)
B. tripartita	*Candida albicans*	All parts[b]	Towers et al. (1977)

[a] Additional references to folk uses found in work cited. [b] Inactive; PA's, polyacetylenes. Compound abbreviations, see Fig. 1.

1986). Morton (1962) reports on the appeal of *B. pilosa* as a food plant. In this connection it should be noted that *B. pilosa* has relatively high levels of niacin, important in diets depending on maize; lysine levels are also relatively high (Oliveira and De Carvalho 1975).

Within the genus *Bidens* few investigations of natural products have focused on compounds other than the polyacetylenes which are characteristic of the Compositae. Only flavonoids and the related anthochlor pigments (chalcones and aurones) have otherwise been reported (see, e.g., Crawford and Stuessy 1981; Ballard 1986). Since there has been considerable recent interest in the bioactivity of polyacetylenes and their sulfur derivatives, thiophenes and thiarubrines, the occurrence of this group of compounds in *Bidens* is emphasized here.

Polyacetylenes, also termed polyines or acetylenes, are found in 19 higher plant families and in basidiomycete fungi (Sörensen 1977). They are most characteristic of members of the Umbelliferae, Compositae, Santalaceae, and related families (Jones and Thaller 1978). Higher plants account for 85% of the known compounds, with the remainder isolated from fungi. Over 725 known acetylenes have been described (Sörensen 1977). In recent years studies on the bioactivity of many of these compounds have shown that some have powerful phototoxic antibiotic activities with potential for application in pest control.

1.2 Controlled Studies of Bioactive Compounds from *Bidens* spp.

Extracts and purified compounds from *Bidens* have been evaluated for both allelopathic potential and for antibiotic activity. Although not concerned with characterized compounds, a study by Kahl (1987) demonstrates the basis for a traditional use of *Bidens pilosa*. The Tarahumara Indians of the northwest Mexican highland not only use the plant for food, as do a number of indigenous peoples, but it is also allowed to grow among the maize fields along with a number of other weedy plants ("quelites") occurring naturally, that are also used for food. Kahl has shown that water extracts of the leaves of young *B. pilosa* and *Lepidium virginicum* plants cause a significant stimulation of seedling growth of the maize. Extracts of older plants do not have the effect.

Several studies have demonstrated allelopathic effects from soil of *Bidens*-infested areas (Bonasera et al. 1979; Meissner et al. 1986). The work of Stevens and Tang (1985) is interesting not only for their results, which showed that the exudate from *B. pilosa* roots are toxic to four agriculturally important plants (lettuce, beans, corn, and sorghum), but also for the methodology employed. The plants are grown in a system such that the exudate from undisturbed roots is passed from the donor plant to the acceptor (test) plants. Campbell et al. (1982) tested α-terthienyl, a polyacetylene derivative from *Tagetes erecta*, and PHT (see Fig. 2 for abbreviations) against four species which could have been natural competitors of *B. pilosa*. All four species were found to be sensitive to both compounds, especially to α-terthienyl. The latter was suggested as a possible natural weed-control agent. A range of polyacetylenes, including PHT and PDE-ol, were tested against marine and freshwater algae and found to be toxic at levels near 1 ppm. It was suggested that the compounds warranted further research as algicides (Arnason et al. 1981a). One study has reported the effects of polyacetylenes from Hawaiian species of *Bidens* on yeast and yeast-like organisms cultured from the host plants (Marchant and Towers 1987).

Several screen-type surveys gave the earliest indications of activity of crude extracts or plant parts against a range of organisms, including bacteria, fungi, and insects (Carlson et al. 1948; Carlson and Douglas 1948; Heal et al. 1950; Pates and Madsen 1955; Farnsworth et al. 1966; Camm et al. 1975). There is now considerably

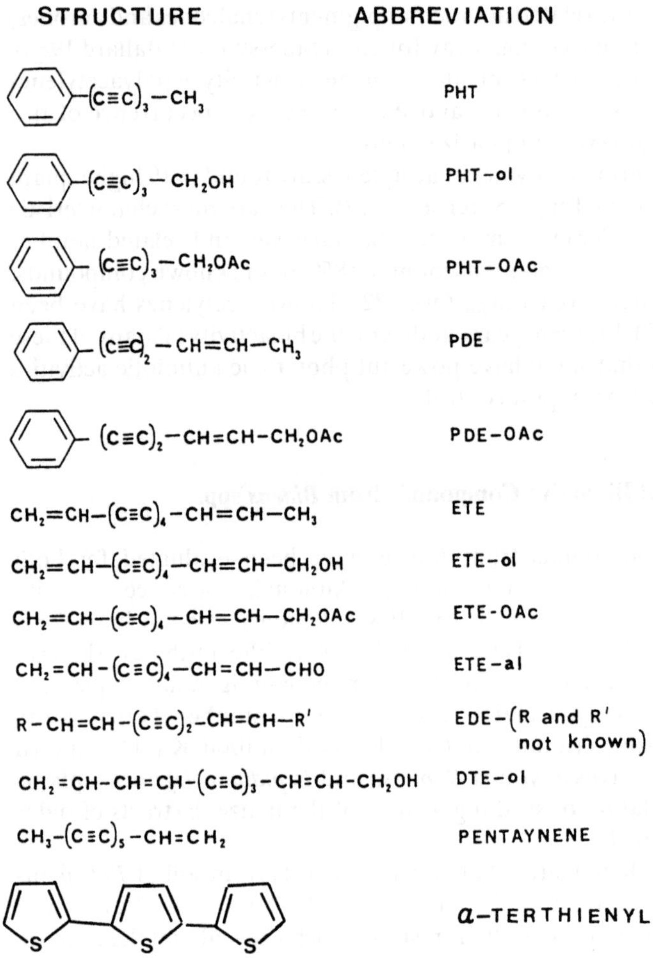

Fig. 2. Compounds discussed: Structures, systematic names, and abbreviations. *PHT*, 1-phenylhepta-1,3,5-triyne; *PHT-ol*, 7-phenylhepta-2-4-6-triyn-1-ol; *PHT-OAc*. 7-phenylhepta-2,4,6-triyn-1-yl acetate; *PDE*, 1-phenylhepta-1,3-diyn-5-ene; *PDE-OAc*, 7-phenylhepta-4,6-diyn-2-ene-1-yl cetate; *ETE*, trideca-1,11-diene-3,5,7,9-tetrayne; *ETE-ol*,trideca-1,11-diene-3,5,7,9-tetrayn-13-ol; *ETE-OAc*, trideca-1,11-diene-3,5,7,9-tetrayn-13-acetate; *ETE-al*, rideca-1,11-diene-3,5,7,9-tetrayne-13-al; *DETE-ol*, trideca-1,3,11-triene-5-7-9-triyn-13-ol; pentaynene, trideca-1-ene-3,5,7,9,11-pentayne; α-terthienyl, α-terthienyl. (From Norton and Towers 1985)

more evidence for the range of organisms sensitive to polyacetylenes found in species of *Bidens*, as shown in Table 1. The mode of action usually involves photoactivation and interaction of the compounds with membranes, although some have activity in the dark (Towers 1980). A review of the mechanism of action can be found in Towers (1980).

Recent reports extend the range of microorganisms (Towers et al. 1977; Arnason et al. 1980; Wat et al. 1979, 1980; Marchant and Towers 1986). Gram-positive bacteria are especially sensitive to photoactivated polyacetylenes because of the lipid in their cell wall. The role of lipid in sensitivity has been further shown by studies with viruses and bacteriophages (Hudson et al. 1986a,b; Warren et al. 1980). PHT has been shown to strongly inhibit germination of macroconidia of *Fusarium culmorum*, a cereal pathogen (Bourque and Arnason 1985). PHT and related polyacetylenes have also shown activity against insects as growth inhibitors (Nakajima and Kawazu 1980), ovicides (Nakajima and Kawazu 1980; Kagan and Chan 1983), and as larvicides (Arnason et al. 1981b).

There is now abundant evidence that the polyacetylenes found throughout most groups of the Compositae, and especially the aromatic polyacetylene PHT found in species of *Bidens*, are toxic to a wide range of organisms ranging from viruses to insects. Table 2 presents a summary of studies which have reported on the polyacetylenes present in species of *Bidens*. Although the ecological role these

Table 2. Species of *Bidens* for which polyacetylene composition has been reported

Species	Reference	
B. andicola	Bohlmann et al. (1983)	The following species are
B. aurea	Bohlmann and Zdero (1975)	in Marchant et al. (1984):
B. bipinnata	Bohlmann et al. (1965)	
	Craveiro et al. (1986)	*B. amplectens*
B. cernua	Bohlmann et al. (1973)	*B. asymmetrica*
B. chrysanthemoides	Bohlmann et al. (1973)	*B. campylotheca*
B. connata	Bohlmann et al. (1964)	*B. cervicata*
	Bohlmann et al. (1965)	*B. conjuncta*
	Bohlmann et al. (1983)	*B. cosmoides*
B. cynapiifolia	Bohlmann et al. (1983)	*B. forbesii*
B. dahlioides	Bohlmann et al. (1965)	*B. hawaiensis*
B. ferulaefolia	Bohlmann et al. (1965)	*B. macrocarpa*
	Bohlmann and Kleine (1964)	*B. mauiensis*
	Jensen and Sörensen (1961)	*B. menziesii*
B. frondosa	Bohlmann et al. (1973)	*B. micrantha*
	Pagani et al. (1972)	*B. molokaiensis*
	Pagani and Romussi (1971)	*B. hillebrandiana*
B. frutelifolia	Bohlmann et al. (1973)	*B. populifolia*
B. grandiflora	Bohlmann et al. (1973)	*B. sandvicensis*
(*Cosmos*)	Sörensen and Sörensen (1958a)	*B. torta*
"	Sörensen and Sörensen (1958b)	*B. valida*
B. graveolens	Bohlmann et al. (1983)	*B. wiebkei*
B. humilis	Bohlmann et al. (1973)	
B. leucantha	Bohlmann et al. (1962)	
	Bohlmann et al. (1965)	
B. maximowicziana	Bohlmann and Zdero (1975)	
B. nitis	Bohlmann et al. (1983)	
B. pilosa	Bohlmann et al. (1964)	
(= *alba*)	Norton and Towers (1985)	
B. radiata	Bohlmann et al. (1973)	
	Jensen and Sörensen (1961)	
B. rubrifolia	Bohlmann et al. (1983)	
B. squarrosa	Bohlmann et al. (1983)	
B. tripartita	Bohlmann et al. (1962)	
B. vulgata	Bohlmann et al. (1973)	
B. warszewiczsiana	Bohlmann et al. (1973)	

compounds play is not clear, it does appear that they are a part of the armory of defensive compounds that plants use against other organisms, and as such may be adaptable to use by people to control some organisms.

2 In Vitro Approaches

To date, the only reported work on in vitro culture of species of *Bidens* is the work of Norton and Towers (1984, 1985, 1986) which has focused on the production of polyacetylenes in transformed and organ cultures. This work is summarized below. Characteristics of plants regenerated from crown gall callus are described in Norton and Towers (1984). Briefly, T-DNA was found to be transmitted through two cycles of sexual reproduction; double transformant galls could be obtained, i.e., producing both nopaline and octopine; and transformed plants differed morphologically in several characteristics from normal plants. The author does not think that there would be any special problem in regeneration of normal *B. pilosa* or *B. alba* plants from callus. A somewhat related plant, *Aspilia mossambicensis*, has proven to be easily propagated in vitro by using 1.0 mg/l BAP and 1.5 mg/l GA3 on MS medium. Root induction in 98% of plants was achieved with no hormones and 0.2% activated charcoal. Root cultures grew best on 0.1 mg/l NAA (Norton et al. 1989). *B. cernua* and *B. connata* both formed good callus cultures on SH + 0.32 mg/l 2,4-D medium.

2.1 Culture for Physiological Studies

2.1.1 Methodology

2.1.1.1 Culture Methods

Agrobacterium tumefaciens strains A208, inducing nopaline-containing galls and A277, inducing octopine-containing galls, were obtained from Dr. M.P. Gordon, U. of Washington, Seattle. Bacterial strains, grown on SH medium (Schenk and Hildebrandt 1972) on which pieces of *B. alba* were cultured, were inoculated into the internodes of plants by syringe. Plants were grown in incubators at 70% relative humidity, 25 °C, and illuminance of 8000 lx for 12 h a day. Organized shoots appeared on A208 tissue about 6 weeks after harvest and were transferred to a 25 °C incubator with 6000-lx illumination and a photoperiod of 16 h. Rooting in regenerants began about 3½ months after culture began. Rooted plants were transferred to sterile potting soil and grown to maturity in a growth chamber with 8000-lx illuminance and 70% relative humidity (Norton and Towers 1984). Double transformants were obtained by inoculating regenerated nopaline-positive plants with A277 bacteria and harvesting the resulting galls. Galls were harvested approximately 3 months after inoculation, sterilized, and grown on SH medium, pH 5.7, containing 3% sucrose and 0.7% agar. No hormones were included in the medium. Filter-sterilized disodium carbenicillin and kanamycin sulfate solutions were added to the medium after autoclaving at concentrations of 0.5 and 0.13 g/l respectively.

After 3 weeks, tumor callus was transferred to medium without antibiotics. Cultures were grown in the dark at 25 °C and transferred every 3 weeks.

Seeds of normal plants were sterilized by rinsing in 70% ethanol for 1 min followed by 10 min in a 10% commerical bleach solution and then rinsed in sterile water several times. Seeds were germinated on moist filter paper and cut into roots, hypocotyl, and cotyledon sections before explanting to SH agar medium. Dry weights were determined on duplicate samples. Callus of normal tissue was grown on medium containing either 1 mg/l 2,4-D or 4 mg/l NAA.

Root cultures were established from single roots arising from either sterile leaf or stem explants of *B. alba* grown on SH agar medium which contained 4.0 mg/l NAA. Stock cultures were stored on SH agar plates without hormones and were kept in the dark at 23 °C. Roots were stored in this way for 4 to 6 months between transfers. When tip segments were explanted for shake culture, a lag phase of about 7 days occurred before vigorous growth started. Roots were also stored in liquid medium without hormones. A single segment was used to start a liquid culture, which was then used to start both experimental and continuing stock cultures. Transfers were made by pulling out, with forceps, a tuft of young roots from the parent clump, and inoculating the flask. Equality of explants was determined visually. Roots were grown for 11 days in 125-ml Erlenmeyer flasks containing 75 ml of liquid SH medium pH 5.7 with 0.5 mg/l NAA. Flasks were stoppered with foam plugs, covered with two layers of paper towel, and sealed at the neck with rubber bands. Incubation was at 25 °C in the dark on a rotary shaker at 125 r.p.m. Medium was autoclaved for 20 min at 15 lb/sq. in. Variations on these conditions are noted. Transformed roots, arising from callus of cultured crown gall cells induced on *B. alba* stems by *A. tumefaciens* strain A208, were grown in the same manner as normal roots. Transformed roots contained nopaline (Norton and Towers 1984). Seedlings of *B. alba* were grown in soil at 25 °C with 8000-lx illumination and a 16-h photoperiod for 6 weeks before harvesting. Seeds of *B. alba* var. *radiata* were obtained from south Florida, U.S.A.; those of *B. pilosa* var. *minor* from Belize.

2.1.1.2 Identification and Quantification of Compounds

One gram of tumor or root tissue, or 5 g of normal tissue, was homogenized in 15 ml methanol. Five ml of water was added followed by 5 ml of petroleum ether (30–60 °C). The mixture was shaken, centrifuged, and the upper phase drawn off. Each homogenate was extracted three times, the pooled extracts concentrated with a stream of nitrogen and redissolved in HPLC grade methanol. The extracts were subjected to HPLC on a Micropak MCH-10 reverse-phase analytical column (30 cm × 4 mm) with water/acetonitrile according to the following program: (time in minutes: % acetonitrile) 0:61, 10:61, 20:100, 24:100, 27:61; (time:flow rate in ml/min) 0:1, 19:1, 20:2, 24:2, 27:1. The detecting wavelength was 287 nm. Compounds were quantitated by comparison with known concentrations of standards and identified by ultraviolet and mass spectra (Norton and Towers 1985). Figure 2 shows the structures, systematic names, and abbreviations of the polyacetylenes identified. The compound designated EDE was not completely identified. The compound ETE-OAc was found in two forms, ETE-OAc (A) and ETE-OAc (B), which are cis/trans isomers; both forms are combined in the tabulations.

2.1.2 Results

2.1.2.1 Background

Bidens alba var. *radiata* and *Bidens pilosa* var. *minor* synthesize polyacetylenes in their roots, stems, and leaves with different patterns for each of these organs. *B. alba* contains three principle polyacetylenes which are easily detected by their UV spectra: PHT (see Fig. 2 for abbreviations), found in leaf, flower, and stem; PDE, found in stem and root; and ETE, found primarily in root. Tissue cultures might afford suitable material for studying the control of polyacetylene synthesis, although Jente (1971) reported very low levels (0.01% of the intact roots) of polyacetylenes in callus of *Centaurea ruthenica*, and Setia (1978) obtained similarly low levels of polyacetylenes in short-term callus cultures of *Tagetes erecta*. Factors which affect polyacetylene synthesis in callus cultures include light (Jente 1971; Ichihara and Noda 1977), types of hormones in the nutrient medium (Setia 1978), and size of cell clumps for suspension cultures (Ketel 1988). In addition, polyacetylenes normally not found in appreciable amounts in the plants may accumulate in the cultured tissue (Jente 1971).

2.1.2.2 Studies on Normal Callus

Callus was easy to establish from explants of *B. alba*, regardless of the geographic source. Seeds from the U.S.A., China, Belize, Africa, Mexico, and other countries showed little variation in adapting to the standard culture method described. However, the product level of nearly all lines (explants from approximately 220 different seedlings) fell below the limit of detection within four to five subcultures when grown on any of eight different media (Norton and Towers 1985). The characteristic UV spectra of the PHT chromophore could be detected in the extract from 0.01 g or less of tissue and in 0.02 g or less or root tissue for the entetraynene (ETE) chromophore. Levels of compound of 0.5% of the intact plant in callus tissue would have been detected. In an attempt to find nutritional or environmental conditions or additives which would enhance or stop the rapid loss of polyacetylene synthesis in explants of normal tissue, the following ingredients or treatments were evaluated for effect on polyacetylene synthesis:

1. Hormones: 2,4-D, NAA, NAA + kinetin, GA3 + kinetin.
2. Sugars and vitamins: inositol, sucrose, sucrose (− light), glucose, ascorbic acid.
3. Primary metabolites: acetate, malate, citrate, combinations of these compounds.
4. Macronutrients: KNO_3, $MgSO_4$, NH_2PO_4; $CaCl_2$, KH_2PO_4.
5. Polyacetylene precursors and related compounds: Oleic acid, linoleic acid, linolenic acid, combinations of all three.
6. Miscellaneous: heavy metal elicitation ($HgCl_2$), UV irradiation, heat and cold shock.

The final result was that none of the treatments induced measurable levels of polyacetylenes or retarded the loss of product. Although changes in texture and growth rate varied considerably with the treatments, none induced product synthesis. Fatty acid supplements resulted in hard, dense callus somewhat like the transformed callus described later, but there was no trace of product in any of the extracts. Green callus was induced by lowering sucrose level to less than 1% and illuminating the callus, but again the results were negative.

It is conceivable that loss of biosynthetic capacity might be irreversible and that attempting to regain synthesis once lost would not be possible. However, safynol, a polyacetylenic phytoalexin, is produced by *Carthamus tinctorius* in response to infection (Ingham 1972). Safynol has been elicited in suspension cultures of *C. tinctorius* by cell wall preparations from two different fungal pathogens (Tietjen and Martern 1981). One of the culture lines of *B. alba* established by us synthesized low levels of PHT in response to a fungal broth preparation (DiCosmo et al. 1982). Clearly, callus and suspension cultures which have stopped making polyacetylenes nevertheless retain the capacity to make them if appropriately treated.

The following points summarize the results from work with normal callus:

1. Media containing high levels of salts (MS, SH, ER, and B5) gave the best callus growth for *Bidens alba*. Maximum growth was obtained with either 1 mg/l 2,4-D or 4 mg/l NAA. Choice of medium did not affect the loss of polyacetylene synthesis in successive cultures.
2. The part of seedlings cultured (root, hypocotyl, or cotyledon) had little effect on the growth of callus, and no effect on polyacetylene synthesis.
3. If *B. alba* seedlings differ in the capacity to continue polyacetylene synthesis in callus, then the fraction of plants which show continued synthesis of product is less than could be detected in a sample of 220 seedlings.
4. Production of polyacetylenes could not be achieved by varying selected nutritional and environmental parameters.

2.1.2.3 Results from Crown Gall Cultures

Background. Although crown gall tumors, and the transformed cultures derived from them, have been studied for many years, there are relatively few reports on secondary metabolism in the intact galls or cultures derived from them. Kado (1976) has reviewed and tabulated recent literature on primary and secondary metabolites of crown gall tumor cells. Older references may be found in Klein (1955) and Krikorian and Steward (1969). Misawa (1977) has reviewed several Japanese patents for the production of secondary products from crown gall cultures including berberine from *Coptis japonica*, saponins and sapogenins from ginseng, and betanin and stevioside from several plants. Among composites, polyacetylenes have been obtained from crown gall cultures of *Matricaria chamomilla* (Reichling et al. 1979), from which two acetylenic dicycloethers were isolated, bioactive thiarubrines A and B, and the associated thiophenes, from *Chaenactis douglasii* (Cosio et al. 1986), and thiophenes from *Tagetes patula* (Norton et al. 1985b; Ketel 1988).

Product Selection Program. The polyacetylenes of *B. alba* are colorless compounds, but their detection is relatively easy, involving a simple extraction and UV spectrophotometry of the extract. However, this method requires from 10 mg to 1 g of material, and the number of such tests that can be run sets a severe limit on the detection of lines with high yield. Soon after a number of transformed octopine callus lines had been established, variation in color of the calli was apparent. Some had dark brown spots in varying amounts, while others were nearly white. Initial work suggested a correlation between amount of color and acetylene content. Fifteen different calli from one line were graded into five groups on the basis of the amount of pigment present. The average product levels for the groups showed a correlation between color and product significant at a $p < 0.01$. Continuous selection of one dark line based on the criterion of pigment – selecting the darkest pieces for one line and only the lightest for the other – led to a convergence of levels for the two groups after 8 months. When a second round of selection, using the same criterion, was performed on the calli having the lowest and highest levels from the first selection, a considerable divergence of yield occurred over 8 months of selection. The level of polyacetylene in the high, pigmented line rose to 0.7 mg/g dry wt., while the unpigmented line fell to 0.1 mg/g dry wt. Composition changed as well: level of EDE-OAc for both lines was 80% of total polyacetylenes; after 30 months, which included the selection intervals, this had dropped to approximately 50% (Norton and Towers 1985).

Long-term changes in product ratios have been reported for thebaine- and protopine-type alkaloids in *Papaver bracteatum* cultures (Zito and Staba 1982), but the reason for such slow changes is not clear. On the basis of the results from roots reported later, the indication would be that the level of cytokinins or the response to cytokinins was declining over time.

Stability of Product Synthesis. Several different lines of transformed callus, including callus from both A277 and A208 gall tissue, were cultured and harvested for analysis at various times after explanting. The results showed that most lines had extremely erratic levels of synthesis varying by as much as several hundred percent over the course of a year, sometimes declining, sometimes increasing. One line from an A208 gall from *B. pilosa* var. *minor*, a related species which cross-hybridizes with *B. alba*, showed a relatively stable, but low level of synthesis, suggesting that this parameter may be determined by the genotype of the plant.

Composition of Callus. Shown in Table 3 is a breakdown of the composition and product levels for the two species of *Bidens* cultured. Results for both normal as well as singly and doubly transformed callus are shown. In situ tumors differ from the adjacent tissue in having low levels of PHT and high levels of PDE-OAc instead of PDE, especially nopaline-type galls. In culture, all types of tumor had similar profiles, which differed from the in situ profiles, the interesting change being in the occurrence of significant amounts of the ETE aldehyde (ETE-al) in all tumor calli. The compound was not found in normal tissue or galls. In addition, DTE-ol and TDE-ol occurred at low, but consistent, levels only in tumor callus. *Bidens pilosa* had a simpler, more restricted range of compounds than *B. alba*; PHT and PDE were both absent from plants and cultures, otherwise tumor callus showed the same trends as for *B. alba*.

Summary. Crown galls induced on *B. alba* by two different strains of *Agrobacteria tumefaciens* were cultured and investigated for polyacetylene production and

Table 3. Comparison of polyacetylene levels and composition of normal plant organs, transformed cultures, and galls of *B. alba* and *B. pilosa*. Values are the average of at least five different samples. Callus values were averaged over 1 year. (Norton and Towers 1985)

Tissue/ species	Total product (mg/g dry wt)	Polyacetylene composition (wt % of total)								
		PHT[a])	PDE-OAc	PDE	ETE-ol	ETE-[b]) OAc	ETE-Al	ETE	EDE	DTE-ol
B. alba										
Leaf	4.6	98.3	–	1.7	–	–	–	–	–	–
Stem	2.3	54.2	2.2	43.5	–	t	–	–	–	–
Root	8.7	5.7	58.2	2.9	0.4	24.2	–	0.8	7.7	t
Oct⁺ stem gall	0.24	7.4	67.7	8.9	3.4	4.4	–	–	6.3	1.9
Nop⁺ stem gall	0.41	7.1	55.8	31.3	0.5	5.3	–	–	t	–
Oct⁺ callus	0.30	–	55.7	–	16.5	10.0	5.2	t	9.1	3.5
Nop⁺ callus	0.25	–	48.1	–	5.8	15.2	14.9	t	11.2	6.8
Oct⁺ -[c])–Nop⁺ callus	0.26	–	51.3	–	7.4	14.7	15.3	t	6.2	6.0
Normal callus	–	–	–	–	–	–	–	–	–	–
B. pilosa										
Leaf	–	–	–	–	–	–	–	–	–	–
Stem	0.1	–	43.2	–	–	56.7	–	t	–	–
Root	3.5	–	50.5	–	0.4	40.1	–	0.4	8.6	–
Nop⁺ callus	0.08	–	77.8	–	4.3	11.8	6.1	–	–	–
Normal callus	–	–	–	–	–	–	–	–	–	–

[a] Includes PHT, PHT-ol and PHT-OAc.
[b] Includes ETE-OAc (A) and (B).
[c] Double transformant.
Symbols, abbreviations: –, not detectable; t, trace; d.w., dry weight.

composition. An effort to use a visual difference in some calli as a guide for selecting those with higher product levels resulted in a line with a higher content but an uncertain relation to the selection criterion. Callus of both *B. alba* and *B. pilosa* grown over several years showed a continuing ability to synthesize polyacetylenes at levels up to 20% of that in cultured roots. However, levels fluctuated widely over time, and composition showed a decrease in the proportion of phenyldiynene acetate (PDE-OAc). Octopine, nopaline, and double transformant callus all showed similarities in composition characterized by high levels of PDE-OAc and the presence of minor compounds not found in plant tissues.

2.1.2.4 Root Cultures

Root cultures, especially hairy root cultures, have recently become the focus of active research after being neglected for many years. One of the drawbacks has been the slow growth rate of most root cultures when grown as excised root tips on agar medium. Hairy roots appear to offer a general approach to overcoming this problem (Flores et al. 1986). However, normal *B. alba* root cultures, as will be shown, have a high growth rate and a relatively stable polyacetylene composition over time and a good yield of product.

Four other composites have been cultured as liquid root cultures for polyacetylene-related compounds: *Chaenactis douglasii* as normal roots (Cosio et al. 1986), as hairy roots (Constabel and Towers 1988); *Aspilia mossambicensis* (Norton et al. 1989), *Tagetes patula* as hairy roots (Flores 1986; Parodi and Fischer 1988), and *Eriophyllum lanatum* (Norton et al. 1985a); the latter cultures were scaled up and grown in 3-l Fernbach flasks to provide quantities of compounds for characterization and studies of bioactivity. Both species produce thiarubrines (dithiacyclohexyl derivatives of polyacetylenes) and the related thiophenes, but in different proportions. In addition, hairy roots of species of the following Compositae grown on agar have been reported: *Ambrosia, Centaurea, Cichorium, Galinsoga, Helianthus,* and *Tagetes* (Mugnier 1988).

Experimental Results: Growth of Cultures. Figure 3 shows the relationship between age of culture and growth and product levels. The polyacetylene level of root cultures, per gram of tissue, decreased during the lag phase and reached a minimum

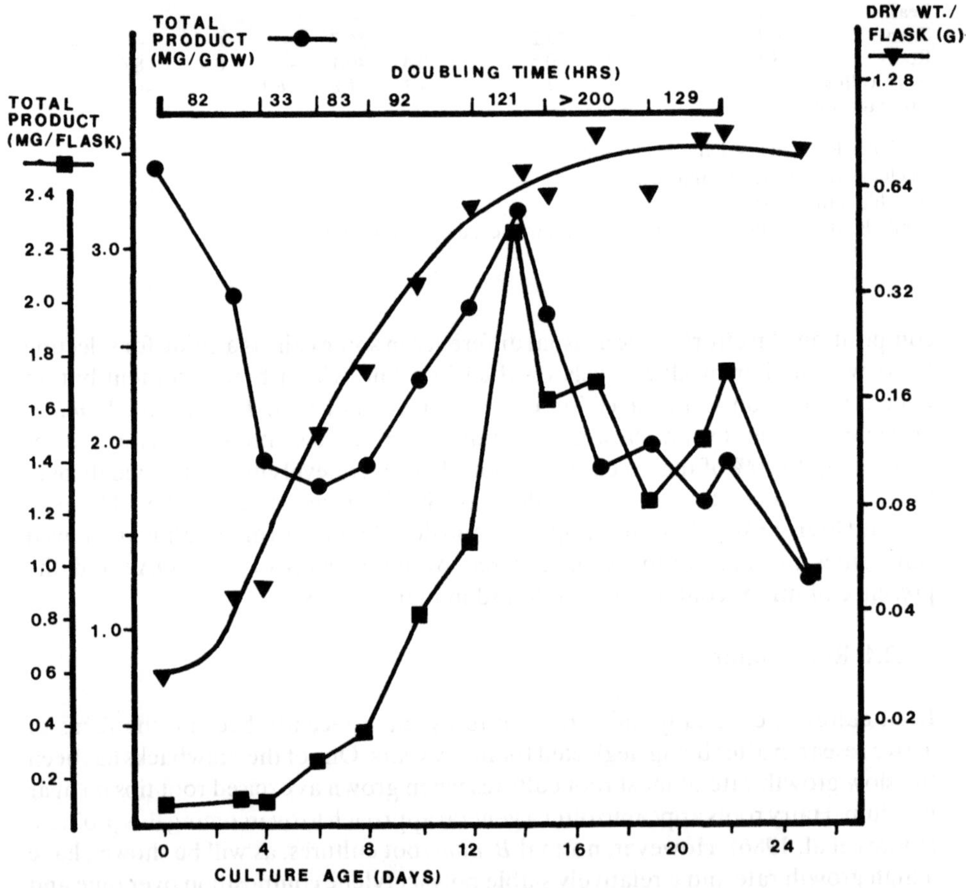

Fig. 3. Effect of culture age on growth and polyacetylenes of *B. alba* root cultures; total product (PA's) per gram root and per flask, doubling time and growth. Dry weight scale is logarithmic. (Norton and Towers 1986)

during the period of most rapid growth (days 4–6), increased during mid-to-late log phase (days 8–14) and declined again as growth ceased. Total acetylenes per flask changed little during early log phase growth but rapidly increased as growth began to slow and decreased as growth began to stop, suggesting that the compounds were being metabolized or excreted into the medium. However, medium levels were typically less than 1 μg per ml of medium.

Polyacetylene Composition. Table 4 shows a comparison of polyacetylene level and composition for *B. alba* seedling roots and dark-grown cultured roots. Intact roots contained a small amount of PHT and PDE which was not found in cultured roots, and which may have been contributed by stem tissue at the top of the roots when they were harvested. The principal difference in composition between the two types of roots was the higher percentage of PDE-OAc in intact roots and a corresponding lower percentage of ETE-OAc.

Table 4. Polyacetylene composition of seedling and cultured roots of *B. alba* grown under standard conditions. (Norton and Towers 1986)

Root type	Polyacetylene composition (wt.% of total)								Total PA's (mg/g dry wt.)
	PHT	PDE-OAc	PDE	EDE	ETE-ol	ETE-OAc		ETE	
						A	B		
Seedling	5.7	58.3	2.9	7.7	0.4	4.9	19.3	0.8	8.7
Cultured	–	38.7	–	7.6	0.7	3.9	47.1	2.0	3.3

Symbol: –, not detectable.

Effect of Temperature. For roots grown at increasing temperature over the range 18–33°C product level decreased at very nearly a constant rate from 4.3 mg/g dry wt. at 18°C to 0.65 mg at 33°C. Growth also showed a nearly linear response to temperature; increasing from 0.05 g dry wt./flask at 18°C to 0.7 g at 30°C, but decreasing sharply at 33°C. Cultures grown at 15 and 35°C showed no appreciable growth at either temperature over the culture period of 11 days. In sum, the overall composition was relatively stable over changes in temperature, while product levels changed sharply.

Effect of Sucrose/Nitrate Ratios. Growth and product increased as sucrose increased from 1.5 to 6%, but both declined sharply at higher concentrations of sucrose. Doubling the sucrose level from 3 to 6% resulted in a 25% increase in product; decreasing nitrate level by 1/2 and 1/4 increased it by 37 and 25% respectively. Similar effects on fat accumulation in fungi have been found. The C:N ratio is reported to be the most important nutritional parameter for lipid production by fungi (Weete 1980).

Effect of Kinetin. The results of growing *B. alba* roots in medium supplemented with kinetin were investigated in auxin-free medium. Low levels of kinetin increased growth moderately, but a 0.135 mg/l growth increased sharply and additional kinetin had little effect on growth; however, total product increased over the expected level of 2.5 mg/g for 11-day cultures, to a maximum of over 5.0 mg/l at the highest level of kinetin (1.2 mg/l). Roots grown at the highest level of kinetin were thicker and shorter than those grown at lower levels. In this respect they resembled roots grown in light, transformed roots, and older roots.

Effect of Auxin. The relationship between NAA concentration and polyacetylene level and growth was also investigated. Overall, growth increased almost linearly from 0.1 g dry wt./flask at 0 NAA to 0.7 g at 1.28 mg/l NAA and total polyacetylene declined from 3.7 mg/g dry wt. at 0 NAA to 1.5 mg at 1.28 mg/l. However, maximum product level occurred at 0.16 mg/l NAA. Changes in the proportion of PDE-OAc were less pronounced than with kinetin, but showed the same trend. In general, roots showed an increase in PDE-OAc formation as either NAA or kinetin levels increased.

Effect of Light and Hormones. Many polyacetylenes are sensitive to light and undergo decomposition or polymerization when concentrated or solubilized in polar solvents, including water. Therefore it was not expected that light would enhance the amount of polyacetylene in root cultures per se, although it might stimulate growth enough to offset losses caused by light. When cultures received 1400 lx continuous illumination or dark with or without 0.5 mg/l NAA, or were grown in light with either 0.1 or 1 mg/l kinetin, two features of the results were of interest: (1) The product ratios of illuminated cultures were quite different from dark-grown roots; the proportion of PDE-OAc was over twice that of dark cultures. (2) Kinetin resulted in two to three times higher total product levels than NAA. Total product level was highest at the higher kinetin level, confirming the effect seen with dark-grown cultures. Comparison of light-grown kinetin cultures with similar dark-grown cultures showed that light reduced the product level to about 1/3 of dark conditions. However, an even sharper decrease was seen in NAA and hormone-free cultures; total product level was less than 1/6 and 1/16 respectively of the dark levels. Kinetin markedly enhanced product level in light-grown cultures relative to NAA and hormone-free media.

Effect of Dedifferentiation. Transformed roots originating from regenerated shoots (Fig. 4A) of transformed callus of nopaline type were used to examine the effect of dedifferentiation on polyacetylene production. Transformed roots (Fig. 4B) were similar to normal roots in types of polyacetylenes present and in the enhancement of growth by NAA. Transformed roots were used because they seemed to be more sensitive to kinetin than normal roots, showing a greater tendency to form callus in the older portions of the roots and to shed cells into the medium. Kinetin was used as the hormone, since earlier experiments had shown it to increase product level at the concentration necessary for dedifferentiation. Table 5 shows the results of these experiments. Two main trends are evident from the data: (1) product level decreased sharply as the organization of the culture decreased; the youngest portion of the root had the highest polyacetylene level and the oldest portion, the lowest level. (2) Composition changed markedly.

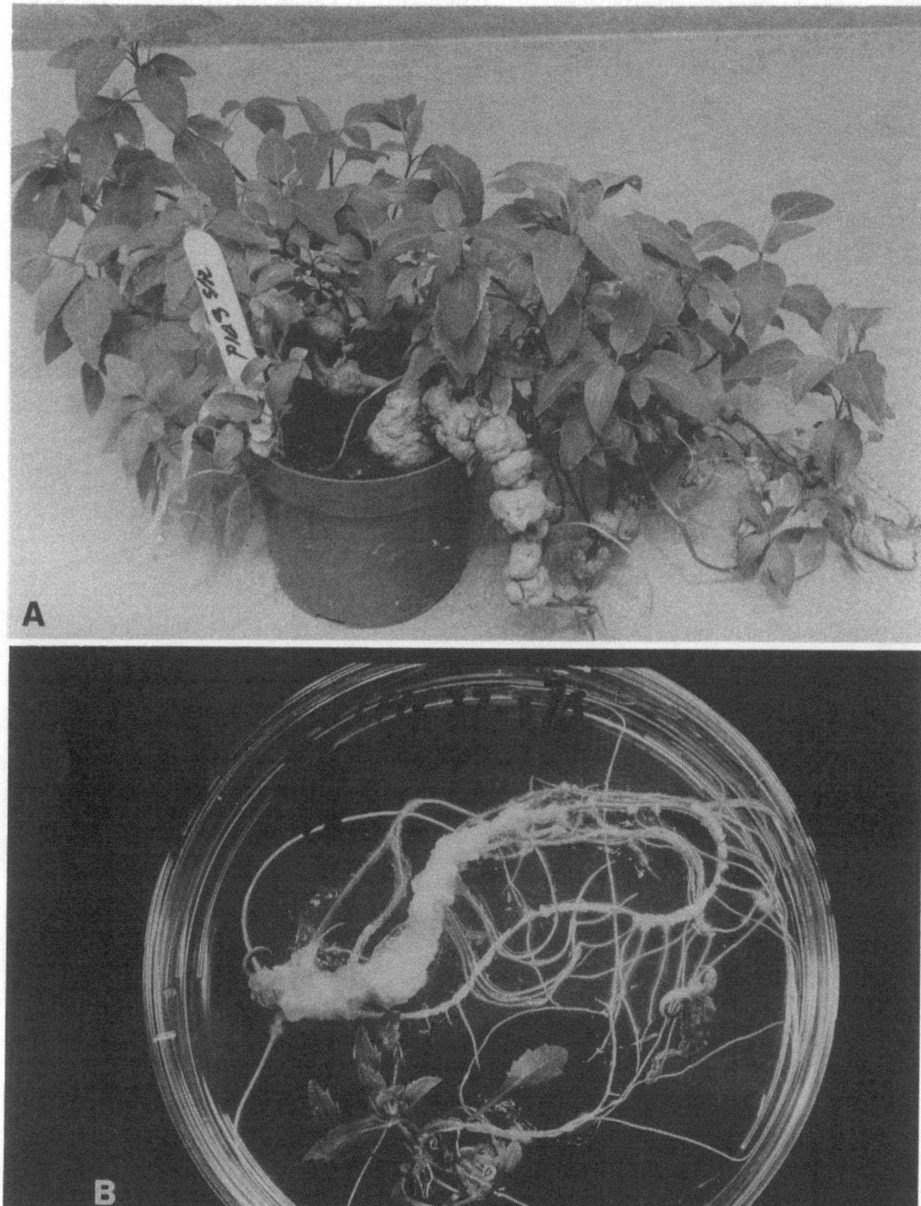

Fig. 4A,B. Regenerated transformed *Bidens alba* var. *radiata* plants. **A** Nearly mature regenerant (nopaline type) showing typical hypertrophy of nodes and lower stem. **B** Seedling from regenerant parent growing on hormone-free SH medium and showing characteristic callusing of root axis

Table 5. Effect of dedifferentiation on cultures of transformed roots. Callusing was promoted by growing roots in liquid medium with 1.5 mg/l kinetin for one passage and transferring the most heavily callused roots to flasks for the next passage. The free cells from the same passage were used as the inoculum for suspension cultures. Root tip was the first 2 cm of root from the tip, middle root was the next 3–4 cm, and the remainder of the root was the oldest root. Roots and suspension cultures were grown under standard conditions except as noted. (Norton and Towers 1986)

Tissue stage	Relative callus	Polyacetylene composition (wt.% of total)							Total PA's (mg/g dry wt.)
		PDE-OAc	EDE	DTE-OAc	PHT	ETE-ol	ETE-OAc		
							A	B	
Root tip	0	47.1	9.2	–	–	0.4	4.3	39.0	2.72
Middle root	0	43.7	9.0	–	–	0.6	5.1	41.6	2.36
Oldest root	+	45.8	9.5	t	t	0.9	6.2	37.7	0.64
Callused[a] root	+ +	28.8	30.2	1.9	3.2	1.7	13.8	20.4	0.49
Callused[a] root	+ + +	19.9	55.4	1.1	2.8	1.4	8.2	11.2	0.06
Suspension[a] culture		31.2	41.4	1.3	3.0	5.4	9.0	8.7	0.02

[a]Grown with 1.5 mg/l kinetin. Symbols: 0, no visible callusing; +, + +, + + + increasing callus formation; –, no detectable compound; t, trace.

None of the effects from the parameters studied could reasonably be extrapolated to give the type of composition found in the callused roots and suspension cultures. The appearance of PHT especially was quite unexpected. The implication is that the organization in roots determines both level and composition and loss of organization not only changes the relative proportions of typical compounds but also allows the synthesis of compounds either normally not present or present at much lower levels.

Stability over Time. Normal roots analyzed bi-weekly over 5 months showed statistically significant changes in composition, total polyacetylenes, and growth. Growth, as measured by dry wt./flask, increased by 50%, amount of total polyacetylenes decreased by 28%, and PDE-OAc increased by 6% relative to ETE-OAc.

Summary. Excised root cultures of *Bidens alba* grown in darkness with 0.5 g/l NAA in liquid SH medium were evaluated for growth, polyacetylene level, and polyacetylene composition as nutritional or environmental conditions were altered. Factors which increased polyacetylenes in roots above the standard level were: late log phase growth, lowered temperature, high levels of kinetin, and high sucrose to nitrate ratios. Factors which decreased polyacetylene levels were: log and stationary phase growth, high temperature, low sucrose to nitrate ratios, high levels of NAA, and illumination. Composition was shifted to increased synthesis of phenyldiynene acetate by high levels of kinetin and illumination. Transformed roots were evaluated for the effect of organization on product level and composition. Young root tips had the highest level of product and the oldest, the basal region of the roots, the lowest. Disorganization of roots, induced by high kinetin levels, resulted in low levels of product with a quite atypical composition, including the synthesis of related compounds not found in the intact plant. As cultures, roots showed a small but statistically significant change in characteristics over time. Nevertheless, *B. alba* roots are a sensitive system for investigating factors affecting polyacetylene synthesis, and lend themselves to growth at scaled-up levels for production of moderate quantities of the polyacetylenes they contain.

3 Conclusions and Prospects

Three aspects of species of *Bidens* of possible relevance to biotechnological and/or biomedical applications are:

1. Species of the plant are troublesome weeds in many parts of the world; however, the same plants find use in folk medicine for a variety of ailments. Whether the preparations are really efficaceous in most of these uses is certainly doubtful, but the folk remedy for thrush, an infection of children caused by *Candida albicans*, is evidently on a sounder basis, since one of the likely components of the plants used, PHT, has been shown in several studies to be toxic to the fungus (Table 1).
2. Controlled studies of a wide variety of polyacetylenes have demonstrated the toxicity of these compounds to a number of pathogenic and nonpathogenic microorganisms. Two compounds, PHT and α-terthienyl, have been tested extensively and found to be lethal to a wide range of organisms at levels that could be practical for application to water areas as mosquito larvicides or other pesticide. Selectivity at the level of microorganisms seems to depend largely on the presence or absence of a lipid component in the cell wall.
3. Two species of *Bidens* are adaptable to producing their characteristic compounds in either crown gall cultures or root cultures. As sources of compounds only the latter can be proposed on any reasonably economical basis. Root cultures of *B. alba* grow rapidly and produce a good yield of the root polyacetylenes. They can also be manipulated to produce compounds not normally found in them but at a sharply decreased yield. Evidently differentiation is necessary for the compartmentalization that leads to efficient synthesis. How to use these cultures? The most immediate use could be for metabolic studies; there are still many questions unanswered about polyacetylene metabolism, not the least of which is the enzymology of the formation of the triple bond. Another possibility is for small-scale production of the major acetylenes produced by the roots; root cultures of *Eriophyllum lanatum* have been used in this way.

Still another use could be biotransformation of polyacetylenes to novel structures normally not occurring in the plant or, possibly, in other plants. These compounds could then be used for toxicity studies. To the author's knowledge, no work has been done on the use of inhibitors of any of the steps of polyacetylene synthesis. Given the toxicity of some of these compounds, it is possible that the right inhibitors could cause the build-up of an intermediate highly toxic to the host plant. This could be the basis for a species-specific herbicide for these plants. Finally, it is interesting that *B. pilosa*, which owes its survival to its ability to rapidly colonize available ground – the qualities which make it a pernicious weed (along with allelopathic compounds), should prove so readily adaptable to callus and root culture.

References

Akobundu IO (1987) Weed science in the tropics: principles and practices. John Wiley & Sons, Chichester New York

Arnason T, Wat C-K, Downum K, Yamamoto E, Graham E, Towers GHN (1980) Photosensitization of *Escherichia coli* and *Saccharomyces cerevisiae* by phenylheptatriyne from *Bidens pilosa*. Can J Microbiol 26:698–705

Arnason T, Stein JR, Graham E, Wat C-H, Towers GHN, Lam J (1981a) Phototoxicity to selected marine and freshwater algae of polyacetylenes from species in the Asteraceae. Can J Bot 59:54–58

Arnason T, Swain T, Wat C-K, Graham EA, Partington S, Towers GHN (1981b) Mosquito larvicidal activity of polyacetylenes from species in the Asteraceae. Biochem Syst Ecol 9:63–68

Arnason JT, Bourque GJ, Madhosingh C, Orr W (1986) Disruption of membrane functions in *Fusarium culmorum* by an acetylenic allelochemical. Biochem Syst Ecol 14:569–574

Ballard R (1986) *Bidens pilosa* complex (Asteraceae) in North and Central America. Am J Bot 73:1452–1465

Bohlmann F, Kleine K-M (1964) Über zwei neue Polyinacetate. Chem Ber 97:1193–1196

Bohlmann F, Zdero C (1975) Ein neues Eugenol-Derivat aus *Bidens aurea* (Ait.) Sherff. Chem Ber 95:1315–1319

Bohlmann F, Arndt C, Bornowski H, Kleine K-M (1962) Über die Polyine der Gattung *Bidens* L. Chem Ber 95:1315–1319

Bohlmann F, Bornowski H, Kleine K-M (1964) Über neue Polyine aus dem Tribus Heliantheae. Chem Ber 97:2135–2138

Bohlmann F, Arndt C, Kleine K-M, Wotschokowsky M (1965) Neue Inhaltstoffe aus *Bidens*-arten. Chem Ber 98:1228–1232

Bohlmann F, Burkhardt F, Zdero C (1973) Naturally occurring acetylenes. Academic Press, New York London

Bohlmann F, Ahmed M, King RM, Robinson H (1983) Acetylenic compounds from *Bidens graveolens*. Phytochemistry 22:1281–1283

Bonasera J, Lynch J, Leck MA (1979) Comparison of the allelopathic potential of four marsh species. Bull Torrey Bot Club 106:217–222

Bourque G, Arnason JT (1985) The photosensitization of the plant pathogen *Fusarium culmorum* by phenylheptatriyne from *Bidens pilosa*. Can J Bot 63:899–902

Bushnell OA, Fukuda M, Makinodan T (1950) The antibacterial properties of some plants found in Hawaii. Pac Sci 4:167–183

Camm EL, Towers GHN, Mitchell JC (1975) UV-mediated antibiotic activity of some Compositae species. Phytochemistry 14:2007–2011

Campbell G, Lambert JDH, Arnason T, Towers GHN (1982) Allelopathic properties of α-terthienyl and phenylheptatriyne, naturally occurring compounds from species of Asteraceae. J Chem Ecol 8:961–972

Carlson HJ, Douglas HG (1948) Screening methods for determining antibiotic activity of higher plants. J Bacteriol 55:235–240

Carlson HJ, Douglas HG, Robertson J (1948) Antibacterial substances separated from plants. J Bacteriol 55:241–248

Constabel CP, Towers GHN (1988) Thiarubrine accumulation in hairy root cultures of *Chaenactis douglassi*. J Plant Physiol 133:67–72

Cosio EG, Norton RA, Towers E, Finlayson AJ, Rodriguez E, Towers GHN (1986) Production of antibiotic thiarubrines by a crown gall tumor line of *Chaenactis douglasii*. J Plant Physiol 124:155–164

Craveiro AA, Andrade CHS, Matos FJA, Alencar JW, Machado MIL (1986) Essential oils from Brazilian Compositae. J Nat Prod 49:361–363

Crawford DJ, Stuessy TF (1981) The taxonomic significance of anthochlors in the subtribe Coreopsidinae (Compositae, Heliantheae). Am J Bot 68:107–117

Degener O (1975) Plants of Hawaii National Parks. Braun-Brumfield, Honolulu

DiCosmo F, Norton R, Towers GHN (1982) Fungal culture-filtrate elicits aromatic polyacetylenes in plant tissue culture. Naturwissenschaften 69:550–551

Duke JA (1986) Isthmian ethnobotanical dictionary, 3rd edn. Scientific Publishers, Jodhpur

Farnsworth NR, Henry LK, Svoboda GH, Blomster RN, Yates MJ, Euler KL (1966) Biological and phytochemical evaluation of plants. I. Biological test procedures and results from two hundred accessions. Lloydia 29:101–122

Fleischer A (1981) Preparations comprising as active ingredients an extract derived from plant of the *Bidens* species or 1-phenylheptatriyne (natural or synthetic). Chem Abstr 94:357

Flores H (1987) Use of plant cells and organ culture in the production of biological chemicals. In: Lebaron HM, Mumma RO, Honeycutt RC, Duesing JH (eds) Biotechnology in agricultural chemistry, ACS Symp Ser 334, Am Chem Soc, Washington DC, pp 66–86

Flores HE, Hoy MW, Pickard JJ (1986) Production of secondary metabolites by normal and transformed root cultures. In: Somers D, Gegenbach BG, Bisboer DD, Hackett WP, Green CE (eds) Proc 6th Int Congr Plant tissue and cell culture. Univ Minn, Minneapolis, p 117

Fong HHS, Farnsworth NR, Henry LK, Svoboda GH, Yates MJ (1972) Biological and phytochemical evaluation of plants. X. Test results from a third two-hundred accessions. Lloydia 35:35–48

Gillett GW (1975) The Diversity and History of Polynesian *Bidens*, Section Campylotheca. HA Lyon Arbor, Univ Hawaii

Goode PM (1989) Edible Plants of Uganda. FAO, Rome

Graham K, Graham EA, Towers GHN (1980) Cercaricidal activity of phenylheptatriyne and α-terthienyl, naturally occurring compounds in species of Asteraceae (Compositae). Can J Zool 11:1955–1958

Heal RE, Rogers EF, Wallace RT, Starnes O (1950) A survey of plants for insecticidal activity. Lloydia 13:89–162

Holm L, Pancho JV, Herberger JP, Plucknett DL (1979) A Geographical Atlas of World Weeds. John Wiley & Sons, New York

Hudson JB, Graham EA, Chan G, Finlayson AJ, Towers GHN (1986a) Comparison of the antiviral effects of naturally occurring thiophenes and polyacetylenes. Planta Med 453–457

Hudson JB, Graham EA, Towers GHN (1986b) Investigation of the antiviral action of the photoactive compound phenylheptatriyne. Photochem Photobiol 43:27–33

Ichihara K-I, Noda M (1977) Distribution and metabolism of polyacetylenes in safflower. Biochem Biophys Acta 487:249–260

Ingham JL (1972) Phytoalexins and other natural products as factors in plant disease resistance. Bot Rev 38:344–394

Jensen SL, Sörensen A (1961) Studies related to naturally occurring acetylene compounds: XXIX. Preliminary investigations in the genus *Bidens*: I. *Bidens radiata* Thuill and *Bidens ferulaefolia* (Jacq.) DC. Acta Chem Scand 15:1885–1981

Jente R (1971) Polyacetylenverbindungen in gewebekulturen von *Centaurea ruthenica* Lam. Tetrahedron 27:4077–4083

Jones ERH, Thaller V (1978) Natural acetylenes. In: Patai S (ed) The Chemistry of the Carbon-Carbon Triple Bond. John Wiley & Sons, New York, pp 621–633

Kado DI (1976) The tumor-inducing substance of *Agrobacterium tumefaciens*. Annu Rev Phytopathol 14:268–308

Kagan J, Chan G (1983) The photoovicidal activity of plant components towards *Drosophila melanogaster*. Experientia 39:402–403

Kahl H (1987) Allelopathic effects in the maize-quelites-agroecosystem of the Tarahumara Indians. J Agron Crop Sci 158:56–64

Ketel KH (1988) Accumulation of thiophenes by cell cultures of *Tagetes patula* and the release of 5-(4-hydroxy-1-butynyl)-2,2′-bithiophene into the medium. Planta Med 400–405

Klein RM, Link GKK (1955) The etiology of crown-gall. Q Rev Biol 30:207–277

Krikorian AD, Steward FC (1969) Biochemical differentiation: the biosynthetic potentialities of growing and quiescent tissue. In: Steward FC (ed) Plant physiology: a treatise, Vol VB. Academic Press, New York London, pp 227–326

Marchant YY, Towers GHN (1986) Phototoxicity of polyacetylene to *Cryptococcus laurentii*. Biochem Syst Ecol 14:565–568

Marchant YY, Towers GHN (1987) Phylloplane fungi of Hawaiian plants and their photosensitivity to polyacetylenes from *Bidens* species. Biochem Syst Ecol 15:9–14

Marchant YY, Ganders FR, Wat C-K, Towers GHN (1984) Polyacetylenes in Hawaiian *Bidens*. Biochem Syst Ecol 12:167–178

Meissner R, Nel PC, Beyers EA (1986) Allelopathic influence of *Tagetes*- and *Bidens*-infested soils on seedling growth of certain crop species. S Afr J Plant Soil 3:176–180

Mirvish SS, Rose EF, Sutherland DM (1979) Studies on the esophagus. II. Enhancement of [3H] thymidine incorporation in the rat esophagus by *Bidens pilosa* (a plant eaten in South Africa) and by croton oil. Cancer Lett 6:159–165

Mirvish SS, Salmasi S, Lawson TA, Pour P, Sutherland D (1985) Test of catechol, tannic acid, *Bidens pilosa*, croton oil, an phorbol for cocarcinogenesis of esophageal tumors induced in rats by methyl-n-amylnitrosamine. J Nat Cancer Inst 74:1283–1290

Misawa M (1977) Production of natural substances by plant cell cultures described in Japanese patents. In: Barz W, Reinhard E, Zenk MH (eds) Plant tissue culture and its bio-technological application. Springer, Berlin Heidelberg New York, pp 17–26

Morton JF (1962) Spanish needles (*Bidens pilosa* L.) as a wild food resource. Econ Bot 16:173–179

Mugnier J (1988) Establishment of new axenic hairy root lines by inoculation with *Agrobacterium rhizogenes*. Plant Cell Rep 7:9–12

Nakajima S, Kawazu K (1980) Insect development inhibitors from *Coreopsis lanceolata* L. Agric Biol Chem 44:1529–1533

Norton RA, Towers GHN (1984) Transmission of nopaline crown gall tumour markers through meiosis in regenerated whole plants of *Bidens alba*. Can J Bot 62:408–413

Norton RA, Towers GHN (1985) Synthesis of polyacetylenes in tumor callus of *Bidens alba*. J Plant Physiol 120:273–283

Norton RA, Towers GHN (1986) Factors affecting synthesis polyacetylenes in root cultures of *Bidens alba*. J Plant Physiol 122:41–53

Norton RA, Finlayson AJ, Towers GHN (1985a) Two dithiacyclohexadiene polyacetylenes from *Chaenactis douglasii* and *Eriophyllum lanatum*. Phytochemistry 24:356–357

Norton RA, Finlayson AJ, Towers GHN (1985b) Thiophene production by crown galls and callus tissues of *Tagetes patula*. Phytochemistry 24:719–722

Norton RA, Huang DQ, Towers GHN, Rodriguez E (1990) In vitro propagation and production of antibiotic polyines by root cultures of *Aspilia mossambicensis*. (submitted)

Oliveira JS, DeCarvalho MD (1975) Nutritional value of some edible leaves used in Mozambique. Econ Bot 29:255–263

Pagani F, Romussi G (1971) New polyinic D-glucoside from *Bidens frondosa* flowers. Phytochemistry 10:2233

Pagani F, Romussi G, Bohlmann F (1972) Notiz über die Struktur des Polyin-Glucosids aus *Bidens frondosa*. Chem Ber 105:3126–3127

Parodi FJ, Fischer NH (1988) Benzofuran and bithiophenes from root cultures of *Tagetes patula*. J Nat Prod 51:594–595

Parry DW, O'Neill CH, Hodson MJ (1986) Opaline silica deposits in the leaves of *Bidens pilosa* L. and their possible significance in cancer. Ann Bot (London) 58:641–647

Pates AL, Madsen GC (1955) Occurrence of antimicrobial substances in chlorophyllose plants growing in Florida. II. Bot Gaz 116:250–261

Reichling J, Beiderbeck R, Becker H (1979) Comparative studies on secondary products from tumors, flowers, herb and roots of *Matricaria chamomilla* L. Planta Med 36:322–332 (in German)

Rose EF, Guillarmod AJ (1974) Plants gathered as foodstuffs by the Transkeian peoples. S Afr Med J 86:1688–1690

Schenk RV, Hildebrandt AC (1972) Medium and techniques for induction and growth of monocotyledonous and dicotyledonous plant cell cultures. Can J Bot 50:199–204

Setia B (1978) Polyacetylen von *Tagetes erecta* and *Tagetes patula* nana; Biosynthese der Theiophenderivate and Polyacetylen von *Tagetes erecta* durch Zellkulturen. Thesis, Univ Münster

Solevilla RC (1984) Research on the analgesic, antipyretic and/or anti-inflammatory activities of the four Philippine medicinal plants – *Ipomoea muricata* (L.) Jacq. (Convolvulaceae), *Bidens pilosa* L. (Compositae), *Calophyllum inophyllum* L. (Guttiferae) and *Tinospora rumphii* Boerl. (Menispermaceae). Acta Manilana: 79–81

Sörensen JS, Sörensen NA (1958a) Studies related to naturally occurring acetylene compounds XXIII. 1-phenylhepta-1:3:5-triyne from *Coreopsis grandflora*, Hogg ex Sweet. Acta Chem Scand 12:765–770

Sörensen JS, Sörensen A (1958b) Ibid. XXIV. 2-phenyl-5(-propynl)-thiophene from the essential oils of *Coreopsis grandiflora*, Hogg ex Sweet. Acta Chem Scand 12:771–776

Sörensen NA (1977) Polyacetylenes and conservation of chemical characters in the compositae. In: Heywood VH, Harborne JB, Turner BL (eds) The biology and chemistry of the compositae, vol 1. Academic Press, New York London, pp 385–433

Stevens GA Jr, Tang C-S (1985) Inhibition of seedling growth of crop species by recirculating root exudates of *Bidens pilosa* L. J Chem Ecol 11:1411–1425

Tietjen K, Matern U (1981) Mode of action of *Alternaria carthami* toxin in safflower (*Carthamus tinctorius* L.). Phytochem Soc N Am Newslett, July, Abstr C-23

Towers GHN (1980) Photosensitizers from plants and their photodynamic action. Prog Phytochem 6:183–202

Towers GHN, Wat C-K, Graham EA, Bandoni RJ, Chang GFQ, Mitchell JC, Lam J (1977) Ultraviolet-mediated antibiotic activity of species of Compositae caused by polyacetylenic compounds. Lloydia 40:487–498

Warren RAJ, Hudson JB, Downum K, Graham EA, Norton R, Towers GHN (1980) Bacteriophages as indicators of the mechanism of action of photosensitizing agents. Photobiochem Photobiophys 1:385–389

Wat C-K, Biswas RK, Graham EA, Bohm L, Towers GHN, Waygood ER (1979) Ultraviolet-mediated cytotoxic activity of phenyl-heptatriyne from *Bidens pilosa* L. J Nat Prod 42:103–111

Wat C-K, Johns T, Towers GHN (1980) Phototoxic and antibiotic activities of plants of the Asteraceae used in folk medicine. J Ethnopharm 2:279–290

Weete JD (1980) Lipid Biochemistry of Fungi and other Organisms. Plenum, New York London

Zito SW, Staba EJ (1982) Thebaine from root cultures of *Papaver bracteatum*. Planta Med 45:53–54

IX *Carthamus tinctorius* L. (Safflower): Production of Vitamin E in Cell Cultures

TSUTOMU FURUYA and TAKAFUMI YOSHIKAWA[1]

1 Introduction

Safflower, *Carthamus tinctorius* L. (family Compositae), consists of tubular florets, which are light red in color. The florets are used for much the same purposes as saffron, the dried stigma of *Crocus sativus* L., and are sometimes admixed with it and occasionally substituted for it (Claus et al. 1970). Although the plant does not exist as a wild species, the same genus, *Carthamus oxyacantha* Bieb., is indigeneous to Central Asia, such as Caucasus, Iran, Afghanistan, and Pakistan, and also cultivated in India. Thus, it is supposed that the *C. tinctorius*, long cultivated in China and Japan as a source of crude drug, was brought from Central Asia. The red florets are widely used as a crude drug in Oriental medicine and natural dye, especially in silk cloth or rouge (Kitamura 1978). A water-soluble yellow dye, called safflower yellow, also used to be extracted from it as well as an alcohol-soluble red dye, saffower carmine, which is carthamin, a chalcone glycoside (Obara and Onodera 1979).

Recently, it has been noted that safflower seed oil contains abundant (over 75%) linoleic acid, an unsaturated fatty acid. Thus, the oil is used as part of the diet for arterial and heart diseases, and for overweight because of its anticholesterol quality (Chiej 1984). In addition, safflower oil is well known as α-tocopherol-rich, like sunflower oil, which is extracted from sunflower (*Helianthus annuus*) seeds (Kanematsu et al. 1983). Tocopherols have the physiological action of vitamin E (Horwitt 1976) and an antioxidant action (Tappel 1974; Diplock 1974), and are used as ingredients of pharmaceutical preparations and as antioxidants in food (Kanno et al. 1970). It has been found that tocopherols may act to prevent a deterioration in physical condition or geriatric disease (Wasserman and Taylor 1972; Janiszowska and Pennock 1976). The tocopherols from vegetable sources consist of eight analogs, i.e., four tocopherols and four tocotrienols, as shown in Fig. 1. Alpha-tocopherol has the strongest vitamin E activity of these. In particular, the physiological activity of (d)-α-tocopherol obtained from natural sources is 1.36 times greater than that of racemic (dl)-α-tocopherol, obtained by chemical synthesis (Katsui 1983). Thus, the production at low cost of tocopherols from a vegetable source in which the physiologically most effective α-tocopherol is present in large amounts is very desirable. Cell culture of safflower, which contains one of the largest amount of α-tocopherol in plants, has therefore been carried out to produce tocopherols, especially α-tocopherol, and other related compounds (Furuya et al. 1987).

[1] School of Pharmaceutical Sciences, Kitasato University, Minato-ku, Tokyo 108, Japan

Biotechnology in Agriculture and Forestry, Vol. 15
Medicinal and Aromatic Plants III (ed. by Y.P.S. Bajaj)

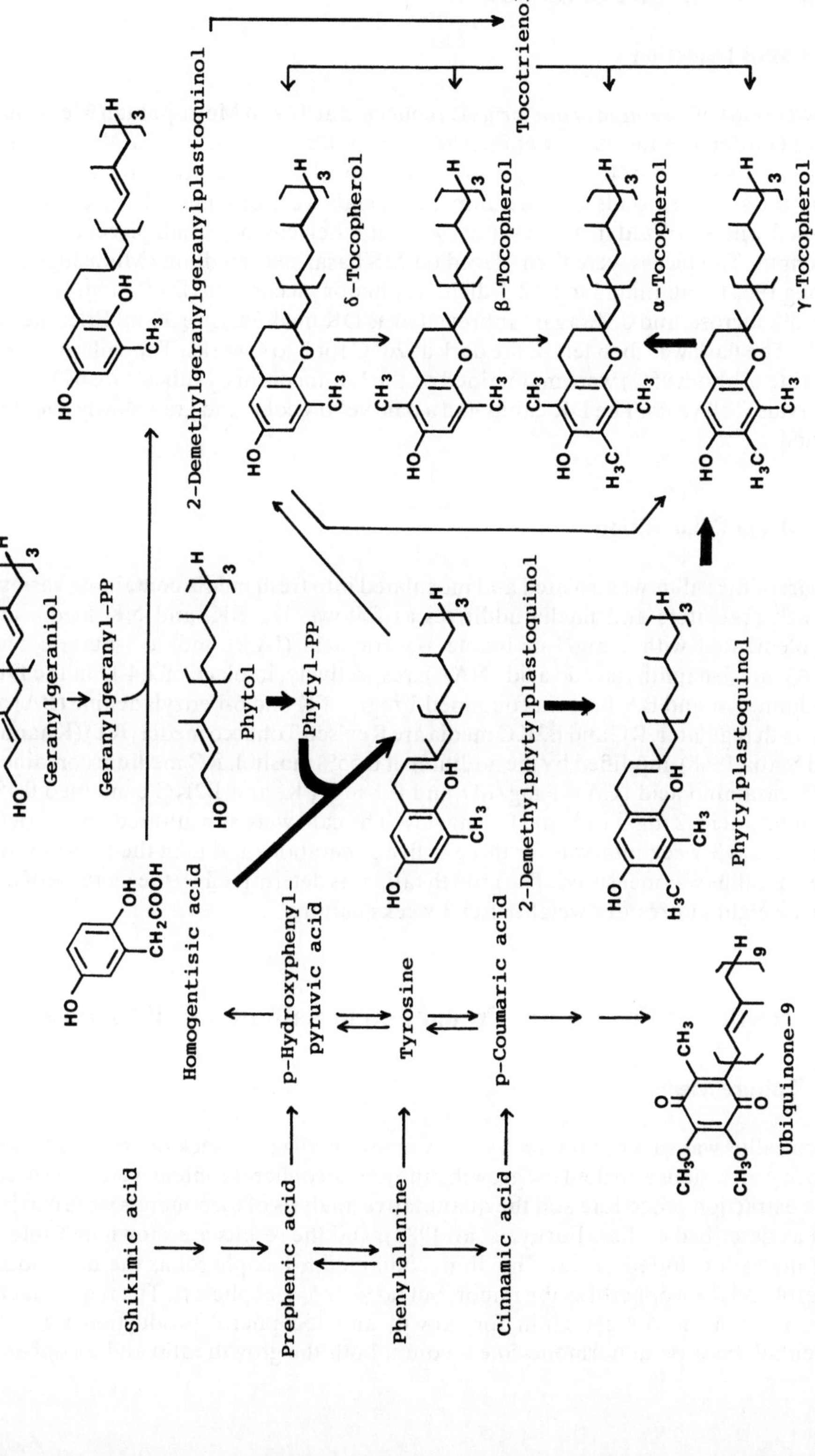

Fig. 1. Probable biosynthetic pathway of α-tocopherol

2 In Vitro Culture of Safflower

2.1 Callus Induction

Flower buds of *Carthamus tinctorius* L., collected at Tokyo Metropolitan Medicinal Plant Garden in June 1976, were sterilized by treatment with 70% EtOH for 1 min and a 10% solution of bleaching powder for 5 to 10 min. The bracts were removed from the sterilized buds and the immature petals were obtained. The petals were washed with sterilized distilled water (×2) and then cut into small pieces ca. 5 mm in length. The pieces were then placed on MS basal agar medium (Murashige and Skoog 1962) containing 1 mg/l 2,4-dichlorophenoxyacetic acid (D), 0.1 mg/l kinetin (K), 3% sucrose, and 0.9% agar (abbreviated as DK medium) in a 50-ml Erlenmeyer flask. The flask was then left in the dark at 26°C for 3 to 4 weeks. The calli (induced at a rate of about 95%) were maintained on the DK medium by subculture of 3-week intervals for 9 years. The DK callus had a pale yellow color and grew slowly over the period.

2.2 Media Composition

A part of the callus was isolated and inoculated into fresh media containing various growth regulators and media additives as follows. IK, BK, and NK media are supplemented with 1 mg/l of indole-3-acetic acid (IAA), indole-3-butyric acid (IBA) and 1-naphthylacetic acid (NAA), respectively, in place of 2,4-D in the DK medium. K 1 and BA 1 media contained 1 mg/l of K and 6-benzyladenine (BA) as a growth regulator. RC and B2KC media are Revised Tobacco media (RT) (Khanna and Staba 1968), modified by the addition of 0.55% inositol. RC medium contained 0.1% casamino acid (CA), 1 mg/l D and 0.1 mg/l K, and B2KC contained 0.1% casamino acid, 2 mg/l IBA and 0.1 mg/l K. The calli were subcultured on a variety of media at 3-week intervals for three to five generations and then the growth ratio of each callus was measured. The growth ratio was determined by the increase of the fresh weight and/or dry weight after 3 weeks culture.

3 Growth and Tocopherol Production in Safflower Cell Cultures

3.1 Culture Media

Each callus was subcultured on a variety of test media at 3-week intervals for three to five generations and then the growth ratio and tocopherol content were measured. The extraction procedure and the quantitative analysis of tocopherols were carried out as described earlier (Furuya et al. 1987), and the results are shown in Table 1. All the calli cultured on each medium contained α-tocopherol as the main tocopherol and β-tocopherol as the minor, but no γ- or δ-tocopherols. The requirement of an auxin and/or cytokinin for growth and tocopherol production was not essential, because in hormone-free medium both the growth ratio and tocopherol

Table 1. Growth ratio and tocopherol contents of safflower calli cultured on various media. The growth ratio was determined by the increase of the fresh wt after 3 weeks culture in 1 g inoculum and 40 ml medium. The values are the quotient of the weight after 3 weeks culture and that of the inoculum. (Furuya et al. 1987)

Medium (mg/l)	Basal	Growth ratio	Tocopherol content	(mg/100 g dry wt)	
			α-	β-	Total
D 1 K 0.1 (DK)	MS	8.85	0.75	0.40	1.15
Basal	MS	8.21	1.60	0.36	1.96
IAA 1 K 0.1 (IK)	MS	4.44	0.55	0.31	0.86
IBA 1 K 0.1 (BK)	MS	6.10	0.57	0.16	0.73
NAA 1 K 0.1 (NK)	MS	8.31	0.46	0.38	0.84
IAA 5 K 0.1 (I5K)	MS	5.82	1.60	0.45	2.05
K 1	MS	8.13	0.43	0.21	0.64
BA 1 (white)	MS	13.57	1.10	0.16	1.26
BA 1 (yellow)	MS	9.53	0.54	0.36	0.90
D 1 K 0.1 CA 0.1 (RC)	RT	8.51	2.68	1.63	4.31
IBA 2 K 0.1 CA 0.1 (white) (B2KC)	RT	10.47	1.83	0.54	2.37
IBA 2 K 0.1 CA 0.1 (yellow) (B2KC)	RT	17.50	3.80	1.15	4.95

Abbreviations: D, 2,4-dichlorophenoxyacetic acid; K, kinetin; IAA, indole 3-acetic acid; IBA, indole 3-butyric acid; NAA, α-naphthylacetic acid; BA, 6-benzyladenine; CA, casamino acid; MS, Murashige and Skoog's basal medium; RT, revised tobacco basal medium.

content were higher than those in the original callus (DK callus). On the other hand, calli cultured on RT medium supplemented with 0.1% casamino acid (CA) showed higher value for the tocopherol content in comparison with those cultured on MS medium. In particular, the callus cultured on RT medium supplemented with 2 mg/l IBA, 0.1 mg/l K and 0.1% CA (abbreviated as B2KC medium) gave the highest values for both the growth ratio (ca. 2 times) and α-tocopherol content (ca. 5 times), when compared with the parent callus (DK medium). The B2KC callus grown on the B2KC medium is deep yellow and has grown vigorously over 3 years.

To investigate the effective components in the B2KC medium, cell suspension cultures were set up for various liquid media. The growth ratio and the tocopherol content were found to be almost the same in both MS and RT basal media. Since the RT medium contained 55 times more inositol (0.55%) than the MS medium, the effects of inositol on growth and tocopherol production were investigated, singly or in combination with casamino acid. The results showed that inositol had least effect when added alone, but in the presence of casamino acid it was the most effective for both growth and tocopherol production (Furuya et al. 1987).

The effect of casamino acid, from 0.1 to 1%, on growth and tocopherol production in safflower cell suspension cultures was investigated in detail. All the cells grew vigorously after transfer to a medium containing casamino acid, except at the 1% level, which was inhibitory to both growth and tocopherol production. However, the addition of 0.1 to 0.5% casamino acid had almost the same effects on both growth, 8 to 9.8 times in ratio, and tocopherol content, 11.4 to 13.4 mg per 100 g dry weight. Thus, using the cells subcultured for several generations in suspension on the B2KC medium in which 0.1% casamino acid is contained, the relationship between growth, as measured by dry weight increase, and tocopherol production was investigated next in a time course experiment (Fig. 2). Cells (4 g) were inoculated into 250 ml medium contained in a liter flask and then shaken on a reciprocating shaker at 78 spm in the dark. The cells were harvested every 5 days, and their growth and the tocopherol content were determined. Maximal growth was achieved 20 days after inoculation, after which the growth rapidly decreased. The tocopherol content of the cells remained almost constant at about 10 mg per 100 g dry wt. in the lag and exponential phases of growth, and then rose rapidly to 28.1 mg per 100 g dry wt. at 25 days, i.e., on the 5th day after maximal growth. Alpha- and β-tocopherols were both produced throughout the culture period, the former accounting for some 73 to 88.6% of the total. No tocopherol was found in the medium throughout the culture period. This pattern in the time course experiment is characteristic of secondary metabolites rather than primary ones, although tocopherol is generally thought of as a primary metabolite, which is essential for the maintenance of life, and widely distributed in plants. This result is, however, in agreement with the observations that tocopherol is stored mainly in the seed of plants and that young fast-growing plants have very little α-tocopherol, whereas slow-growing plants contain quite large amounts.

Each tocopherol (α-, β-, γ- and δ-) was identified by TLC using co-chromatography and HPLC using co-injection. Further confirmation of the identity of α-tocopherol, the main product of the safflower cells, was identified in comparison with the authentic sample by means of mass spectrometry of a HPLC-purified sample (6.1 mg) of α-tocopherol.

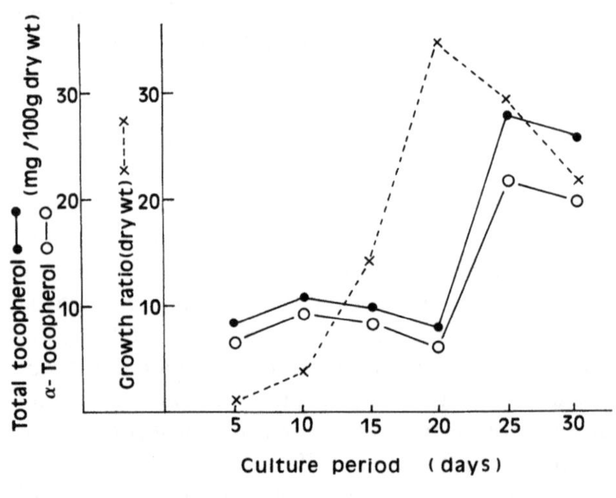

Fig. 2. Time course of the growth and tocopherol production of safflower cells grown in a suspension culture on B2KC medium. Growth ratio was determined by the increase in the dry weight of the cells, lyophilized after culturing for each time in 4 g inoculum in 250 ml medium in 1-l flask. The values are the quotient of the weight after every culture and that of the inoculum. Tocopherol content was determined by the method described in the text. The values are the average of duplicate estimations in every 2 or 4 flasks

3.2 Enhancement of Tocopherol Production by Biosynthetic Precursors

The enhancement of tocopherol production by administration of potential precursors of the tocopherol molecule was investigated by incubating the test compounds with safflower cell suspension cultures. Callus (ca. 10 g) from 3-week-old cultures was inoculated into 250 ml liquid medium (in 1-l Erlenmeyer flasks) and incubated at 26 °C for 1 week. Two-ml aliquots of the precursors dissolved in 20% Tween 80 solution at 12.5 mg/ml were administered to the cell suspensions under sterile conditions, filtered through the Millex-GS filter (Millipore Co. Ltd.). The precursors tested were L-phenylalanine, D-, L-tyrosine, p-coumaric acid, p-hydroxyphenylpyruvic acid, and homogentisic acid as the precursors of the aromatic ring (so-called chromanol nuclei) and geranylgeraniol, phytol, and isophytol as the possible precursors of the side chain. After culturing for 2 weeks, the cells and media were separately extracted with n-hexane, and the amount of each tocopherol was determined by HPLC. The results are shown in Fig. 3.

From the results it was clearly shown that the precursors of the aromatic ring, such as L-phenylalanine, D- or L-tyrosine (Whistance et al. 1967), p-coumaric acid and p-hydroxyphenylpyruvic acid (Whistance and Threlfall 1971) had no effect on tocopherol production, except in the case of homogentisic acid (Whistance and Threlfall 1970; Peake et al. 1970), which had a slight stimulatory effect. Homogentisic acid is a very effective precursor in tocopherol biosynthesis in broken chloroplast systems (Soll and Schultz 1980; Soll et al. 1980a; Fiedler et al. 1982), but in the safflower cultured cells it was only slightly effective. These results suggest that homogentisic acid, a hydrophilic compound, has difficulty in penetrating the cytoplasmic membrane of the cultured cell, in comparison with the broken chloroplast system. Furthermore, homogentisic acid is rapidly oxidized, probably before being incorporated into the cell. This was evident from the fact that the cells and medium turned brown immediately after administering homogentisic acid to the cell suspension.

On the other hand, phytol, a potential side-chain precursor, markedly increased the tocopherol content of the cells (45.43 mg/100 g dry wt. as the total amount), whereas geranylgeraniol was ineffective. Although the cells produced α-tocopherol as the major component and β-tocopherol as the minor, the administration of phytol led to the accumulation of γ- and δ-tocopherols, as well as α- and β-tocopherols. These facts suggest that the biosynthesis of α-tocopherol proceeds actively by way of γ- and/or δ-tocopherols in the presence of added phytol, that is by the tocopherol pathway. As geranylgeraniol, a potential precursor of phytyl pyrophosphate needed for tocopherol biosynthesis (Soll and Schultz 1981), had no effect on tocopherol biosynthesis, it was clearly demonstrated that in safflower cells there is no direct route from geranylgeraniol via geranylgeranyl pyrophosphate to tocopherol, i.e. that is the tocotrienol pathway (Wellburn 1970), and that the route from geranylgeraniol via phytol and/or geranylgeranyl pyrophosphate to phytyl pyrophosphate may be rate-limiting (Block et al. 1980). Moreover, as shown in the administration experiment of phytol in combination with homogentisic acid (Furuya et al. 1987), both the high concentration of homogenistic acid and its long-term biotransformation were inhibitory. The reasons for these effects are not known. Further

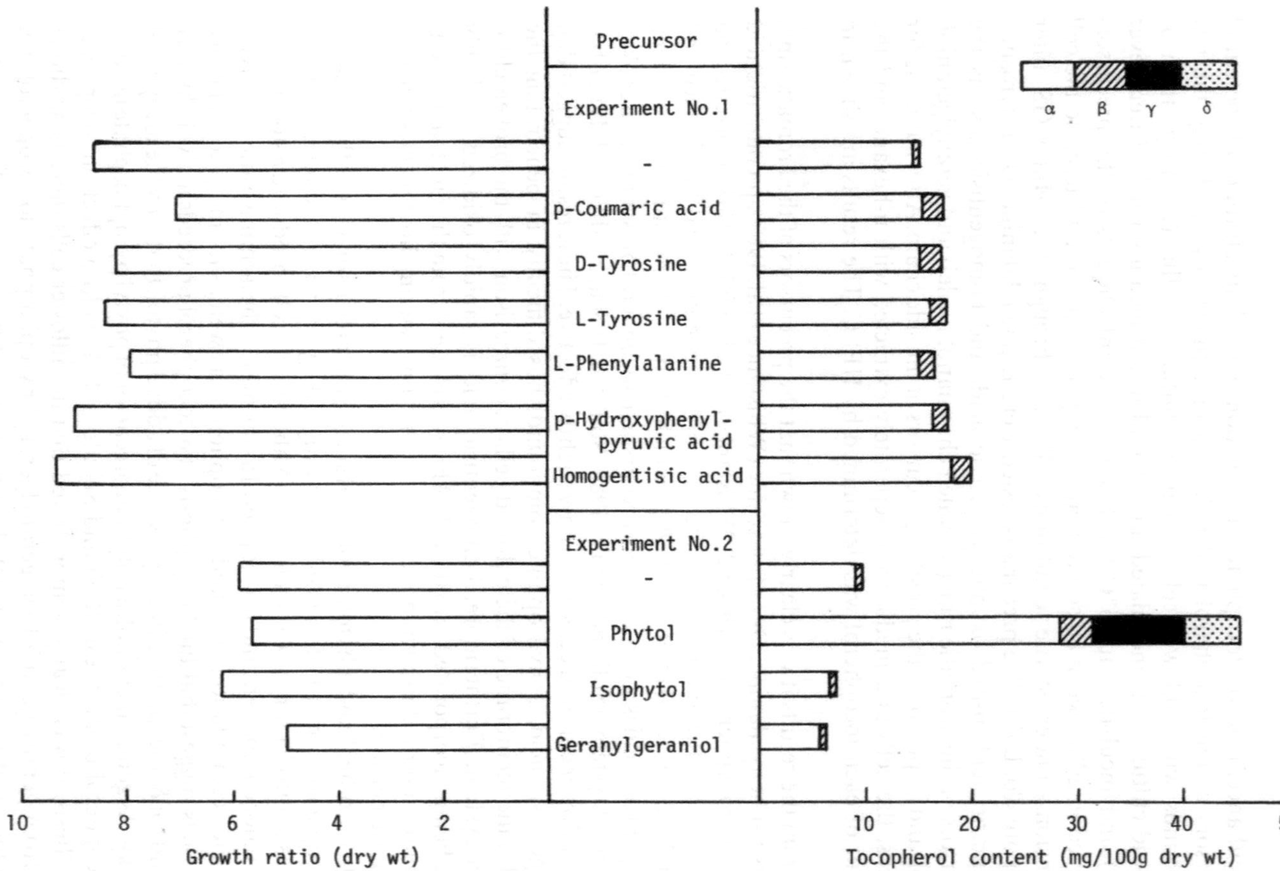

Fig. 3. Effects of various precursors on the tocopherol production of safflower cultured cells. Each precursor, 100 ppm, was administered on the 1-week culture and subsequently incubated for 2 weeks. Experiment No. 1 and No. 2 were carried out separately under the same conditions

experiments are required in which homogentisic acid is administered to a cell-free system and/or in the presence of an antioxidant reagent, such as ascorbic acid.

4 Metabolism of Tocopherols

4.1 Accumulation and Metabolism of Tocopherols in Safflower Cells

It was particularly interesting that not only α- and β-tocopherols, the normal tocopherol constituents of safflower cells, but also γ- and δ-tocopherols were produced in large amounts in the cells upon administration of phytol. This result suggests that the biosynthesis of α-tocopherol proceeds actively by way of γ- and/or δ-tocopherols in the presence of phytol. Soll and Schultz (1980a) have reported that the product of the condensation of homogentisic acid and phytyl pyrophosphate, 2-demethylphytyl-plastoquinol in Fig. 1, is methylated to form phytylplastoquinol, which is cyclized to γ-tocopherol, which is then methylated to produce α-tocopherol. Delta-tocopherol, which is sometimes found and is thought to be formed as a result of a side reaction, can be methylated to form β- and/or γ-tocopherols, and then α-tocopherol. Marshall et al. (1985) also arrived at the same conclusions in their re-examination of tocopherol biosynthesis. Thus, in order to confirm these facts and further investigate the ability of the safflower cells to accumulate the tocopherols, 10-mg aliquots of α-, β-, γ-, and δ-tocopherols were administered to the cell suspension cultured in 250 ml medium per liter flask for 2 weeks. After culturing each 1 week, the cells and media were separately extracted with n-hexane and the amounts of each tocopherol were determined as above. The results are shown in Table 2. No α-tocopherol and little γ-tocopherol was found in the medium even 1 week after the administration. This suggests that almost all of α- and γ-tocopherols administered in vitro were incorporated into the cells. Nevertheless, α-tocopherol level in the cells was the same as that with no administration, i.e., the control level.

Table 2. Accumulation and metabolism of α-, β-, γ-, and δ-tocopherols in safflower cultured cells. Ten mg of each tocopherol was administered to 2-week-old culture which was then incubated for a further period of 1 week. (Unpubl.)

Tocopherol	Cell or medium	Tocopherol content (μg/flask) α-	β-	γ-	δ-
No addition	Cell	152.3	1.1	0.8	–
	Medium	–	–	–	–
α-	Cell	171.8	1.7	1.3	–
	Medium	0.9	0.4	–	–
β-	Cell	149.6	138.3	1.4	–
	Medium	1.4	141.9	–	–
γ-	Cell	159.1	1.0	77.4	–
	Medium	0.7	0.8	23.6	–
δ-	Cell	138.5	47.3	0.9	402.8
	Medium	–	0.4	–	524.1

It appeared thereby that α-tocopherol was metabolized rapidly in the cells or medium, and that the cells had a constant pool size of α-tocopherol. This was also proved by the fact that α-tocopherol content in the cell was held constant in all experiments. On the other hand, both β- and δ-tocopherols were recovered in large quantities from the cells fed each compound as well as the medium, thus indicating that both β- and δ-tocopherols are difficult to metabolize in the safflower cells. Moreover, it is noteworthy that the level of β-tocopherol increased only in the cells of the cultures fed δ-tocopherol. From these results, it was clearly demonstrated that in the safflower cells α-tocopherol is biosynthesized mainly according to the route shown with a thick line in Fig. 1, i.e., via γ-tocopherol from phytylplastoquinol. This was also strongly supported by the results of enzymatic studies as described below.

4.2 Enzymatic Studies of Tocopherol Biosynthesis

S-Adenosylmethionine: γ-tocopherol methyltransferase in the cells was confirmed by the enzyme experiment carried out according to the modified method in *Capsicum* chloroplasts (Camara 1985a,b). The plastid fraction obtained from safflower B2KC cells, which is most likely proto-chloroplast fraction in the safflower cells, was assayed to S-adenosylmethionine: γ-tocopherol methyltransferase (Camara 1985b).

This enzyme had high activity to γ-tocopherol, about 2.7% incorporation to the initial radioactivity in [^{14}C]SAM, but almost no activity to α-, β-, and δ-tocopherols by the same assay method except the use of each tocopherol instead of γ-tocopherol. This also suggests that the biosynthetic route from γ- to α-tocopherol must be the main route of α-tocopherol biosynthesis, and that from δ- via β- and/or γ- to α-tocopherols may be a minor one. Both the metabolic and enzymatic experiment of tocopherols in safflower cells, thus strongly supported the theory of Soll and Schultz (1980) related to α-tocopherol biosynthesis.

5 Increase of Tocopherol Content by Greening of Cells

Alpha- and gamma-tocopherols were rapidly metabolized and were never accumulated in the cells at levels higher than the maximum pool size, which must vary with the age of the cells. Therefore to bring about an increase in α-tocopherol content of the cells, it will be necessary to enlarge the pool size in the cell. It is known that tocopherols are biosynthesized both inside and outside the chloroplast (Soll et al. 1980b). In fact, the tocopherol content in plant leaves is much higher in green tissue than in yellow tissue (Janiszowska and Pennoch 1976). The greening-up of cells, therefore, may be a prerequisite for an increase of tocopherol content. The safflower cells used in this experiment never became green under illumination. If a chloroplast could be formed by any method, an increase in the tocopherol production of the safflower cells would be expected.

Thus, the greening of safflower cells was tried next. As a result, new callus, named Ca-2 callus, was induced from seedling of *Carthamus tinctorius* on the DK medium (Yoshikawa unpubl.), which is the same medium as described in the case from the flower bud (Furuya et al. 1987). Immediately after callus induction, the Ca-2 callus was transferred onto the B2KC medium, which is optimal for growth and tocopherol production in the DK callus, and then cultured under illumination of 7000 lx 16 h light and 8 h dark at 26°C. The Ca-2 callus became slightly green under the illumination. The tocopherol content of the Ca-2 callus, which was cultured static and in suspension under the same conditions as above, was compared with those of the DK and B2KC calli with and without phytol administration in the dark and light. The results in Table 3 show that the Ca-2 callus grown in the dark produced tocopherols in higher amounts than the DK callus, but lower than the B2KC callus. On the contrary, the Ca-2 green callus grown under illumination produced tocopherol 3.3 times higher than the B2KC callus. In particular, the content of α-tocopherol in Ca-2 green cells, 34.17 mg per 100 g dry wt. (82.6% in total), was higher than that of B2KC cells, 21.8 mg (75.2%). In the B2KC cells, the total tocopherol increased about five times by phytol addition, while in the Ca-2 cells it was not so much, up to 1.8 times. Moreover, it is most interesting that Ca-2 green cells produced γ- and δ-tocopherols, as well as α- and β-tocopherols, without administerring phytol, whereas in the B2KC cell γ- and δ-tocopherols were produced at first by exogenously adding phytol. These facts suggest that the precursor, phytol, is supplied endogenously in the Ca-2 green cells. This coincides with the fact that phytol is a component of chlorophyll, which must be formed in the green cells by the illumination. Although the B2KC cell produced up to 144.28 mg in total tocopherol by the administration of phytol, the α-tocopherol content was very low, 37.78 mg (26.2%). On the other hand, the Ca-2 green cell produced α-tocopherol to 60.75 mg (82.6%), in spite of the low content of total tocopherol,

Table 3. Tocopherol contents in various safflower cultured strains. DK and B2KC strains are of the same origin derived from flower bud, and both strains have no ability to grow under illumination. Ca-2 strain is slightly green under illumination at 7000 lx for 16 h a day. (unpubl.)

Strain	Phytol	Illumination	Tocopherol content (mg/100 g dry wt.)				
			α-	β-	γ-	δ-	Total
Static culture							
DK	–	–	0.75	0.40	–	–	1.15
B2KC	–	–	3.80	1.15	–	–	4.95
Ca-2	–	–	1.09	0.64	–	–	1.73
	–	+	13.14	2.07	1.12	0.08	16.41
Suspension culture							
DK	–	–	1.05	0.45	–	–	1.50
B2KC	–	–	21.80	7.04	–	0.14	28.98
	+	–	37.78	17.26	29.28	59.96	144.28
Ca-2	–	–	6.51	1.20	0.07	–	7.78
	+	–	22.28	1.22	5.54	0.74	29.78
Ca-2 green	–	+	34.17	2.89	3.24	0.38	40.68
	+	+	60.75	1.54	10.14	1.11	73.54

73.54 mg, which was half of that of the B2KC cell in the total amount. Therefore, the production system using Ca-2 green cell has a good prospect of developing the production of α-tocopherol by cell culture.

6 Production of Other Compounds

6.1 Ubiquinone-9

Ubiquinone is a vitamin-like reactive compound, which plays a role as H^+ transfer in the electron-transferring enzymatic system, and is widely contained in plants. The number of isoprene units in the side chain is different in various species. It has been reported that safflower cultured cells contain ubiquinone-9 (Hagimori et al. 1978). Ubiquinone-9, which is a by-product of α-tocopherol, as shown in Fig. 1, was isolated from the safflower jar fermentor cultured cells and then identified. The ubiquinone-9 (15.4 mg), which was purified by a silica gel chromatography and HPLC, was identified by TLC (co-chromatography), HPLC (co-injection) and MS spectrometry. $C_{54}H_{82}O_4$, 794.6196, was found by high resolution MS, and calculated to be 794.6212. The mass spectrum of ubiquinone-9 showed as major fragments m/z 795 (M^+, 7%), 235 (30) and 197 (100), to be due to the characteristic fragmentation of ubiquinone-9 molecules, and is consistent with those of the authentic sample.

6.2 Fatty Acids

It is well known that safflower seed oil contains abundant linoleic acid, an unsaturated fatty acid, as well as tocopherols and ubiquinone. A typical sample contains linoleic acid, 75%, oleic acid, 18%, and a mixture of saturated acids totaling about 6%. The claims that ingestion of quantities of polysaturated fatty acids in various forms results in a decreased blood cholesterol level are based on insubstantial evidence. Thus the fatty acid contents in various safflower cultured cells were determined by GLC. The 3-week-cultured cells were collected by filtration with a nylon cloth, freeze dried, and crushed. The resulting powder was extracted with n-hexane, as described for tocopherols. The extract was resolved to 25 ml with $CHCl_3$ and then subjected to GLC, the results are shown in Table 4. All the cultured safflower cells contained 43.8 to 53.5% linoleic acid ($C_{18:2}$) in total as the main fatty acid as well as safflower oil from the seeds, 75.5%. The linoleic acid content in the cultured cells was the highest in the DK callus, 1.55 g per 100 g dry weight, and next in BA 1-Y, 1.14 g. It has been reported that in various vegetable oils, linoleic and linolenic acid contents in fatty acids related closely to tocopherol contents (Kanematsu et al. 1983). However, in safflower cultured cells there was no relationship between fatty acids and tocopherol contents, because the high tocopherol-producing line, B2KC, did not show so much high fatty acid content and further there was no increase in the fatty acid amounts on the addition of phytol. In the dry cell base, the fatty acids contents in the cultured cells were lower, one twentieth to one sixth, than that in the seed, whereas in the extract base those in the cells were almost the same as in the seed.

Table 4. Fatty acid contents of various safflower strains and seeds. (unpubl.)

Callus (seed)	Culture condition	Fatty acids (g/100 g dry wt.) C_{16}	C_{18}	$C_{18:1}$	$C_{18:2}$	$C_{18:3}$	Others	Total
DK	Static	0.69	0.03	0.13	1.55	0.33	0.18	2.91
BA 1-W	Static	0.33	0.02	0.04	0.63	0.17	0.09	1.28
BA 1-Y	Static	0.64	0.04	0.09	1.14	0.23	0.11	2.25
B2KC	Static	0.21	0.02	0.03	0.38	0.09	0.06	0.79
B2KC	Suspension	0.34	0.02	0.02	0.56	0.21	0.08	1.23
B2KC + Phytol		0.40	0.04	0.02	0.67	0.17	0.14	1.44
Seed		1.11	0.32	1.83	13.54	0.04	1.09	17.93

Abbreviations: BA 1, 6-benzyladenine 1 mg/l; W, white; Y, yellow; C_{16}, palmitic acid; C_{18}, stearic acid; $C_{18:1}$, oleic acid; $C_{18:2}$, linoleic acid; $C_{18:3}$, linolenic acid.

7 Conclusions

The florets of safflower, *Carthamus tinctorius*, which contain a red pigment, carthamin, a chalcone glycoside, have been used as a crude drug and a natural dye since ancient times in the Orient. In our experiments, however, the cultured cells induced from flower buds did not contain any red pigment. On the other hand, safflower oil has been recently noted as α-tocopherol-rich, having the highest vitamin E activity, and rich in linoleic acid, an unsaturated fatty acid. Thus, research was conducted for the purpose of the production of these compounds by plant cell culture. At first we investigated mainly the cultured cells of sunflower, safflower, soybean, etc., whose original plant contained a high amount of tocopherols. It was shown from the results that only safflower cultured cells produced a remarkable amount of tocopherols. Next, a better cell line, named the B2KC strain, in terms of both growth rate and tocopherol content, was selected using various growth regulators and media additives. This was a high-growth and high-tocopherol-producing strain, ca. two times in growth ratio and 4.3 times in tocopherol production in comparison with the original strain, DK cells. Further, the increase of tocopherol contents was tried by administering various biosynthetic precursors. As shown in Fig. 3, tocopherol production was effectively stimulated by phytol, the precursor of the side chain, and the amount reached up to five times, 144.28 mg per 100 g dry cells in maximum, although the increase was predominant in γ- and δ-tocopherols. However, homogentisic acid, the most closely related precursor of chromanol nuclei, had almost no effect. As a result, the effective incorporation of homogentisic acid into the cell remained as the most important subject for increase of the tocopherol content.

Furthermore, the amount of α-tocopherol produced by safflower cell culture showed an almost constant value in each strain and never exceeded this level. This suggests that the pool size of α-tocopherol in the cell is always constant in the same culture strain. Further, it has been reported that in spinach and lettuce, α-tocopherol is biosynthesized in the isolated chloroplast, hence we considered the formation of chloroplast essential for the further increase of α-tocopherol content. Thus a new strain, named Ca-2, induced from the seedling and cultured under illumination, resulted in slight greening. Although in this strain the increase on phytol addition was lower, 1.8 times, than that in the DK strain, five times, most of it, 82.6% in total tocopherol, was α-tocopherol, i.e., 60.75 mg per 100 g dry cell (Table 3).

In conclusion, in the present study, the best medium and conditions for the establishment of saffower callus and the production of vitamin E are as follows. (1) Safflower callus is induced at a high frequency on Murashige and Skoog's medium supplemented with 1 mg/l 2,4-D and 0.1 mg/l K from seedling or flower bud. (2) Callus is cultured on RT medium containing 2 mg/l IBA, 0.1 mg/l K and 0.1% casamino acid, under the dark and/or illumination (7000 lx) and subcultured at 3-week intervals. (3) The maximum production of tocopherols is obtained by the addition of 100 mg/l phytol on cell suspension culture precultured for 1 week, followed by incubation for a further 2 weeks.

Finally, the tocopherol amounts produced by our experiments were compared with those in seeds or vegetable oils. The results showed that tocopherol content per n-hexane extract from the safflower cultured cells was ten times higher than those from seeds and oil. Hence, it was shown that the oil produced by safflower cell culture is α-tocopherol-rich with a good quality, and further that the oil also contains abundant linoleic acid. Safflower cell cultures, therefore, have a good prospect of developing the new method of tocopherol production.

References

Block MA, Joyard J, Douce R (1980) Site of synthesis of geranylgeraniol derivatives in intact spinach chloroplasts. Biochim Biophys Acta 631:210–219

Camara B (1985a) Carotene synthesis in *Capsicum* chromoplasts. Meth Enzymol 110:244–253

Camara B (1985b) S-Adenosylmethionine: γ-tocopherol methyltransferase (*Capsicum* chlomoplasts). Meth Enzymol 111:544–547

Chiej R (1984) In: Mondadori A (ed) The Macdonald encyclopedia of medicinal plants. Macdonald, London, Entry No. 70

Claus EP, Tyler VE, Brady LR (1970) (eds) Pharmacognosy. Kimpton, London p 129

Diplock AT (1974) Possible stabilizing effect of vitamin E on microsomal, membrane-bound, selenide-containing proteins and drug-metabolizing systems. Am J Clin Natrit 27:995–1004

Fiedler E, Soll J, Schultz G (1982) The formation of homogentisate in the biosynthesis of tocopherol and plastquinone in spinach chloroplasts. Planta 155:511–515

Furuya T, Yoshikawa T, Kimura T, Kaneko H (1987) Production of tocopherols by cell culture of safflower. Phytochemistry 26:2741–2747

Hagimori M, Matsumoto T, Noguchi M (1978) Isolation and identification of ubiquinone 9 from cultured cells of safflower (*Carthamus tinctorius* L.). Agric Biol Chem 42:499–500

Horwitt MK (1976) Vitamin E: a reexamination. Am J Clin Nutrit 29:328–337

Janiszowska W, Pennock JF (1976) The biochemistry of vitamin E in plants. Vitamin Horm 34:77–105

Kanematsu H, Ushigusa T, Maruyama T, Niiya I, Fumoto D, Toyoda T, Kawaguchi Y, Matsumoto T (1983) Comparison of tocopherol contents in crude and refined edible vegetable oils and fats by high performance liquid chromatography. Yukagaku 32:122–126 (in Japanese)

Kanno C, Hayashi M, Yamauchi K, Tsugo T (1970) Antioxidant effect of tocopherols on autoxidation of milk fat, pt 1. Antioxidant activity of tocopherols in fatty acid methylester of milk fat. Agric Biol Chem 34:878–885

Katsui G (1983) Recent trend of natural tocopherols. Yukagaku 32:411–417 (in Japanese)

Khanna P, Staba EJ (1968) Antimicrobials from plant tissue cultures. Lloydia 31:180–189

Kitamura S (1978) Spermatophyta 1. In: Itoh M (ed) The Asahi encyclopedia of plants of the world. Asahi, Tokyo, p 27 (in Japanese)

Marshall PS, Morris SR, Threlfall DR (1985) Biosynthesis of tocopherols: A re-examination of the biosynthesis and metabolism of 2-methyl-6-phytyl-1,4-benzoquinol. Phytochemistry 24:1705–1711

Murashige T, Skoog F (1962) A revised medium for rapid growth and bioassays with tobacco tissue cultures. Physiol Plant 15:473–497

Obara H, Onodera J (1979) Structure of carthamin. Chem Lett 201–204

Peake IR, Audley BG, Pennock JF (1970) Some studies on the biosynthesis of tocochromanols. Biochem J 119:58 p

Soll J, Schultz G (1980) 2-Methyl-6-phytylquinol and 2,3-dimethyl-5-phytylquinol as precursors of tocopherol synthesis in spinach chloroplasts. Phytochemistry 19:215–218

Soll J, Schultz G (1981) Phytol synthesis from geranylgeraniol in spinach chloroplasts. Biochem Biophys Res Commun 99:907–912

Soll J, Kemmerlitz M, Schultz G (1980a) Tocopherol and plastoquinone synthesis in spinach chloroplasts subfractions. Arch Biochem Biophys 204:544–550

Soll J, Douce R, Schultz G (1980b) Site of biosynthesis of α-tocopherol in spinach chloroplasts. FEBS Lett 112:243–246

Tappel AL (1974) Selenium-glutathione peroxidase and vitamin E. Am J Clin Nutrit 27:960–965

Wasserman RH, Taylor AN (1972) Metabolic roles of fat-soluble vitamin D, E, and K. Annu Rev Biochem 41:179–202

Wellburn AR (1970) Studies on the biosynthesis of the tocopherols in higher plants. Phytochemistry 9:743–748

Whistance GR, Threlfall DR (1970) Biosynthesis of phytoquinones: Homogentisic acid; a precursor of plastoquinones, tocopherols and α-tocopherolquinone in higher plants, green algae and blue-green algae. Biochem J 117:593–600

Whistance GR, Threlfall DR (1971) Incorporation of shikimic acid into isoprenoid quinones and chromanols by maize shoots. Phytochemistry 10:1533–1538

Whistance GR, Threlfall DR, Goodwin TW (1967) Observations on the biosynthesis of phytoterpenoid quinone and chromanol nuclei. Biochem J 105:145–154

X *Chamomilla recutita* (L.) Rauschert (Camomile): In Vitro Culture and the Production of Secondary Metabolites

J. Reichling[1] and R. Beiderbeck[2]

1 General Account

1.1 Habit and Distribution of the Plant

According to recent studies in plant systematics, the valid name of the medically used camomile is *Chamomilla recutita* (L.) Rauschert. Linné used the name "*Matricaria recutita*", but the plant has become generally known as "*Matricaria chamomilla*".

Chamomilla recutita is a member of the family of the Asteraceae. It is an annual herbaceous plant which flowers from May to August and reaches a height of 20–60 cm. It carries leaves with a pinnate structure and inflorescences of 10–20 mm diameter which are located terminally, or in the leaf axils supported by pedicels of 5–10 cm length (see Fig. 1). The inflorescences (capitula) consist of numerous yellow tubular florets in the center surrounded by a circle of white ligulate florets with all florets imbricated on a hollow inflorescence base. The flowers, and especially the inferior ovaries, are rich in excretory glands containing the valuable essential oil. The fruit is an achene. The original area of distribution was southern and eastern Europe and the Near East. From there the camomile has spread through most of Europe, and today is found from Scandinavia in the north to the Ural mountain range in the east. Its area of distribution has extended further to Iran, Afghanistan, Pakistan, India, China, and Japan on the one hand, and to Egypt, East Africa, and the Congo basin on the other. It was introduced into the Americas (U.S.A., Mexico, Peru, Argentina, Brazil), and Australia, probably together with shipments of grain. In keeping with its modest demands on the prevailing soil conditions (loamy, salty, sandy soils are tolerated) the camomile grows on fallows, dumps, village paths, grasslands, grain fields, and vineyards (Reichling 1983; Schilcher 1987).

1.2 Medically Important Compounds of the Camomile Plant

Several groups of compounds of the camomile have been shown to be of medical importance.

[1] Institut für Pharmazentische Biologie, Im Neuenheimer Feld 364, W-6900 Heidelberg, FRG
[2] Botanisches Institut der Universität, Im Neuenheimer Feld 360, W-6900 Heidelberg, FRG

Biotechnology in Agriculture and Forestry, Vol. 15
Medicinal and Aromatic Plants III (ed. by Y.P.S. Bajaj)

Fig. 1. Flower heads of *Chamomilla recutita* (L.) Rauschert

1.2.1 Essential Oil

The essential oil of the wild camomile is accumulated in various excretory domains such as excretory glands, schizogenic oil ducts, and oil cells. It is composed mainly of terpenes and polyines (see Fig. 2). Some compounds are restricted to special organs (see Table 1), while others can be detected in almost all parts of the plant, e.g., trans-α-farnesene, trans-β-farnesene, spathulenol, or the cis/trans-en-in-dicycloethers (Reichling 1983; Reichling et al. 1984; Carle and Isaac 1985; Schilcher 1987).

The terpenes (–)-α-bisabolol, (–)-α-bisabolonoxide, the (–)-α-bisabololoxides, and the sesquiterpene-lactone matricine are the most important compounds

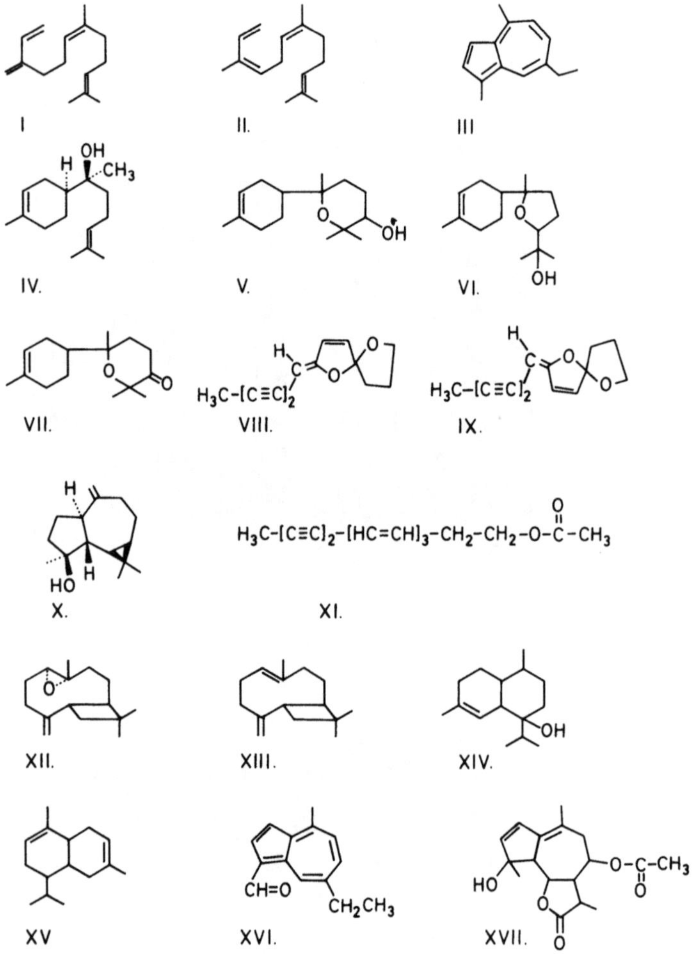

Fig. 2. Terpenes and polyines in the essential oil of camomile. *I* trans-β-farnesene; *II* trans-α-farnesene; *III* chamazulene; *IV* (–) α-bisabolol; *V* (–)-α-bisabololoxide A; *VI* (–)-α-bisabololoxide B; *VII* (–)-α-bisabolonoxide; *VIII* cis-en-in-dicycloether; *IX* trans-en-in-dicycloether; *X* spathulenol; *XI* chamomillaester; *XII* caryophyllenepoxide; *XIII* β-caryophyllene; *XIV* chamomillol; *XV* cadinene; *XVI* chamavioline; *XVII* matricine

regarding the pharmaceutical quality of the camomile and are found in the excretory glands of the flower heads only. Additionally, these contain some more sesquiterpenes (cadinene, α-cubebene, calamene, α-muurolene, chamaviolene) and low amounts of monoterpenes, e.g., γ-terpenene and Δ^3-carene (Motl et al. 1978, 1983; Reichling et al. 1979a,b, 1984). If the flower heads are subjected to steam distillation, generally an essential oil of blue color (0.4–1% dry wt.) is obtained. The blue color is caused by its content of the sesquiterpenehydrocarbon chamazulene, which is generated from matricine during the steam distillation procedure.

In the essential oil of the root (0.09–0.16% dry wt.) the sesquiterpenes caryophyllene-epoxide, cis-caryophyllene, and β-caryophyllene, and also the two

Table 1. Occurrence of well-known sesquiterpenes and polyines in flower heads, shoots, and roots of camomile

Flower head	Shoot	Root
Sesquiterpenes	Sesquiterpenes	Sesquiterpenes
Chamazulene	trans-β-Farnesene	trans-β-Farnesene
trans-β-Farnesene	trans-α-Farnesene	trans-α-Farnesene
trans-α-Farnesene	Spathulenol	Spathulenol
Cadinene		β-Caryophyllene
α-Cubebene		cis-Caryophyllene
Calamene		Caryophyllenepoxide
α-Muurolene		Chamomillol
Chamavioline		
Spathulenol		
(−)-α-Bisabolol		
(−)-α-Bisabolonoxide		
(−)-α-Bisabololoxide A		
(−)-α-Bisabololoxide B		
(−)-α-Bisabololoxide C		
Polyines	Polyines	Polyines
cis/trans-en-in--Dicycloether	cis/trans-en-in--Dicycloether	cis/trans-en-in--Dicycloether
		Chamomillaester I
		Chamomillaester II

isomeric polyines, chamomillaester I and II are present (see Fig. 2, Table 1). The essential oil is deposited in schizogenic excretory ducts and in excretory cells (oil cells, idioblasts) of the root; the oil cells are restricted to the inner layers of the root cortex (Reichling et al. 1979b, 1983, 1984).

Allergenic effects of the camomile have been described often in the past. A sesquiterpene lactone with an exocyclic methylene group was thought to be the responsible allergen. This compound has the trivial name of anthecotulide (Hausen et al. 1984; Carle and Isaac 1985). As could be demonstrated recently, this substance is present in commercial samples of the bisabololoxide B-type of *Chamomilla recutita* in trace amounts at best. Previous reports demonstrating the presence of great amounts of anthecotulide in the camomile have proven erroneous due to the mistaken identity of *Anthemis cotula* for camomile (Hausen et al. 1984).

1.2.2 Flavonoids

Flavonoids and flavonoid glycosides are found in all organs of the camomile except the roots. In the vegetative parts aboveground only traces of flavonoids are detected, whereas the flower heads contain considerable amounts (1–2% dry wt.) of flavones, flavonols, and flavonoid glycosides (Reichling 1983). The flavonoids of the flower heads can be classified into five groups of increasing polarity (Reichling 1983; Carle and Isaac 1985):

a) Methoxylated Flavones and Flavonols

Jaceidin, chrysosplenol, eupatolitin, spinacetin, axillarin, eupalitin, 6-methoxykämpferol, chrysoeriol, isorhamnetin, patuletin, chrysosplenetin (see Fig. 3);

b) Hydroxylated flavones and flavonols

Luteolin, quercetin, apigenin (see Fig. 3);

c) Acetylated flavone-monoglycosides

Apigenin-7-0-(6-0-acetyl)-glycoside, apigenin-7-0-(2-0-acetyl)-glycoside;

d) Flavone- and flavonol-monoglycosides

Apigenin-7-0-glucoside, luteolin-7-0-glucoside, quercetin-7-0-glucoside, quercetin-3-0-galactoside, patuletin-7-0-glucoside, isorhamnetin-7-0-glucoside;

e) Flavone- and flavonol-diglycosides

Apigenin-7-0-(6-0-apiosyl)-glucoside, apigenin-7-0-rutinoside, apigenin-7-0-neohesperidoside, luteolin-7-0-rutinoside, quercetin-3-0-rutinoside.

Parts of the flavonoids are distributed inside the flower heads with organ or tissue specificity: for example, apigenin-glycosides are lacking in the tubular florets, whereas in the ligulate florets the quercetin derivatives cannot be found. Generally, the presence of highly methylated flavonoids is restricted to the flower heads (Reichling 1983). A recent paper demonstrate that apigenin does not actually occur in living camomile flower heads (Schreiber et al. 1990). Apigenin is the result of post-harvest process converting apigenin-glycosides into the aglycon. So, apigenin could be found in considerable amounts in the drug-matricariae flos.

1.2.3 Coumarins and Phenolic Acids

The coumarins herniarine (0.04–0.1% dry wt.) and umbelliferone (0.006–0.02% dry wt.) and in smaller amounts the phenolic acids, syringic acid, vanillic acid, anisic acid, and coffeic acid (see Fig. 4) are found in the flower heads but not in other parts of the plant (Reichling et al. 1979a,b; Schilcher 1987).

1.2.4 Mucilages

At the surfaces of the inferior ovaries slime cells are located which contain high amounts (up to 10% dry wt.) of camomile mucilage composed of complex carbohydrates. After hydrolysis glucose, fructose, galactose, arabinose, xylose, rhamnose, glucosamine, and galacturonic acid have been identified (Carle and Isaac 1985).

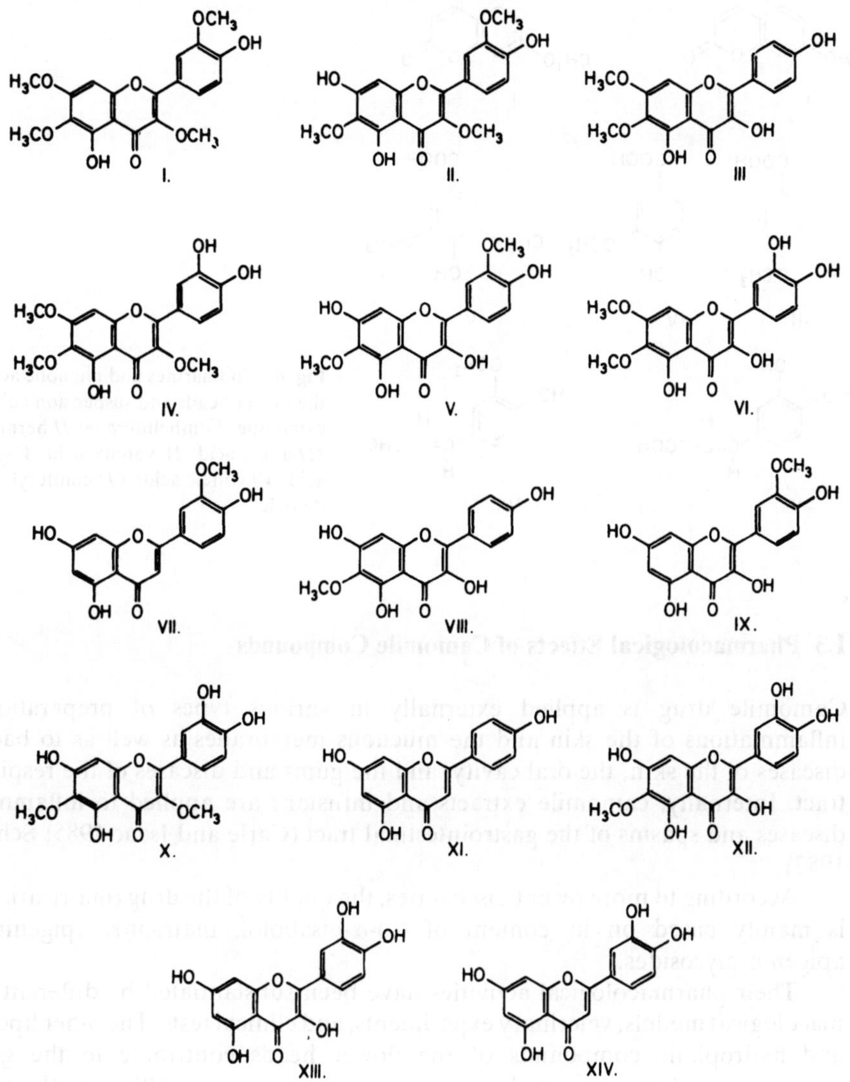

Fig. 3. Flavones and flavonols in the flower heads of camomile. *I* chrysosplenetin; *II* jaceidin; *III* eupalitin; *IV* chrysosplenol; *V* spinacetin; *VI* eupatolitin; *VII* chrysoeriol; *VIII* 6-methoxykämpferol; *IX* isorhamnetin; *X* axillarin; *XI* apigenin; *XII* patuletin; *XIII* quercetin; *XIV* luteolin

On the basis of the organ-specific distribution of the medically important compounds (see below) only the flower heads are used in the fresh or dried state (camomile drug). By far the greatest portion of the camomile drug is utilized by the cosmetic industry and by private households to make camomile infusions; a smaller portion is used by the pharmaceutical industry. In the Federal Republic of Germany, for example, about 290 camomile-containing formulations are registered (Carle and Isaac 1985; Schilcher 1987).

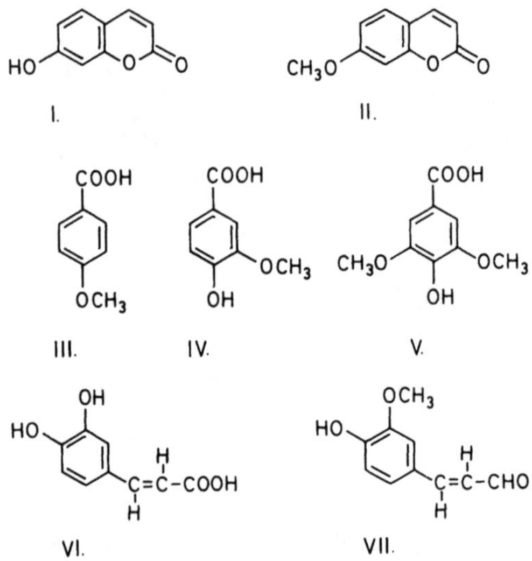

Fig. 4. Coumarines and phenolic acids in the flower heads and suspension cultures of camomile. *I* umbelliferone; *II* herniarine; *III* anisic acid; *IV* vanillic acid; *V* syringic acid; *VI* coffeic acid; *VII* coniferyl aldehyde

1.3 Pharmacological Effects of Camomile Compounds

Camomile drug is applied externally in various types of preparations to inflammations of the skin and the mucuous membranes as well as to bacterial diseases of the skin, the oral cavity, and the gums and diseases of the respiratory tract. Internally, camomile extracts and infusions are applied to inflammatory diseases and spasms of the gastrointestinal tract (Carle and Isaac 1985; Schilcher 1987).

According to more recent discoveries, the quality of the drug (matricariae flos) is mainly based on its content of (–)-α-bisabolol, matricine, apigenin and apigenin-glycosides.

Their pharmacological activities have been substantiated by different pharmacological models, veterinary experiments, and clinical tests. The other lipophilic and hydrophilic compounds of the flower heads contribute to the general effectiveness of the camomile extracts and preparations, although their pharmacological activity is clearly less obvious (Jakovlev et al. 1979; Szelenyi et al. 1979; Achterrath-Tuckermann et al. 1980; Della Loggia 1985).

As far as the main compounds are concerned, antiphlogistic, antimicrobial, ulcusprotective and musculotropically spasmolytic activities have been proven. For example, (–)-α-bisabolol shows a pronounced antiphlogistic activity, as do matricine (or chamazulene, respectively) and, less pronounced, the other bisaboloids (Jakovlev et al. 1979; Szelenyi et al. 1979). Moreover, recent studies have shown that apigenin also has an antiphlogistic activity if applied topically (Della Loggia 1985). The effectiveness of (–)-α-bisabolol, chamazulene, and apigenin is comparable to that of several synthetic antiphlogistics, e.g., salicylamide, indomethacin.

Beyond it (–)-α-bisabolol shows a papaverine-like musculotropically spasmolytic activity; it is twice as efficient as bisabololoxides and cis/trans en-in-dicycloethers. The flavonoids apigenin, patuletin, luteolin, and quercetin also have pronounced musculotropically spasmolytic effects. Apigenin is significantly more potent than papaverine (3.29 times); the other flavonoids are less active than papaverine (Achterrath-Tuckermann et al. 1980).

To summarize, it is clear that the lipophilic as well as the hydrophilic compounds contribute to the therapeutic importance of camomile. Standardized hydroalcoholic extract preparations therefore represent the optimal form of camomile prescriptions.

1.4 Conventional Practices of Camomile Propagation and Camomile Demand in the World Market

According to the present state of knowledge for therapeutical purposes, a camomile plant type with high contents of (–)-α-bisabolol, cis/trans en-in-dicycloether, matricine, and flavonoids (with a high portion of apigenin-glycosides) is required. The steadily increasing need for high quality drugs cannot be provided by plant collection in the wilderness. For several years the systematic and controlled cultivation of camomile types with reproducible pharmaceutical quality has been promoted worldwide. Today well-bred camomiles are cultivated in Argentina, Bulgaria, Egypt, and Hungary; smaller amounts are grown in Czechoslovakia, Germany, and Spain. They give a drug yield of 500–800 kg per hectare; the drug obtained contains about 1% (dry wt.) essential oil with a high proportion of (–)-α-bisabolol (200–500 mg%) and chamazulene (150–240 mg%) (Franz and Kirsch 1985).

The annual harvest world-wide is estimated to be 6000-7000 t of camomile drug. In the Federal Republic of Germany alone 3000 t of camomile drug are imported annually, representing a value of about $11 million. In 1987 1 kg of high quality camomile oil cost $1700–2300 on the world market (Carle et al. 1987).

The various commercial camomile drugs still differ widely as far as the total content of active compounds and the compound spectrum are concerned (Table 2). This is partly a consequence of the camomile variety cultured. Four or five different chemical types of cultivars are known which are each characterized by the predominance of a different compound of the essential oil spectrum, e.g., bisabolol type, bisabolonoxide type, bisabololoxide A type, bisabololoxide B type. On the other hand, the compound content and pattern of the drug are influenced by daylength, culture and fertilization methods, seasonal weather fluctuation, and the developmental state of the flower heads at harvest time (Reichling 1983; Schilcher 1987). In the coming years progress in camomile breeding and improvements in culture methods should result in a supply with more standardized camomile drugs. Breeding programs to increase homogeneously all groups of medically important compounds, especially the apigenin-glycoside content, need to be designed. Since the demand for camomile drugs of high quality is expected to increase in the future, additional efforts concerning breeding programs should repay on a long-term basis.

Table 2. Quantitative determination of essential oil (mg oil/100 g dry wt.) and essential oil components (% of essential oil) in various commercial camomile drugs

Flower heads (drug)	Content of essential oil mg oil/100 g dry wt.	Farnesene %	(−)-α-Bisabolo-loxide B %	(−)-α-Bisabo-lonoxide %	(−)-α-Bisabolol %	Chama-zulene %	(−)-α-Bisabolo-loxide A %	cis/trans en-in-Dicyclo-ether %
Bulgaria	590	11.2	1.5	66.8	Traces	Traces	13.6	7.5
Mexico	992	4.2	5.8	7.6	3.5	2.0	68	9.1
Bohemia	760	5.1	5.8	7.3	2.5	9.7	60.4	9.5
Bohemia	936	3.1	5.8	5.7	2.2	9.8	64.0	10.1
Egypt	902	3.4	8.6	5.5	2.3	3.5	65.5	11.5
Egypt	803	4.3	8.5	7.6	4.4	3.4	61.5	10.4

2 Biotechnological Approaches

In previous years very few research groups have been concerned with in vitro cultures of camomile. Mainly the problems of the formation and accumulation of essential oil have been studied using callus cultures (on agar surfaces) and suspension cultures.

Table 3. In vitro studies on *Chamomilla recutita* (L.) Rauschert

Inoculum	Medium	Growth response	Reference
Root, stem,	MS+2, 4-D(1)+kin(1);	Callus culture	Szöke et al. (1977)
inflorescence	MS+CM	Callus culture	Szöke et al. (1977)
Inflorescence	MS+2, 4-D(1)+kin(1);	Callus culture	Szöke et al. (1978)
–	MS+CM	Callus culture	Szöke et al. (1979)
–	LS+NAA(0.5)+kin(0.5)	Callus culture	Cellarova et al. (1982)
–	LS+kin(0.1)	Shoot/root formation	Cellarova et al. (1984)
–	LS+2, 4-D(1)	Callus culture	Cellarova et al. (1984)
–	LS+NAA(2)+kin(2)	Callus culture	Cellarova et al. (1986)
Crown gall of shoot	MS(hormone-free)	Suspension culture	Beiderbeck (1982)
–	MS(hormone-free)	Suspension culture	Knoop and Beiderbeck (1983)
–	MS(hormone-free)	Suspension culture, plating	Knoop and Beiderbeck (1985)
Shoot	MS+NAA(0.5)	Callus culture	Reichling and Becker (1976)
–	MS(hormone-free)	Suspension culture	Bisson (1983)
–	MS+NAA(2)+kin(1)	Callus culture	Bisson (1983)
–	MS(hormone-free)	Suspension culture	Bisson et al. (1983)
–	MS(hormone-free)	Suspension culture	Becker et al. (1984)
–	MS+NAA(5)+kin(2.5)	Callus culture	Reichling et al. (1984)

Abbreviations: NAA = naphthylacetic acid; 2,4-D = 2.4-dichlorophenoxyacetic acid; BA = benzyladenine; kin = kinetin; MS = Murashige and Skoog (1962); CM = coconut milk; LS = Linsmaier and Skoog (1965).

2.1 Callus Cultures

Reichling et al. (1984) isolated a callus culture (BK2/p) from surface sterilized shoot segments of the camomile variety BK2. This culture consisted of relatively firm, nodular cell aggregates (Fig. 5) when cultivated in continuous light on a modified MS medium (Murashige and Skoog 1962) supplemented with the phytohormones NAA (5 mg/l) and kinetin (2.5 mg/l). The cell aggregates consisted mainly of parenchyma cells; but besides numerous tracheary elements also excretory cells (oil cells) were detected, exclusively containing the complete complement of essential oil of the culture (0.2% of culture dry weight). The following compounds of this oil could be identified (% of essential oil): trans-α-farnesene (5%), trans-β-farnesene (1%), spathulenol (3%), caryophyllene epoxide (1%), cis- en-in dicycloether (19%), and the isomeric chamomillaesters (35%) (see Fig. 2). This compound pattern is almost identical to the substance pattern found in the camomile root (see Table 1). The bisaboloides and matricine which are characteristic compounds of the

Fig. 5. Callus culture (type BK2/p) on solid medium (surface culture) initiated from camomile (type BK2) (3 years old)

inflorescence could not be detected. Szöke et al. (1977, 1978, 1979) established callus cultures from root, shoot, and flower head tissues and cultivated them in permanent darkness on a modified MS medium supplemented with 2,4-D (1 mg/l) and kinetin (1 mg/l). Callus cultures of flower head origin yielded the highest growth rates and had the highest content of essential oil (about 0.06% of dry weight) of all cultures established. The essential oil of these cultures contained trace amounts of (−)-α-bisabolol, (−)-α-bisabololoxides A and B, trans-β-farnesene, spathulenol, and the sesquiterpenoid alcohol chamomillol, which was first identified by Reichling et al. (1983). Polyines have not been accumulated. Cultivation in continuous light for 2 years resulted in the production of essential oil, with chamomillol being the main compound (Szöke et al. 1977, 1979). The accumulation of essential oil by dark-grown cultures changed considerably after 2,4-D and kinetin had been replaced by 10% coconut milk. The total content of essential oil was increased to more than 0.2% dry wt. as a consequence of the induced formation of oil reservoirs which accumulated the terpenes as described for the BK2/p culture (Szöke et al. 1978, 1979; Reichling et al. 1984).

The differences in the compound patterns produced by different cultures in different laboratories under a variety of conditions cannot at present be explained. Even a combination of conditions thought favorable to the essential oil production (culture strain BK2/p, permanent darkness, 10% coconut milk) does not lead to a biosynthesis program including the flower-specific compounds. None of the callus cultures described is capable of synthesizing the sesquiterpene lacton matricine and BK2/p never produces bisaboloids (Bisson 1983).

2.2 Suspension Cultures

Starting with the callus culture BK2/p, a suspension culture (BK2/sub) was established consisting of small globular cell aggregates of intensively green color (see Fig. 6). This suspension was cultured in 12:12 h fluorescent light-dark alternations in a modified MS medium (Bisson 1983; Bisson et al. 1983; Becker et al. 1984). Since it grew well even in the absence of phytohormones, involuntarily a habituated cell culture had been selected. From the very beginning this culture produced much less essential oil than the mother culture BK2/p. After 3 years of subculturing in liquid medium it accumulated merely 0.014% essential oil per dry weight of culture. Of the characteristic compounds of the camomile oil only low amounts of cis/trans-en-in-dicycloether could be detected during routine checks (Bisson 1983; Bisson et al. 1983; Becker et al. 1984). Histochemical studies did not give any indication of the presence of excretory ducts, oil idioblasts, or other possible reservoirs of essential oil in the suspension culture. Variations of light quality (blue and red light), light intensity (500–2500 lx), composition of the nutrient medium (e.g., addition of 10% coconut milk) did not result in changes of the compound pattern (Bisson 1983).

Fig. 6. Suspension culture (BK2/sub) initiated from callus culture (type BK2/p) (2 years old)

2.3 Two-Phase Culture

A comparison between cell cultures and the situation inside the intact plant may provide insight as to the reasons of "nonproduction" of secondary substances in unorganized suspension cultures. In higher amounts most of the volatile terpenes are toxic to the cellular metabolism (Reichling 1983). These compounds are, therefore, accumulated in special intra- and extracellular compartments (excretory reservoirs) and in this way separated from the active metabolism of the cells. In contrast, suspension cultures (e.g., BK2/sub) lack excretory reservoirs and consequently an extraplasmatic accumulation of terpenes is no longer possible. It is conceivable that in this situation a continued synthesis of terpenes is suppressed by negative feed-back mechanisms. Alternatively, it is conceivable that the terpenes are synthesized continuously but are removed from the cytoplasma by one of the following mechanisms: by the formation of water-soluble derivatives (terpene-glycosides) and accumulation of these derivatives in the vacuoles; by excretion of the lipophilic terpenes into the surrounding aqueous medium and the gas phase. The first hypothesis was tested by treating a water-extract of a suspension culture (BK2/sub) by acid hydrolysis or by digestion with β-glucosidase, respectively. These treatments did not indicate the presence of water-soluble sesquiterpene derivatives in the cultured cells. The second hypothesis was more difficult to investigate. The volatile terpenes, when released into the medium, could have been subject to degradative processes (e.g., oxidative reactions) or have disappeared from the medium by evaporation. In both cases the addition to the culture medium of an accumulation reservoir may favor the accumulation and conservation of lipophilic compounds in the cell environment and make them accessible to detection and analysis.

These considerations led to the development of a two-phase culture system for the accumulation of lipophilic compounds from camomile plant suspension cultures (Beiderbeck 1982; Bisson et al. 1983). Such a system consists of the cells and the aqueous nutrient medium as one phase and a water-insoluble liquid or solid compound as a second phase. This second phase has to match a variety of conditions to be suitable for suspension cultures (Bisson et al. 1983; Becker et al. 1984; Beiderbeck and Knoop 1987). As second phases Miglyol 812 (Dynamit Nobel) and RP8 (LiChroprep, Merck) proved suitable. Miglyol 812 is a water-insoluble triglyceride composed of fatty acids with 8–10 carbon atoms, and RP8 is a modified silica gel with C-8 side chains.

2.3.1 Growth of Suspension Cultures in the Presence of Lipophilic Phases

The presence of 2.5 ml Miglyol 812 or 0.5 g RP8 per 25–50 ml culture medium did not impair the growth characteristics of normal or tumorous camomile cultures, e.g., fresh weight increase, cell form, cell color (Fig. 7) (Bisson 1983).

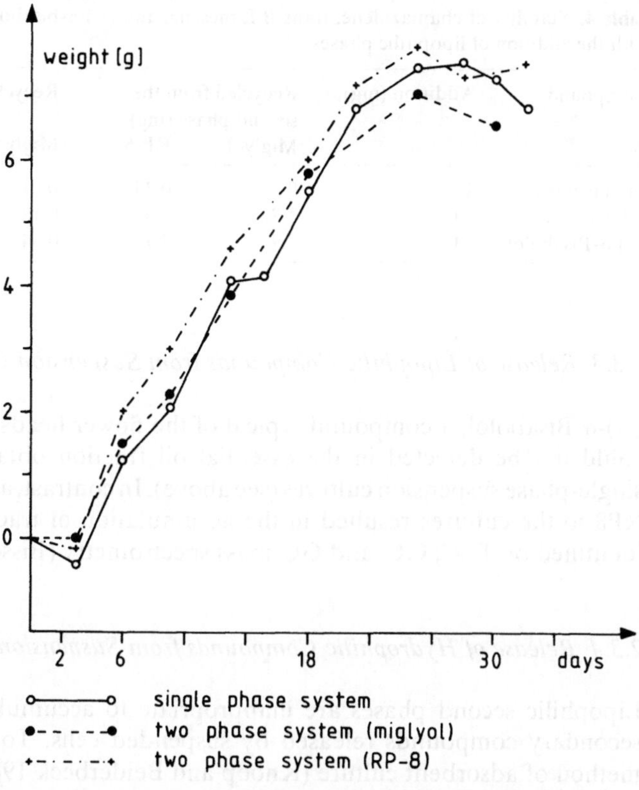

Fig. 7. Growth curve of suspension culture: BK2/sub (fresh weight increase); comparison of single phase and two phase systems

2.3.2 Ability of Lipophilic Phases to Accumulate Lipophilic Compounds from the Liquid Medium

To test the accumulating and preserving capabilities of the second phases, mixtures of 50 ml aqueous medium with 2.5 ml Miglyol or 0.5 g RP8, respectively, were supplied with 1 mg chamazulene, 2 mg trans-β-farnesene, or 5 mg (–)-α-bisabolol and incubated on a shaker for 25 days under the usual culture conditions. The recovery of the secondary compounds and the distribution between the aqueous and the lipophilic phases were subsequently determined (see Table 4). The secondary compounds used were almost completely recovered in the lipophilic second phases, and were accumulated and preserved during the incubation period of 25 days. In such experiments, Miglyol proved a more efficient accumulation phase than RP8. The lipophilic compounds could, however, be recovered from RP8 by simple elution with ethanol or other solvents, whereas steam distillation with its inconveniences was necessary to recover the secondary compounds from Miglyol 812 (Bisson 1983; Becker et al. 1984).

Table 4. Stability of chamazulene, trans-β-farnesene, and (−)-α-bisabolol in liquid nutrient medium with the addition of lipophilic phases

Compound	Addition (mg)	Recycled from the second phase (mg)		Recycled from the medium (mg)	
		Miglyol	RP-8	Miglyol system	RP-8 system
Chamazulene	1.0	0.6	0.34	0.01	0
β-Farnesene	2.0	1.72	1.4	0.18	0
(−)-α-Bisabolol	5.0	3.9	3.1	0.34	0

2.3.3 Release of Lipophilic Compounds from Suspension Cultures

(−)-α-Bisabolol, a compound typical of the flower heads of *Chamomilla recutita*, could not be detected in the essential oil fraction obtained from unorganized single-phase suspension cultures (see above). In contrast, addition of Miglyol 812 or RP8 to the cultures resulted in the accumulation of traces of (−)-α-bisabolol as identified by TLC, GC, and GC-mass spectrometry (Bisson 1983).

2.3.4 Release of Hydrophilic Compounds from Suspension Cultures

Lipophilic second phases are inappropriate to accumulate or bind hydrophilic secondary compounds released by suspended cells. To serve this purpose, the method of adsorbent culture (Knoop and Beiderbeck 1983) was developed using activated charcoal as a second phase. Twenty-five ml camomile suspension culture were supplied with 0.2 g activated charcoal. After 7 days of culture, the charcoal was separated from the cells and the culture medium, washed and extracted by means of ethanol/1M KOH (4 + 1). A great variety of secondary compounds could be eluted from the charcoal, one of which was identified as coniferyl aldehyde (see Fig. 4) by means of TLC, and 300-MHz H-NMR.

2.4 Crown Gall Tumors and Tumor Suspension Cultures

Since it had been demonstrated that unorganized suspension cultures of camomile are hardly able to produce sesquiterpenes in vitro characteristic of the intact plant, experiments were undertaken to obtain variant cell cultures with modified metabolic characters (regulation variants). A possible source of such variants is made available by the genetic transformation of plants by infection with virulent strains of the tumor-inducing soil bacterium *Agrobacterium tumefaciens*. During the infection process, a portion of the bacterial virulence plasmid (Ti plasmid), namely the T-DNA, is transferred into a host cell and integrated into the host's nuclear genome leading to the formation of a tumor cell with a permanently changed metabolism. The metabolic changes best studied are the establishment of the opine production, of a new pathway of auxin synthesis (Weiler and Schroeder 1987; Reichling and Beiderbeck 1988) and of cytokinin synthesis. The tumor cells can be isolated, and in contrast to normal, nontransformed cells cultured in vitro using hormone-free media.

It is conceivable that the transformation of a plant cell into a tumor cell, with all its accompanying alterations in cell metabolism, might also influence the production of secondary compounds either directly by special gene products coded for on the T-DNA or indirectly as a consequence of changed phytohormone levels of the transformed cells (Reichling and Beiderbeck 1988). Young shoots of camomile plants were thus infected with *Agrobacterium* strains (Reichling et al. 1979b). The resulting tumors were unorganized and consisted of meristematic and parenchymatous cells and tracheary elements. Excretory reservoirs have never been found. The essential oil fraction of 4-week-old tumors was compared to that of normal flower heads, shoots, and roots of corresponding age. The composition of the tumor oil was to a large extent identical with that of the shoots: trans-β-farnesene, trans-α-farnesene and the cis/trans-en-in-dicycloethers were identified. In addition, the essential oil of the tumor tissues contained some polyines lacking in the essential oil of the shoots, e.g., the isomeric chamomillaesters.

Tumor formation thus influences essential oil metabolism by mechanisms not clearly explained. However, the complete biochemical properties of the camomile inflorescence are not attained by the tumor cells: matricine, the bisaboloids, and the flavonoids characteristic of the flower heads could not be detected in tumor cells. The phenylcarbonic acids of the camomile plant are again present in tumor cells (see Fig. 4). Crown gall suspension cultures of camomile behaved similarly to nontransformed suspension cultures as far as the production of essential oils and flavonoids are concerned. Any influence of tumor transformation on the secondary compound production seemed to be leveled out by the conditions of suspension culture (Bisson et al. 1983; Becker et al. 1984).

2.5 Selection of Cell Lines with Modified Compound Production

Any cell suspension culture consists of a cell population which is more or less heterogenic with respect to chromosome number, cell size, antibiotic resistance, and certainly secondary compound synthesis (Reichling 1985).

A cell culture with a low total level of secondary compound production generally consists of a high portion of cells without any measurable production at all, accompanied by a low portion of variant cells with a higher production capacity (Ellis 1982). This can be as high as or even higher than that of excretory cells inside the intact plant. If the secondary compounds considered can be detected either by their color or their fluorescence, the selection of cell strains with different production is a comparably simple task. Selection of variants producing unusual levels of compounds not detectable by visual inspection is a rather difficult undertaking. Methods based on immunological techniques (Zenk et al. 1977) have been applied successfully. But these methods are complicated, expensive, and not applicable to any problem and in any laboratory. A relatively simple method for the selection of cell variants with a secondary compound production superior to the production of the average cell has been devised using camomile and tobacco cell cultures: the adsorbent filter technique (Knoop and Beiderbeck 1985). Using this technique, cell aggregates of camomile tumor cells could be recovered from mixtures with tobacco cells on the basis of their secondary compound production. Additionally, the suitability of the method was demonstrated by the selection of stable tobacco cell lines with variant compound production (Thumann et al. 1987).

2.6 Organogenesis and Micropropagation of Camomile

Cellarova et al. (1982, 1984) studied the organogenesis of callus cultures of *Chamomilla recutita*. From top parts of young receptacles explanted to LS medium (Linsmaier and Skoog 1965) supplemented with 1 mg/l 2,4-D slowly grown compact callus cultures were obtained. Twelve-months-old calli were transferred to the same basal medium with 0.1 mg/l kinetin and without 2,4-D where they started to regenerate shoots. Transfer to the hormone-free LS medium resulted in root formation. From regenerant plants thus obtained clonal propagation was started under sterile conditions: Cuttings were prepared consisting of stem segments with leaves and cultivated on hormone-free LS medium. After a transfer to sand with hydroponic solution, the plants obtained could be grown in soil without difficulties. Since generally callus cultures show a high degree of cytological variability, and consequently the plants regenerated from callus cultures via organogenesis have different chromosome numbers, Cellarova et al. (1984) studied the cytological status of callus cultures, regenerant plants and individual members of micropropagated clones of *Chamomilla recutita*. In callus cultures they could demonstrate considerable chromosomal instability: besides a high proportion of diploid cells ($2n = 18$), 33–48% of the cells showed deviations from the diploid chromosome number; haploid, tetraploid, and aneuploid cells were present. Root tips of regenerant plants reflected the chromosome instability present in the calli. The occurrence of diploid metaphases was approximately the same as that in callus cultures. If such regenerants were cloned in vitro the percentage of diploid metaphases among the clone members significantly increased. This indicates the possibility of elimination of aneuploid cells and cells with different ploidy levels during the cloning procedure.

The experiments of Cellarova et al. constitute the basis for future selection and micropropagation of camomile plants with desired phenotypes. A high content of essential oil or compounds of this oil [e.g., (−)-α-bisabolol and chamazulene], resistance to pests and diseases, and increased tolerance of environmental stress are among such characters. Experiments with respect to salt tolerance have been performed. On LS medium supplemented with different concentrations of mineral salts nonselected callus cultures tolerated high concentrations of NaCl (85 mM), KNO_3 (2.4 mM), $CaCl_2$ (9.5 mM), and $MgSO_4$ (2.2 mM) (Cellarova et al. 1986). Plants regenerated from such salt-adapted calli and from control calli did not differ with respect to salt tolerances, phenotype and growth. So, without further selection an extension of the area where camomile can be cultivated into areas of salt-enriched soils seems possible.

2.7 Nutrient Media for Callus Induction, Callus Growth, and Micropropagation

Callus induction

LS + 1 mg/l 2,4-D	(Cellarova et al. 1982)
MS + 0.5 mg/l NAA	(Reichling and Becker 1976)

Callus growth

MS + 0.5 mg/l NAA	(Reichling and Becker 1976)
MS + 5 mg/l NAA + 2.5 mg/l kinetin	(Reichling et al. 1984)
MS + 1 mg/l 2,4-D + 1 mg/l kinetin	(Szöke et al. 1979)
LS + 0.5 mg/l NAA + 0.5 mg/l kinetin	(Cellarova et al. 1982)

Micropropagation

LS + 0.1 mg/l kinetin	(Cellarova et al. 1984)

In undifferentiated callus or suspension cultures of camomile, a specific correlation between a certain phytohormone combination and/or concentration and the accumulation of secondary compounds could never be demonstrated. If specific oil cavities were differentiated, secondary compounds like terpenes could be detected in the in vitro cultures (Reichling et al. 1984).

3 Conclusions and Prospects

The few papers concerned with the secondary compound production by cell cultures of *Chamomilla recutita* have yielded very interesting results with respect to the formation and accumulation of volatile compounds and with respect to techniques for manipulating such formation. Up to the present it has, however, proved impossible to establish a callus or suspension culture which continuously produces or releases high yields of terpenes of therapeutical importance, e.g., matricine and (−)-α-bisabolol. In plants the accumulation of these compounds is restricted to the flower heads and consequently underlies tissue-specific regulations, the mechanisms of which are barely understood. At the same time many experiments have shown that cell cultures of camomile do not synthesize flavonoids (Szöke et al. 1978; Reichling et al. 1979b, 1984), the other class of compounds of therapeutical importance. At the present, no indication is thus available which points to a forthcoming replacement of agricultural production of camomile drug by in vitro culture (Becker and Reichling 1985). For the foreseeable future, hydroalcoholic extracts of high quality camomile varieties produced by agriculture seem to be the only basis of therapeutically valuable remedies.

References

Achterrath-Tuckermann U, Kunde R, Flaskamp E, Isaac O, Thiemer K (1980) Pharmacological investigations with compounds of camomile. V. Investigations on the spasmolytic effect of compounds of camomile and kamillosan on the isolated Guinea pig ileum. Planta Med 39:38–50

Becker H, Reichling J (1985) Can cell cultures of camomile replace field grown plants? In: Fac Farm Univ Camomile in industrial and pharmaceutical use. Eigenverlag, Triest, pp 90–100

Becker H, Reichling J, Bisson W, Herold S (1984) Two phase culture – a new method to yield lipophilic secondary products from plant suspension cultures. In: Eur Fed Biotechnol (ed) 3rd Eur Congr Biotechnology, vol 1. Chemie, Weinheim, pp 209–213

Beiderbeck R (1982) Zweiphasenkultur – ein Weg zur Isolierung lipophiler Substanzen aus pflanzlichen Suspensionskulturen. Z Pflanzenphysiol 108:27–30

Beiderbeck R, Knoop B (1987) Two-phase culture. In: Constabel F, Vasil IK (eds) Cell culture and somatic cell genetics of plants, vol 4. Academic Press, New York London, pp 255–266

Bisson W (1983) Untersuchungen zur Bildung, Zusammensetzung und Akkumulation von ätherischem Öl in Zellkulturen von *Matricaria chamomilla L.* unter besonderer Berücksichtigung systemfremder Akkumulationsorte (Zweiphasensystem). Diss, Univ Heidelberg

Bisson W, Beiderbeck R, Reichling J (1983) Produktion ätherischer Öle durch Zellsuspensionen der Kamille in einem Zweiphasensystem. Planta Med 47:164–168

Carle R, Isaac O (1985) Fortschritte in der Kamillenforschung in den Jahren 1974 bis 1984. Dtsch Apotheker Z 125 Suppl 1:2–8

Carle R, Fleischhauer J, Fehr D (1987) Qualitätsbeurteilung von Kamillenölen. Dtsch Apotheker Z 127:2451–2457

Cellarova E, Grelakova K, Repcak M, Honcariv R (1982) Morphogenesis in callus tissue cultures of some *Matricaria* and *Achillea* species. Biol Plant 24:430–433

Cellarova E, Rychlova M, Honcariv R (1984) Cytological instability in *Matricaria chamomilla L.* tissue cultures. Herba Hung 23:37–51

Cellarova E, Repcakova K, Honcariv R (1986) Salt tolerance of *Chamomilla recutita* (L.) Rauschert tissue cultures. Biol Plant 28:275–279

Della Loggia R (1985) Lokale antiphlogistische Wirkung der Kamillenflavone. Dtsch Apotheker Z 125 Suppl 1:9–118

Dölle B, Carle R, Müller W (1985) Flavonoidbestimmung in Kamillenextraktpräparaten. Dtsch Apotheker Z 125 Suppl 1:14–19

Ellis BE (1982) Chemical production patterns at the single cell level. IAPTC Newslett 38:13–16

Franz Ch, Kirsch C (1985) Neuere Ergebnisse der Kamillenzüchtung. Dtsch Apotheker Z 125 Suppl 1:20–23

Hausen BM, Busker E, Carle R (1984) Über das Sensibilisierungsvermögen von Compositenarten. Planta Med 50:229–234

Jakovlev V, Isaac O, Thiemer K, Kunde R (1979) Pharmacological investigations with compounds of camomile. II. New investigations on the antiphlogistic effect of (–)-α-bisabolol and bisabololoxides. Planta Med 35:125–140

Knoop B, Beiderbeck R (1983) Adsorbenskultur – ein Weg zur Steigerung der Sekundärstoffproduktion in pflanzlichen Suspensionskulturen. Z Naturforsch 38c:488–486

Knoop B, Beiderbeck R (1985) Adsorbent filter – a tool for the selection of plant suspension culture cells producing secondary substances. Z Naturforsch 40c:297–300

Linsmaier E, Skoog F (1965) Organic growth factor requirements of tobacco tissue cultures. Physiol Plant 18:100–127

Motl O, Repcak M, Sedmera P (1978) Weitere Bestandteile des Kamillenöls, 2. Mitt. Arch Pharm 311:75–76

Motl O, Repcak M, Budesinsky, Ubik K (1983) Weitere Bestandteile des Kamillenöls, 3. Mitt. Arch Pharm 316:908–912

Murashige T, Skoog F (1962) A revised medium for rapid growth and bioassays with tobacco tissue culture. Physiol Plant 15:473–497

Reichling J (1985) Pflanzenzellkulturen in Forschung und Praxis, pt 2. GIT Fachz Lab 9:854–867
Reichling J, Beiderbeck R (1988) Pflanzenzellkulturen in Forschung und Praxis, pt 3. Sekundärstoffakkumulation in transformierten in vitro Kulturen. GIT Fachz Lab 5:466–473
Reichling J, Becker H (1976) Calluskulturen von *Matricaria chamomilla*. Planta Med 30:258–268
Reichling J, Becker H, Exner J, Dräger PD (1979a) Vergleichende Untersuchungen verschiedener Handelsmuster von *Matricariae flos*. Pharm Z 124:1998–2005
Reichling J, Beiderbeck R, Becker H (1979b) Vergleichende Untersuchungen über sekundäre Inhaltsstoffe bei Pflanzentumoren, Blüte, Kraut und Wurzeln von *Matricaria chamomilla* L. Planta Med 36:322–332
Reichling J, Bisson W, Becker H, Schilling G (1983) Zusammensetzung und Akkumulation des ätherischen Öls in *Matricariae Radix*, 2. Mitt. Z Naturforsch 38c:159–161
Reichling J, Bisson W, Becker H (1984) Vergleichende Untersuchungen zur Bildung und Akkumulation von ätherischem Öl in der intakten Pflanze und in Calluskulturen von *Matricaria chamomilla*. Planta Med 51:334–337
Schilcher H (1987) Die Kamille. Wiss Verlagsges, Stuttgart
Schreiber A, Carle R, Reinhard E (1990) On the accumulation of apigenin in chamomile flowers. Planta Med 56:179–181
Szelenyi J, Isaac O, Thiemer K (1979) Pharmacological experiments with compounds of camomile. III. Experimental studies of the ulceroprotective effect of camomile. Planta Med 35:218–227
Szöke E, Kuzovkina JN, Verzar-Petri G, Smirnov AM (1977) Cultivation of wild camomile tissues. Fiziol Rast 24:832–840
Szöke E, Kuzovkina JN, Verzar-Petri G, Shavarda AL (1978) The influence of coconut-milk on the synthesis of essential oils in plant tissue cultures. In: Proc 18th Hung Annu Meet. Biochemistry, Budapest, pp 189–190
Szöke E, Shavarda AL, Verzar-Petri G, Kuzovkina JN (1979) Effects of growth regulators and light on essential oil synthesis in camomile tissues cultures. Herba Hung 18:7–19
Thumann J, Knoop B, Beiderbeck R (1987) The separation of 2 cell strains with different secondary substance production by means of an improved adsorbent filter technique. Biol Plant 29:422–424
Weiler EW, Schroeder J (1987) Hormone genes and crown-gall disease. Trends Biochem Sci 12:271–275
Zenk MH, El Shagi H, Arens H, Stöckigt J, Weiler EW, Deus B (1977) Formation of the indole alkaloids serpentine and ajmalicine in cell suspension cultures of *Catharanthus roseus*. In: Barz W, Reinhard E, Zenk MH (eds) Plant tissue culture and its biotechnological application. Springer, Berlin Heidelberg New York, pp 27–43

XI *Carum carvi* L. (Caraway): In Vitro Culture, Embryogenesis, and the Production of Aromatic Compounds

M. Furmanowa, D. Sowińska, and A. Pietrosiuk[1]

1 General Account

The genus *Carum* Rup. ex Linn. Syst. ed I. (1735) from the family Umbelliferae (= Apiaceae) comprises 195 species. In Europe five species grow: *Carum carvi* L., *C. heldreichii* Boiss., *C. multiflorum* (Sibth. et Sm.) Boiss., *C. rigidulum* (Viv.) Koch. ex DC., and *C. verticillatum* (L.) Koch. From the economical point of view, the most important is caraway, *Carum carvi* L. Sp. Pl., 263 (1753), known also as *Carum aromaticum* Salisb., *C. decussatum* Gilib., *C. officinale* S.F. Gray, and *C. careum* Bub. (Index Kewensis 1895–1974).

Caraway is biennial herb which grows up to 150 cm, indigenous to Europe and Asia, and widely cultivated in many countries for its aromatic fruits. Leaves are two- to three-pinnate, lobes 3 to 25 mm linear-lanceloate or linear. The lowest leaf segments are at least twice as long as wide. Petals whitish or pink. Fruits 3 to 6 mm, ovoid, 3 to 3.5 times as long as wide, with low rounded ridges, smelling strongly after crushing, 2n = 20 (Tutin 1968).

1.1 Medicinal Components

Carum carvi L., caraway, is known as a spice and a medicinal herb. Fruits of this plant are used as flavoring spice in various foods. Dried fruits are used for preparing a stimulating tea, they are also a mild stomachic.

Caraway fruits contain essential oil (localized in vittae) from 3 to 7%, 20% of fixed oil (i.e., nonvolatile oil) and proteins. The main component of the essential oil is the ketone carvone from 50 to 60%, the terpene d-limonene from 40 to 50%, and small amounts of other constituents, including carveol and dihydrocarvone (Fig. 1). Salveson and Baerheim Svendsen (1976), using gas liquid chromatography, presented the detailed composition of caraway oil. Embong et al. (1977) described the 12 main constituents of the essential oil obtained from plants growing in Canada. Dijkstra and Speckmann (1980) studied autotetraploids of caraway. In these plants the yield of essential oil in fruits was 35.6% higher when compared to diploid plants.

Caraway oil possesses carminative properties, is often used to correct nausea and griping (Tyler et al. 1977; Lewis 1977). Lewis (1977), discussing the problem of allergy, mentioned carvone as a sensitizing substance and for this reason classified

[1] Department of Biology and Pharmaceutical Botany, Medical Academy, 02–097 Warsaw, Banacha 1, Poland

Biotechnology in Agriculture and Forestry, Vol. 15
Medicinal and Aromatic Plants III (ed. by Y.P.S. Bajaj)

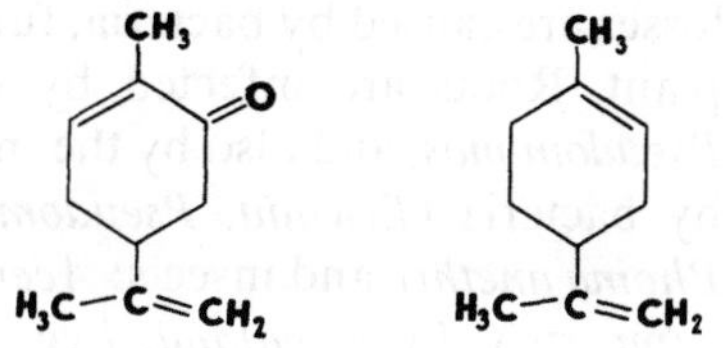

Fig. 1. Structure of carvone and limonene

Carum carvi among the plants causing contact dermatitis. Because of this allergic potential essential oil cannot be used as a flavor for toothpaste.

Weiss (1982) classified the fruits of *Carum carvi*, together with fruits of *Foeniculum vulgare* and *Anethum graveolens*, as the most important plant raw stuffs with carminative properties. Lutomski et al. (1987) mentioned Fructus Carvi as one of the components of four prescriptions for making plant mixtures. Three of these compositions are used as Remedia Carminativa, one as Remedium Digestivum. Oleum Carvi is also a component of a drug used as Remedium Corminativum.

1.2 Conventional Propagation

Caraway has great nutritional requirements and therefore the cultivation of this plant is not easy. There are many factors, which influence the dynamics of growth and the crop of caraway. Formanowicz (1969) observed that a significant dependence of the germination capability on moisture content is found in caraway seeds. Bocheńska and Kozłowski (1969) studied the fluctuations in essential oil, water content, and germination capability of *C. carvi* L. fruits, during ripening and storing. These authors found that seeds of high germination capability, above 80%, and of high essential oil content, within 5 to 7%, can be obtained only when the weather during the ripening season is warm (daily temperature about 18°C), with not more than 400 mm of rainfall. Okoniewska (1977) tested various mutagenic agents for application in caraway. As a result of this work, two mutants were selected, scrub line 211, after irradiation by fast neutrons, and line 298, with large fruits obtained by chemical mutagen treatment. Kordana et al. (1983) found that the increase of mineral fertilization level accelerates the rate of the plant mass production, enhances the mass of fruits but does not affect the content of essential oil. Caraway has a high demand for nitrogen and potassium but lower for calcium and phosphorus. Węglarz (1983a,b) studied the effect of fertilization, soil moisture, time of sowing, and amount of seeds per ha on the development and cropping of caraway. It was found that these factors significantly influenced the number of leaves, root mass, the number of flowering plants, and the crop itself. A favorable growth, development, and crop of caraway were observed in the case of an early sowing in the spring, a low amount of sowing seeds (5 kg/ha), and a high level of mineral fertilization.

Apart from agricultural factors which affect cropping of *Carum carvi*, the second problem which also must be taken into account is that caraway is often attacked by a large number of diseases, that cause a decrease in fruit yield. Heavy losses are caused by bacteria, fungi, and insects, which can damage all parts of the plant. Roots are infected by various species of bacteria: *Fusarium, Erwinia, Pseudomonas*, and also by the insect, *Psila rosae*. Diseases of the stems are caused by bacteria (*Erwinia, Pseudomonas, Xantomonas*), fungi (*Centrospora acerina, Phoma anethi*), and insects: *Aceria carvi*, which deforms leaves and umbels; *Lygus campestris, Lygus calami, Lygus lucorum attaceka*, the young stems, leaves, and umbels; species of *Cavariella* and *Aphis subdue*, the higher part of stems; *Lasioptera carophila* damages the umbels. *Depressaria nervosa* feeds on flowers and fruits of caraway (Czabajska 1984).

So far, there are only few reports on vegetative propagation of caraway in vitro. In this chapter progress on micropropagation of this plant, using various methods of regeneration of plant from explants in vitro, is reviewed.

2 In Vitro Approaches

2.1 Review of Tissue Culture (Table 1)

Bajaj et al. (1988) reviewed the biotechnology of the micropropagation of medicinal and aromatic plants, and the species described were propagated from existing meristem or from nonmeristematic tissues. Among these plants *Carum carvi* is mentioned.

The first observations were made by Ammirato (1973, 1974, 1977, 1978), who studied the dependence of somatic embryo development on medium composition (growth regulators). Ammirato (1973, 1974) drew attention to the abnormal structure of embryoids formed on White's medium with sucrose (2% wt/vol), NAA (2 mg/l) and casein hydrolysate (200 mg/l). ABA added to the culture at noninhibitory levels permitted somatic embryo maturation, but inhibited abnormal proliferations, including initiation of accessory and adventitious embryos or callus. The normal development of caraway somatic embryo can be maintained in the medium by altering the relative levels of ZEA, GA_3, and ABA (Ammirato 1977). Later (Ammirato 1978) it was shown that purine analog 8-azaguanine (AZG), an inhibitor of cytokinin activity, also increased the frequency of normal embryos from cultured caraway cells. In the concentration 10^{-6} to 10^{-7}M, AZG allowed embryo growth from the globular to the heart-shaped stage and the maturation of cotyledons, but inhibited the inception of accessory embryos, malformed or callused cotyledons, and precocious germination. Schäfer et al. (1986) induced somatic embryos in caraway callus culture, using petioles of young leaves as explants.

Ammirato (1983) described the various aspects of somatic embryogenesis and presented a list of species in which this natural phenomenon has been observed. The recent problems connected with embryogenesis and production of synthetic seeds (artificial seeds) were discussed by Sluis and Walker (1985).

Fujii et al. (1987) discussed the role of artificial seeds for plant propagation. These authors mentioned some crops that may benefit from an artificial seed delivery system, that is the plants in which true seed costs are high or seeds are not used for propagation due to genetic instability or are attacked by various diseases.

Table 1. Differentiation and regeneration of plants of *Carum carvi* in tissue culture

Explant	Growth regulators used in the first passage and in the next ones	Response	Reference
Petiole	White + NAA	Abnormal types of somatic embryos	Ammirato (1973)
Petiole	White + NAA → MS auxin-free → MS + ABA	Normal types of somatic embryos	Ammirato (1974)
Petiole	MS + ZEA, GA_3, ABA	Normal somatic embryos	Ammirato (1977)
Petiole	MS + AZG	Normal somatic embryos	Ammirato (1978)
Shoot tip of seedling	MSN + IBA + AS	Plantlets	Sowińska (1982)
Shoot tip of seedling	MSN + IBA + BA → MSN + IAA → MSN + IBA + AS	Plantlets	Sowińska (1982)
Hypocotyl	MS, NN, White → MSN + NAA + SA + BA → MSN + IAA	Plantlets	Żelazek (1982)
Hypocotyl	MSN + NOA ↘ MSN + 2,4-D ↗ MSN auxin-free → MSN + IBA + AS	Plantlets	Furmanowa et al. (1984a)
Hypocotyl	MSN + 2,4-D → MSN + IAA → MSN + IAA	Plantlets	Furmanowa et al. (1984b)
Hypocotyl	MSN + 2,4-D → MSN auxin-free → MSN + IBA + AS	Plantlets	Furmanowa et al. (1986)
Petiole, seedling		Somatic embryos	Schäfer et al. (1986)
Hypocotyl	MSN + NOA → MSN auxin-free → 1/2 MSN auxin-free	Plantlets	Furmanowa et al. (1989)

Redenbaugh et al. (1986b) mentioned two species from the umbellifers: celery, and carrot, whose embryoids have been encapsulated in alginate beads.

The research work which has been carried on in the past years indicated that somatic embryogenesis can be used as the fastest method for micropropagation of caraway and can be applied in practice. However, this technique requires many manipulations of hormonal balance in the medium. Our work (Furmanowa et al. 1984a,b, 1986, 1989) is summarized here.

2.2 Aromatic Compounds in In Vitro Cultures

The production of essential oil and its components in tissue culture is still not satisfactory. Plant tissue culture in many cases lost the ability to synthesize the components of essential oils typical for intact plants. This may be caused by lowering the level of cell differentiation necessary for the formation of secretory structures. Kuzovkina et al. (1975), discussing the problem of essential oils in the culture of isolated plant tissues, paid attention to the close connection between the biosynthetic property of the tissues cultured and their morphology. For the study of the biogenesis of essential oils in tissue culture, it is very important to know the physiological dependences and primary and secondary metabolism in intact plant and in tissue culture. Staba (1980) mentioned only four species in which the production of essential oils in tissue culture has been studied, and none of them belonged to the Umbelliferae (= Apiaceae) family. Among the many plants which contain essential oil, most studies are on members of the Labiatae, in which the biosynthesis of the components of essential oil in tissue culture was observed. Fewer studies concern the plants of the umbellifers. Sardesai and Tipnis (1969) found geraniol in tissue culture of *Coriandrum sativum*. Becker (1970) studied the formation of volatile substances in plant tissue cultures of: *Pimpinella anisum*, *Foeniculum vulgare* subsp. *capillaceum* var. *azoricum*, and *Levisticum officinale*. The steam distillates of *Pimpinella anisum* roots and leaves differentiated in tissue culture showed a substance which was already known from the entire plant, but not in the same quantitative proportion. Becker (1970) also discussed the dependence of the accumulation of volatile compounds on the degree of differentiation. Collin and Watts (1986) found that suspension culture of celery (*Apium graveolens*) produced phthalides (flavor compounds) and limonene (terpenoid – a component of essential oil) during the exponential phase. The cultures lost the ability to synthesize the flavor compounds after a prolonged period of subculture.

Mulder-Krieger et al. (1988) presented a review on the production of essential oils and flavors in plant cell and tissue cultures. In this publication six species from the Umbelliferae (Apiaceae) family were mentioned. The most interesting was the production of the phthalides compounds (components of essential oil) by suspension with embryoids of *Apium graveolens*. In suspension culture of this species phthalides and monoterpene compounds were also found. Monoterpene compounds were produced by callus of *Coriandrum sativum*, *Daucus carota*, *Foeniculum vulgare*, *Levisticum officinale*, and *Pimpinella anisum*.

2.3 Establishment of Tissue Culture

2.3.1 Regeneration of Plants by Embryogenesis

To induce callus formation of caraway (*Carum carvi* L.), Furmanowa et al. (1984a,b, 1986, 1989) used fruits received from the Institute of Medicinal Plants in Poznań (Poland). The fruits were surface sterilized by a 15-min treatment with 10% sodium hypochlorite solution, and then washed three times with sterile water. Fruits germinated at 26°C in the dark. After 10 days of germination, 0.5- to 1-cm-long hypocotyl segments were excised and placed on media which contained Murashige and Skoog (1962) salts with Nitsch and Nitsch (1969) vitamins (MSN) supplemented with 3% sucrose and auxin at various concentrations: 0.05, 0.1, 0.2, 0.3, 0.4, 0.5, 0.8, 1 mg/l 2,4-D or 0.01, 0.1, 1, 10 mg/l NOA. The pH was adjusted to 5.6 before adding the agar (0.5% Phytagar) and autoclaving at 120°C for 15 min. Callus was cultured at 26°C in the dark. After 2 to 6 weeks of growth, callus was transferred to MSN liquid medium supplemented separately with 2,4-D or NOA or without auxin. The suspension culture was kept on a rotary shaker (110 rpm) at 26°C under photoperiods of 12 h/day with light intensity of 2000 lx. One part of the callus was transferred for the same period onto the MSN solid media containing IAA at four concentrations: 0.05, 0.1, 0.2, and 0.4 mg/l. The cluster of embryoids formed in suspension culture and on solid medium were fixed in formalin, passed through an ethanol series and embedded in paraffin. Sections (10 mm thick) were stained with safranin and fast green, and studied under the light microscope.

Somatic embryos maturated in the 2nd passage were inoculated on solid MSN medium in three modification: (1) with IBA 0.5 mg/l and AS 10 mg/l (2) with 0.2 mg/l IAA or (3) on solid half-strength MSN medium without growth regulators. Whole plantlets which developed on these media were first transferred to flower pots and later to the field.

The results of the various treatments were recorded and the number of plants obtained from one explant was counted. Each treatment, consisting of ten replicates, was repeated twice.

From among four media tested for caraway callus culture the MSN medium was chosen as the best for callus initiation and growth (Żelazek 1982). The results show that the presence of 2,4-D or NOA are essential to callus induction and proembryoid initiation. Depending on NOA concentration in solid MSN primary medium, different results in the secondary suspension culture were obtained (Furmanowa et al. 1989). The fastest growth of callus was observed on solid MSN medium supplemented with 1 mg/l NOA. Yellow, compact callus was obtained after only 2 weeks of growth. On MSN medium with 10 mg/l NOA callus was achieved later, after 4 weeks of culture. The concentration of 2,4-D with the most stimulating effects on callus formation was 0.3 mg/l. Somatic embryos were developed in suspension culture in MSN medium when 2,4-D or NOA had been removed (Fig. 2). The presence of these auxins in the medium of the second passage prevented maturation of caraway embryoids, and only little growth of callus was observed. Similarly, embryogenesis of *Carum carvi* after removal of auxin (NAA) was obtained by Ammirato (1973, 1974, 1977). The majority of species of the

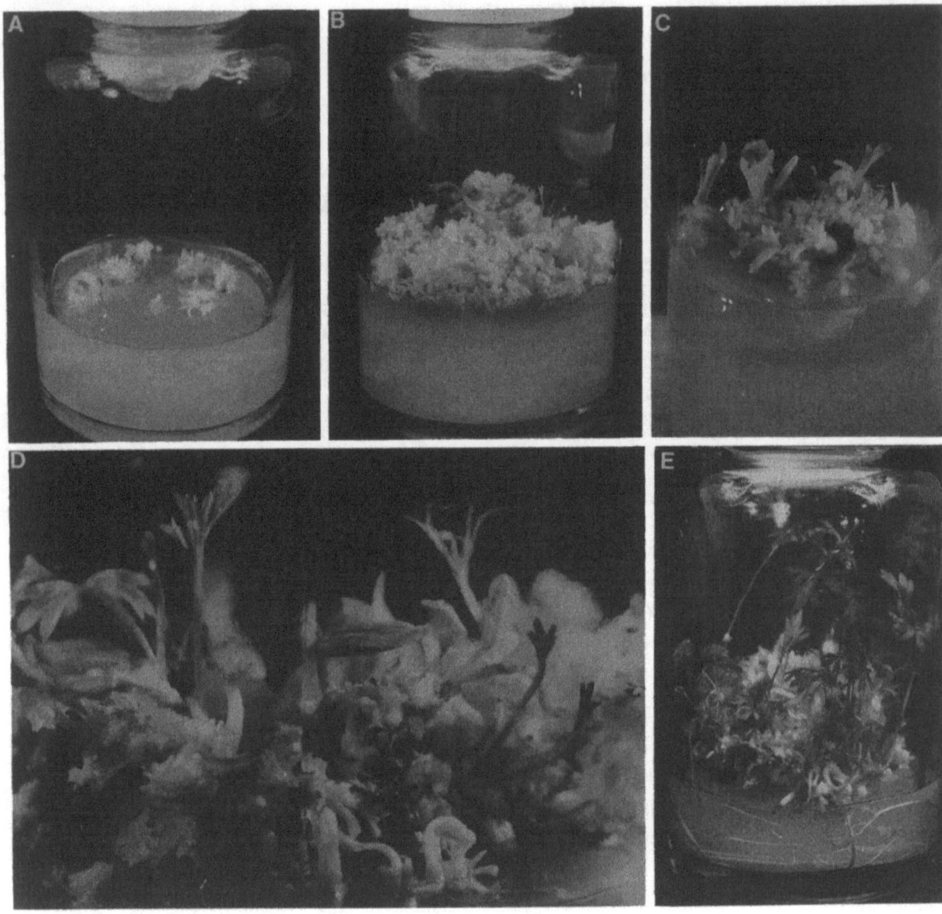

Fig. 2A-E. Different stages of caraway plantlets regenerated from somatic embryos in the 3rd passage on MSN solid medium with 0.5 mg/l IBA and 10 mg/l AS: after **A** 1 week, **B** 2 weeks, **C** 3 weeks, **D** 4 weeks, **E** 6 weeks. In the primary culture, callus was induced on MSN solid medium with 0.3 mg/l 2,4-D and subcultured into MSN liquid medium free of auxin

Umbelliferae have produced somatic embryos when the cells are removed to a medium free of growth regulators (Ammirato 1983).

In our study it was noticed that NOA at 1 and 10 mg/l greatly induced the formation of caraway somatic embryos, which developed in auxin-free MSN medium. NOA at 0.1 mg/l stimulated somatic embryo formation only slightly, but root formation strongly. NOA at 0.01 mg/l had no effect on somatic embryogenesis but stimulated the formation of roots (Fig. 3). The results are in agreement with those of Kamada and Harada (1979), who observed the effects of various growth regulators on somatic embryogenesis and root formation in carrot tissue culture. They showed that NOA stimulated root formation at lower concentrations (0.001, 0.01, 0.1 mg/l) and embryo formation at higher concentrations (0.1 to 10 mg/l).

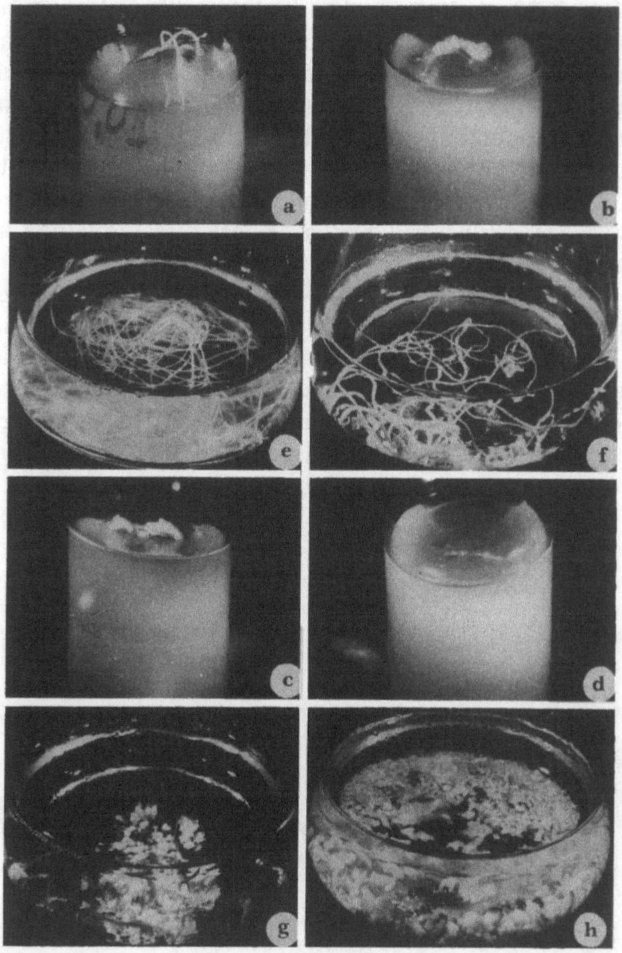

Fig. 3a-h. The effect of various concentration of NOA on caraway somatic embryos development. The tubes show MSN primary culture containing NOA: **a** 0.01 mg/l, **b** 0.1 mg/l, **c** 1 mg/l, **d** 10 mg/l. Erlenmeyer flasks shows secondary suspension culture in auxin-free MSN medium. **e** Roots developed from primary explant which was plated on medium with 0.01 mg/l NOA, no somatic embryos were observed. **f** Roots and some somatic embryos developed from callus growing previously on medium with 0.1 mg/l NOA. **g** Somatic embryos and only some roots developed from callus growing earlier on medium with 1 mg/l NOA. **h** Numerous somatic embryos developed from callus growing on medium with 10 mg/l NOA, no roots were formed

A very interesting difference was found between using NOA in the higher two doses. On MSN medium with 1 mg/l NOA, callus was formed after 2 weeks, but it was less embryogenic than callus which was obtained after 4 weeks of culture on medium with 10 mg/l NOA. Time of callus formation influenced the time of plantlet development and rate of micropropagation. When the culture of callus was initiated on MSN medium with 1 mg/l NOA after 2 months, 2000 plantlets could be obtained from one explant. However, when it was inoculated on the medium with 10 mg/l

NOA, 3000 plantlets could be obtained, but later, after 3 months. The lesser embryogenic capacity was observed in callus which was initiated on the MSN medium with 2,4-D, embryoids developed after transferring callus to liquid MSN medium without auxin. When callus growing on medium with 2,4-D was transferred onto solid medium containing 0.2 mg/l IAA, the differentiation of cells was visible after 1 week (Furmanowa et al. 1984b). In this callus some adventitious and malformed embryos were sometimes observed. Apart from indirect initiation of embryogenesis via callus intermediary direct initiation from differentiated tissue was also observed.

During 5 weeks of caraway liquid culture three developmental stages of embryoids were observed (globular, heart, and cotyledonary). Histological studies of somatic embryos were done at 5 weeks of suspension culture. Most of the embryoids were of cotyledonary form (Fig. 4). The essential compounds were not found in the embryoids and embryogenic callus. For these analyses the same steam distillation method was used as for distillation of the essential oil from the fruits of *Carum carvi* (Farmakopea Polska – The Polish Pharmacopoeia 1970). On transfer to pots, the plants required high humidity, which was provided by covering the plants with glass beakers for 1 to 2 weeks. Regenerated plants were transferred to the greenhouse, and then to the field (Fig. 5B,C).

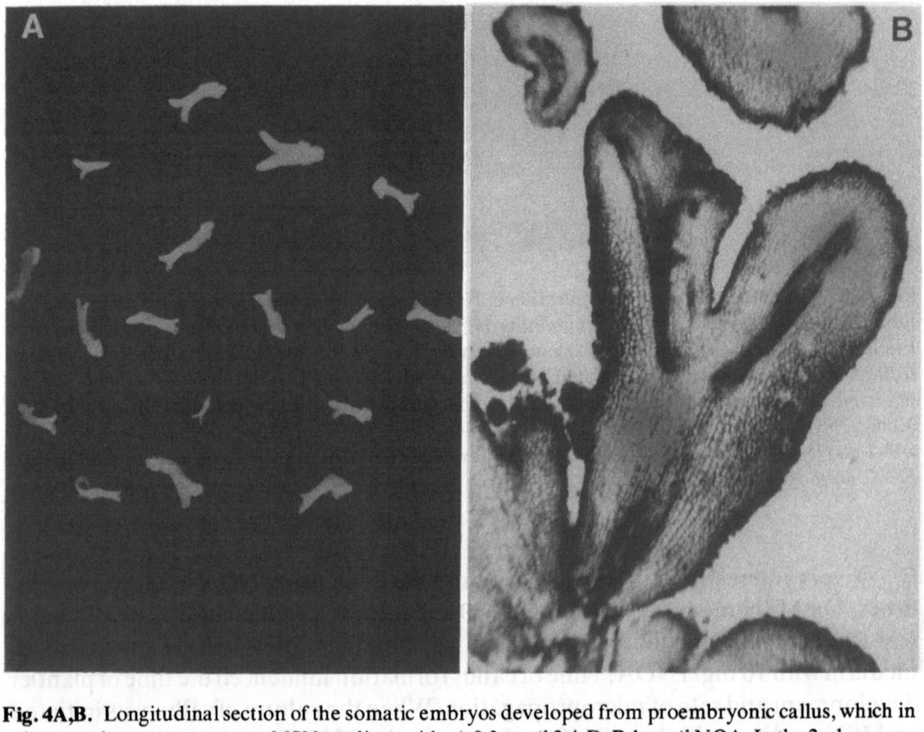

Fig. 4A,B. Longitudinal section of the somatic embryos developed from proembryonic callus, which in primary culture was grown on MSN medium with: **A** 0.3 mg/l 2,4-D, **B** 1 mg/l NOA. In the 2nd passage MSN medium without auxin was used

Fig. 5A-C. Caraway plants regenerated in vitro after transferring to the field: **A** Blooming plant obtained from shoot tip in the first year of vegetation. **B** Plants regenerated from somatic embryos in the first year of vegetation in the field. Two big rosettes of leaves, which are thick and deep green (as compared with the intact plants) are seen. **C** Plants regenerated from somatic embryos in the second year of vegetation. All plants are in bloom, 95% of plants survived after transferring to the field

2.3.2 Regeneration of Plants by Shoot-Tip Culture

The second method which can also be used in the micropropagation of caraway is the initiation of culture from existing meristem. In this technique the regenerated plants should be genetically identical with the donor plants.

Complete caraway plants were produced from shoot tips derived from 7-day-old seedlings. Using 1 to 2 mm shoot tip of seedling, the multiplication was obtained on MSN solid medium with 0.1 mg/l IBA and 1 mg/l of BA. The shoots which developed from adventitious buds were transferred to MSN solid medium supplemented with 0.2 mg/l IAA for rooting. Rooted shoots growing in the next passage on MSN solid medium with 0.5 mg/l IBA and 10 mg/l AS multiplied, and gave rise to adventitious plantlets, which after separation could be transferred to the flower pot and later to the field (Fig. 5A). Using this method, the rate of micropropagation is rather low – only 16 plants from one shoot tip; the development of plants took 2 months, but what is very important, 75% of the plants which were transplanted to the ground bloomed and bore fruits in the first year of vegetation. The plantlets can also be developed from shoot tips placed on filter paper bridges, dipped in liquid MSN medium with 0.5 mg/l IBA and 10 mg/l AS. After 3 to 4 weeks of growth of one passage, the plantlets 3 to 5 cm long (green part) could be transferred to the pots. This method of regeneration gave the lowest rate of micropropagation – one plant from one shoot tip. Some of these plants, after transfer to the field, bloomed in the first year of vegetation (Sowińska 1982).

3 Plant Development in the Field

Plantlets regenerated in vitro were transferred to the field in the Botanical Garden of the Institute of Medicinal Plants in Poznań (Fig. 5A,B,C). One part was transplanted in spring, and the second in the autumn. These plants were compared with the plants which developed from seeds. It was found that plants regenerated in vitro in the first year of vegetation have a big rosette of leaves which were bigger, thicker, and greener than the leaves of the plants propagated by the conventional method (Fig. 5B). A part of the plants, which developed in vitro from shoot tips of seedling, were transferred to the soil in the spring, bloomed and bore fruit in the first year of vegetation, at the end of September. In the second year of vegetation all the plants under the same condition were up to 100 cm high, all developed flowers (Fig. 5C). The fruits were collected in the autumn and biologically evaluated. The results are presented in Table 2. In comparable experiments there are not many biological differences between the plants. The most interesting from the pharmaceutical point of view is the content of essential oil, which is the highest (5.8%) in the fruits from plants regenerated in tissue culture and transferred to the soil in spring. In fruits used as control the yield of essential oil was 4.9%. For this purpose the steam distillation method was used (Farmakopea Polska – The Polish Pharmacopoeia 1970).

Table 2. Biological evaluation of caraway fruits and yield of essential oil from plants regenerated by embryogenesis from callus growing on MSN medium with NOA

Fruits of caraway	Humidity (%)	Weight of 1000 fruits (g)	Size of fruits length (mm)	Width (mm)	Germination capability after 21 days (%)	Content of essential oil (%)
Used as control (sowing in the spring)	9.0	2.58	5.1	1.1	80	4.9
Regenerated in tissue culture (transferred to the soil in spring)	9.6	1.88	4.6	1.0	67	5.8
Used as control (sowing in the autumn)	10.4	2.81	5.1	1.0	51	5.8
Regenerated in tissue culture (transferred to the soil in the autumn)	11.3	2.96	4.6	1.0	62	5.0

4 Chromosome Analysis

The chromosome number of the developed plants was determined in the root tips. The roots were fixed in 3:1 ethanol-acetic acid and stained with orcein (Gerlach 1972). There was no change in the chromosome number in plants regenerated by somatic embryogenesis. Metaphase cells from root tips of regenerated plants had the normal 2n = 20 chromosomes (Fig. 6). These results are in agreement with the observation of several authors.

The problem of genetic stability in plants micropropagated by somatic embryogenesis was discussed by Ammirato (1983). Some examples given by him showed that regenerated plants (mainly *Daucus carota*) have no changes in the chromosome set. On the other hand, sometimes somaclonal variation that has arisen spontaneously can be observed. From the pharmaceutical point of view it is very essential to keep the genetic stability of the plants.

5 Preliminary Reports of Encapsulation of Embryoids and Shoot Tips

The fast propagation of *Carum carvi* L. by embryoids and shoot tips stimulated the use of these organs for encapsulation. Somatic seeds were obtained only in two species of the Umbelliferae family: *Daucus carota* L. (Kitto and Janick 1985a,b) and *Apium graveolens* L. (Redenbaugh et al. 1986a).

Embryoids of caraway were encapsulated in calcium alginate beads to produce somatic (artificial) seeds. The same method was used to encapsulate apical shoot buds. Somatic embryos or apical buds were mixed in sodium alginate (in the range of 2–6%) and dropped individually into 50 mM solution of $CaCl_2$. The best results

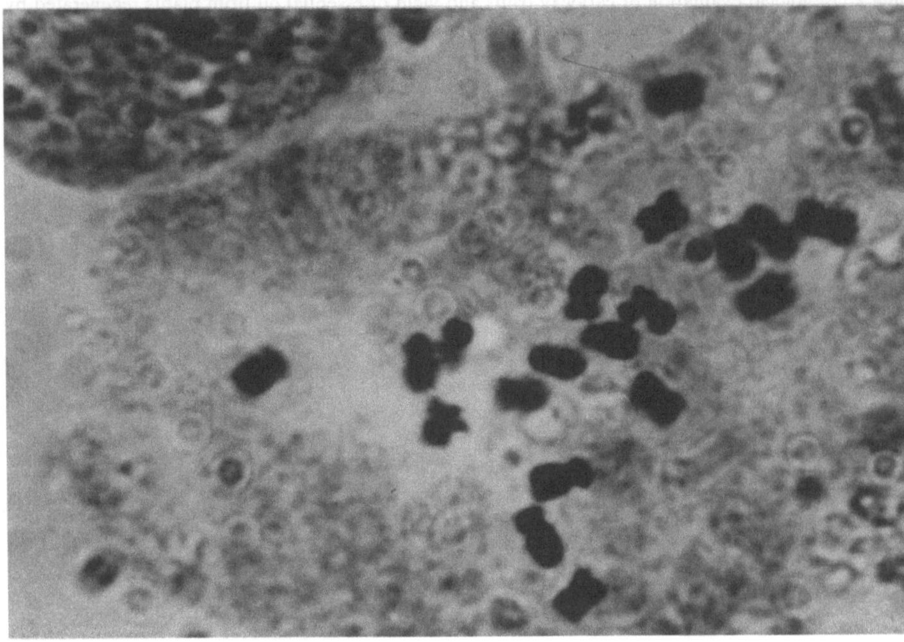

Fig. 6. Chromosomes in root tip cells of young caraway plantlet regenerated in vitro from somatic embryo. Normal diploid cells with 2n = 20 chromosomes are observed. (All photographs were made by I. Rudnicki from experimental materials, in the Department of Biology and Pharmaceutical Botany, Medical Academy in Warsaw)

were obtained with 4.5% alginate. A coating of calcium alginate formed around the somatic embryos or apical shoot buds after 30-min incubation in $CaCl_2$. The beads were 0.5 cm in diameter (Fig. 7). Somatic seeds and encapsulated buds developed into plants after transferring onto SH solid medium (Fig. 8).

6 Protocol

Fast regeneration of plants via somatic embryogenesis

1. Initiation of callus cultures from hypocotyl segments 0.5 to 1 cm long on MSN medium supplemented with 1 mg/l NOA.
2. After 4 weeks subculture 0.5 to 1 g of callus growing in the dark into 100 ml of MSN liquid medium free of NOA and place on a rotary shaker. Growth conditions: 12 h photoperiod, low intensity light – 2000 lx, temp. 26°C, agitation 110 rpm.
3. After 3 weeks of culture clusters of somatic embryos developed.
4. Transfer of somatic embryos with embryogenic callus to 60 ml of auxin-free half-strength MSN solid medium for development of plantlets. Green cotyledons of somatic embryos visible after 2 weeks.
5. Plantlets 5 to 7 cm long in green parts transferred to pots and acclimated for 2 weeks by covering with glass beakers.

Fig. 7. Calcium alginate beads with shoot buds (0.5 cm in diameter)

Fig. 8. Plantlet developed from encapsulated shoot buds on SH medium without growth regulators

7 Conclusions and Prospects

Many of the experiments done on micropropagation of caraway indicated that this method should find practical application. A high rate of micropropagation can be obtained in vitro via somatic embryogenesis. The number of caraway plants produced through somatic embryogenesis was significantly higher as compared to the micropropagation from existing meristem in seedling shoot tip.

It is expected that in vitro techniques will help to solve some of the problems concerning improvements of caraway and its breeding. In future, more attention should be paid to encapsulation of somatic embryos in gel-like substances as synthetic seeds, which could be comparable to dry seeds from the point of view of planting equipment. Micropropagation of medicinal and aromatic plants, from the Umbelliferae (= Apiaceae) family, especially caraway, appears a very promising method.

References

Ammirato PV (1973) Some effects of abscisic acid on the development of embryos from caraway cells in suspension culture. Am J Bot (Suppl)60:22–23

Ammirato PV (1974) The effects of abscisic acid on the development of somatic embryos from cells of caraway (*Carum carvi* L.). Bot Gaz 135:328–337

Ammirato PV (1977) Hormonal control of somatic embryo development from cultured cells of caraway. Plant Physiol 59:579–586

Ammirato PV (1978) The effects of 8-azaguanine on the development of somatic embryos from cultured caraway cells. Plant Physiol 61 (suppl):46

Ammirato PV (1983) Embryogenesis. In: Evans DA, Sharp WR, Ammirato PV, Yamada Y (eds) Handbook of plant cell culture, vol 1. Macmillan, New York, pp 82–123

Bajaj YPS, Furmanowa M, Olszowska O (1988) Biotechnology of the micropropagation of medicinal and aromatic plants. In: Bajaj YPS (ed) Biotechnology in agriculture and forestry, vol 4: Medicinal and aromatic plants I. Springer, Berlin Heidelberg New York Tokyo, pp 60–103

Becker H (1970) Studies on the formation of volatile substances in plant tissue cultures. Biochem Physiol Pflanzen 161:425–441

Bocheńska I, Kozłowski J (1969) Fluctuations in essential oil and water content in a germination capability of *Carum carvi* L. fruits, during ripening and storing. Herba Pol 15:251–263

Collin HA, Watts MJ (1986) Flavour production in celery tissue cultures. In: Somers DA, Gengenbach BG, Biesboer DD, Hackett WP, Green CE (eds) 6th Int Congr Plant tissue and cell culture, Minneapolis, Minn, USA, August 3–8, Abstr, p 118

Czabajska W (1984) Choroby i szkodniki wystepujace na kminku zwyczajnym (Diseases and insects which attack the caraway plant). Wiadom Ziel 24:3–7

Dijkstra H, Speckmann GJ (1980) Autotetraploids in caraway (*Carum carvi* L.) for the increase of the aetheric oil content of the seed. Euphytica 29:89–96

Embong MB, Hadziyer D, Molnar S (1977) Essential oils from spices grown in Alberta - caraway oil (*Carum carvi*). Can J Plant Sci 57

Farmakopea Polska (The Polish Pharmacopoeia) (1970) 4th edn, vol 2. Warszawa, Państwowy Zakład Wydawnictw Lekarskich, pp 38–41

Formanowicz H (1969) The influence of the humidity of seeds on their viability. Herba Pol 15:46–55

Fujii JAP, Slade DT, Redenbaugh K, Walker KA (1987) Artificial seeds for plant propagation. Trends Biotechnol 5:335–339

Furmanowa M, Olędzka H, Sowińska D (1984a) Regeneration of plants by embryogenesis with callus cultures of *Carum carvi* L. J Plant Physiol 115:209–210

Furmanowa M, Sowińska D, Olędzka H (1984b) Influence of several growth regulators on em-

bryogenesis in tissue culture of *Carum carvi* L. In: Novák FJ, Havel L, Dolezel J (eds) Plant tissue and cell culture application to crop improvement. Czech Acad Sci, Prague, pp 159–160
Furmanowa M, Sowińska D, Olędzka H (1986) Regeneracija rastienij putem embriogeneza iz kallusa tmina obyknowiennogo (*Carum carvi* L.). In: Butenko RG (ed) Kultura Kletok Rastenij i Biotechnołogia. Nauka, Moscow, pp 127–128
Furmanowa M, Pietrosiuk A, Sowińska D (1989) Wpływ kwasu β-naftoksyoctowego na rozwój somatycznych zarodków *Carum carvi* L. w kulturze in vitro. Regeneracja roślin i ocena biologiczna owoców (Influence of β-naftoxyacetic acid on somatic embryogenesis in tissue culture of *Carum carvi* L. Plant regeneration and biological estimate of fruits). Acta Pol Pharmaceut 46:280–287
Gerlach D (1972) Zarys Mikrotechniki Botanicznej (The base of botanical microtechnology). PWRiL, Warszawa
Index Kewensis Plantarum Phanerogamarum, tomus I (1895) Suppl 1–15 (1906–1974). Clarendon, Oxford
Kamada H, Harada H (1979) Studies on the organogenesis in carrot tissue culture. I. Effects of growth regulators on somatic embryogenesis and root formation. Z Pflanzenphysiol 91:255–266
Kitto SL, Janick J (1985a) Production of synthetic seeds by encapsulating asexual embryos of carrot. J Am Soc Hort Sci 110(2):277–282
Kitto SL, Janick J (1985b) Hardening treatments increase survival of synthetically-coated asexual embryos of carrot. J Am Soc Hort Sci 110(2):283–286
Kordana S, Leśniewska S, Golcz L (1983) Nutritional requirements of *Carum carvi* L. Herba Pol 29:27–38
Kuzovkina IN, Kuznetsova GA, Smirnov AM (1975) Essential oils in the culture of isolated plant tissues. Referat J Ser Biol 3:377–381
Lewis WH (1977) Medical botany. Plants affecting man's health. Wiley Interscience, New York London Sydney Toronto
Lutomski J, Alkiewicz J, Maiwald L (1987) Phytotherapie in der Praxis. Ein Abriβ für Ärzte und Apotheker. Hippokrates, Stuttgart
Mulder-Krieger T, Verpoorte R, Baerheim Svendsen A, Scheffer JJC (1988) Production of essential oils and flavours in plant cell and tissue cultures. A review. Plant Cell Tissue Organ Cult 13:85–154
Murashige T, Skoog F (1962) A revised medium for rapid growth and bioassays with tobacco tissue cultures. Physiol Plant 15:473–497
Nitsch JP, Nitsch C (1969) Haploid plants from pollen grains. Science 163:85–87
Okoniewska J (1977) Testing of various mutagenic agents for application in caraway (*Carum carvi* L.). Hodowla Roślin, Aklimatyz Nasienn 21:63–86
Redenbaugh K, Paasch BD, Nichol JW, Kossler ME, Viss PR, Walker KA (1986a) Somatic seeds: encapsulation of asexual plant embryos. Bio/Technology 4:797–801
Redenbaugh K, Viss P, Slade D, Fujii J (1986b) Scale-up artificial seeds. In: Somers DA, Gengenbach BG, Biesboer DD, Hackett WP, Green CE (eds) 6th Int Congr Plant tissue and cell culture, Minneapolis, Minn, USA, August 3–8, Abstr, p 132
Salveson AA, Baerheim Swendsen A (1976) Gas liquid chromatographic separation and identification of the constituents of caraway seed oil. Planta Med 30:93–96
Sardesai DL, Tipnis HP (1969) Production of flavouring principles by tissue culture of *Coriandrum sativum*. Curr Sci 38:545
Schäfer F, Li Tianran, Neumann KH (1986) Petiole explants and intact seedlings as sources of somatic embryos of some Apiaceae (*Daucus carota*, *Anethum graveolens*, *Pimpinella anisum*, *Coriandrum sativum*, *Petroselinum crispum*, *Carum carvi*). In: Somers DA, Gengenbach BG, Biesboer DD, Hackett WP, Green CE (eds) 6th Int Congr Plant tissue and cell culture, Minneapolis, Minn, USA, August 3–8, Abstr, p 346
Sluis CJ, Walker KA (1985) Commercialization of Plant Tissue Culture Propagation. IAPTC Newslett 47:2–12
Sowińska D (1982) Regeneracja *Carum carvi* L. w kulturze tkankowej in vitro (*Carum carvi* L. regeneration by tissue culture in vitro). M Sc Thesis, Dep Pharm Bot, Med Acad, Warsaw
Staba EJ (1980) Plant tissue culture as a source of biochemicals. CRC, Boca Raton, p 248
Tutin TG (1968) *Carum* L. In: Tutin TG, Haywood VH, Burges NA, Moore DM, Valentine DH, Walters SM, Webb DA (eds) Flora Europaea, vol 2: Rosaceae to Umbelliferae. Univ Press, Cambridge, p 354
Tyler VE, Brady LR, Robbers JE (1977) Pharmacognosy, 7th edn. Lea & Febiger, Philadelphia
Węglarz Z (1983a) Effect of agricultural agents on the transformation of *Carum carvi* L. from vegetative

to generative phase. III. Effect of time or sowing amount of seeds per ha and fertilization level on the development and cropping of caraway. Herba Pol 29:103–111

Węglarz Z (1983b) Effect of agricultural agents on transition of *Carum carvi* L. from vegetative to generative phase. II. Effect of fertilization and soil moisture on the development and cropping of *Carum carvi* L. Herba Pol 29:21–26

Weiss RF (1982) Lehrbuch der Phytotherapie, 5th edn. Hippokrates, Stuttgart

Żelazek M (1982) Próby rozmnåzania *Carum carvi* L. przez kultury tkankowe in vitro (Preliminary reports on in vitro vegetative propagation of caraway). M Sc Thesis, Dep Pharm Bot, Med Acad, Warsaw

XII *Citrus* spp.: In Vitro Culture and the Production of Naringin and Limonin

R.L. Mansell and C.A. McIntosh[1]

1 General Account

1.1 Distribution and Importance of *Citrus*

Citrus is cultivated in many tropical and subtropical regions of the world. Those areas with the highest level of production include parts of the United States (Florida, California, Texas, and Arizona), Asia (Japan, China, India, and other areas), Central America (Mexico), South America (Brazil, Argentina, Chile, Trinidad, and Tobago), the Mediterranean (Spain, Italy, Greece, Cyprus, Israel, and other areas), South Africa, and Australia. These regions help define a "citrus belt" which is contained between 35° N and 35° S (Ziegler and Wolfe 1975). Some of these areas produce citrus on a large scale and export fruit, while others produce fruit mainly for local consumption.

Basically, there are four major commercial species of *Citrus*, each of which contains different cultivars which are grown in different areas of the world depending upon the climate. *Citrus sinensis* includes the sweet oranges, and the major cultivars include Hamlin, Navel, Valencia, and Pineapple oranges. The major cultivars of *C. paradisi* (grapefruit) are Duncan, Marsh, Thompson Pink, and Ruby. The most common lemons (*C. limon*) include the Bearss, Meyer, and Ponderosa cultivars, while the most common *C. aurantifolia* variety is the Tahiti or Persian lime.

The nutritive value of citrus has been investigated and well documented (Araujo 1977; Baig and Cerda 1980; Ting 1980). These studies have focused on the availability of vitamins, especially vitamin C, as well as on levels of pectins, fiber, calcium, and other desirable components. Therapeutic effects of citrus bioflavonoids (Robbins 1980) and of flavonoids in general (Cody et al. 1986) have also been investigated and the reader is referred to these sources for comprehensive reviews of this topic.

Fruit taste, nutrition, and the development of methods for production of frozen concentrated juice have been responsible for the development of a large citrus industry in many areas of the "citrus belt". In this type of industry, fruit color and ease in peeling (for the fresh fruit market) or processing (canning and beverage industry) are also important considerations. These factors, in addition to cold hardiness and disease resistance, have dictated much of the selection process in the propagation of citrus.

[1]Department of Biology, University of South Florida, Tampa, Fl 336201, USA

Biotechnology in Agriculture and Forestry, Vol. 15
Medicinal and Aromatic Plants III (ed. by Y.P.S. Bajaj)

1.2 Propagation of *Citrus*

Most citrus varieties can be propagated by seeds, and this was a popular method during the 18th and 19th centuries in Europe and the United States (see Ziegler and Wolfe 1975 for a comprehensive review). While this method can result in large, productive trees, it has a major drawback in that there is a relatively long juvenility period (up to 10 years) before the trees start producing fruit. In addition, most high quality varieties are not resistant to soil microbes, and therefore they are readily attacked by these pathogens. Neither of these negative features is acceptable or desirable in a commercial operation.

The most actively used method of propagation today is by the process of budding or grafting. These procedures were practiced in Europe during the 16th and 17th centuries, and regained popularity in the late 1800's in both Europe and the U.S. (Ziegler and Wolfe 1975). Budding involves the insertion of a scion bud under the bark of the stock so that the cambial tissues of the two samples contact each other. This results in a union of the cambia and in the wood itself. The advantages of this method over seedling propagation are tremendous since the resulting plant shows little juvenility, and therefore results in the earlier production of fruit. Secondly, the grower can select a rootstock with specifically desired characteristics (disease-resistant, drought-resistant, etc.) and most plants show uniform fruit quality and seasonal development (Ziegler and Wolfe 1975). Most rootstocks are chosen on the basis of their cold hardiness and disease resistance, while budstocks are chosen on the basis of fruit quality.

1.3 Secondary Compounds in *Citrus*

There are a wide variety of secondary compounds found in *Citrus*. These include phenolic compounds such as flavonoids, benzoic acids, cinnamic acids, and coumarins as well as terpenoids such as carotenoids and limonoids. Research on the benzoic and cinnamic acids of citrus have focused on physiological as well as antimicrobial studies (Feldman and Hanks 1965; Ismail and Brown 1975; Waterman and Grundon 1983) and will not be discussed here. Coumarins and coumarin glycosides are being utilized in chemotaxonomy (McHale et al. 1987 and references therein; Wu 1988) and also will not be discussed here. Most recently, research efforts have focused mainly on flavonoids and limonoids in citrus, and the work on these two groups of compounds will be discussed in greater detail.

1.3.1 Flavonoids

These are, by far, the most studied group of phenolic compounds in *Citrus* of the many compounds which have been reported. The predominant class is the flavanones and these are the major compounds found in both vegetative parts as well as in the fruit. For example, naringin (5,7,4′-trihydroxyflavanone 7-0-beta-neohesperidoside; Fig. 1) is the most abundant flavonoid in grapefruit and this single compound can account for up to 70% of the dry weight of small green fruit

(Kesterson and Hendrickson 1953). Concentrations as high as 94,000 ppm (micrograms/g fr. wt.) have been reported in young shoot tips (Jourdan et al. 1985). Flavones are also found, although in lesser amounts, and anthocyanins have been found only in the "blood orange". The isolation, identification, and distribution of flavonoids in the genus *Citrus* have been extensively reviewed (Kefford and Chandler 1970; Nishura et al. 1971; Horowitz and Gentili 1977; Waterman and Grundon 1983).

Of all the flavonoids the most interesting group, from an economic standpoint, are the flavanones. These compounds are important to the citrus-processing industry due to their flavor qualities and they are of interest to the pharmaceutical industry and to medical researchers as they are reported to have anti-inflammatory and tumor growth-inhibitory activity (Cody et al. 1986).

The naturally occurring flavanones are present in glycoside forms with 7-0-rutinoside (e.g., narirutin and hesperidin) and 7-0-neohesperidoside (e.g., naringin and neohesperidin) substitutions being common (Fig. 1). The type and position of the sugar substitution has a profound effect on the taste of the compounds, and therefore upon the fruit and processed fruit product which contain them (Rouseff 1980; Horowitz 1986). The 7-0-rutinosides (alpha 1,6 rhamnoglucosides) are tasteless, whereas the 7-0-neohesperidosides (alpha 1,2 rhamnoglucosides) are bitter. The taste threshold of naringin can be as low as 20 ppm (micrograms/ml) (Mansell and Weiler 1980). An increase in the oxidation of the flavanone (e.g., to a flavone) results in compounds which are nonbitter and the degree of ring substitutions (OH, OMe) on the flavanone C-15 ring system modulates the bitter properties of the compound (Rouseff 1980 and references therein).

The relationship between naringin content and grapefruit juice acceptability was first established by Guadagni and associates (Maier 1983 and references therein) although the presence of other juice parameters (e.g., acid concentration, sugar concentration, limonin concentration) greatly complicates the assignment of a quality score on the basis of naringin content alone (Barros et al. 1981; Mansell and McIntosh unpubl.; McIntosh et al. 1988).

Dihydrochalcone derivaties of the bitter flavanone neohesperidosides also have interesting taste properties. In contrast to the bitterness of their flavanone counterparts, naringin dihydrochalcone and neohesperidin dihydrochalcone are sweet and the latter compound is reported to be 250 times sweeter than sucrose (Horowitz 1986). These dihydrochalcones reduce bitterness when added to citrus juices and it has been suggested that they act by interfering with the bitter taste receptors (Horowitz 1986). This interference with bitterness perception, coupled with their inherent sweetness, suggests that they might be useful as additives to bitter drugs or as taste-enhancing agents in the food industry. At present, naringin is being used in the food and beverage industry as an additive to products in which a "tonic" or bitter flavor is desired (Maier 1983; Horowitz 1986).

The flavanone hesperidin (Fig. 1) is a tasteless compound and as such does not affect the flavor of processed citrus products; however, the high level of this compound in oranges does cause a problem in the citrus industry. Due to the unique chemistry of this compound, it readily precipitates out of orange juice concentrate upon storage and the resulting juice, which now has excessive hesperidin crystals, will receive a downgraded quality score due to visual changes which are apparent

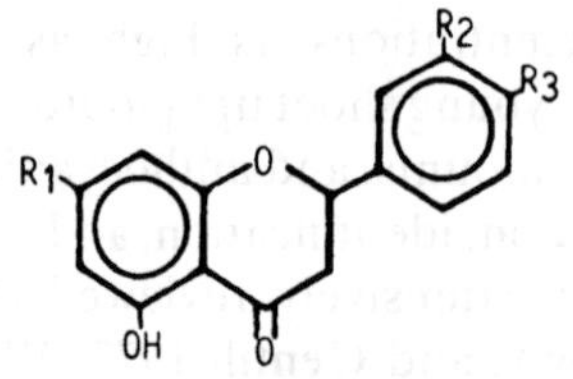

Flavanone

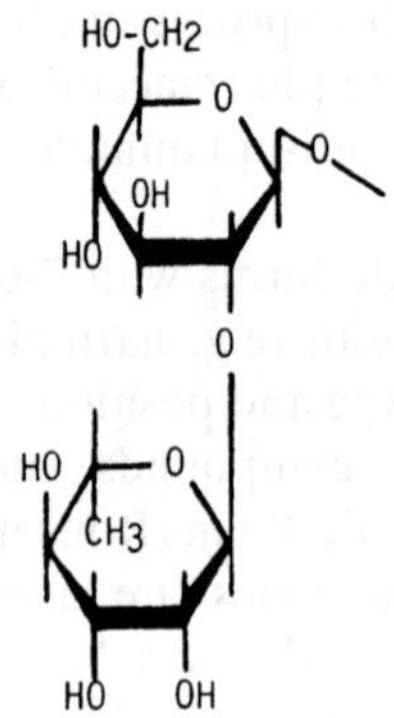

Neohesperidose

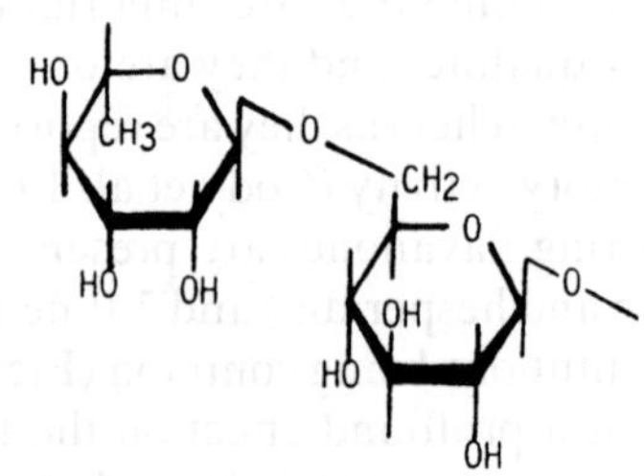

Rutinose

Compound	R_1	R_2	R_3
Naringenin	OH	H	OH
Prunin	O-glucose	H	OH
Naringin	neohesperidose	H	OH
Narirutin	rutinose	H	OH
Hesperitin	OH	OH	OCH_3
Hesperitin 7-glucoside	O-glucose	OH	OCH_3
Hesperidin	rutinose	OH	OCH_3
Neohesperidin	neohesperidose	OH	OCH_3

Fig. 1. Structures of the flavanones discussed in this chapter

(Rouseff 1980). Hesperidin also coats the juice extractors, and this results in loss of efficiency of concentrate production. In spite of these esthetic and processing problems, hesperidin is being actively studied as a bioflavonoid with positive effect in humans (Robbins 1980; Cody et al. 1986).

1.3.2 Limonoids

Limonoids are terpenoids and thus have a very different biosynthetic origin from that of the flavonoids. Limonin ($C_{26}H_{30}O_8$) itself is an intensely bitter triterpenoid dilactone compound, even more bitter than quinine (Maier 1983), and is found throughout the genus (Fig. 2). Thus citrus is impacted by two unrelated bitter compounds each of which has its own properties as well as being able to act synergistically. Bitterness in grapefruit is due to the presence of both limonin and naringin; however, the bitterness in oranges is due primarily to the presence of limonin (Maier et al. 1980; Rouseff and Nagy 1982).

A unique feature of limonin is demonstrated by the fact that it is responsible for the problem of "delayed bitterness", a character which is strongly exhibited by navel and shamouti oranges (see Maier et al. 1980; Maier 1983; Dreyer 1983 for reviews). Delayed bitterness is due to the slow conversion of a nonbitter compound, limonoate A-ring monolactone (Fig. 2), to limonin after fruits have been juiced or processed. The conversion process is exacerbated by the juice acids or by heating. Recent reports on the presence of limonoid glycosides in *Citrus* (Hasegawa et al. 1988) provide another pool for limonin since these compounds might be deglycosylated during extraction and processing, and thus could also contribute to bitterness. At present, however, there is no information available on the taste properties of the glycosides or on their chemical conversion to known bitter compounds.

Much of the recent research on limonin has focused on methods for the debittering of citrus juices including the use of cellulose adsorbents and microbial enzymes (e.g., Chandler and Johnson 1977; Hasegawa et al. 1983). Unfortunately, not much is known about the biosynthesis of this compound or of the physiological and environmental factors which control and regulate biosynthesis and accumulation.

In addition to the impact of limonin and other limonoids on the citrus industry, the potential insect antifeedant activities of these compounds have also been explored (Dreyer 1983; Nakatani et al. 1985; and references therein). While several different limonoids have been shown to have pronounced negative effects on insect feeding activities, limonin itself does not seem to possess strong antifeedant activities, at least in those insects studied to date.

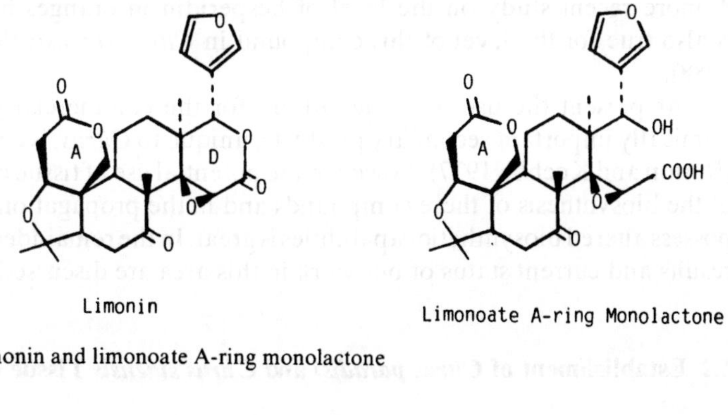

Fig. 2. Structures of limonin and limonoate A-ring monolactone

2 In Vitro Approaches

The use of plant tissue culture in the study of secondary compound biosynthesis is of great interest and a great deal of research has been directed toward the use of plant tissue culture to isolate cell lines which produce atypical levels of these secondary metabolites (Zenk et al. 1977; Zenk 1978). Studies from many different plants have shown that the level of production of secondary compounds virtually ceases during culture initiation and callus growth and resumes only upon organogenesis (Hiraoka and Tabata 1974; Tabata et al. 1978, 1974, 1972; Barthe et al. 1987; Wink 1987).

2.1 Citrus Tissue Culture

Research on tissue culture of *Citrus* has been actively pursued for the past 25 years and the early work in this area and its potential application in the industry has been comprehensively reviewed (Button and Kochba 1977). The emphasis of most of the recent research has centered on embryogenetic and regeneration control, protoplast culture, and on the isolation of lines which are resistant to herbicides or which are salt-tolerant (Spiegel-Roy and Kochba 1980; Spiegel-Roy et al. 1983; Kobayashi et al. 1985; Moore 1985; Barlass and Skene 1986; Vardi and Galun 1989). The use of tissue culture in the study of the growth and development of fruit tissue is also being explored (Kordan 1959, 1977; Tisserat and Galletta 1987).

Relatively little work has been done on the analysis of secondary compounds by citrus tissue cultures. Flavonoid-like material has been detected in lemon juice vesicle cultures (Kordan and Morgenstern 1962), trace amounts of highly methylated flavones have been found in cultures of *Citrus* peel (Brunet and Ibrahim 1973), and we have recently published results obtained from an analysis of limonin and naringin in citrus tissue cultures (Barthe et al. 1987). Earlier work in our laboratory has shown that the level of limonin and naringin in different organs and tissues of grapefruit is variable and dependent upon the developmental state and tissue age (McIntosh et al. 1982; McIntosh and Mansell 1983; Jourdan et al. 1985). A more recent study on the level of hesperidin in oranges has shown that this is also true for the level of this compound in *Citrus sinensis* tissues (Barthe et al. 1988).

At present the use of tissue culture for the commercial production of economically important secondary products, unique to *Citrus*, is currently impractical (Button and Kochba 1977); however, the potential use of tissue cultures in the study of the biosynthesis of these compounds and in the propagation of cell lines which possess altered biosynthetic capabilities is great. In the remainder of this chapter, the results and current status of our work in this area are discussed.

2.2 Establishment of *Citrus paradisi* and *Citrus sinensis* Tissue Cultures

Callus cultures are initiated as in Barthe et al. (1987). Cultures originating from fruit tissues are obtained by surface sterilization of the fruit, dissection, and removal of the desired tissue. Culture is done on Murashige and Skoog basal medium as modified by Murashige and Tucker (1969) (MST) and supplemented with 2,4-dichlorophenoxyacetic acid, kinetin, and coconut milk. Cultures originating from seedling tissues are obtained by growing aseptic seedlings on 1% agar and the explants are cultured as above.

Induction of shoots from the callus is accomplished using a modification of the procedure of Chaturvedi and Mitra (1974). Callus is placed on MST medium supplemented with benzyladenine, naphthaleneacetic acid, and malt extract and grown on a 16 h light/8 h dark regime. Regenerated shoots begin to appear after 3 weeks.

For the initiation of suspension cultures, approximately 20 g (fr. wt.) of callus is pulverized into a fine slurry of cells and then transferred into a flask containing 50 ml of liquid MST medium. Cultures are shaken at 120 r.p.m. on a rotary shaker at 28 °C (Barthe 1984).

2.3 Extraction and Analysis of Tissue and Callus

For those samples in which only the limonin and naringin will be analyzed, the tissue is extracted by refluxing in methanol. The extracts are diluted with water, and the limonin and naringin are analyzed using radioimmunoassay (RIA) (Weiler and Mansell 1980; Jourdan et al. 1983). A similar method is used for the extraction and analysis of hesperidin (except that the samples are diluted with dimethylformamide) (Barthe et al. 1988); however, the RIA for this compound has not, as yet, been utilized to analyze tissue cultures.

For those samples in which limonoate A-ring monolactone will be analyzed in addition to limonin and naringin, the tissue is extracted in Tris buffer (pH 8) and aliquots are diluted with water or with acid to catalyze lactonization of the D-ring of limonoate A-ring monolactone (to which the antibody does not bind), and then both are assayed. The difference between the total limonin (acid-diluted extract) and the endogenous limonin (water-diluted extract) equals the amount of limonoate A-ring monolactone (McIntosh and Mansell 1983).

2.4 Naringin and Limonin Levels in Citrus Tissue Cultures

2.4.1 Comparison of Levels in Explants and Callus

Table 1 shows the results obtained from a study of the levels of naringin and limonin in different *Citrus paradisi* (grapefruit, cv. Duncan) tissues and in the callus subsequently obtained from explants of these tissues (Barthe et al. 1987). The albedo contained the highest concentration of both limonin and naringin in the fruit tissues analyzed. In the seedlings, the highest level of limonin was in the cotyledon and the highest level of naringin was in the young leaves. These trends agree with those found in previous studies (McIntosh et al. 1982;Jourdan et al. 1985).

The callus derived from explants of these tissues show much lower levels of these compounds ranging from 0.2 ppm naringin in callus from cotyledon to 3 ppm in leaf-derived callus; the limonin levels were 0.1 ppm in fruit tissue- and stem-

Table 1. Concentration of bitter principles in random, pooled samples of differentiated tissues and in unorganized, 5-month-old callus derived from such tissue of Duncan grapefruit. (Barthe et al. 1987)

Plant part	Total limonin (ppm)	Naringin (ppm)
Fruit		
Flavedo	12	3500
Callus	0.1	0.3
Albedo	30	6200
Callus	0.1	0.4
Juice vesicles	9	200
Callus	0.1	0.5
Seedling		
Leaf	650	5800
Callus	4.0	3.0
Stem	600	450
Callus	0.1	1.5
Cotyledon	1200	10
Callus	1.0	0.2

derived callus and were 4 ppm in leaf-derived callus. Limonoate A-ring monolactone levels were not determined in this analysis.

The limonin, limonoate A-ring monolactone, and naringin levels were determined in the stem tissues and derived callus from four different *Citrus* spp., Duncan and Thompson grapefruit, and Parson Brown and Valencia orange (Table 2; cf. Barthe et al. 1987). Once again, the callus contained substantially reduced levels of each of these compounds as compared to the original explants. The range of concentration of these compounds was very large (up to a factor of 10 or more) in the callus as well as in the stem tissue.

The limonin and naringin content in multiple callus pieces derived from a single albedo explant also showed considerable variability (Fig. 3). The average limonin concentration in the 21 colonies was 0.4 ppm with a range of 0.14–1 ppm. The average naringin concentration was 27.6 ppm with a range of 3–99 ppm. This indicates that the variation in limonin and naringin content in callus cultures may not be solely due to the variation between the explants themselves but may also be due to differences within each individual explant or even to epigenetic factors due to culture (micro-environmental?) conditions. At this time there is no information available which would clarify this point.

2.4.2 Changes in Limonin and Naringin Levels During Callus Initiation and Growth

The above results lead us to the following question – what are the kinetics of this decrease in limonin and naringin levels as explants are induced to produce callus?

Table 2. Levels of limonin, limonoate A-ring monolactone, and naringin in stem explants and derived callus. (Barthe et al. 1987)

Sample	n	Limonin (ppm) Ave ± SD (range)	A-ring lactone (ppm) Ave ± SD (range)	Total (ppm) Ave ± SD (range)	n	Naringin (ppm) Ave ± SD (range)
Duncan stems	10	66 ± 63	495 ± 369	561 ± 376	3	36 ± 14
		(23 – 220)	(46 – 1056)	(69 – 1121)		(22 – 50)
Callus	6	1.1 ± 1.2	4.4 ± 4.3	5.3 ± 5.4	12	0.6 ± 0.4
		(0.2 – 3.1)	(0.8 – 11)	(1.0 – 14)		(0.1 – 1.5)
Thompson stems	6	108 ± 53	776 ± 385	874 ± 427	5	41 ± 30
		(52 – 176)	(592 – 1400)	(489 – 1555)		(7.8 – 88)
Calus	5	0.7 ± 0.6	2.0 ± 1.6	2.6 ± 1.7	10	0.4 ± 0.3
		(0.3 – 1.7)	(0.6 – 4.7)	(0.9 – 5.3)		(0.1 – 1.0)
Parson Brown	10	111 ± 125	427 ± 385	538 ± 393	9	2.8 ± 0.3
stems		(25 – 397)	(42 – 1249)	(68 – 1382)		(1.7 – 5.9)
Callus	3	0.3 ± 0.1	0.7 ± 0.4	1.0 ± 0.5	6	0.3 ± 0.3
		(0.1 – 0.4)	(0.5 – 1.1)	(0.6 – 1.6)		(0.1 – 0.9)
Valencia stems	9	138 ± 111	642 ± 435	779 ± 448	9	2.2 ± 0.8
		(41 – 312)	(75 – 1308)	(152 – 1460)		(1.0 – 3.5)
Callus	3	0.3 ± 0.2	2.3 ± 2.4	2.6 ± 2.2	6	0.3 ± 0.1
		(0.1 – 0.5)	(0.3 – 4.9	(0.6 – 4.9)		(0.2 – 0.4)

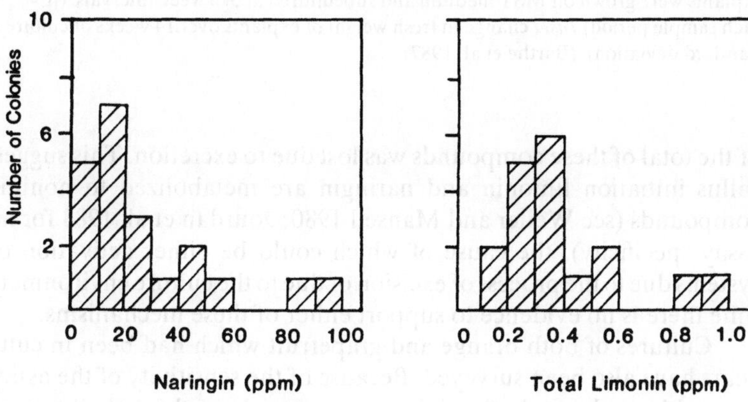

Fig. 3. Levels of naringin and total limonin in 21 callus colonies derived from a single Duncan grapefruit albedo explant after 11 months of culture. Total limonin reflects the level of endogenous limonin plus limonoate A-ring monolactone. (Barthe et al. 1987)

Figure 4 shows the changes in total limonin (endogenous limonin plus limonoate A-ring monolactone) and naringin during the onset of callus growth of Duncan grapefruit albedo. There was a rapid decrease in the levels of these compounds during the first 3 weeks of culture followed by a more gradual decline. By the 42nd week of culture, the naringin level was only 0.1% of the original concentration and the limonin level was 0.06%. While the decrease in the concentration of these compounds is due in part to dilution from an increase in fresh weight as a result of callus production, this alone cannot explain the concomitant drop in the total amount per explant. Also, analysis of the culture medium showed that less than 1.5%

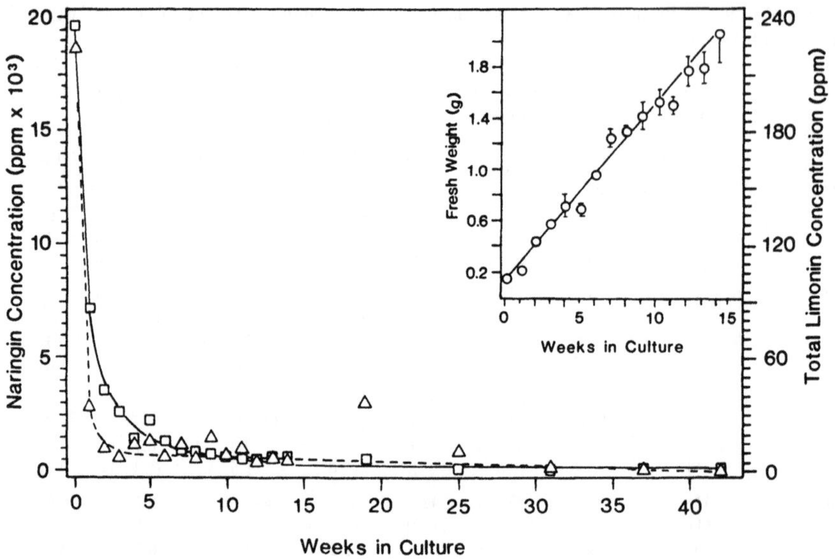

Fig. 4. Changes in the levels of naringin and total limonin during culture of Duncan grapefruit albedo. Total limonin reflects the level of endogenous limonin plus limonoate A-ring monolactone. Albedo tissue explants were grown on MST medium and subcultured at 5–6 week intervals. (n = 7 to 20 explants for each sample period) *Inset* change in fresh weight of explants over 14 weeks of culture (n = 16, mean ± standard deviation). (Barthe et al. 1987)

of the total of these compounds was lost due to excretion. This suggests that during callus initiation limonin and naringin are metabolized to nonimmunoreactive compounds (see Weiler and Mansell 1980; Jourdan et al. 1983 for information on assay specificity), the cause of which could be either activation of degradative systems due to the process of excision or due to the culture environment itself. At this time there is no evidence to support either of these mechanisms.

Cultures of both orange and grapefruit which had been in culture for over 5 years have also been surveyed. Because of the sensitivity of the assay methods, we were able to detect both of these compounds in the ppb (parts per billion i.e., nanograms compound/g culture sample) range. Both limonin and naringin were still measurable in these cultures even after 5 years of subculturing (Fig. 5). This would suggest that there is a constitutive low level of biosynthesis of these compounds still being expressed in the callus cultures; however, accumulation analogous to that in intact tissues does not occur.

2.4.3 Analysis of Liquid Suspension Cultures

The limonin and naringin levels in liquid suspension cultures derived from Marsh grapefruit hypocotyl were monitored throughout subculturing and growth (Barthe 1984). These cultures had a lag phase of 7–10 days after subculturing followed by a rapid growth phase which leveled off after 20 days (Fig. 6). Both the limonin and

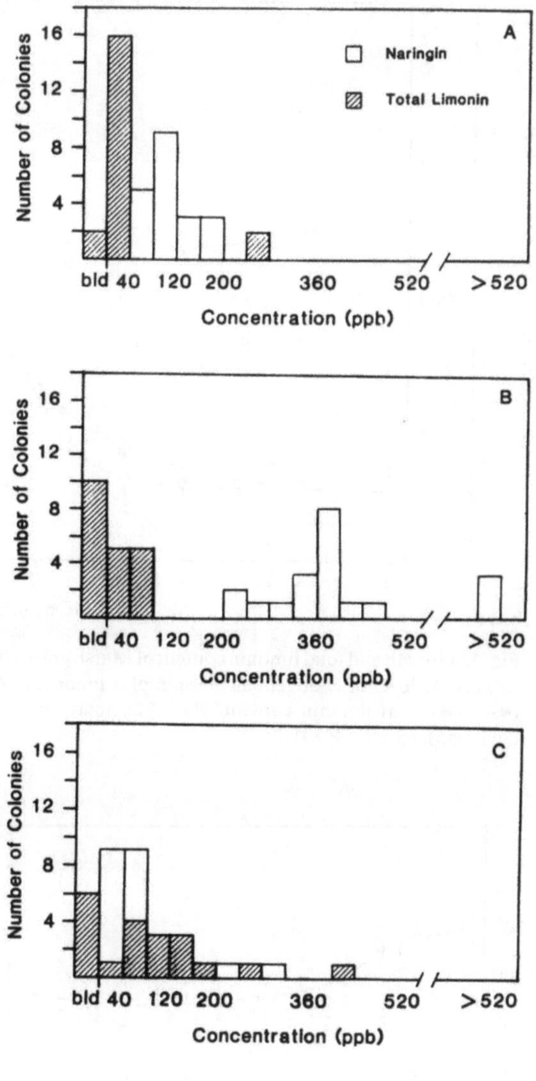

Fig. 5A-C. The distribution of naringin and total limonin in *Citrus* callus cultures derived from different tissue sources. **A** Marsh grapefruit hypocotyl-derived callus, 5 years in culture. **B** Thompson grapefruit albedo-derived callus, 1.5 years in culture. **C** Thompson grapefruit juice vesicle-derived callus, 1.5 years in culture. (n = 20; *bld* = below the limits of detection of the assay). (Barthe et al. 1987)

naringin levels paralleled the increase in biomass of the cultures, thus the concentration of these compounds in the cultures remained relatively constant (Figs. 6,7). These data support the callus study showing that there is a constitutive low level of biosynthesis of these compounds under culture conditions.

Lewinsohn et al. (1986, 1989) investigated the glucosylation of exogenous naringenin and hesperitin by suspension cultures of Duncan grapefruit. They found that the cultures were able to glucosylate both flavanone aglycones to their respective 7-0-glucosides (prunin and hesperitin 7-0-glucoside; Fig. 1) even though de novo synthesis of the flavanone C-15 ring structure could not be detected. No rhamnosylating activity could be detected. These cultures had been derived from 3-year-old callus and thus they may have lost the ability to express the rhamno-

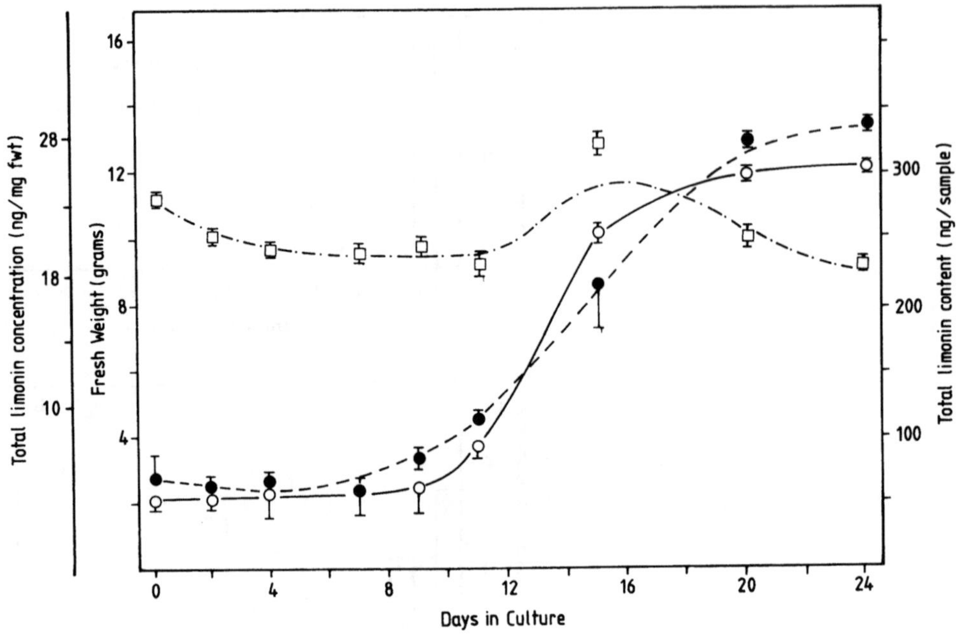

Fig. 6. Growth and total limonin content of Marsh grapefruit derived suspension cultures. Total limonin reflects the level of endogenous limonin plus limonoate A-ring monolactone. (○——○) fresh weight; (●----●) total limonin content; (□----□) total limonin concentration. (n = 3, mean ± standard deviation). (Barthe 1984)

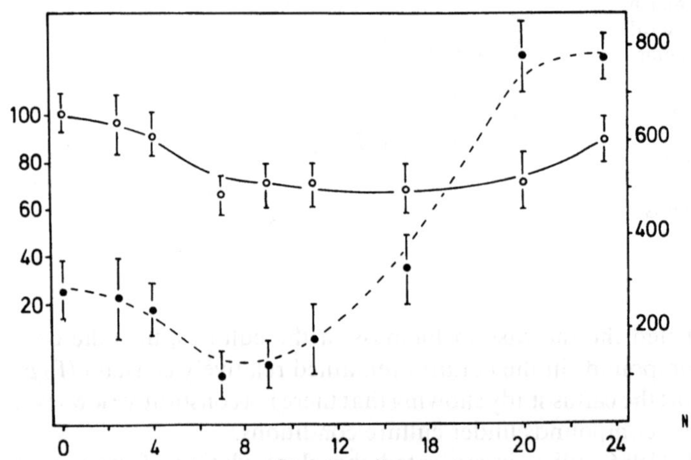

Fig. 7. Naringin content of Marsh grapefruit derived suspension cultures. (○——○) naringin concentration; (●----●) naringin content. (n = 3, mean ± standard deviation). (Barthe 1984)

sylating activity. Another possible explanation of the lack of detection of naringin synthesis in this study may lie in the method of analysis itself. Lewinsohn et al. (1986) used a thin layer chromatographic method for analysis of flavanones, which is more than 1000 times less sensitive than the radioimmunoassay employed in our laboratory. It would be very important to analyze the culture media containing the biotransformed flavanones using the immunological methods.

In addition, the endogenous levels of the compounds were not measured in this biotransformation study, therefore it is not possible to speculate about the effect of exogenous precursors on endogenous levels. It is also possible that the glucosyl transferase activity could be excreted from the cells into the medium under these conditions thereby making the resultant exogenous glycoside production independent of internal cellular processes. It is important to note that neither prunin nor hesperitin 7-0-glucoside have been reported to be present in citrus tissues; however, the demonstration of the ability of grapefruit suspension cultures to biotransform naringenin and hesperitin into these glucosides suggests that these compounds could be precursors in the biosynthetic pathway leading to flavanone rhamnoglucosides in *Citrus* (Lewinsohn et al. 1986, 1989). It would appear that endogenous concentrations of these putative intermediates are below the detection levels of normal flavonoid quantification methods and therefore it will be important to reexamine *Citrus* plants with immunoassays.

2.4.4 *Naringin, Limonin, and Limonoate A-Ring Monolactone Levels During Shoot Regeneration*

Duncan, Marsh, and Thompson pink grapefruit callus (hypocotyl-derived) began to develop buds after 3 weeks on regeneration medium and ultimately shoots were produced. The callus contained a total limonin concentration of 3.6 ppm and a naringin concentration of 7.4 ppm (Barthe et al. 1987). Newly produced Duncan buds contained an average of 40 ppm limonin and 148 ppm naringin, and regenerated shoots contained even higher levels of these compounds – 77 ppm limonin and 1051 ppm naringin (Barthe et al. 1987). Ninety-five percent of the total limonin found was present as the A-ring monolactone (Table 3). When the levels of limonin

Table 3. Limonin, limonoate A-ring monolactone, and naringin levels during the development of shoots from Duncan grapefruit callus. (Barthe et al. 1987)

Sample	n	Limonin (ppm) Ave ± SD (range)	A-ring lactone (ppm) Ave ± SD (range)	Total (ppm) Ave ± SD (range)	n	Naringin (ppm) Ave ± SD (range)
Duncan callus[a]	10	0.8 ± 0.7 (0.1 – 2.0)	6.6 ± 6.7 (1.6 – 21)	7.4 ± 7.3 (1.8 – 23)	9	3.6 ± 3.2 (0.7 – 8.8)
E1 (buds)	14	2.1 ± 1.4 (0.3 – 5.3)	38 ± 22 (4.1 – 69)	40 ± 23 (7.5 – 73)	15	148 ± 113 (17 – 357)
E2 (shoots)	3	4.0 ± 3.1 (1.5 – 7.5)	73 ± 59 (32 – 141)	77 ± 58 (40 – 144)	2	1051 ± 115 (970 – 1132)

[a] On regeneration medium.

and naringin were measured in a single grapefruit stem section through callus initiation and shoot regeneration (Table 4), it was again found that there is a dramatic decrease in the naringin and limonin content during initiation and growth of callus, and a subsequent increase upon shoot regeneration. These data indicate that there is an increase in the production and/or accumulation of these compounds upon organogenesis. It is not clear at this time, however, whether this increased level of limonin and naringin is due to the hormonal treatment, the differentiation process and accompanying metabolic changes which result from this treatment, or both.

Table 4. Limonin, limonoate A-ring monolactone, and naringin levels in a single Thompson grapefruit stem explant through callus formation and shoot production. (Barthe et al. 1987)

Stem[a] -------> callus----> ---------- -----------> Regenerated Shoot

Sample	Limonin (ppm)	A-ring lactone (ppm)	Total (ppm)	Naringin (ppm)
Stem	155	1400	1555	45
Callus	0.5	1.1	1.5	0.2
	0.3	0.6	0.9	0.1
Regenerated shoot	0.9	7.2	8.2	504

[a] n = 1, Thompson grapefruit.

2.5 Conclusions and Directions for Further Research

The inability of callus cultures to accumulate naringin and limonin coupled with the maintenance of very low levels in actively growing cultures would seem to point to the presence of a balanced metabolic system, enzyme levels that are low, or controlled so that their activities are low. The increased levels of the two compounds found in organoid areas (buds) and regenerated shoots suggest that limonin and naringin synthesis and net accumulation is tied to morphological development. Whether it is the production of organized tissues (e.g., leaf) or the other hormonal or metabolic activities which accompany these structures that regulate limonin and naringin synthesis and accumulation is not clear. We are currently investigating the effect of hormonal levels in the medium and of culture conditions on the production of limonin and naringin in grapefruit callus cultures.

Several questions about the biosynthesis of limonin and naringin are unanswered. Do the increased levels of limonin which are observed in young developing tissue and during organogenesis represent de novo synthesis of limonin or conversion of precursors? Do the increased levels of naringin under these same conditions represent glycosylation of a precursor pool (e.g., naringenin, prunin)? Are the increased levels of naringin due to an isomerization of the glycoside moiety from a rutinose, which does not bind to the antibodies used in the RIA, to a neohesperidose (i.e., conversion of narirutin to naringin) which does bind to the antibodies? Is there an increased synthesis of the C-15 ring structure along with the glycosylation, or a mobilization of "reserves" of the C-15 ring nucleus?

Current efforts in our laboratory to clarify these points include the development of immunoassays for both naringenin and prunin. Such assays will enable us to test for the presence of these compounds in tissue cultures and, if found, to follow levels during callus initiation and shoot regeneration. In addition, it will be possible to test for these compounds during the growth and development of *Citrus* and thus to monitor developmental aspects of the control of naringin biosynthesis.

We are also working on the isolation and characterization of the glycosyl transferase(s) involved in naringin biosynthesis in grapefruit seedlings, since both activities should be present in intact tissues as well as tissue cultures. To date, the flavanone-specific 7-0-glucosyl transferase activity in grapefruit seedlings has been purified and characterized (McIntosh and Mansell 1990; McIntosh et al. 1990). Once the glycosylating enzymes have been characterized in seedlings, their levels and activity will be studied in cultures from callus initiation through the process of shoot regeneration. Cultures are also being tested for levels of activity of the enzymes involved in the biosynthesis of the C-15 flavonoid ring. Information from this work may provide the basic information necessary for an investigation of the control of the activities of these enzymes at the cellular and at the gene level.

It is apparent that tissue culture has the potential to be a valuable tool in the study of the biosynthesis of citrus bitter compounds. As this process is clarified, it will be possible to consider and evaluate techniques to suppress or enhance this synthesis in plants. Such insight would greatly aid any efforts to use molecular techniques in an attempt to improve this crop.

The screening of plants regenerated from callus culture for individuals with altered limonin- and/or naringin-producing potential could also be conducted. Shoots with desired biosynthetic capacities could be used in a propagation program. It would be important to monitor levels through the propagation phase and in the resultant plant so as to determine if this capacity will hold through these processes. The determination of the relationship between the levels produced in the regenerated shoots and the levels eventually found in the fruits produced by the plants obtained after budding of these shoots would be crucial to an organized breeding effort.

3 Protocol

We have cultured citrus under a variety of conditions as have others (e.g., Murashige and Tucker 1969). We use the basic MS salts supplemented with 10 mg/l thiamine, 10 mg/l pyridoxine, 5 mg/l nicotinic acid, 2 mg/l glycine, 100 mg/l inositol, 50 g/l sucrose, 0.66 mg/l 2,3-diclorophenoxyacetic acid, 1.32 mg/l kinetin, 100 ml/l coconut milk, and 8.5 g/l agar, pH 5.8 for callus culture. Tissues are maintained at 28 °C in the dark for maximal callus growth.

For regeneration of buds and shoots from callus cultures, a modification of the procedure of Chaturvedi and Mitra (1974) is used. Callus tissue is placed on a medium consisting of the basic MS salts supplemented with 10 mg/l thiamine, 10 mg/l pyridoxine, 5 mg/l nicotinic acid, 2 mg/l glycine, 100 mg/l inositol, 50 g/l sucrose, 5 mg/l benzyladenine, 0.1 mg/l naphthaleneacetic acid, 500 mg/l malt extract, and 8.5 g/l agar, pH 5.8. Incubation is under either a 16 h light/8 h dark regime or under continuous light at 28 °C. Regenerated shoots begin to appear after 3 weeks.

References

Araujo PE (1977) Role of citrus fruit in human nutrition. In: Nagy S, Shaw PE, Veldhuis MK (eds) Citrus science and technology. Avi, Westport, Conn, pp 1-32

Baig MM, Cerda JJ (1980) Studies on the role of citrus in health and disease. In: Nagy S, Attaway JA (eds) Citrus nutrition and quality. ACS Symp Ser 143, Am Chem Soc, Washington, DC, pp 5-42

Barlass M, Skene KGM (1986) Citrus (*Citrus* species). In: Bajaj YPS (ed) Biotechnology in agriculture and forestry, vol 1: Trees I, Springer, Berlin Heidelberg New York Tokyo, pp 207-219

Barros SM, Carter RD, Fellers PJ, Ting SV, Mansell RL (1981) Effect of finishing variations on quality and yield of juice from frozen oranges. Proc Fla State Hortic Soc 94:276-279

Barthe GA (1984) Levels of naringin and limonin in cell cultures of *Citrus* sp. MA Thesis, Univ S Fl, Tampa, Fl 33620

Barthe GA, Jourdan PS, McIntosh CA, Mansell RL (1987) Naringin and limonin production in callus cultures and regenerated shoots from *Citrus* sp. J Plant Physiol 127:55-65

Barthe GA, Jourdan PS, McIntosh CA, Mansell RL (1988) Radio-immunoassay for the quantitative determination of Hesperidin and Analysis of Its Distribution in *Citrus sinensis*. Phytochemistry 27:249-254

Brunet G, Ibrahim RK (1973) Tissue culture of citrus peel and its potential for flavonoid synthesis. Z. Pflanzenphysiol 69:152-162

Button J, Kochba J (1977) Tissue culture in the citrus industry. In: Reinert J, Bajaj YPS (eds) Applied and fundamental aspects of plant cell, tissue, and organ culture. Springer, Berlin Heidelberg New York, pp 70-92

Chandler BV, Johnson RL (1977) Cellulose acetate as a selective sorbent for limonin in orange juice. J Sci Food Agric 28:875-884

Chaturvedi HC, Mitra GC (1974) Clonal propagation of citrus from somatic callus cultures. Hortscience 9:118-120

Cody V, Middleton E Jr, Harborne JB (eds) (1986) Progress in clinical and biological research, vol 213:Plant flavonoids in biology and medicine. Liss, New York

Dreyer DL (1983) Limonoids of the Rutaceae. In: Waterman PG, Grundon MF (eds) Chemistry and chemical taxonomy of the rutales. Phytochem Soc Eur Symp Ser 22. Academic Press, New York London, pp 215-246

Feldman AW, Hanks RW (1965) Phenolic compounds in roots and leaves of four citrus cultivars. Nature (London) 207:985-986

Hasegawa S, Pelton VA, Bennett RD (1983) Metabolism of limonoids by *Arthrobacter globiformis* II. Basis for a practical means of reducing the limonin content of orange juice by immobilized cells. J Agric Food Chem 31:1002-1004

Hasegawa S, Herman Z, Bennett RD (1988) Limonoid glucosides in citrus. In: Proc 1988 Subtropical technology Conf USDA/ARS, Lake Alfred, Fl

Hiraoka N, Tabata M (1974) Alkaloid production by plants regenerated from cultured cells of *Datura innoxia*. Phytochemistry 13:1671-1675

Horowitz RM (1986) Taste effects of flavonoids. In: Cody V, Middleton E Jr, Harborne JB (eds) Progress in clinical and biological research, vol 213:Plant flavonoids in biology and medicine. Liss, New York, pp 163-175

Horowitz RM, Gentili B (1977) Flavonoid constituents of *Citrus*. In: Nagy S, Shaw PE, Veldhuis MK (eds) Citrus science and technology. Avi, Westport, Conn, pp 397-426

Ismail MA, Brown GE (1975) Phenolic content during healing of Valencia orange peel under high humidity. J Am Soc Hortic Sci 100:249-251

Jourdan PS, Weiler EW, Mansell RL (1983) Radioimmunoassay for naringin and related flavanone-7-neohesperidosides using a tritiated tracer. J Agric Food Chem 31:1249-1255

Jourdan PS, McIntosh CA, Mansell RL (1985) Naringin levels in citrus tissues II. Quantitative distribution of naringin in *Citrus paradisi* Macfad. Plant Physiol 77:903-908

Kefford JF, Chandler BV (1970) The chemical constituents of citrus fruits. Advances in Food Research, Suppl 2. Academic Press, New York London

Kesterson JW, Hendrickson R (1953) Naringin, a bitter principle of grapefruit. Fla Agric Exp Stn Tech Bull 511:1-35

Kobayashi S, Ikeda I, Uchimiya H (1985) Conditions for high frequency embryogenesis from orange (*Citrus sinensis* Osb.) protoplasts. Plant Cell Tiss Org Cult 4:249–259
Kordan HA (1959) Proliferation of excised juice vesicles of lemon in vitro. Science 129:779–780
Kordan HA (1977) Sucrose activation of mitosis in lemon fruit explants. (*Citrus limon* L.) Experientia 33:924–925
Kordan HA, Morgenstern L (1962) Flavonoid production by mature citrus fruit tissue proliferating in vitro. Nature (London) 195:163–14
Lewinsohn E, Beman E, Mazur Y, Gressel J (1986) Glucosylation of exogenous flavanones by grapefruit (*Citrus paradisi*) cell cultures. Phytochemistry 25:2531–2535
Lewinsohn E, Berman E, Mazur Y, Gressel J (1989) (7)-Glucosylation and (1–6) rhamnosylation of exogenous flavanones by undifferentiated *Citrus* cell cultures. Plant Sci 61:23–28
Maier VPS (1983) Chemistry and significance of selected *Citrus* limonoids and flavonoids. In: Waterman PG, Grundon MF (eds) Chemistry and chemical taxonomy of the Rutales. Phytochem Soc Eur Symp Ser 22. Academic Press, New York London, pp 319–342
Maier VPS, Hasegawa S, Bennett RD, Echols L (1980) Limonin and limonoids: chemistry, biochemistry, and juice bitterness. In: Nagy S, Attaway JA (eds) Citrus nutrition and quality. ACS Symp Ser 143, Am Chem Soc, Washington, DC, pp 63–82
Mansell RL, Weiler EW (1980) Immunological tests for the evaluation of citrus quality. In: Nagy S, Attaway JA (eds) Citrus nutrition and quality. ACS Symp Ser 143, Am Chem Soc, Washington, DC, pp 341–359
McHale D, Khopkar PP, Sheridan JB (1987) Coumarin glycosides from *Citrus* flavedo. Phytochemistry 26:2547–2549
McIntosh CA, Mansell RL (1983) Distribution of limonin during the growth and development of leaves and branches of *Citrus paradisi*. J Agric Food Chem 33:319–325
McIntosh CA, Mansell RL, Rouseff RL (1982) Distribution of limonin in the fruit tissues of nine grapefruit cultivars. J Agric Food Chem 30:689–692
McIntosh CA, Mansell RL, Barros S (1988) Relationship between limonin concentration and naringin concentration in commercial strength grapefruit juice. J Food Sci 52:1734–1735
McIntosh CA, Mansell RL (1990) Biosynthesis of naringin in *Citrus paradisi*; UDP-Glucosyltransferase activity in grapefruit seedlings. Phytochemistry (in press)
McIntosh CA, Latchinian L, Mansell RL (1990) Demonstration of flavanone-0-7-glucosyltransferase activity and its resolution from other glucosyltransferase activities in seedlings of *Citrus paradisi*. (submitted)
Moore GA (1985) Factors affecting in vitro embryogenesis from undeveloped ovules of mature *Citrus* fruit. J Am Soc Hortic Sci 110:66–70
Murashige T, Tucker DPH (1969) Growth factor requirements of citrus tissue culture. In: Proc 1st Int Citrus Symp 3:1155–1161
Nakatani M, Iwashita T, Naoki H, Hase T (1985) Structure of a limonoid antifeedant from *Trichilia roka*. Phytochemistry 24:195–196
Nishura M, Kamiya S, Esaki S (1971) Flavonoids in citrus and related genera, pt 3. Flavonoid pattern and citrus taxonomy. Agr Biol Chem 35:1691–1706
Robbins RC (1980) Medicinal and nutritional aspects of citrus bioflavonoids. In: Nagy S, Attaway JA (eds) Citrus nutrition and quality. ACS Symp Ser 143, Am Chem Soc, Washington, DC, pp 43–62
Rouseff RL (1980) Flavonoids and citrus quality. In: Nagy S, Attaway JA (eds) Citrus nutrition and quality. ACS Symp Ser 143, Am Chem Soc, Washington, DC, pp 83–108
Rouseff RL, Nagy S (1982) Distribution of limonoids in citrus seeds. Phytochemistry 21:85–90
Spiegel-Roy P, Kochba J (1980) Embryogenesis in *Citrus* tissue cultures. In: Fiechler A (ed) Advances in biochemical engineering, vol 16: Plant cell cultures I. Springer, Berlin Heidelberg New York, pp 27–48
Spiegel-Roy P, Kochba J, Saad S (1983) Selection for tolerance to 2,4-dichlorophenoxyacetic acid in ovular callus of orange (*Citrus sinensis*). Z Pflanzenphysiol 109:41–48
Tabata M, Yamamoto H, Hiraoka N, Konoshima M (1972) Organization and alkaloid production in tissue cultures of *Scopolia parviflora*. Phytochemistry 11:949–955
Tabata M, Muzukami H, Hiraoka N, Konoshima M (1974) Pigment formation in callus cultures of *Lithospermum erythrorhizon*. Phytochemistry 13:927–932
Tabata M, Ogino T, Yoshioka K, Yoshikawa N, Hiraoka N (1978) Selection of cell lines with higher yield

of secondary products. In: Thorpe TA (ed) Frontiers of plant tissue culture. Univ Press, Calgary, Can, pp 213-222
Ting SV (1980) Nutrients and nutrition of citrus fruits. In: Nagy S, Attaway JA (eds) Citrus nutrition and quality. ACS Symp Ser 143, Am Chem Soc, Washington, DC, pp 1-24
Tisserat B, Galletta PD (1987) In vitro culture of lemon juice vesicles. Plant Cell Tiss Org Cult 11:81-95
Vardi A, Galun E (1989) Isolation and culture of *Citrus* protoplasts. In: Bajaj YPS (ed) Biotechnology in agriculture and forestry, vol 8: Plant protoplasts and genetic engineering I. Springer,. Berlin Heidelberg New York Tokyo, pp 147-159
Waterman PG, Grundon MF (eds) (1983) Chemistry and chemical taxonomy of the Rutales. Phytochem Soc Eur Symp Ser 22. Academic Press, New York London
Weiler EW, Mansell RL (1980) Radioimmunoassay for limonin using a tritiated tracer. J Agric Food Chem 28:543-545
Wink M (1987) Why do lupin cell cultures fail to produce alkaloids in large quantities? Plant Cell Tiss Org Cult 8:103-111
Wu T-S (1988) Alkaloids and coumarins of *Citrus grandis*. Phytochemistry 27:3717-3718
Zenk MH (1978) The impact of plant cell culture on industry. In: Thorpe TA (ed) Frontiers of plant tissue culture. Univ Press, Calgary, Can, pp 1-14
Zenk MH, El-Shagi H, Arens H, Stockigt J, Weiler EW, Deus B (1977) Formation of the indole alkaloids serpentine and ajmalicine in cell suspension cultures of *Catharanthus roseus*. In: Barz W, Reinhard E, Zenk MH (eds) Plant tissue culture and its biotechnological application. Springer, Berlin Heidelberg New York, pp 27-43
Ziegler L, Wolfe HS (1975) Citrus growing in Florida. Univ Press, Gainesville, Fla

XIII *Cymbopogon* Spreng. (Aromatic Grasses): In Vitro Culture, Regeneration, and the Production of Essential Oils

H.L. Sreenath[1,2] and K.S. Jagadishchandra[1]

1. Introduction

1.1 Distribution and Importance of the Plant

Cymbopogon Spreng. is one of the major aromatic plant genera, belonging to the tribe Andropogoneae of the family Poaceae. The number of species recognized by different workers in this genus varies from 55 to over 100 (Bor 1960; Chase and Niles 1962; Soenarko 1977). With the exception of the cultivated and introduced species, *Cymbopogon* occurs only in the Old World tropics and subtropics. Like other members of the tribe Andropogoneae, *Cymbopogon* is adapted to hot moist conditions. The species are more or less evenly distributed in the area, but several rather diffuse centers of diversity such as Indo-China, India, East Africa, and Queensland can be recognized (Soenarko 1977). The species of *Cymbopogon* are either densely or loosely tufted plants ranging in height from 20 cm to 3 m. They are mostly perennial; culms of most species are erect and unbranched (Fig. 1).

Most of the species of *Cymbopogon* can be clearly distinguished from the related genera in the tribe by their aromatic smell. The species like *C. nardus* (L.) Rendle, *C. winterianus* Jowitt, *C. flexuosus* (Nees) Wats., *C. citratus* (DC.) Stapf and *C. martinii* (Roxb.) Wats. have been cultivated for a long time for their essential oils (Guenther 1950; Bor 1960; Jagadishchandra 1975a,b).

Lemongrass oil, citronella oil, palmarosa oil, and gingergrass oil are the names of commercially important essential oils extracted from the genus. In addition to the main aromatic constituents listed in Table 1, more than 50 other aromatic components have been reported to be present in the oils in varying concentrations (Guenther 1948, 1950; Thapa et al. 1971; Sobti et al. 1982; Maheshwari and Mohan 1985). These essential oils have traditionally been regarded as large-volume oils used in a wide variety of perfumery products and to a much smaller extent in flavoring. They are used either as perfumery oils per se or as sources of individual chemical constituents (Table 1 and Fig. 2) or their derivatives, which are themselves important materials in the perfumery or flavoring industries.

Lemongrass oil per se is used as an ingredient of aerosol deodorants, floor polishes, household detergents, and a whole range of domestic and industrial products in which a pleasant fresh lemon fragrance is desired, whether in its own

[1]Department of Botany, University of Mysore, Manasagangotri, Mysore-570 006, India
[2]Present Address: Tissue Culture Division, Central Coffee Research Institute, Coffee Research Station-577 117, Chikmagalur District, Karnataka, India

Biotechnology in Agriculture and Forestry, Vol. 15
Medicinal and Aromatic Plants III (ed. by Y.P.S. Bajaj)

Fig. 1. Mature plant of *Cymbopogon winterianus*. The scale is 1 m long

right or as a mask for the unpleasant odors of certain ingredients. From citral, the major constituent of lemongrass oil, ionones (β-ionone, methyl ionone, etc.), an important group of perfumery isolates and finally vitamin A and E are manufactured (Guenther 1948, 1950, Anonymous 1950, Jagadishchandra 1975a, Robbins 1983).

The range of products in which citronella oil is used for its fresh fragrance is very large and includes household and toilet soaps, many types of detergents, various household and industrial polishes, cleaning compounds, aerosols, bath preparations, and so on. Geraniol and citronellal are important isolates of citronella oil. Citronellal is important as a source of hydroxy-citronellal, which is of great importance in its own right, possessing a fragrance reminiscent of lily-of-the-valley. From hydroxy-citronellal it is also possible to manufacture 1-menthol. Geraniol

Table 1. List of important aromatic species of *Cymbopogon*, their essential oils and major aromatic isolates

Name of oil	Species	Chromosome no. (2n)	Common name	Wild or cultivated	Major aromatic isolates (%)
Lemongrass oil					
East Indian type	*C. flexuosus* (Nees) Wats.	20+2B, 40+6B's, 60	East Indian Lemongrass	Wild and cultivated	Citral (65–85)
	C. pendulus (Nees) Wats.	30, 60	North Indian lemongrass	Wild and cultivated	Citral (75–83)
West Indian type	*C. citratus* Wats.	40, 60	West Indian lemongrass	Cultivated	Citral (54–87)
Citronella oil					
Ceylon type	*C. nardus* (L.) Rendle var. *nardus*	40	Ceylon citronella	Cultivated	Geraniol, (50–60) Citronellal (15–20)
	C. nardus (L.) Rendle var. *confertiflorus*	20	Wild citronella	Wild	Geraniol (35–40)
Java type	*C. winterianus* Jowitt	20	Java citronella	Cultivated	Citronellal, (32–45) Geraniol, (12–18) Citronellol (11–15)
Palmarosa oil	*C. martinii* (Roxb.) Wats. var. *motia*	20+1–2B's	Rosha grass or rusa grass	Wild and cultivated	Geraniol (88–93)
Gingergrass oil	*C. martinii* (Roxb.) Wats. var. *sofia*	40	Gingergrass	Wild	Geraniol, (39–48) perillyl alcohol
Karnkusa oil	*C. jwarancusa* (Jones) Schult.	20	Khavigrass	Wild	Piperitone (60–70)

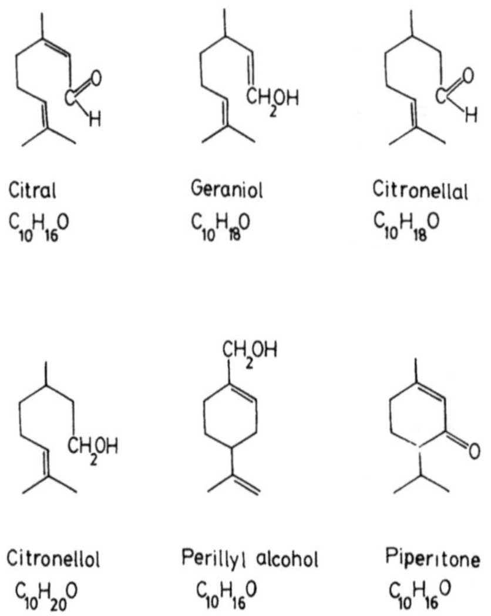

Fig. 2. Chemical structure of important aromatic isolates of essential oils of *Cymbopogon* species

and its derivative citronellol (the latter can also be manufactured from citronellal) both possess fragrances reminiscent of rose and are in fact known as rose alcohols. They are both of considerable importance in an extensive range of perfumery products. Derivatives of citronellol include esters such as formate and acetate, the former having applications in floral composition of rose and lily-of-the-valley type, and the latter when a more fruity character is required. Other products from geraniol and citronellol include nerol, laevo citronellol etc., each of which has a special use in different branches of perfumery (Anonymous 1950; Thapa et al. 1971; Jagadishchandra 1975a; Gupta and Jain 1978; Robbins 1983).

Palmarosa oil is rich in geraniol and is an important raw material in perfumery, extensively used for imparting a rose-like aroma to wide range of soaps, cosmetic products, toiletry goods, tobacco products, and many others. Geraniol isolated from palmarosa oil can also be used to synthesize the perfumery compounds mentioned in the previous paragraph (Anonymous 1950; Thapa et al. 1971; Jagadishchandra 1975a; Gupta and Jain 1978).

Gingergrass oil is closely related to palmarosa oil and in it perillyl alcohol exists along with geraniol. Gingergrass oil is used as a cheap substitute for palmarosa oil. Perillyl alcohol has a powerful warm woody and herbaceous floral odor, more like linalol and terpineol. It blends well with green floral notes in narcissus and lily, imparting a petal-like effect to rose perfumes. Perillyl acetate, having an odor reminiscent of dill and spearmint, can be synthesized from perillyl alcohol (Zutchi 1982).

C. jwarancusa yields an aromatic pale yellow oil possessing a minty odor. The oil has been used medicinally for all kinds of fever (Bor 1960). Piperitone, which is the major constituent of the oil (Saeed et al. 1978), has recently been used as an asthmolytic. Further, it can be synthetically converted into menthol and thymol, with great pharmaceutical utility (Thapa et al. 1971; Jagadishchandra 1975a).

Several *Cymbopogon* species are also used in native medicine. The essential oil and infusions of these aromatic grasses are used to treat a number of human diseases like leprosy, gout, rheumatism, sprain, coughs, cold, fever, and various stomach troubles including cholera (Kirtikar and Basu 1935; Anonymous 1950; Nadkarni 1954; Bor 1960; Chopra and Handa 1961; Soenarko 1977; Suri et al. 1978).

1.2 Production and Demand for Essential Oils on the World Market

The major producers and the main consumers on the world market of important essential oils of *Cymbopogon* species are given in Tables 2 and 3, respectively.

Lemongrass oil, which ranks among the most important essential oils in terms of quantities used, has been available in two basic forms, namely, "East Indian" oil, produced in South Asia, and "West Indian" oil, produced in Central and South America, parts of Africa, Indo-China, and the islands of the Indian Ocean. The volume of lemongrass oil traded internationally has declined considerably over the years from approaching 1500 tonnes in the late 1960's to under 500 tonnes at the beginning of the 1980's. Of this export total, Guatemala accounts for 50% on average, India for 35–40%, and China for most of the remainder (Robbins 1983).

Citronella oil ranks alongside lemongrass as one of the most widely used of all the natural essential oils, world consumption amounting to several thousand tonnes a few years back. The Ceylon type of citronella oil, produced solely in Sri Lanka,

Table 2. Major producers of important essential oils of *Cymbopogon* species. (Anonymous 1985)

Name of essential oil	Country	Quantity produced (in tonnes)	Total quantity (in tonnes)
Citronella oil	China	1200	2300
	Indonesia	550	
	Argentina	150	
	Sri Lanka	120	
	India	120	
	Taiwan	50	
	Brazil	40	
	Guatemala		
	Colombia		
Lemongrass oil	India	120	310
	China	120	
	Guatemala	40	
	Brazil	25	
	Sri Lanka		
	Haiti		
	USSR		
Palmarosa oil	India	35	55
	Brazil	10	
	Paraguay	8	
	Madagascar		
	Comoros		
	Guatemala		
	Indonesia		

Table 3. Annual consumption of lemongrass oil and citronella oil in importing countries at the end of 1970's. (Robbins 1983)

Country	Quantity (tonnes)	
	Lemongrass oil	Citronella oil
USA	140	550
UK	60–70	110
France	35–40	140
FRG	10–20	100
Netherlands	10	150
Switzerland	10	50–100
Japan	35	190
USSR	145	–
Mexico	–	130
Spain, Italy, and Belgium	–	80–90 (total)
Eastern Europe (including USSR)	–	80
Other countries	20–40	150–160 (minimum)

accounts for a very small proportion of world production and exports, the bulk of which is accounted for by the Java type of citronella. At the beginning of the 1980's, the total trade of citronella oil was around 1750 tonnes/a. Of this total, Indonesia and China each contributed between 40 and 45%, Taiwan and Guatemala each a little under 3%, Sri Lanka 6%, and Brazil 2%.

In recent years, severe competition has arisen for important essential oils of *Cymbopogon* species in the form of synthetic substrates mostly derived from turpentine (via α-pinene and β-pinene) and petroleum-derived hydrocarbons like acetylene. Lemongrass oil and citronella oil are also facing severe competition from natural sources, namely, oil of *Litsea cubeba* and oil of *Eucalyptus citriodora*, respectively. The current world demand for the essential oils of *Cymbopogon* species is met by the existing producers and the situation is likely to remain the same in the near future (Robbins 1983).

1.3 Conventional Practices for Propagation and Improvement

Species like *C. winterianus*, *C. citratus*, and *C. nardus*, which are known in cultivation only, are erratic in flowering and seed setting (Soenarko 1977; Sreenath and Jagadishchandra 1985–86) and they are conventionally propagated vegetatively by means of slips (root stocks). *C. flexuosus* and *C. martinii*, which show good flowering and seed setting, are grown from seeds in the nursery before transplanting, although often propagated by vegetative methods (Guenther 1950; Bor 1960; Jagadishchandra 1975a; Atal and Kapur 1982).

Cymbopogon species are naturally cross-pollinated and highly heterozygous. There have been various reports of natural hybridization between different species of *Cymbopogon* (Bor 1960; Gupta 1971; Jagadishchandra 1975b; Soenarko 1977). Earlier improvement work in *Cymbopogon* species was based on selection and introduction from the natural populations (Guenther 1950; Bor 1960; Jagadishchandra 1975a,b; Gupta and Jain 1978; Atal and Kapur 1982). However, in recent years, scientists have resorted to controlled hybridization followed by selection in the progeny for desired gene combinations. This has given encouraging results in developing many improved strains with improved oil quality and strains with altogether new essential oil components different from the traditional strains (Sobti et al. 1982). Recently, some economically important strains have been developed in species like *C. pendulus* (Steud.) Wats., *C. khasianus* (Munro ex Hack.) Bor, *C. jwarancusa* (Jones) Schult. and introduced to cultivation on a commercial scale (Sobti et al. 1982). Attempts are also being made to induce variability in *Cymbopogon* species through induced mutations (Choudhary and Kaul 1980). The natural diversity present in *Cymbopogon* species is decreasing at an alarming rate, due to deforestation and overexploitation. The long-term objectives of *Cymbopogon* improvement cannot be realized unless a lot of genetic variation is generated and conserved. This could, however, be possible, if the conventional practices are combined with some of the recently evolved biotechnology methods (See Bajaj 1988).

2 In Vitro Approaches

2.1 Review of Tissue Culture in *Cymbopogon*

The tissue culture studies on *Cymbopogon* species started with our report on callus induction and morphogenesis in some commercially important species of the genus (Sreenath and Jagadishchandra 1980a,b). Since then, a number of reports have appeared on callus induction and maintenance, morphology and histology of the callus, morphogenesis and plant regeneration, cytology of callus cultures and regenerated plants, and somaclonal variation in regenerants in commercially important species of *Cymbopogon*. A summary of the tissue culture work done in *Cymbopogon* species is presented in Table 4.

2.2 Sterilization and Preparation of Explants

Different plant parts like mature caryopses, entire seedlings, seedling sections, mature plant parts like rhizome, leaf, culm, and inflorescence are used as explants and the procedure of explant preparation and sterilization have been described (Sreenath and Jagadishchandra 1980a,b, 1988a,b,c, 1989; Jagadishchandra and Sreenath 1982, 1987). In all cases, the sterilant used is 0.1% mercuric chloride and the duration of the treatment varies from 5–15 min depending on the explant. After treatment with the sterilant, the explants are thoroughly washed with sterile distilled water and blotted before transferring to the culture medium.

The caryopses are sterilized for 5 min. The surface sterilized caryopses are cultured for 1 week on 1% water agar under sterile conditions to obtain seedlings. These seedlings are then transferred entire to the culture medium, and root and

Table 4. Summary of in vitro studies on *Cymbopogon*

Species	Ploidy	Explants and/or response	Medium	Reference
C. flexuosus *C. martinii* var. *motia* *C. jwarancusa*	Diploids	Callus induction from seeds, and differentiation	MS + 2,4-D or NAA for callus induction, MS + Kin or NAA for plant regeneration	Sreenath and Jagadishchandra (1980a)
C. nardus var. *nardus*	Tetraploid	Callus induction and growth from seeds, entire seedlings and seedling sections	MS + 2,4-D or NAA for callus induction, MS without growth regulator or MS with Kin for plant regeneration	Sreenath and Jagadishchandra (1980b)
C. nardus var. *confertiflorus*	Diploid			
C. martinii var. *motia*	Diploid	Regeneration of plants from seed callus from 2-year-old cultures	Callus induction and maintenance on MS + 2,4-D, plant regeneration on auxin-free MS medium	Sreenath and Jagadishchandra (1981)
C. winterianus	Diploid	Callus induction and plant regeneration from immature inflorescence explants	Callus induction and maintenance on MS + 2,4-D, plant regeneration on auxin-free MS medium	Sreenath and Jagadishchandra (1982, 1989)
C. flexuosus *C. nardus* var. *nardus* *C. nardus* var. *confertiflorus* *C. citratus* *C. winterianus* *C. jwarancusa*	Diploids to hexaploids	Callus induction from seeds, seedlings, root, inflorescence, and rhizome explants, plant regeneration through somatic embryogenesis	Callus induction on MS + 2,4-D morphogenesis on auxin-free MS or MS with IAA or NAA	Jagadishchandra and Sreenath (1982)
Cymbopogon species	Diploids to hexaploids	Nuclear cytology of callus, induction of euploid changes in long-term cultures	Callus induction and maintenance on MS + 2,4-D	Sreenath and Jagadishchandra (1983)

C. nardus var. *confertiflorus* *C. nardus* var. *nardus* *C. flexuosus* *C. winterianus* *C. citratus* *C. martinii* var. *motia* *C. jwarancusa*	Diploids to hexaploids	Morphohistological studies of callus, morphological differences of embryogenic and nonembryogenic callus	Callus induction and maintenance on MS + 2,4-D or MS + NAA supplemented with BAP or Kin	Sreenath and Jagadishchandra (1988a)
C. nardus var. *confertiflorus* *C. nardus* var *nardus* *C. martinii* var *motia* *C. winterianus* *C. citratus*	Diploids to hexaploids	Rhizome culture for direct sprouting of axillary buds and callus induction	Direct sprouting of axillary buds on MS, MS + BAP or MS + Kin with IAA, callus induction on MS + 2,4-D	Jagadishchandra andSreenath (1987)
C. flexuosus	Diploids to hexaploids	Nuclear cytology of the donor plants, calli, and regenerated plants, polyploidy in long-term cultures	MS + 2,4-D for callus induction and maintenance, plant regeneration in auxin-free MS medium	Sreenath and Jagadishchandra (1988b)
C. jwarancusa	Diploid	Cytology of the calli induced from seeds and inflorescence explants and the plants regenerated from them, euploid changes in long-term cultures, regeneration of autotetraploids	MS + 2,4-D for callus induction and maintenance, plant regeneration in auxin-free MS medium	Sreenath and Jagadishchandra (1987)
C. martinii var. *motia*	Diploids with B chromosomes	Instability of B chromosomes in cultured tissues and regeneration of plants without Bs from inflorescence culture of B-containing plants	MS + 2,4-D for callus induction and maintenance, plant regeneration in auxin-free MS medium	Sreenath and Jagadishchandra (1988c)
C. flexuosus *C. nardus*	Diploids to	Control of germination of somatic embryos by manipula-	MS + 2,4-D with BAP or Kin for	Sreenath and Jagadishchandra (1988d)

(*to be continued*)

Table 4. *(continued)*

Species	Ploidy	Explants and/or response	Medium	Reference
C. martinii *C. winterianus* *C. citratus*	hexaploids	ting sucrose concentration in the medium	embryogenic calli, MS basal medium with 9% sucrose for germination of somatic embryos	
C. winterianus	Diploid	Somaclonal variants in regenerated plants from calli of leaf sheath and release of high yielding strains	MS + 2,4-D + Kin for callus induction and proliferation, MS + IAA + Kin for plant regeneration through shoot buds	Mathur et al. (1988a,b, 1989)
C. martinii	–	High frequency plant regeneration by somatic embryogenesis and organogenesis from nodal segments	MS and LS media with 2,4-D and Kin for callus maintenance; MS with Kin, BAP, biotin, Ca-pantothenate and NAA for organogenesis LS with Kin, BAP, and IAA for embryogenesis	Baruah and Bordoloi (1989)

shoot sections separately. Roots collected from the field-grown plants are sterilized for 5 min and cultured. Leaf explants are prepared as described below. Shoots are collected from the adult plants, outermost leaves and debris are removed and then sterilized for 10 min. Then the whorl of leaves is cut into 4–5-mm-long explants, starting at the level of the shoot meristem and going above. The rhizomes are collected from the field-grown mature plants, the shoot portions at the level of apical meristem, and the roots at their bases are cut off. Then the rhizomes are cleaned carefully by removing the old leaf bases, scale leaves, soil etc. without damaging the axillary buds. The rhizomes are washed thoroughly in running tap water for 10 min and then sterilized for 15 min. The culms which consist of basal vegetative and the terminal inflorescence portions are collected from mature plants during the flowering season. The culms are cut into 15–20-mm-long explants and sterilized for 10 min.

2.3 Establishment of Embryogenic Callus and Suspension Cultures

2.3.1 Callus Cultures

Murashige and Skoog (1962) semi-solid medium devoid of any growth regulator is the basal medium and is designated MS. In various experiments, the basal medium was supplemented with IAA, NAA, 2,4-D, BAP, and Kin in various combinations, which play an important role in inducing embryogenic callus (Table 5). 2,4-D is essential for induction of embryogenic callus. Addition of BAP or Kin to 2,4-D-containing medium is beneficial for inducing embryogenic callus. NAA can also induce callus from certain explants, but the calli thus induced are nonembryogenic. The source of explants is also important in inducing embryogenic callus. Mature embryo, seedling mesocotyl and shoot sections, and young inflorescence and leaf

Table 5. Role of growth regulators in inducing embryogenic callus in *Cymbopogon* species. (Compiled from published and unpublished results of the authors)

Growth regulators	Type of response
MS (No growth regulator)	No callus
MS + IAA (1–50 mg/l)	No callus
MS + NAA (1–5 mg/l)	Nonembryogenic callus
MS + IAA + NAA (1–5 mg/l each)	Nonembryogenic callus
MS + 2,4-D (1–5 mg/l)	Embryogenic callus
MS + Kin (1–5 mg/l)	No callus
MS + BAP (1–5 mg/l)	No callus
MS + 2,4-D (1–5 mg/l) + BAP (1 mg/l)	Embryogenic callus
MS + 2,4-D (1–5 mg/l) + Kin (1 mg/l)	Embryogenic callus
MS + IAA (1–5 mg/l) + BAP (1 mg/l)	No callus
MS + IAA (1–5 mg/l) + Kin (1 mg/l)	No callus
MS + NAA + Kin (1 mg/l each)	Nonembryogenic callus
MS + NAA + BAP (1 mg/l each)	Nonembryogenic callus

Table 6. Explants capable of induction of embryogenic and nonembryogenic callus in *Cymbopogon* species. (Compiled from the published and unpublished results of the authors)

Species	Ploidy	Explant	
		Embryogenic calli	Nonembryogenic calli
C. flexuosus	Diploid	Mature embryo, seedling mesocotyl and shoot section, inflorescence, leaf	Rhizome, root, vegetative culm
	Tetraploid with 6B's	– " –	– " –
	Hexaploid	Inflorescence, leaf	– " –
C. nardus var. *confertiflorus*	Diploid	Mature embryo, seedling mesocotyl and shoot section, inflorescence, leaf	– " –
C. nardus var. *nardus*	Tetraploid	– " –	– " –
C. martinii var. *motia*	Diploid with and without B's	– " –	
C. martinii var. *sofia*	Tetraploid	– " –	
C. winterianus	Diploid	Leaf, inflorescence	Rhizome, root, vegetative culm
C. citratus	Hexaploid	Leaf	Rhizome, root
C. jwarancusa	Diploid	Mature embryo, seedling mesocotyl and shoot sections, inflorescence, leaf	Rhizome, root, vegetative culm
C. flexuosus × *C. martinii* hybrid	Diploid	Leaf, inflorescence	

explants of mature plants give rise to embryogenic calli; the rhizome, root, and vegetative culms give rise to nonembryogenic callus (Table 6).

From the embryos, callus is initiated from the mesocotyl region (Fig. 3A) after 20–25 days of culture. From the intact seedlings, callus is induced from the shoot, mesocotyl, and root portions (Fig. 3B). The shoot sections of seedlings readily produce callus. The root explants of seedlings and adult plants produce no callus when separated from shoot portions, implying that some important substance is provided by the shoots for callus induction in roots. When once induced, the root callus can also be isolated and subcultured. A 3-cm-long basal region of the two or three innermost leaves is most responsive for callus induction. Callus tissue is initiated mainly at the cut ends of leaf explants (Fig. 3D). Explants from immature and elongating vegetative culms give good response and the percentage of response varies from species to species. Presence of an axillary bud in the explant is also beneficial for callus induction. Species, like the diploid *C. winterianus*, the tetraploid *C. nardus*, and the hexaploid *C. flexuosus*, possessing fleshy culms with numerous intercalary meristematic and parenchymatous tissues, showed a high percentage of callus induction from vegetative culm explants. The explants from young unemerged inflorescences, in which floral primordia are differentiated, but at very early stages of development, responded well (Fig. 3C). In inflorescence explants also, the percentage of response varies from species to species. The response of mature rhizomes is better than that of the immature ones. Further, entire rhizomes

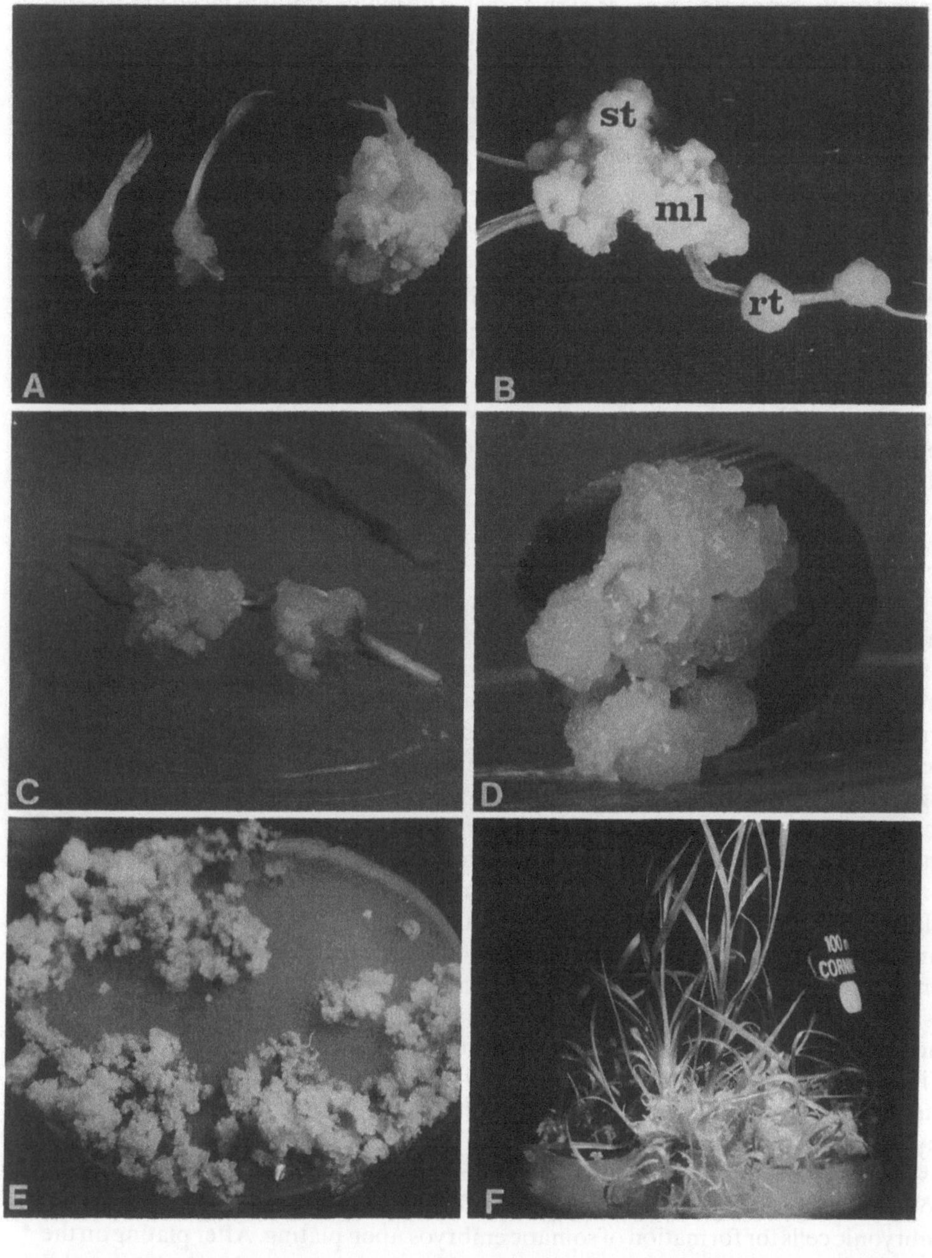

Fig. 3A-F. Induction of embryogenic callus from different explants and plant regeneration in *Cymbopogon* species. **A** Different stages of embryogenic callus induction from caryopses in *C. martinii* var. *motia* (Jagadishchandra and Sreenath 1982). **B** Callus induction from different parts of entire seedling in *C. nardus* (*st* shoot; *ml* mesocotyl; *rt* root). (Sreenath and Jagadishchandra 1980b). **C** Callus induction from immature inflorescence explants in *C. martinii* var. *motia*. **D** Callus induction from leaf explant in *C. martinii* × *C. flexuosus* hybrid. **E** Subcultured embryogenic callus of *C. martinii* var. *motia*. **F** Regenerated plantlets of *C. citratus* from leaf callus. (Sreenath and Jagadishchandra 1988d)

respond well, compared to pieces of rhizomes. The major problem faced in rhizome culture is fungal contamination (Jagadishchandra and Sreenath 1987).

Morphological differences are observed between embryogenic and nonembryogenic callus (Sreenath and Jagadishchandra 1988a). The embryogenic callus is nodular, hard, yellow, and opaque (Fig. 3E). The nonembryogenic callus is non-nodular, soft, whitish, and translucent. The microtome section of the embryogenic callus exhibits two zones, viz., a central core and peripheral (nodular) region. The core is composed of tracheary elements and parenchymatous cells. The peripheral region has a meristematic zone bounded on either side by parenchyma. The nonembryogenic callus shows loosely arranged parenchymatous cells, ranging from isodiametric to elongated. The meristematic zone is not clearly demarcated in this type of callus; only a few rather irregular patches of meristematic cells are present deeply embedded in the bulky parenchymatous zone. Identification, selection, and culture of the embryogenic callus at an early stage and its preferential maintenance are critical in retaining the long-term morphogenetic potential of such cultures.

2.3.2. Suspension Cultures

Embryogenic calli obtained from cultured embryos, leaf, and inflorescence segments are used for induction of embryogenic suspensions in *Cymbopogon* species (Sreenath and Jagadishchandra unpubl.). The calli initiated on MS + 2,4-D + BAP (1 mg/l each) were teased apart into small pieces and any nonembryogenic callus present was discarded. To obtain a fine suspension culture, it is of prime importance that as far as possible, a friable callus is used initially. The friability of the callus increases if it is maintained on a semi-solid medium for two or three passages. The embryogenic callus is then transferred into a liquid medium having the same composition. Generally, 250-ml Erlenmeyer flasks containing 50 ml medium with callus fragments are placed in a shaker at 100 r.p.m. at 27°C under 12-h light and dark cycles. This gives rise to a suspension of friable cell colonies, which develop very rapidly. After a few subcultures a stable embryogenic suspension culture results. Such a culture is composed of some elongated thick-walled nondividing cells and a large population of small richly cytoplasmic starch-filled embryogenic cells (Fig. 4A,B). The suspension is maintained by subculture at 1:5 dilution at 10–15-day intervals. Such suspensions grow rapidly, are finely dispersed, and do not contain any organized meristems or meristemoids. Depending on the planned use of the suspension, it may be maintained in a condition in which somatic embryos up to the globular or the early scutellar stage are formed, or it may consist entirely of embryonic cells for formation of somatic embryos upon plating. After plating on the agar medium of the same composition, callus colonies are obtained from cell suspensions. These callus colonies differentiate into plantlets upon transferring to the regeneration medium described in the next section.

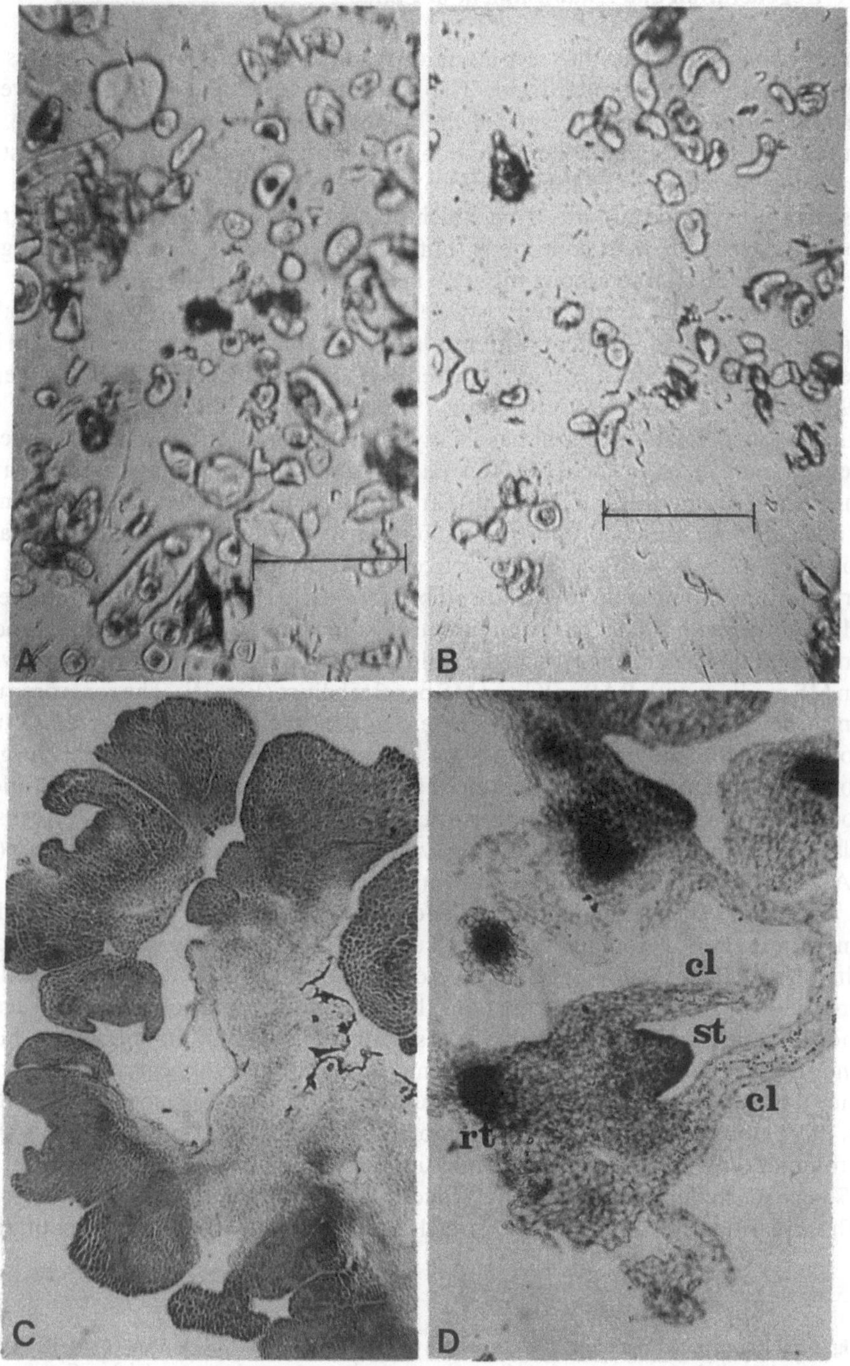

Fig. 4. A,B Cells from embryogenic suspension culture of *C. citratus*. *Bar* = 150 μm. **C,D** Somatic embryos of *C. martinii* var. *motia* (*rt* root; *cl* coleoptile; *st* shoot; *sc* scutellum). (Jagadishchandra and Sreenath 1982)

2.4 Regeneration and Establishment of Plants

In *Cymbopogon* species the technique of regeneration of plants from callus cultures is fairly well standardized. The studies have revealed that the developmental and physiological stages of the donor tissues, source of the explant, and suitable growth regulator combinations are most important in initiating stable embryogenic cultures (Tables 5 and 6). All species mentioned in Table 6 give rise to embryogenic cultures from which plants have been regenerated and grown to maturity under field conditions, but differences in the vigor of embryogenic competence are found among different species (Sreenath 1983).

The embryogenic calli can be isolated and subcultured at 4-6-week intervals on media with 2,4-D for further proliferation. The established cultures exhibited a high degree of embryogenic activity for over 3 years (Jagadishchandra and Sreenath 1982; Sreenath and Jagadishchandra 1988a). Embryogenic callus ready for subculture on a fresh medium has many somatic embryos at various stages of development on its surface (Fig. 4C). Early visible stages of somatic embryos are in the form of fine nodules on calli. These are usually clustered together at high density. Under appropriate conditions these globular structures differentiate into mature somatic embryos with scutellum, coleoptile, and coleorhiza (Fig. 4D). The embryonic development up to the late globular or early coleoptile stage continues on MS medium with 2,4-D. The somatic embryos germinate precociously and produce green plantlets after 2 months on the callus induction medium itself or after 2 weeks on MS medium with BAP or Kin or MS basal medium (Fig. 3F). The roots produced on these media are very slender. These plantlets are transferred to culture tubes containing MS medium with IAA or MS medium with full-strength inorganic compounds but without any organic substances, including sucrose, for additional rooting. Formation of fresh roots started within 1 week, and after 4-6 weeks the plantlets formed four to six tillers with a vigorous root system and several leaves (Fig. 5A). They are transferred to the soil with more than 95% survival.

Recently, the effect of different concentrations of sucrose in the basal medium on germination of somatic embryos has been reported (Sreenath and Jagadishchandra 1988d). At low levels of sucrose (3%), the somatic embryos germinate precociously before attaining the typical grass embryo morphology. At 6% sucrose, the germination of somatic embryos is slightly delayed and the shoots are shorter and stronger. At 9% sucrose, maturation of somatic embryos occurred rapidly and they develop to sizes similar to those found in seeds. Clearly recognizable features of these embryos include a well-developed scutellum and an embryonic axis with a prominent coleoptile. Embryo germination is delayed up to 20 days on medium with 9% sucrose, but the resulting shoots are compact and hardy, with a good root system. These plantlets can be transferred directly to soil, without an additional rooting step.

Fig. 5. A Regenerated plantlets of *C. flexuosus* (diploid) rooted on MS + IAA (1 mg/l). **B** Parent (*extreme left*), diploid regenerant (*middle*) and tetraploid regenerant (*extreme right*) in *C. flexuosus*. **C** Direct shoot formation by sprouting of axillary bud by node culture in *C. martinii* var. *motia*. (Jagadishchandra and Sreenath 1982). **D** Direct shoot formation by sprouting of axillary buds by rhizome culture in *C. winterianus*. (Jagadishchandra and Sreenath 1987)

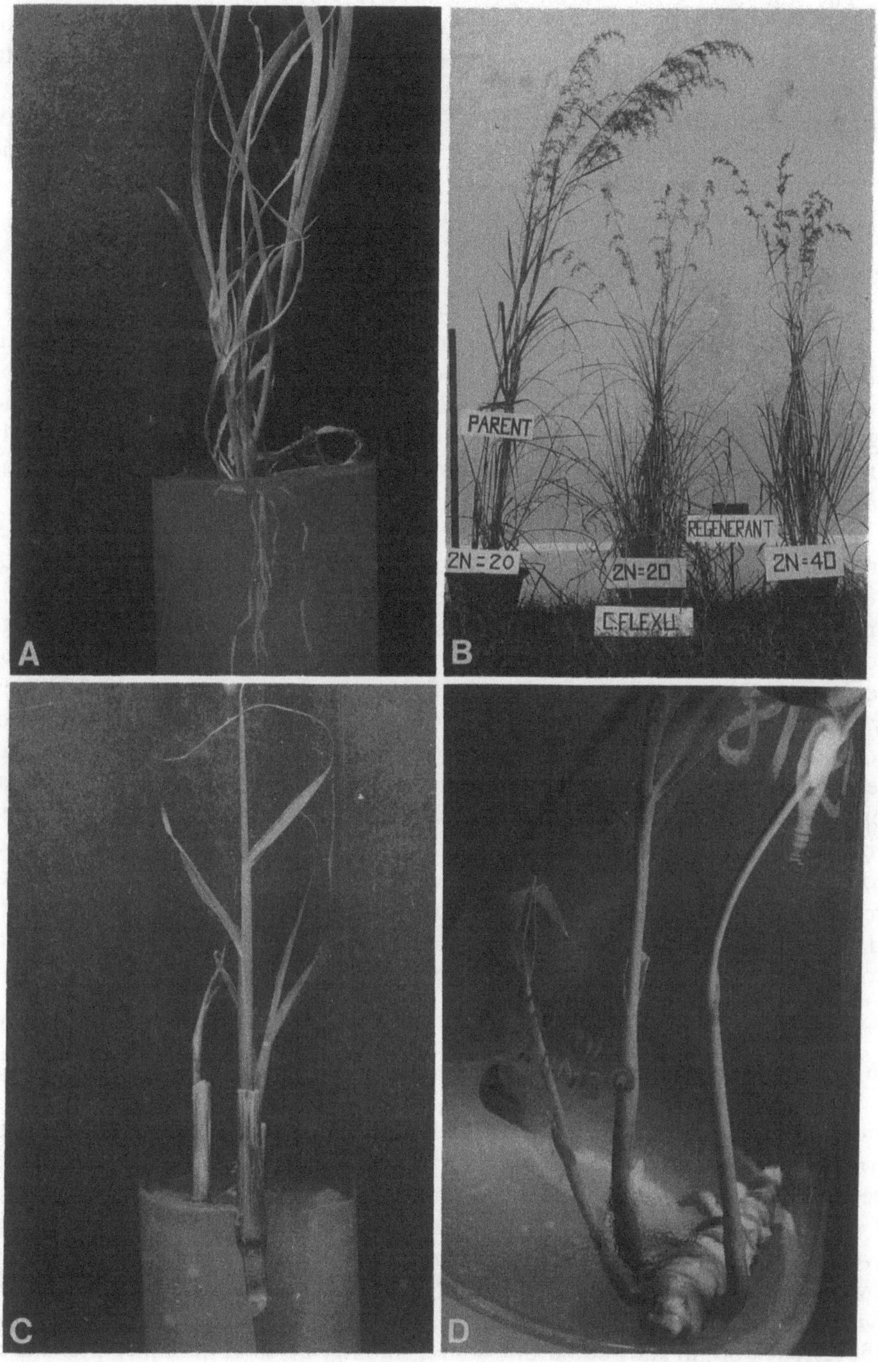
A
B
PARENT
2N=20
2N=20
REGENERANT
2N=40
C.FLEXU.
C
D

The studies in *Cymbopogon* species have revealed that use of high osmotic medium in the form of an elevated sucrose level could promote full maturation of somatic embryos, preventing their precocious germination. It would be easier to handle these mature embryos and their germination could be further delayed by partial desiccation and storage at 5–10°C. This is a step towards the development of synthetic seed technology in this genus by encapsulating the quiescent somatic embryos in a suitable matrix.

2.5 Direct Regeneration by Sprouting of Axillary Buds

As described in the previous sections, plant regeneration through callus culture is quite common in *Cymbopogon* species. However, callus cultures are reported to be associated with genetic instability (Larkin and Scowcroft 1981; Sreenath and Jagadishchandra 1987, 1988b). Hence, direct development of the shoots, entirely eliminating the callus stage, is desired for clonal propagation purposes, where the genetic stability is most important.

In *Cymbopogon* species the dormant axillary buds of culms and the rhizomes can be made to develop directly into shoots (Jagadishchandra and Sreenath 1982, 1987; Sreenath 1983). These studies reveal that the cytokinins, Kin and BAP, in combination with IAA, can trigger preformed axillary buds of culms to develop into shoots, and the percentage of response varies in different species. *Cymbopogon* species produce many tillers. Each such tiller can produce culms with 10–20 nodes. By culturing these nodal explants (Fig. 5C) with axillary buds, a large number of shoots can easily be produced. *Cymbopogon* species are typically rhizomatous and the rhizomes are short stems, underground or subaerial, responsible for perennation. *Cymbopogon* is conventionally propagated by planting the separated rhizomes, also called slips or root stocks. Rhizome culture and subsequent direct shoot regeneration (Fig. 5D) can enhance the rate of this vegetative method of propagation by many times. The technique of inducing preformed axillary buds of rhizomes and culms directly into shoots is of great importance in the rapid clonal propagation of elite plants of *Cymbopogon* species.

2.6 Cytogenetic Studies of Callus and Regenerated Plants

There are a few reports on the cytogenetics of callus and regenerated plants of *Cymbopogon* species from our laboratory (Jagadishchandra and Sreenath 1982; Sreenath 1983; Sreenath and Jagadishchandra 1983, 1987, 1988b).

The nuclear cytology of cultured somatic tissues of *Cymbopogon* species revealed the relative stability of chromosome number. After a long period of culture, however, euploid changes were induced (Jagadishchandra and Sreenath 1982). In *Cymbopogon jwarancusa*, a diploid species with $2n = 20$, calli of seed and inflorescence origin showed chromosomal stability up to 12 months, after which tetraploid cells with $2n = 40$ resulted in the calli (Fig. 6E). At the end of 30 months of subculture, further polyploidization resulted in a few octoploid cells with $2n = 80$. Both diploid and autotetraploid plants have been regenerated from the

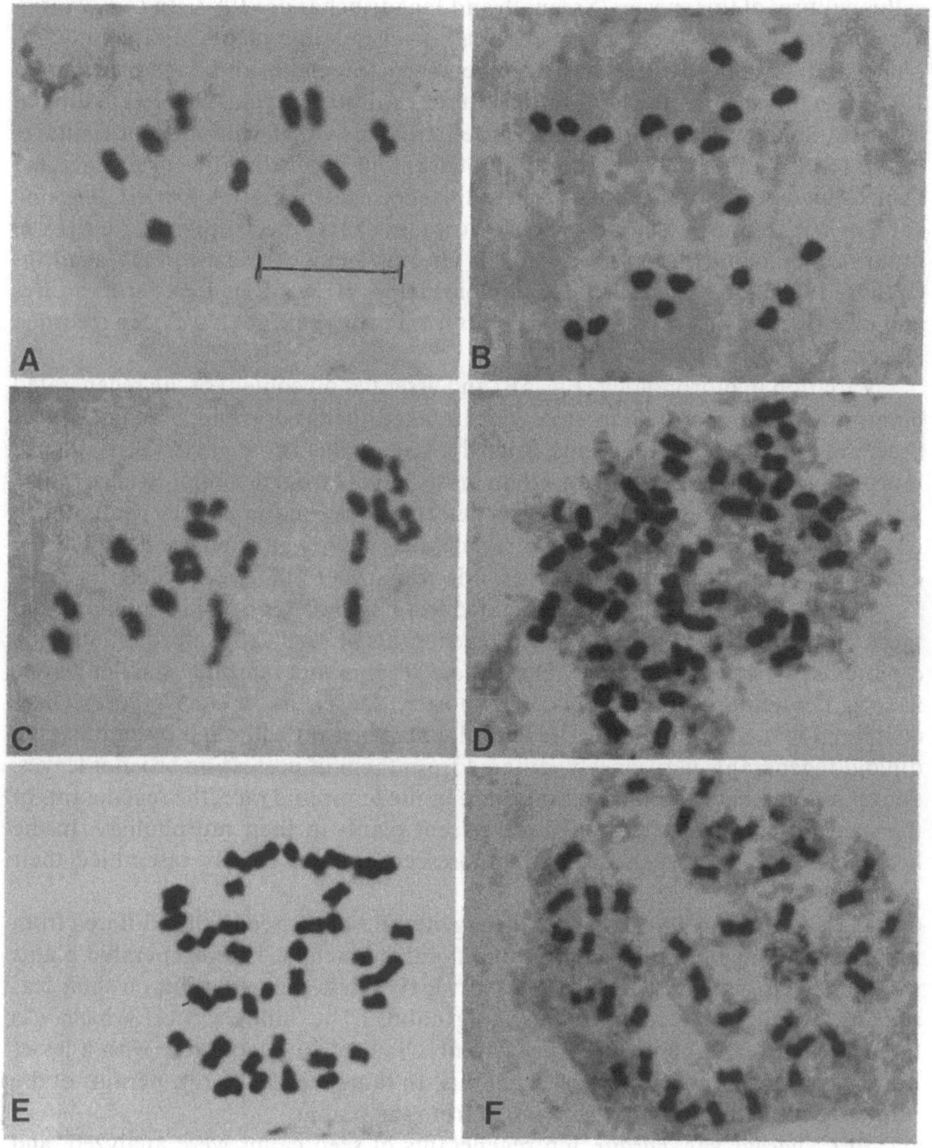

Fig. 6A-F. Nuclear cytology of callus cells and regenerated plants of *Cymbopogon* species. **A,B** Normal metaphase I and anaphase I showing 10 bivalents and 10-10 segregation, respectively, in diploid regenerants of *C. flexuosus*. **C** Metaphase I in an autotetraploid regenerant derived from diploid donor plant, showing bivalent and multivalent formation, in *C. flexuosus*. **D** Hyperpolyploid callus cell showing 80 A's, derived from the tetraploid race with 40 A's + 6 B's, in *C. flexuosus*. **E** Tetraploid callus cell with 40 chromosomes derived from diploid donor plant of *C. jwarancusa* with 2n = 20. **F** Tetraploid callus cell with 40 chromosomes derived from diploid donor plant of *C. martinii* var. *motia* with 2n = 20 (**A,C** Sreenath and Jagadishchandra 1988b; **E** Sreenath and Jagadishchandra 1987; **F** Jagadishchandra and Sreenath 1982). *Bar* = 10 μm

callus cultures of this species (Sreenath and Jagadishchandra 1987). In *C. flexuosus*, karyomorphology of the donor plants, callus-regenerated plants of diploid, tetraploid, and hexaploid races has been reported (Sreenath and Jagadishchandra 1988b). In this species, the long-term cultures exhibited euploid changes in diploid and tetraploid races. B chromosomes are eliminated from the inflorescence cultures of the tetraploid race. Diploid and autotetraploid plants have been regenerated from callus cultures of the diploid race of this species also (Fig. 6A-D). In *C. martinii* var. *motia*, plants without B chromosomes have been regenerated from inflorescence culture of plants with 2B chromosomes (Sreenath and Jagadishchandra 1988c). In this species also, euploid changes (Fig. 6F) in long-term cultures and regeneration of some asynaptic plants from callus cultures have been recorded (Sreenath 1983).

Variation in morphology (Fig. 5B) and essential oil content of the regenerated plants has been reported in this genus (Jagadishchandra and Sreenath 1982; Sreenath 1983). The regenerants from the seed callus show greater variation in morphology from the donor plants than those derived from the inflorescence callus. This is presumably related to the highly heterozygous condition of the genus due to out-breeding. The characters of the regenerants mainly affected are height of the culm, size and pigmentation of leaves, tillering, size of inflorescence, etc.

In *C. flexuosus*, plants are regenerated from diploid, tetraploid, and hexaploid races. In the diploid race, nine of ten regenerants of seed callus were found to be diploids and one was tetraploid. The diploid regenerants exhibited smaller leaves, less pigmentation, shorter culms, and shorter, denser, more erect inflorescences compared to the parent plant. The tetraploid regenerant, although resembling the diploid regenerants very closely, differed from them in possessing broader leaves, denser inflorescences, and larger racemes. In the tetraploid race, the regenerants of the seed callus closely resembled the parent plants in their morphology. In the hexaploid race, the regenerants of inflorescence callus closely resembled their parent plants.

In *C. nardus* var. *confertiflorus*, the regenerants of the seed callus differed from the parent plants in both vegetative and floral characters. The regenerated plants were less vigorous and had narrow, short leaves, less pigmentation on their leaf sheaths, with greatly reduced flowering culms. The inflorescence, which was typically dense and compact in the parent, changed to a loose one with a lesser number of raceme pairs in the regenerants. In this species, the regenerants of the inflorescence callus did not differ much from the parent.

In *C. nardus* var. *nardus*, ten regenerants of seed callus were analyzed, and showed a higher rate of tillering, thinner leaves with meager pigmentation, and a less waxy coating. However, the flowering culms of the regenerants were similar to those of the parent plants but were occasionally reduced in height, with a lax panicle. In this species more than ten regenerants of the inflorescence callus analyzed were morphologically almost identical to the parent plants.

In *C. winterianus*, more than ten regenerants of the inflorescence callus analyzed showed normal vegetative features and closely resembled the parent plants (Sreenath and Jagadishchandra 1989).

In *C. martinii* var. *motia*, the regenerants of the seed callus as well as inflorescence callus of two chromosomal races, one with 2B's ($2n = 20+2B$) and another without B's (2n = 20), were normal and identical to the parent plants in both vegetative and floral characters (Sreenath and Jagadishchandra 1988c).

In *C. jwarancusa*, 15 regenerants of the seed calli resembled one another closely, but all of them differed drastically from the parent plant. The regenerants were much more vigorous, with a high rate of tillering (50–60 tillers), as against 10–20 tillers of the parent. The leaves of the regenerants were coarse with macrohairs on the upper surface. The flowering was very poor and occurred only once in a year, in contrast to the profuse flowering (three to four times) of the parent. The inflorescence was green and had smaller racemes and spikelets, resembling that of *C. flexuosus*. White hairs, which are characteristically present at the base of bisexual spikelets of the parent plant, were, however, absent in the regenerants. The vegetative characters and the odor of the regenerants were a mixture of *C. jwarancusa* and *C. flexuosus*. The drastic changes observed in the regenerants of the seed callus of *C. jwarancusa* are attributed to hybridization between *C. jwarancusa* and *C. flexuosus*. Spontaneous hybridization is common in *Cymbopogon* species (Soenarko 1977). In *C. jwarancusa*, more than ten regenerants of inflorescence callus analyzed resembled the parent plant in leaf shape, texture, odor, etc., but all the regenerants had short culms and small leaves with a meager amount of wax on the inflorescence. The hairs of the bisexual florets were also reduced. In this species, regenerants from the inflorescence callus flowered within a year, while those of the seed callus flowered only after 1½–2 years. The regenerants of leaf callus were similar to the regenerants of inflorescence callus.

In *C. citratus* more than 50 regenerants of the leaf callus were analyzed, and showed great variation in morphology, such as, leaf size and texture, number of tillers, bush diameter, etc. About 50% of the regenerants showed significant morphological differences from the donor plant.

In the regenerants from leaf sheath callus of *C. winterianus*, a wide range of variation in seven yield-contributing agronomic traits, viz., herbage yield, tiller number, bush diameter, length and area of the longest leaf, fresh and dry matter ratio, and the essential oil content has been reported (Mathur et al. 1988a,b).

2.7 Improvement of Essential Oil

The essential oils of *Cymbopogon* species comprise monoterpenoids (Fig. 2 and Table 1). There are two basic techniques for extracting essential oils from plant material, viz., hydro/steam distillation and solvent extraction. In *Cymbopogon* species the essential oils are extracted from the aerial parts, especially foliage, through hydro/steam distillation.

There is the possibility of assessing the use of in vitro techniques for improvement of essential oil at two levels, namely, (1) callus or cell culture level and (2) at the whole plant level (regenerants). The relative importance of the two possibilities in *Cymbopogon* species is discussed here.

2.7.1 Callus and Cell Cultures

Plant cell cultures have become increasingly significant in the field of biotechnology as a potential source of commercially useful compounds (Misawa 1985; Stafford et al. 1986). The product levels of several culture systems seem to be high enough to warrant industrial production (Berlin 1988; see also Bajaj 1988). Indeed, shikonins are the first natural compounds to be produced in large fermentors in Japan (Fujita et al. 1982). There are two important problems that have to be overcome for in vitro production of useful compounds. These are (1) selection of specific cells that produce high amounts of the desired compounds and (2) the development of an adequate culture medium for the production of such useful compounds.

The presence of even minute quantities of aromatic constituents can be rapidly detected by smelling the squeezed plant parts. The odor expressed is characteristic of the species and is dependent on the aromatic constituents of the essential oil. By experience, one can assess the quality of the essential oil fairly well through the olfactory sense. The callus of *Cymbopogon* species, when squeezed and smelled, does not exude any smell of the essential oil. The hydrodistillation of the callus using Clevenger's apparatus also fails to yield any essential oil in the genus. However, as soon as the embryogenic callus differentiates green shoots, these develop the essential oil with the characteristic smell of the species (Sreenath 1983). The reformation of secondary products in redifferentiated tissue cultures indicates that cells in undifferentiated cultures still retain the capacity to synthesize secondary products, but the path of synthesis is blocked in some way.

The essential oils in plants accumulate in specialized tissues, such as glands. Most of the work on essential oil production in tissue cultures has shown that synthesis of the complete oil or flavor occurs only when the cultures have been morphologically differentiated to produce shoots and roots, or the culture contains specialized cells. In all the aromatic plants like *Ruta graveolens*, *Perilla*, cocoa, mint, and camomile, where essential oil is reported in cultures, a level of cell or organ differentiation was found to be necessary before synthesis and accumulation of the secondary product occurs. This suggests that cells need to attain a specific stage of differentiation before the secondary pathways are activated (Collins and Watts 1983).

Many economically important secondary metabolites are found only in traces or not at all in rapidly growing cultured cells (Berlin 1986). Today, the use of abiotic and biotic elicitors is believed to be a primary tool for the improvement of the yields of low-level or absent products in culture systems (DiCosmo and Misawa 1985). Past results and the failure to achieve a significant breakthrough in improving product levels of very important secondary metabolites have changed the attitude of many researchers in the field towards acknowledging the need of greater biochemical and molecular knowledge of secondary metabolism (Misawa 1985; Berlin 1986; Stafford et al. 1986). The biosynthesis of secondary products is strictly regulated during plant development and in cultured cells.

2.7.2 *Whole Plant Level*

The tissue culture technique is useful for the recovery of variants through somaclonal variation. In the commercially important species of *Cymbopogon*, the oil content of the regenerants was analyzed and compared with that of donor plants (Sreenath 1983). In *C. flexuosus, C. nardus, C. winterianus, C. martinii* var. *motia*, and *C. jwarancusa*, the regenerants show variation in oil content, i.e., some a decrease and others an increase. In *C. winterianus*, 25 randomly selected regenerants were analyzed for essential oil content and they showed significant variation. About 60% of the plants had a considerable increase in oil content. The average oil content of the donor plant was 1.7% and the highest oil content recorded in the regenerated plants was 2.5%, the range being from 1.5 to 2.5% (Sreenath and Jagadishchandra 1989). The odor of the regenerants of seed calli of *C. jwarancusa* shows an admixture of *C. jwarancusa* and *C. flexuosus*, but in all other species the oil of the regenerants shows the same odor as that of donor plants.

Variation with respect to the relative composition of the six major constituents of the essential oil pool, viz., citronellal, citronellol, geraniol, citronellylacetate, geranylacetate, and elemol has been recorded in regenerants of *C. winterianus* (Mathur et al. 1988a,b). These authors found that correlation analysis between agronomic parameters revealed a strong negative correlation between herbage yield and oil content. However, plants with more than threefold increase in herbage yield as well as nearly 1.5-fold increase in oil content over the donor parent could be recovered amongst the callus derivatives. Also, plants could be screened that have improved oil quality, possessing high amounts of desirable constituents like citronellol, citronellal, and geraniol, together with a sevenfold decrease in the undesirable component – elemol. After assessing the stability of the selected variants in row-replicated trials, five superior strains were carried on to multilocational stability and agronomic trials, with subsequent multiplication.

3 Protocol

Plant regeneration through somatic embryogenesis in *Cymbopogon* spp.

1. Surface sterilize caryopses, leaf base whorls, and immature inflorescences in 0.1% mercuric chloride for 5–10 min. Wash thoroughly in sterile distilled water.
2. Cut leaf bases into 4–5-mm-long explants and inflorescences into 15–20-mm-long explants. Use entire caryopses as explants.
3. Transfer sterile explants to callus induction medium to obtain primary cultures. The best medium for induction of embryogenic callus is MS medium supplemented with 2,4-D and BAP (1 mg/l each). Incubate under 8-h photoperiod at low light intensity. Embryogenic calli are obtained in 4–6 weeks. They are nodular, hard, yellow, and opaque.
4. Transfer embryogenic calli to medium of the same composition for maintenance and multiplication. The subcultured calli may be called secondary cultures. Thus the embryogenic calli may be maintained for more than 3 years.
5. Transfer small pieces (50–100 mg) of primary and secondary cultures to MS medium with 9% sucrose, but without growth regulator, for maturation and germination of somatic embryos into plantlets. Incubate under 12-h photoperiod of higher light intensity. Plantlets with four to six leaves are produced in 6–8 weeks.

6. Transfer the regenerated plantlets to MS inorganic medium with or without IAA (1 mg/l) for additional rooting and further shoot growth. In 4 weeks, six to eight additional roots are produced.
7. Transfer the rooted plantlets to the soil. About 95% of the plantlets survive in the soil.

4 Conclusion and Prospects

Essential oil of *Cymbopogon* species is the source of many commercially important aromatic compounds and the quality of the oil is determined by the high percentage of the desirable aromatic constituents and low percentage of undesirable constituents. The major breeding objectives in this genus are: increased herbage and oil yield, high percentage of desired aromatic constituents in the oil, adaptation to wider agroclimatic conditions, resistance to diseases, pests, and stress conditions. It is envisaged that during the next few years biotechnology will play a very important role in the generation of genetic variability and improvement of *Cymbopogon* species.

The development of cell culture methodology for *Cymbopogon* species has advanced to the point of reliable callus induction and subculture for important species. The embryogenic cultures regenerate plants via somatic embryogenesis that can be maintained for more than 3 years. Suspension cultures have been obtained and successfully plated to retrieve callus and regenerate plants. Callus and cell cultures can be used for isolation of mutants that are resistant/tolerant to diseases, salinity, herbicides, pests, cold, drought, and other stresses. Plantlets raised from callus cultures of this genus have shown variations in chromosome number, morphology of plants, and quality and composition of essential oil. High yielding strains of java citronella and lemongrass have been developed througn induction and screening of somaclonal variations. On the basis of stability/agronomic trials of citronella strains over locations, a variety CIMAP-BIO-13 has been released for commercial cultivation. In addition to high initial establishment (80–90%), the variety shows a 50% increase in total herb yield and oil content (Mathur et al. 1989). Thus it is clear that callus cultures are useful for producing useful variants. Direct sprouting of axillary buds to produce single and multiple shoots through node and rhizome culture can be used for rapid clonal propagation of selected varieties.

Acknowledgements. The authors thank the University Grants Commission, New Delhi for financial assistance and the University of Mysore for facilities.

References

Anon (1950) The Wealth of India, vol 2. CSIR, New Delhi, pp 411–418
Anon (1985) Total world production of important essential oils (1984). Pafai J 7,4:21–27
Atal CK, Kapur BM (eds) (1982) Cultivation and utilization of aromatic plants. Reg Res Lab. CSIR, Jammu-Tawi
Bajaj YPS (ed) (1988) Biotechnology in agriculture and forestry, vol 4: Medicinal and aromatic plants I. Springer, Berlin Heidelberg New York Tokyo

Baruah A, Bordoloi DN (1989) High frequency plant regeneration of *Cymbopogon martinii* (Roxb.) Wats. by somatic embryogenesis and organogenesis. Plant Cell Rep 8:483–485

Berlin J (1986) Secondary products from plant cell cultures. In: Rehm HJ, Reed G (eds) Biotechnology, vol 4. VCH, Weinheim, pp 630–658

Berlin J (1988) Approaches to altering regulatory controls of secondary pathways in cultured cells. In: Hanover JW, Keatley DE (eds) Genetic manipulation of woody plants. Plenum, New York, pp 353–364

Bor NL (1960) The grasses of Burma, Ceylon, India and Pakistan, Pergamon, New York, London

Chase A, Niles CD (1962) Index to grass species, Hall, Mass, pp 510–516

Chopra IC, Handa KL (1961) Review on Indian medicinal and allied plants, vol 60. ICAR, New Delhi

Choudhary DK, Kaul BL (1980) Effect of ionizing radiation on *Cymbopogon winterianus* Jowitt. Indian Perfumer 24:79–81

Collins HA, Watts M (1983) Flavor production in culture. In: Evans DA, Sharp WR, Ammirato PV, Yamada Y (eds) Handbook of plant cell culture, vol 1. Macmillan, New York, pp 729–747

DiCosmo F, Misawa M (1985) Eliciting secondary metabolism in plant cell cultures. Trends Biotechnol 3:318–322

Fujita Y, Tabata M, Nishi A, Yamada Y (1982) New medium and production of secondary compounds with the two-staged culture method. In: Fujiwara A (ed) Plant tissue culture 1982. Maruzen, Tokyo, pp 399–400

Guenther E (ed) (1948) The essential oils, vol 1. Van Nostrand, Princeton, NJ, pp 15–77

Guenther E (1950) The essential oils, vol 4. Van Nostrand, Princeton, NJ, pp 3–178

Gupta BK (1971) A note on the occurrence of natural hybrids in Indian cymbopogons. Plant Sci 3:120–121

Gupta BK, Jain N (1978) Cultivation and utilization of genus *Cymbopogon* in India. Indian Perfumer 22:55–68

Jagadishchandra KS (1975a) Recent studies on *Cymbopogon* Spreng. (aromatic grasses) with special reference to Indian taxa: cultivation and ecology: a review. J Plantat Crops 3:1–5

Jagadishchandra KS (1975b) Recent studies on *Cymbopogon* Spreng. (aromatic grasses) with special reference to Indian taxa: taxonomy, cytogenetics, chemistry and scope. J Plantat Crops 2:43–57

Jagadishchandra KS, Sreenath HL (1982) In vitro culture and morphogenetic studies in some species of *Cymbopogon* Spreng. (aromatic grasses). In: Fujiwara A (ed) Plant tissue culture 1982. Maruzen, Tokyo, pp 703–704

Jagadishchandra KS, Sreenath HL (1987) In vitro culture of rhizome (root stock) in *Cymbopogon* Spreng. and *Vetiveria* Bory. In: Reddy GM (ed) Recent advances in plant cell and tissue culture of economically important plants. Osmania Univ Press, Hyderabad, pp 199–208

Janaki Ammal EK, Gupta BK (1978) The aromatic grasses of India – an appraisal. Indian J For: 19–21

Kirtikar KR, Basu BD (1935) Indian medicinal plants, vol 4. Lalit Mohan Basu, Allahabad, pp 2675–2683

Larkin PJ, Scowcroft WR (1981) Somaclonal variation – a novel source of variability from cell cultures for plant improvement. Theor Appl Genet 60:197–214

Maheshwari ML, Mohan J (1985) Geranyl formate and other esters in palmarosa oil. Pafai J 7, 3:21–26

Mathur AK, Ahuja PS, Pandey B, Kukreja AK, Mathur A (1988a) Development of superior strains of an aromatic grass – *Cymbopogon winterianus* Jowitt – through somaclonal variation. In: Int Conf Research in plant science and its relevance to future, Abstr, Univ Press, Delhi, p 160

Mathur AK, Ahuja PS, Pandey B, Kukreja AK, Mandal S (1988b) Screening and evaluation of somaclonal variations for quantitative and qualitative traits in an aromatic grass, *Cymbopogon winterianus* Jowitt. Plant Breeding 101:321–334

Mathur AK, Ahuja PS, Kukreja AK, Pandey B, Mathur A (1989) Progress in the application of tissue culture for the improvement of aromatic grasses. In: XIII plant tissue culture Conf Abstr. Univ Press, North-Eastern Hill University, Shillong, p 31

Misawa M (1985) Production of useful metabolites. Adv Biochem Eng Biotechnol 31:59–88

Murashige T, Skoog F (1962) A revised medium for rapid growth and bioassays with tobacco tissue cultures. Physiol Plant 15:473–497

Nadkarni KM (1954) Indian materia medica. Popular Book Depot, Bombay

Robbins SRJ (1983) Selected markets for the essential oils of lemongrass, citronella and eucalyptus. Trop Prod Inst, London

Saeed T, Sandra PJ, Verzele MJE (1978) Constituents of the essential oil of *Cymbopogon jwarancusa*. Phytochemistry 17:1433–1434

Sobti SN, Verma V, Rao BL (1982) Scope for development of new cultivars of cymbopogons as a source of terpene chemicals. In: Atal CK, Kapur BM (eds) Cultivation and utilization of aromatic plants. RRL, CSIR, Jammu-Tawi, pp 302–307

Soenarko S (1977) The genus *Cymbopogon* Sprengel (Gramineae). Reinwardtia 9:225–375

Sreenath HL (1983) Cytogenetic and tissue culture studies in some commercially important species of aromatic grasses. PhD Thesis, Mysore Univ, Mysore

Sreenath HL, Jagadishchandra KS (1980a) The role of growth regulators in the tissue culture and differentiation of *Cymbopogon* Spreng. In: Rao PS, Heble MR, Chadha MS (eds) Plant tissue culture, genetic manipulation and somatic hybridization of plant cells. BARC, Bombay, pp 328–334

Sreenath HL, Jagadishchandra KS (1980b) Callus induction and growth in two varieties of *Cymbopogon nardus* (L) Rendle (Ceylon citronella). Curr Sci 49:437–438

Sreenath HL, Jagadishchandra KS (1981) Regeneration of plants from the tissue cultures of palmarosa (*Cymbopogon martinii* var. motia). p 11. In: 6th All India Plant tissue culture Conf, Abstr Dep Bot, Univ Poona

Sreenath HL, Jagadishchandra KS (1982) Clonal propagation of Java citronella (*Cymbopogon winterianus* Jowitt) by tissue culture. In: Proc 69th Sess Indian Science Congr, Abstr. Indian Sci Congr Assoc, Calcutta, pp 95–96

Sreenath HL, Jagadishchandra KS (1983) Studies on nuclear cytology of the callus cells and in vitro clones of some aromatic grasses. In: Proc 70th Sess Indian Science Congr, Abstr. Indian Sci Congr Assoc, Calcutta, p 141

Sreenath HL, Jagadishchandra KS (1985–86) Studies on flowering, pollen, seed setting and essential oil yield in commercially important species of *Cymbopogon* Spreng. (aromatic grasses). J Mysore Univ Sci B 30:10–16

Sreenath HL, Jagadishchandra KS (1987) Nuclear cytology of Khavi grass [*Cymbopogon jwarancusa* (Jones) Schult.] under in vivo and in vitro system. Cell Chr Res 10:93–99

Sreenath HL, Jagadishchandra KS (1988a) Morphohistological studies of callus tissues in *Cymbopogon* Spreng. In: Maheshwari Devi H, Rao PN, Rao KVM (eds) Recent advances in developmental morphology of crop plants. Andhra Univ Press Waltair, pp 343–352

Sreenath HL, Jagadishchandra KS (1988b) Cytology of the donor plants, calli and regenerated plants in different ploid races of lemongrass (*Cymbopogon flexuosus*). In: Bir SS, Chatha GS (eds) Glimpses of cytogenetics in India, vol 1. Soc Cytol Genet India, pp 104–112

Sreenath HL, Jagadishchandra KS (1988c) In vivo and in vitro instability of B chromosomes in palmarosa grass (*Cymbopogon martinii* var. motia). Genome 30:966–973

Sreenath HL, Jagadishchandra KS (1988d) Plant regeneration through somatic embryogenesis and progress towards the development of artificial seeds in *Cymbopogon* (aromatic grasses). In: Int Conf Research in plant science and its relevance to future, Abstr. Univ Press, Delhi, p 180

Sreenath HL, Jagadishchandra KS (1989) Somatic embryogenesis and plant regeneration from inflorescence culture of Java citronella (*Cymbopogon winterianus* Jowitt) Ann Bot (London) 64:211–215

Stafford A, Morris P, Fowler MW (1986) Plant cell biotechnology – a perspective. Enzyme Microb Technol 8:578–587

Suri RK, Jain N, Gupta BK (1978) Medicinal potentialities and economic importance of some aromatic grasses of genus *Cymbopogon* Spreng. Indian Drugs Pharm Ind 13:7–12

Thapa RK, Bradu BL, Vashist VN, Atal CK (1971) Screening of *Cymbopogon* species for useful constituents. Flavour India 2:49–51

Zutchi NL (1982) Essential oils-isolates and semisynthetics. In: Atal CK, Kapur BM (eds) Cultivation and utilization of aromatic plants. RRL, CSIR, Jammu-Tawi, pp 38–89

XIV *Dysosma pleiantha* (Hance) Woodson: Somatic Embryogenesis and Plant Regeneration

WEI-CHIN CHANG[1]

1 Introduction

Dysosma comprises seven species distributed from central and southern China to Taiwan (Ying 1979). Among them, *Dysosma pleiantha* (Hance) Woodson (Berberidaceae), highly prized by the mountain tribes of Taiwan for its medicinal properties, is a herbaceous, rhizomatous perennial (Ying 1979) (Fig. 1). It is an interesting species discovered by Watters in Taiwan in 1881 and described by Hance in 1883 as *Podophyllum pleianthum* (Kumazawa 1936). It was later introduced as the synonyms *P. versipelle* Hance, *P. onzoi* Hayata, etc. (Kumazawa 1936). Woodson (1928) compared the floral and vegetative features of *P. pleiantha* and those of other species of *Podophyllum* in herbarium material, and indicated some features which are quite different, and suggested this herb as *D. pleiantha* (Hance) Woodson. This plant, distributed from Himalayan districts to Taiwan through the mountainous parts of China, was rarely cultivated in any of the European and American botanical gardens until earlier in this century. So, contrary to its related species *Podophyllum peltatum* L. and *P. Emodi* Wall., its morphological and histological studies have been rather neglected, although some descriptions were given by Kumazawa (1930, 1936), Ying (1979), and Terabayashi (1983). Studies on its heteromorphic karyotype have been documented (Chen et al. 1975).

Several natural chemicals have been isolated from this herb (Shibata et al. 1962; Chang et al. 1975; Jackson and Dewick 1985). These Table 1 have included podopyllotoxin (*1*), 4'-demethylpodophyllotoxin (*2*), desoxypodophyllotoxin (*3*), 4'-demethyldesoxypodophyllutoxin (*4*), podophyllotoxone (*5*), 4'-demethylpodophyllotoxone (*6*), isopicropodophyllone (*7*) and 4'-demethylisopicropodophyllone (*8*). Among them are several liganas previously known to have potentially useful antitumor activities (Hartwell and Schrecker 1958; Shibata et al. 1962). Those present in *Podophyllum* species have now been chemically modified to produce the clinically useful drugs etoposide and teniposide. Etoposide is of particular value in the treatment of testicular cancer and small-cell lung cancer, and is now produced from podophyllotoxin (Jackson and Dewick 1985). This upsurge in pharmacological activities and demand for large quantities of rhizomes for studies and folk medicine make this species one of the most highly priced herbs in the local herb market in Taiwan.

Natural regeneration of *Dysosma* is either by seed or by sprout from the rhizomes, but due to its long juvenile phase (5–7 years) and low quantity in fruit

[1]Institute of Botany, Academia Sinica, Taipei, Taiwan 11529, Republic of China

Biotechnology in Agriculture and Forestry, Vol. 15
Medicinal and Aromatic Plants III (ed. by Y.P.S. Bajaj)

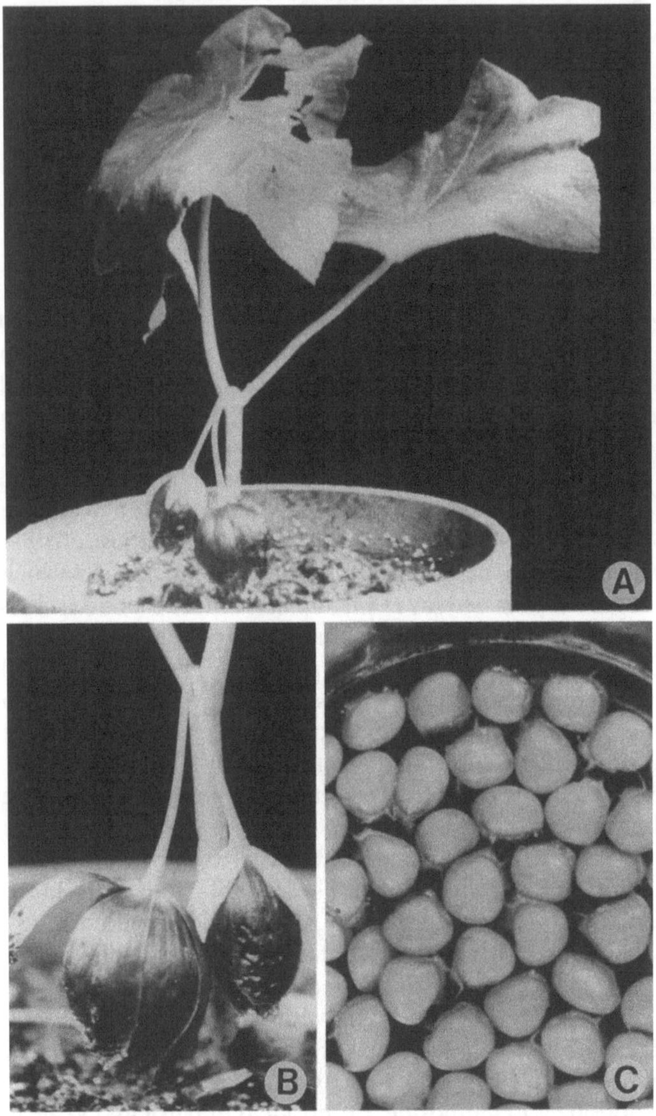

Fig. 1A-C. *Dysosma pleiantha* (Hence) Woodson. **A** A flower-bearing wide accession. **B** Two flower buds. **C** Mature seeds (about 5 mm in size)

Table 1. Lignans found in *Dysosma pleiantha*

Lignan	References
Podophyllotoxin (1)	Shibata et al. (1962) Jackson and Dewick (1985)
4′-Demethylpodophyllotoxin (2)	Jackson and Dewick (1985)
Desoxypodophylloptoxin (3)	Shibata et al. (1962) Chang et al. (1975) Jackson and Dewick (1985)
4′-Demethyldexopodophyllotoxin (4)	Jackson and Dewick (1985)
Podophyllotoxone (5)	Jackson and Dewick (1985)
4′-Demethylpodophyllotoxone (6)	Jackson and Dewick (1985)
Isopicropodophyllone (7)	Chang et al. (1975) Jackson and Dewick (1985)
4′-Demethylisopicropodophyllone (8)	Jackson and Dewick (1985)

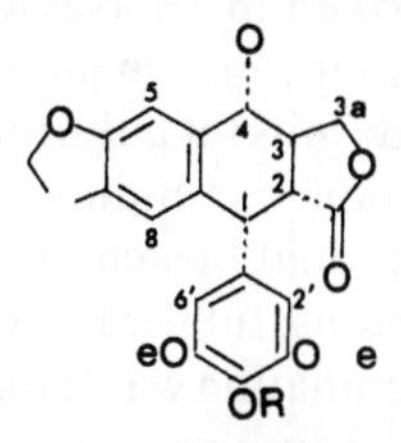

1 R=Me
2 R=H

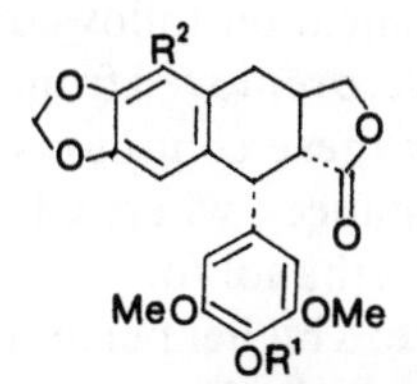

3 R^1=Me, R^2=H
4 R^1=H, R^2=H

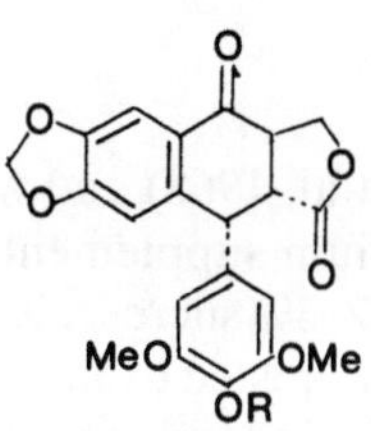

5 R=Me
6 R=H

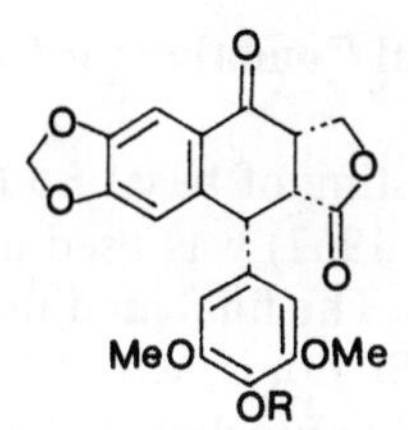

7 R=Me
8 R=H

setting, poor seed variability, and slow seedling growth, propagation through seeds is not satisfactory. Propagation by rhizomes is also very slow. Thus, in vitro culture was undertaken in the attempt to develop a more rapid method for multiplication for urgent needs.

2 Tissue Culture Studies

Data concerning in vitro culture of *Dysosma* are scanty. Besides our work, as documented in two reports (Chuang and Chang 1987a,b), no other papers have been found so far. While somatic embryogenesis and plant regeneration have been described, there are no reports on cultures of embryo, anther and protoplast, and suspension culture. Thus the major method of genetic modification, cell culture for supplying large cell mass for pharmacological studies, has yet to be developed for *Dysosma.*

2.1 Plant Material and Methods of Sterilization

The material used was local accessions obtained from their natural habitat on the central highland of Taiwan, and grown in the glasshouse under natural conditions. Vegetative tissues (leaves, petioles, stems, and rhizomes) and reproductive tissues (immature seeds and mature zygotic embryos) were used for establishing in vitro culture. All the tissues were surface sterilized for 3 min in 75% ethanol followed by one rinse in sterile distilled water. The materials were then soaked in 1% NaOCl (Chlorox) for 3 min with ultrasonication followed by rinsing four times in sterile distilled water. Immature seeds were removed from the sterilized fruits for culturing without sterilizing again. Under these conditions, contamination was limited to 5–15% in most cases. In some instances, where Chlorox caused rapid bleaching of the leaves, sterilization with 70% ethanol for 3–5 min was found useful for leaves. Rhizomes were difficult to sterilize; a higher percentage of contamination was found in the culture of rhizome tissue than in those of other materials.

2.2 Culture Medium and Physical Condition for Culture

The solid culture medium consisting of basal B5 (Gamborg et al. 1968) and MS medium (Murashige and Skoog 1962) was used as basal medium supplemented with various hormonal additions. The final medium contained 2–3% sucrose, 0.7% (w/v) agar (No. 7002, Sigma Co. St. Louis, Mo. U.S.A.), and pH adjusted to 5.8 with 0.1 N NaOH. Media were usually autoclaved at 120°C for 15 min, and stored at room temperature until use. The cultures were maintained either at 25 + 1°C in the dark or under 16/8 h light/dark regime at 27 + 1°C.

2.3 Callus Culture and Somatic Embryogenesis

Young leaves of dormant buds were cut into segments of 5 × 6 mm as the explants. Between 1 and 2 weeks after inoculating on MS medium supplemented with 2,4-dichlorophenoxyacetic aicd (2,4-D) 1–2 mg/l, cultured young leaf segments exhibited embryogenic callus, direct embryoid formation, or a combination of both (Fig. 2A). The presence of 2,4-D was essential for callus proliferation. Initiation of the embryogenic callus was similar in young leaf explants cultured on MS or B5 medium. Subculturing of these primary embryogenic calli in the same media resulted in numerous developing embryoids over the surface of callus (Fig. 2B). On medium without 2,4-D or with lower 2,4-D concentration (0.1–0.5 mg/l) they formed embryogenic callus from which many more embryoids could be recovered. Embryoids were snow-white or slightly green when exposed to light, and morphogenetically resembled zygotic embryos with well-developed cotyledons and epicotyl apex (Fig. 2C). Some embryoids, however, formed flattened or fasciate cotyledons. When these young leaf-derived embryoids were transferred onto either MS or B5 medium plus benyzladenine 1 mg/l and gibberellic acid 1 mg/l for 3 to 4 weeks, more than 90% germinated into individual plantlets (Fig. 2D) which could be established in soil. Some embryoids showed a slightly disorganized proliferation and multiple shoot and root development.

In the case of pith explants of rhizomes, embryogenic callus initiation and direct embryoid formation were observed within 6 to 10 days after inoculation onto MS medium with 1 mg/l 2,4-D in darkness (Fig. 2E). The frequency of callus initiation from pith tissues was 80%. About 20% incubated embryos turned brown without signs of callus initiation. Subculturing of these rhizome-derived calli on MS or half-strength MS medium supplemented with 0.1–1 mg/l 2,4-D resulted in spontaneous embryoid formation (Fig. 2F). Direct embryoids were also formed from the adventive roots which occurred on the rhizome-derived callus (Fig. 2G).

The frequency of embryoids resulting in plants seems to depend on the quality and quantity of embryoids formed on the calli. Although rhizome-derived calli produced embryoids in a higher proportion, many were aberrant, such as fused cotyledons, translucent and albino plantlets, and failed to establish in peat moss or soil. The young leaf-derived calli gave a higher production rate of plantlets through embryoid formation as they produced more embryoids per callus, and generated more secondary and tertiary embryoids on subsequent subculturing. Repeating subculturing of both calli on low 2,4-D resulted in the continued growth of white embryogenic tissue without any apparent loss in the totipotency for 17 months, as long as the tissue was maintained in darkness.

Segments of stems and petioles of young leaves of dormant buds cultured on MS medium supplemented with 1 mg/l 2,4-D showed growth of callus all over the surfaces of the segments after 2 weeks of culture. Callus maintained in the same medium began to form embryoids after 4 weeks of subculture. On the MS medium supplemented with 1 mg/l benzyladenine (BA) and 1 mg/l gibberellic acid (GA_3) these embryoids swelled in size 20 to 30 times but failed to germinate into plantlets.

In the case of reproductive tissues, initially, immature seeds cultured either in the dark or under a 16/8 h light-dark regime produced a dark, growth-inhibiting exudate in the medium. Frequent transfers to fresh medium decreased the effects of the exudation. However, all the incubated seeds turned brown without any sign of growth. Subsequent transfers to fresh media also failed to support any callus initiation. The seed coats had to be removed or wounded by cutting at random with a razor, as no callus proliferation occurred from seeds with intact seed coats under the culture conditions and with the compositions of media employed. Embryos excised from immature seeds also failed to respond to all the media tested.

Callus initiation and proliferation from cultured immature seeds with wounded seed coats was very slow. The presence of 2,4-D was essential for callus proliferation. Initiation of callus was similar in seeds cultured on MS or B5 medium. After 1 month in culture on medium containing 1 mg/l 2,4-D, seeds commenced swelling, particularly in the wounded regions. By the end of the second month, callus had begun to protrude from the wounded seed coats (Fig. 3A). Callus proliferation was observed only in the vicinity of the wound. Uncallused immature seed tissue either

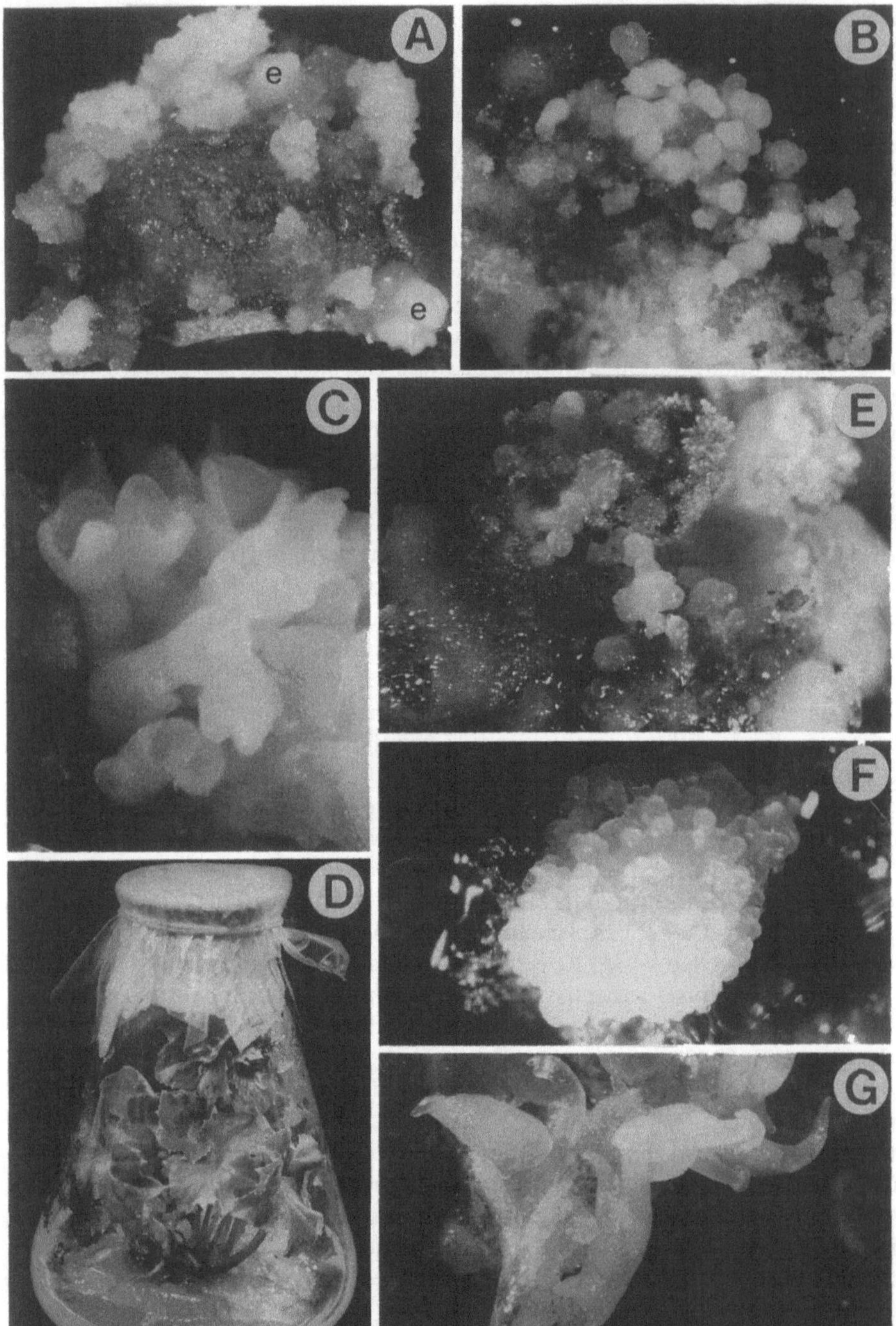
A
e
e
B
C
E
D
F
G

became necrotic or showed no signs of growth. Subculture of the primary callus in the same medium resulted in the formation of a compact dark brown globular mass. Frequent subculture in fresh medium decreased the browning, and white globular structures protruded from the surface of the browning callus. Prolonged culturing of the callus in the dark resulted in spontaneous embryoid proliferation all over the surface of callus (Fig. 3B). Embryoids were snow-white or slightly green when exposed to light, and morphologically resembled zygotic embryos with well-developed cotyledons and epicotyl apex except for their broad attachment to the parent callus and conspicuous absence of constricted root axes. Some, however, formed flattened or fasciate cotyledons (Fig. 3C).

2.4 Regeneration of Plants

Transfer of the rhizome-derived embryoids and immature seed-derived embryoids onto either MS or B5 medium plus 1 mg/l BA and 1 mg/l GA_3 for 3 weeks gave rise to plantlets (Figs. 2D, 3D), 75% of which were grown to full-sized normal plants after transfer to soil. No significant difference was observed on germination and growth of plantlets on the media based on MS or B5 minerals. The combination of 1 mg/l BA and 1 mg/l GA_3 proved to be most efficient in supporting embryoid germination and subsequent normal plantlet development. In addition to dicotyledonous embryoids, aberrant forms of embryoids were observed. Embryoids with multiple cotyledons, secondary embryoids, and multiple shoots (Fig. 3E) developed from a single embryoid were all observed. The frequency of these unusual forms varied with experiment and explant piece. Reduction in the formation of aberrant embryoids is being attempted by manipulation of the composition of the medium.

The dosage-response studies on effects of 2,4-D on growth of embryoid-forming callus cultures and embryoid proliferation showed an optima in a range of 0.1–0.5 mg/l. With progressively increased concentrations of 2,4-D over 1 mg/l to 5 mg/l in either MS or B5 medium, less embryoids were formed. In the presence of 2,4-D (0.1–0.5 mg/l) there was no significant difference in embryoid-forming ability between MS and B5 media (Table 2). Repeated subculturing of such tissues on 0.1 mg/l 2,4-D resulted in continued growth of white embryogenic tissue without any apparent loss in regeneration capacity for more than 3 years, as long as the tissue was maintained in darkness. Under illumination (16/h light-dark regime) the embryogenic callus turned brown and lost its totipotency within 1 month.

Fig. 2A-G. In vitro regeneration of *Dysosma pleiantha* via somatic embryogenesis of vegetative tissue-derived calli. **A** Embryogenic callus derived from young leaf explants. Several embryoids (**e**) directly formed on the cut edges of the leaf explants. **B** Embryoid proliferation over the surface of the embryogenic callus on MS medium plus 1 mg/l 2,4-D. **C** Embryoid proliferation on the subcultured leaf callus on MS medium without plant growth substance. **D** A plantlet derived from a leaf callus embryoid on MS medium plus benzyladenine and gibberellic acid, 1 mg/l each. **E** Direct embryoid formation and embryogenic callus proliferation occurred on the surface of rhizome-derived callus. **F** Cluster of embryoids on the surface of rhizome-derived callus on MS medium plus 1 mg/l 2,4-D. **G** Direct embryoids protruding from the surface of adventive roots generated from rhizome callus. (Chuang and Chang 1987b)

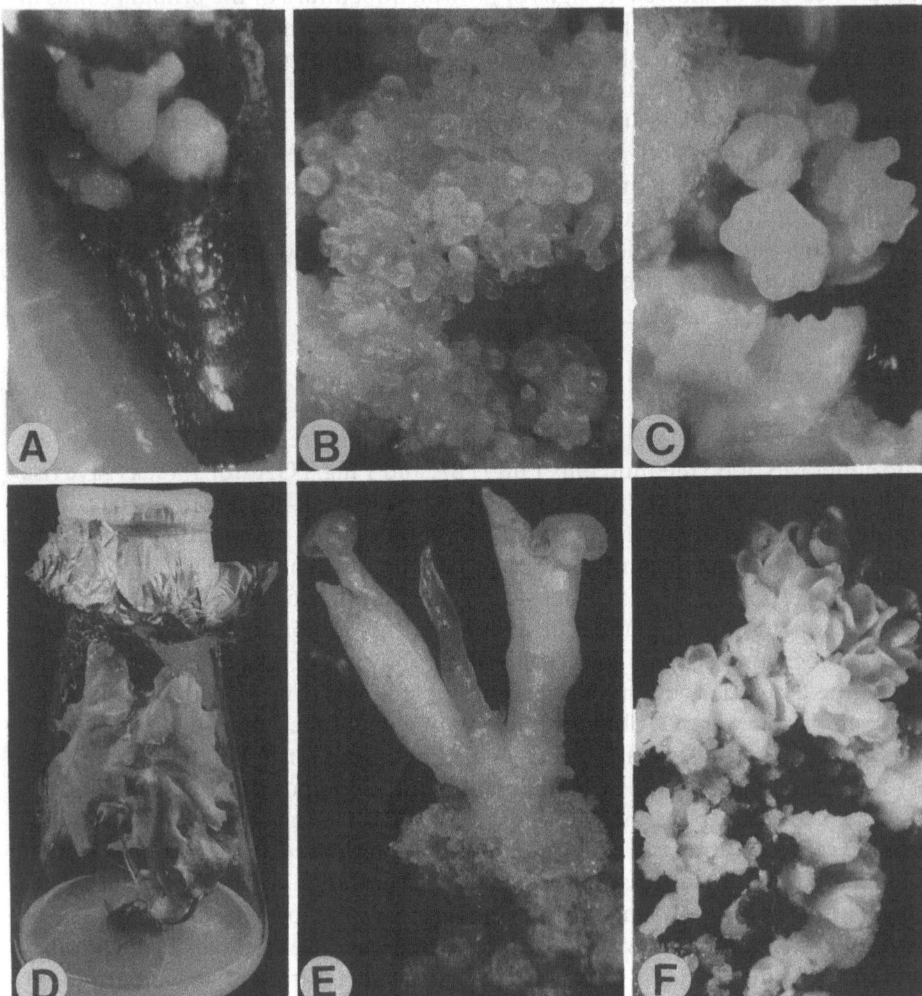

Fig. 3A-F. Plant regeneration via embryoid formation in reproductive tissue-derived callus cultures of *Dysosma pleiantha*. **A** Initiation of callus from the wounded area of an immature seed. **B** Spontaneous embryoid proliferation all over the callus derived from immature seed. **C** Snow-white well-developed embryoids with flattened or distorted cotyledons. **D** Regenerated plantlet. **E** Multiple shoots developed from a single embryoid. **F** Embryoid formed in callus derived from isolated mature zygotic embryos. (Chuang and Chang 1987a)

In the case of mature zygotic embryos, callus initiation was observed within 6 to 10 days on MS medium containing 1 mg/l 2,4-D in darkness. The frequency of callus initiation from embryos was 95%. About 5% turned brown without any signs of callus initiation. Subculturing of the embryo-derived calli on MS media or half-strength MS media supplemented with 0.1 mg/l 2,4-D resulted in spontaneous embryoid formation (Fig. 3F). This embryogenic callus could be subcultured and

Table 2. Effect of 2,4-D on the growth of embryoid-forming callus and embryoid proliferation in the cultures of *Dyososma pleiantha* derived from immature seeds (Chang unpubl.)

Basal medium	Concentration of 2,4-D (mg/l)	Callusing of embryoid--forming callus	Embryoid proliferation embryoid/culture
MS	0.01	+++	7
	0.1	++++	25
	0.5	+++++	27
	1	+++	5
	2.5	+++	1
	5	+	0
B5	0.01	+++	7
	0.1	+++	24
	0.5	++++	25
	1	++	3
	2.5	++	0
	5	++	0

Each treatment has five replicated cultures. Each culture contain one piece of embryoid-forming callus area on 30 ml agar-gelled nutrient medium in a 20 mm x 15 cm test tubes.

maintained on the medium containing 0.5 mg/l 2,4-D or auxin-free medium for over 3 years without loss of totipotency. Under illumination, most of the subcultured calli turned brown, and no morphogenesis occurred. Embryoids germinated either on MS or B5 medium plus 1 mg/l BA and 1 mg/l GA_3, and 80% plants have been established in soil and grown to maturity. No significant difference in embryoid germination was found between MS and B5 minerals, and both BA and GA_3 were required for embryoid germination.

3 Conclusion and Prospects

Explants from young leaves, rhizome pith tissue, and immature seeds are excellent material for production of embryoids in tissue culture of *D. pleiantha*. We recommend their maintenance at 25 ± 1 °C in the dark on MS basal medium plus 1 mg/l 2,4-D as the best medium for establishment of callus cultures and forming embryogenic callus. Once the embryogenic calli have been isolated, embryoid proliferated optimally in MS plus 0.1–0.5 mg/l 2,4-D in the condition as for callusing. Plant regeneration occurred best in a medium based on MS or B5 plus benzyladenine 1 mg/l and gibberellic acid 1 mg/l for 3 to 4 weeks kept under 16/8 h light/dark regime at 27 ± 1 °C.

To the present, dozens of plants have been established under glasshouse condition. The stability of the clone raised so far, with respect to the medicinally valuable components, is under assessment. If it is confirmed that embryoid-derived plants are homogeneous and genetically similar to mother plants, our protocols for micropropagation through somatic embryogenesis could replace the conventional methods for mass propagation, provided that the cost is compatible.

References

Chang FC, Chiang CK, Aiyar VN (1975) Isopicropodophyllone from *Podophyllum pleianthum*. Phytochemistry 14:1440

Chen C, Huang YC, Pi CP (1975) Analysis of heteromorphic karyotype in *Dysosma pleiantha* Hance. Mem Coll Agric Natl Taiwan Univ 17, 1:39–42

Chuang MJ, Chang WC (1987a) Somatic embryogenesis and plant regeneration in callus culture derived from immature seeds and mature zygotic embryos of *Dysosma pleiantha* (Hance) Woodson. Plant Cell Rep 6:484–485

Chuang MJ, Chang WC (1987b) Embryoid formation and plant regeneration in callus culture derived from vegetative tissues of *Dysosma pleiantha* (Hance) Woodson. J Plant Physiol 128:279–283

Gamborg OL, Miller RA, Ojima L (1968) Nutrient requirement of suspension culture of soybean root cells. Exp Cell Res 50:151–158

Hartwell J, Schrecker AW (1958) The chemistry of *Podophyllum*. Forscher Chem Org Nat 15:83–166

Jackson DE, Dewick PM (1985) Tumour-inhibitory aryltaetralin lignans from *Podophyllum pleianthum*. Phytochemistry 24:2407–2409

Kumazawa M (1930) Morphology and biology of *Glancidium palmatum* SIEB et ZUCC. with note on affinities to the allied genera *Hydrastis*, *Podophyllum* and *Diphylleia*. J Fac Sci Imp Univ Tokyo Sec 3 Bot 2:345–380

Kumazawa M (1936) *Podophyllum pleianthum* Hance. A morphological study, with supplementary notes on allied plants. Bot Mag (Tokyo) 50:268–276

Murahsige T, Skoog F (1962) A revised medium for rapid growth and bioassays with tobacco tissue cultures. Physiol Plant 15:473–497

Shibata S, Murata T, Fujita M (1962) Studies on the constituents of Japanese and Chinese drugs VI. On the constituents of rhizome and root of *Phodophyllum pleianthum* Hance. Yakugaku Zasshi 82:777–779

Terabayashi S (1983) Studies in the morphology and systematics of Berberidaceae VI. Floral anatomy of *Diphylleia* Michx., *Podophyllum* L. and *Dysosma* Woodson. Acta Phytotax Geobot 34:27–47

Ying TS (1979) On *Dysosma* Woodson and *Sinopodophyllum* Ying, gen. nov. of the Berberidanceae. Acta Phytotax Sin 17:15–23

Woodson RE (1928) *Dysosma*: a new genus of Barberidaceae. Ann Missour Bot Garden 15 (described in Kumazawa 1936)

XV *Euphorbia* spp.: In Vitro Culture and Production of Anthocyanins

YOSHIKAZU YAMAMOTO[1]

1 Introduction

1.1 Distribution and Importance of *Euphorbia*

The genus *Euphorbia* consists of 2000 to 3000 species which grow in semi-arid lands throughout the world. The latex from these species is highly vesicanting; it irritates the skin and mucous membranes, and is used as a fish poison and an insecticide. Its biological functions have been reported to be caused by the fatty acid esters of diterpene alcohols (Furstenberger and Hacker 1977). Calvin (1979) has spearheaded a drive to develop *Euphorbia* species as future sources of gasoline, and has propagated *E. tirucalli* (African milk bush) and *E. lathyris* (gopherwood) on a large scale. His group has reported that chemical constituents of *Euphorbia* species are mainly triterpene alcohols (Nishimura et al. 1977).

2 In Vitro Approaches

2.1 Review of Tissue Culture on *Euphorbia*

There are a number of reports on tissue culture of nine species of *Euphorbia*: *E. characias* (Hardy et al. 1986 and 1987), *E. cyparissias* (Lee and Starratt 1971), *E. esula* (Lee and Starratt 1971), *E. lathyris* (Tideman and Hawker 1982; Ripley and Preece 1986), *E. millii* (Yamamoto et al. 1981, 1982, 1983), *E. peplus* (Tideman and Hawker 1982), *E. pulcherrima* (Langhe et al. 1974), *E. tannesis* (Tideman and Hawker 1982), and *E. tirucalli* (Biesboer and Mahlberg 1979; Yamamoto et al. 1981; Ohyama et al. 1984). Secondary metabolites produced in cultured *Euphorbia* cells were fatty acids (*E. esula* and *E. millii*), phytosterols (*E. esula*, *E. millii*, and *E. tirucalli*), triterpenes (*E. tirucalli*), and anthocyanin (*E. millii*). These compounds have usually been found in cultured plant cells. There has been no report on isolation of characteristic biological active substances, alkaloids, and diterpene esters, from cultured *Euphorbia* cells.

[1]Research Center, Nippon Paint Co., Ltd., Neyagawa, Osaka 572, Japan

Biotechnology in Agriculture and Forestry, Vol. 15
Medicinal and Aromatic Plants III (ed. by Y.P.S. Bajaj)

2.2 Establishment of Tissue Culture and Cell Suspension

2.2.1 Callus Induction

Sterilized samples of young shoots of *Euphorbia tirucalli* Desf., leaves of *E. millii* Ch. des Moulins, or stems of *E. lathyris* L. were placed on Murashige and Skoog's basal solution with 2% (w/v) sucrose, 0.1–10 ppm 2,4-D or NAA and 0.2% (w/v) natural supplement; malt extract (ME), yeast extract or casein hydrolysate (Yamamoto et al. 1981). The cultures were kept under fluorescent light. The following four calli were obtained: a yellowish white *E. tirucalli* callus, two distinct *E. millii* calli; a yellow callus and a varied-color callus, and a yellowish white *E. lathyris* callus. They grew well and could be subcultured for 10 years on these media.

Fig. 1. *Euphorbia millii* plant at the flowering stage

2.2.2 Selection of Cultured Euphorbia millii Cells that Produced High Level of Anthocyanin

Usually plant cell cultures produce only small amounts of secondary metabolites. Recently, however, cell strains containing amounts of secondary metabolites greater than those found in intact plants have been isolated by clonal selection. There are two questions in respect to clonal selection (Yamamoto et al. 1982): (1) How can we determine whether the high productivity for secondary metabolites in selected cell strains is stable? (2) How long must we continue clonal selection in order to obtain stable cell strains that produce secondary metabolites? I continuously selected cell strains that showed high contents of an anthocyanin from cultured *E. millii* cells and studied the stability of their productivity for this pigment with statistical and cell-pedigree analyses (Yamamoto et al. 1982).

An outline of the selection method, the cell-aggregate selection, is shown in Fig. 2. The color of original *E. millii* calli which were induced from leaves was mottled red and white. These calli were cut into 128 segments (cell mass, ca. 3 mm). Each segment was coded and placed on agar medium in a sectioned Petri dish. The agar medium consisted of Murashige and Skoog's basal solution: 10^{-6} M 2,4-D, 0.2% (w/v) malt extract, 2% (w/v) sucrose, and 0.8% (w/v) agar. The segments were cultured under continuous fluorescent light for 10 days. Each of the nine segments was cut into two cell aggregates; one (D_1) for subculture and the other (D_2) for quantitative analysis of the pigment. From the analysis of D_2, the reddest D_1 cell aggregate from each Petri dish was selected. These selected cell aggregates were cut into several segments (cell mass, ca. 3 mm). All these segments were coded and transplanted onto fresh medium in a nine-section Petri dish. The code of each segment was put into the data file of a computer together with the code of the mother

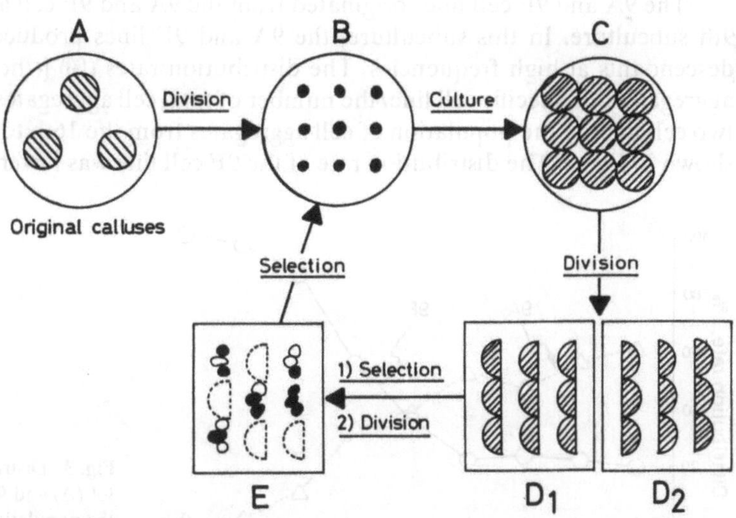

Fig. 2. Outline of the selection method. ○ = unselected segment; ● = selected segment; □ = unselected aggregate; ○ = grown aggregate. (Yamamoto et al. 1982)

cell aggregate. Segments selected at each transplantation were cultured on the same medium under the above conditions. This selection procedure was repeated 28 times. The cell aggregates established were subcultured for about 6 years after selection under the same conditions.

Pedigree analysis has been used in breeding programs of garden plants. The knowledge of parentage obtained by pedigree tracing can serve as an aid when we assign genotypes, and give an indication of the degree of genetic diversity in breeding populations. Only two reports of pedigree analysis by Bell et al. (1980) and McCreight and Favila (1980) have appeared in horticultural science. They searched the pedigree records for pear with a computer program written in COBOL and for muskmelon with a program written in BASIC, and confirmed that computer tracing is a valuable tool for pedigree analysis.

The author collected the net data for parentage code, anthocyanin content, and fresh weight of 1588 entries from the 7th to 28th passages. The net data obtained in each selection procedure were rearranged in the most convenient format, and entered in our computerized system, CELSEL. The CELSEL system consists of a basic subsystem (MASTER) and three practicable subsystems (LIST, TREE, and ROOT). Programs contained in these subsystems were written in DSM-11 as implemented for the PDP 11/34 system, because this language was found to be convenient for handling our data files which showed tree structure (Yamamoto et al. 1983). To my knowledge, CELSEL was the first computer system to be used for the pedigree analysis of cultured plant cells.

Pedigree tracing of cultured *E. millii* cells was accomplished by two distinct repetitive search routines that were contained in the TREE and ROOT subsystems. The TREE routine traced the pedigree from ancestry, but the ROOT routine traced it from progeny. The ROOT output also showed that many of the members that produced relatively higher levels of anthocyanin were derived from one ancestor in a population at each passage.

The 9A and 9F cell lines originated from the 9A and 9F cell aggregates of the 9th subculture. In this subculture, the 9A and 9F lines produced pigment-rich descendants at high frequencies. The distribution rates (%) [(the number of cell aggregates of a specific cell line/the number of total cell aggregates) × 100] of these two cell lines in the population of cell aggregates from the 16th to 24th passages is shown in Fig. 3. The distribution rate of the 9F cell line was lower than that of the

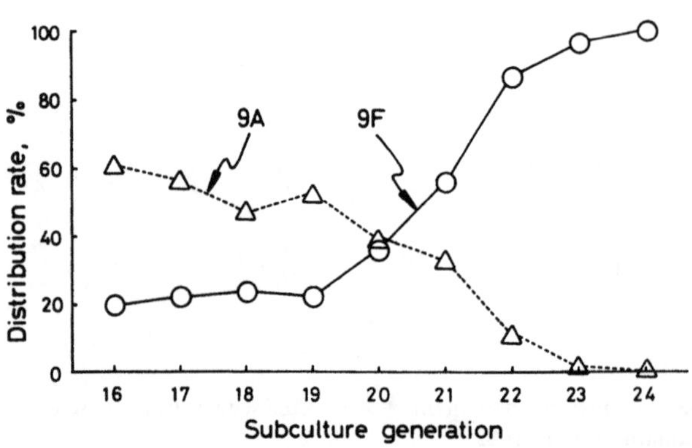

Fig. 3. Distribution rates of the *9A* (△) and *9F* (○) cell lines in the population of cell aggregates from 16th to 22nd passages. (Yamamoto et al. 1982)

9A cell line at the 16th subculture, but equalled it at the 20th subculture. This rate gradually increased and reached 100% at the 24th subculture. This result shows that all the aggregates after the 24th passage were derived from the 9F cell aggregate. Figure 4 shows the histogram for the frequency distributions of cell aggregates from the 16th to 22nd passages. It could also be shown that the 9F cell line was composed of the group of cell aggregates that indicated high pigment contents in the histograms after the 18th passage.

The mean and maximum values of pigment content [absorbance at 530 nm/fresh weight (mg) of the cell aggregate] in the frequency distribution of cell aggregates from the 16th to 28th passages were plotted against the pigment in Fig. 5. The mean value (C) increased nearly threefold from the 16th to 22nd passages. I

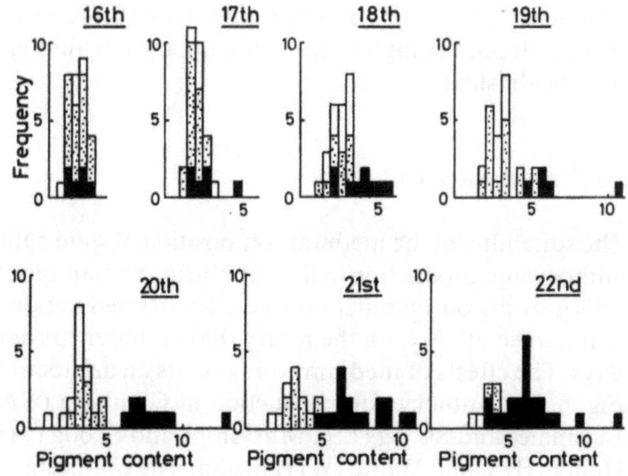

Fig. 4. Frequency distributions of cell aggregates of the 9A (□) and 9F (■) cell lines in histograms of the population of cell aggregates with various pigment contents from the 16th to 22nd passages. □ = cell aggregates other than the 9A and 9F cell lines. (Yamamoto et al. 1982)

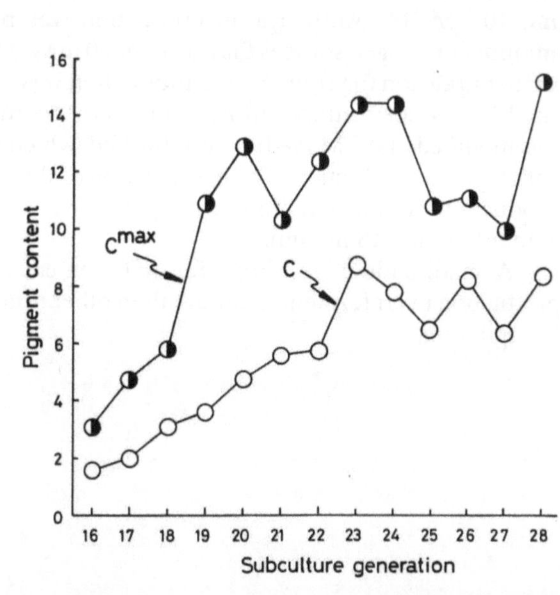

Fig. 5. Trends of the two characteristic terms; the mean value (*C*) and maximum value (C^{max}) in the frequency distribution of cell aggregates with various pigment contents from the 16th to 28th passages. (Yamamoto et al. 1982)

found that the mean value (C) for the anthocyanin content in cell aggregates of *E. millii* became stable after 24 clonal selections.

If the mean value for the content of secondary metabolites is stable in a population of cell aggregates chosen by successive clonal selection, and if all the cell aggregates are derived from one cell aggregate, they should consist of cells with high and stable productivity for secondary metabolites. These results show that a cell strain containing a high and stable pigment content from *E. millii* callus could be cultured after 24 successive clonal selections. I believe that during early passages cultured cells are heterogeneous in their production of pigments and that they contain high pigment-producing variant cells. Results show that more cell aggregates of the 9F cell line could be selected than of the 9A cell line at each subculture, and that the 9F cell aggregate had high pigment-producing variant cells.

Anthocyanin contents (%DW) of *E. millii* cells were 1.32 (high anthocyanin-producing cells), 0.05 (original cells), 0.28 (flowers), and < 0.01 (leaves). A strain that produced the highest level of anthocyanin among various cells of *E. millii* was thus established.

2.2.3 Suspension Culture

The suitability of the medium composition of solid culture was also investigated for anthocyanin production in liquid culture (Yamamoto et al. 1989). Cultured *E. millii* cells growing on agar medium were transferred to liquid medium. These cells were maintained at 26°C on the rotary shaker under continuous fluorescent light for 10 days. The effects of medium components on anthocyanin production are shown in Fig. 6. The anthocyanin production in Gamborg (GA) medium is higher than in Linsmaier and Skoog (LS), Murashige and Skoog (MS), Nitsch and Nitsch (NN), Heller (HE), and White (WH) media. The optimum concentration of sucrose was between 5 and 7%. The optimum combination of phytohormones is 10^{-6} M 2,4-D and 10^{-8} M BA. Anthocyanin production was promoted more at 30 mM total amount of nitrogen sources than at 15 or 60 mM. The optimum ratio of ammonium ion to nitrate ion for anthocyanin production was 1:16. High concentrations of Fe^{2+} and SO_4^{2-} reduced anthocyanin production. According to these results, GA medium was modified and EM medium established, which had the optimal composition for anthocyanin production. The composition of the EM medium is shown in Table 1. Production of anthocyanin (32 mg per liter per day) in EM medium is 4.5 times higher than in MS medium.

A small paddle with holes for agitation caused better results in anthocyanin production in jar fermentor culture than other rotators. *E. millii* cells were very frail for agitation.

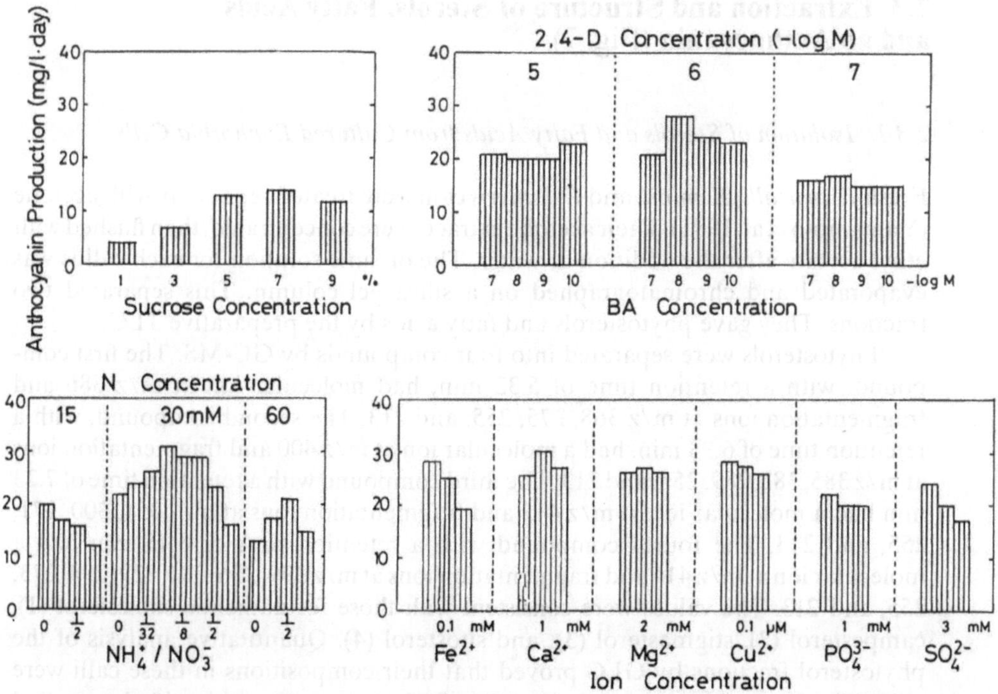

Fig. 6. Effects of medium components on anthocyanin production in suspension cultures of *E. millii* cells. Cultures were maintained on the rotary shaker under continuous fluorescent light at 26 °C for 10 days

Table 1. Components of EM medium. (Yamamoto et al. 1989)

KNO_3	2648 mg/l
NH_4NO_3	141
$MgSO_4 \cdot 7H_2O$	500
$CaCl_2 \cdot 2H_2O$	150
KCl	264
$NaH_2PO_4 \cdot 2H_2O$	150
$FeSO_4 \cdot 7H_2O$	13.9
Na_2EDTA	18.7
$MnSO_4 \cdot 4H_2O$	10
$ZnSO_4 \cdot 7H_2O$	2
$CuSO_4 \cdot 5H_2O$	0.025
$CoCl_2 \cdot 6H_2O$	0.025
Ki	0.75
H_3BO_3	3
$Na_2MoO_4 \cdot 2H_2O$	0.25
Inositol	100
Nicotinic acid	0.5
Pyridoxine·HCl	0.5
Thiamine·HCl	1
Glycine	2
Sucrose	50,000
2,4-D	10^{-6} M
BA	10^{-8} M

2.3 Extraction and Structure of Sterols, Fatty Acids and an Anthocyanin (Fig. 7)

2.3.1. Isolation of Sterols and Fatty Acids from Cultured Euphorbia Cells

Fresh *E. tirucalli, E. millii*, and *E. lathyris* calli were treated separately with acetone (Yamamoto et al. 1981). Their acetone extracts were concentrated, then flushed with ethyl acetate after the addition of water. The organic solution for each callus was evaporated and chromatographed on a silica gel column. This separated two fractions. They gave phytosterols and fatty acids by the preparative TLC.

Phytosterols were separated into four compounds by GC-MS. The first compound, with a retention time of 5.32 min, had molecular ion at m/z 386 and fragmentation ions at m/z 368, 275, 255, and 213. The second compound, with a retention time of 6.93 min, had a molecular ion at m/z 400 and fragmentation ions at m/z 385, 382, 289, 255, and 213. The third compound with a retention time of 7.23 min had a molecular ion at m/z 412 and fragmentation ions at m/z 351, 300, 271, 255, and 213. The fourth compound with a retention time of 8.45 min had a molecular ion at m/z 414 and fragmentation ions at m/z 399, 396, 381, 329, 294, 273, 255, and 213. The values were consistent with those for authentic cholesterol (1), campesterol (2), stigmasterol (3), and sitosterol (4). Quantitative analysis of the phytosterol fractions by GLC proved that their compositions in these calli were alike. The fatty acids were methyl esterified. These esters were identified as methyl esters of palmitic and linoleic acids by comparing their GLC's with those of authentic samples.

Lee and Starratt (1971) reported that the major lipid components in 2,4-D cultures of *E. esula* tissues were triglycerides, palmitic acid, and sitosterol. Except for the data on triglycerides, their results are consistent with the above. Biesboer and Mahlberg (1979) reported that callus of *E. tirucalli* tissues induced on medium containing IAA produced the triterpene alcohols; our 2,4-D calli produced phytosterols. The use of different auxins probably explains the discrepancy in the results.

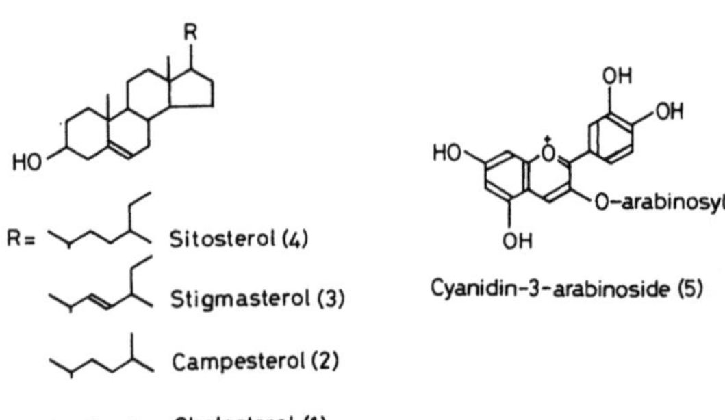

Fig. 7. Structures of various secondary metabolites found in *Euphorbia* species. (Yamamoto et al. 1981)

2.3.2 *Isolation of Anthocyanin from Cultured E. millii Cells*

Anthocyanin, a natural colorant, is mainly contained in flower petals and fruit peels. It has been used to color food stuff for many years. However, it has proved difficult to obtain a stable and abundant supply of this colorant from natural sources. There have been many reports on anthocyanin production in cultured plant cells, but most of the anthocyanins produced in these cells have not been identified. The reason is that anthocyanins are very labile to heat and to exposure to oxygen or light.

Freeze dried cells of *E. millii* were soaked in MeOH containing HCl. After the concentration of the MeOH solution and the addition of water, the solution was washed with ethyl acetate. Next, the aqueous solution was flushed with 1-butanol. The BuOH extract was run twice through preparative HPLC. The reddish eluate was concentrated to give a reddish powder, an anthocyanin (EM-1). EM-1 was hydrolyzed with HCl/MeOH in order to obtain its aglycon and sugar. The MeOH solutions of the aglycon and authentic anthocyanidins and the aqueous solutions of the sugar and authentic sugars were subjected to Avicel TLC and HPLC. By comparison with their Rf and Rt values, the aglycon and sugar were identified as cyanidin and arabinose. The NMR data of EM-1 suggested that EM-1 was cyanidin-arabinoside. SIMS spectral data indicated that EM-1 had a molecular ion at m/z 419 and a fragmentation ion at m/z 287, where there was a peak of the molecular ion of cyanidin formed by the elimination of a pentose from EM-1. The UV spectrum of EM-1 had two λmax at 531 and 282 nm, $E_{440}/E_{max} = 21\%$, and $E_{UVmax}/E_{max} = 56\%$. These values were not consistent with those of cyanidin-7-arabinoside (Sun and Francis 1967). These results showed that EM-1 was cyanidin-3-arabinoside.

Cultured plant cells produce many anthocyanins: cyanidin-3-glucoside [*Vitis* sp. (Yamakawa et al. 1983), *Populus* sp. (Matsumoto et al. 1973) and *Crinum defixum* (Mathur et al. 1982)], peonidin-3-glucoside (*Vitis* sp.), pelalgonidin-3-glucoside (*Crinum defixum*), cyanidin-3-(6″-malonylglucoside) [*Centaurea cyanus* (Kakegawa et al. 1987)], peonidin-3,5-diglucoside (*Vitis* sp.), malvidin-3-(p-cuomaroyl)-rutinoside-5-glucoside [*Petunia hybrida* (Colijn et al. 1981)], and petunidin-3-(p-coumaroyl)-rutinoside-5-glucoside (*Petunia hybrida*). All anthocyanins from these cells are glucosides; ours was the first report of isolation of anthocyanidin arabinoside in cultured cells.

The reddish extract with HCl/MeOH from dried flowers was subjected to Avicel TLC and HPLC. The chromatographic results showed that Rf and Rt values were identical with those of EM-1. There have been some reports on the isolation of cyanidin-3-arabinoside from intact plants: fruits of cranberry, apple, cacao, blueberry, and flowers of *Rhododendron*. This is the first isolation of cyanidin-3-arabinoside from the genus *Euphorbia*.

2.4 Commercial Aspects and Exploitation

Cultured *E. millii* cells produced only one compound as anthocyanin. Therefore, silk and cotton yarns dyed by this anthocyanin showed an extremely clear color. Many variations of this color appeared on using various methods and compounds for its processing. This anthocyanin was thought to be very valuable for dyeing textiles.

The yellow-brown fraction from the BuOH extract of cultured *E. millii* cells that produced a high level of anthocyanin showed a vasodepressor activity. One possibility may be to use this constituent as a material for medicines.

3 Conclusions and Prospects

The genus *Euphorbia* consists of 2000 to 3000 species; however, there are only nine species which have been cultured in vitro, and calli were induced from three species. The lipid constituents of *E. tirucalli*, *E. millii*, and *E. lathyris* calli were sitosterol, stigmasterol, campesterol, palmitic acid, and linoleic acid. *E. millii* callus produced an anthocyanin, cyanidin-3-arabinoside. A strain of cultured *E. millii* cells by cell aggregate selection produced a 27 times higher level of anthocyanin than the original cells. Statistical and cell pedigree analyses using a computerized system proved that this cell strain has a stable production of this anthocyanin. The optimum composition of the medium was investigated on anthocyanin production in *E. millii* suspension culture. Anthocyanin production (32 mg per liter per day) in modified Gamborg's medium (EM medium) is 4.5 times higher than that in MS medium. This strain was cultured in a 14-l jar fermentor. There are two current investigations as to whether the constituents of cultured *E. millii* cells can be applied as dyes and medicines. Useful compounds can certainly be found in cultured cells of the genus *Euphorbia*.

The best conditions for the establishment of cultures and the production of anthocyanins were as follows: callus induction medium containing malt extract and 2,4-D, cell aggregate selection for isolating a variant strain, modified medium (EM medium) for anthocyanin production, and mild agitation for fermentor culture.

References

Bell RL, Janick J, Blake RC (1980) Computer tracing of pedigrees. Hortscience 15:303–304

Biesboer DD, Mahlberg PG (1979) The effect of medium modification and selected precursors on sterol production by short-term callus cultures of *Euphorbia tirucalli*. J Nat Prod 42:648–657

Calvin M (1979) Petroleum plantations for fuel and materials. Bioscience 29:533–538

Colijn CM, Jonsson LMV, Schram AW, Kool AJ (1981) Synthesis of malvidin and petunidin in pigmented tissue cultures of *Petunia hybrida*. Protoplasma 107:63–68

Furstenberger G, Hacker E (1977) New highly irritant *Euphorbia* factors from latex of *Euphorbia tirucalli*. Experientia 33:986–988

Hardy T, Chaumont D, Brunel L, Gudin C (1986) Photoautotrophic suspension cultures I-Obtention of photoautotrophic cultures from *Euphorbia characias* L. J Plant Physiol 128:11–19

Hardy T, Chaumont D, Wessinger ME, Bournat P (1987) Photoautotrophic suspension cultures II-Transition from photoautotrophic to photoautotrophic growth. J Plant Physiol 130:351–361

Kakegawa K, Kaneko Y, Hattori E, Koike K, Takeda K (1987) Cell cultures of *Centaurea cyanus* produce malonated anthocyanin in UV light. Phytochemistry 26:2261–2263

Langhe EDE, Debergh P, Rijk RV (1974) In vitro cultures as a method of vegetative propagation of *Euphorbia pulcherrima*. Z Pflanzenphysiol 71:271–274

Lee TT, Starratt AN (1971) Growth substances requirements and major lipid constituents of tissue cultures of *Euphorbia esula* and *E. cyparissias*. Can J Bot 50:723–726

Mathur JMS, Maiid R, Bhaskaren S (1982) Phytochrome mediated biosynthesis of pigments in lily *Callistephus chinensis* (China aster). Ind J Exp Biol 20, 99

Matsumoto T, Nishida K, Noguchi M, Tamaki E (1973) Isolation and identification of an anthocyanin from the cell suspension culture of poplar. Agric Biol Chem 34:1110–1114

McCreight JD, Favila V (1980) Computer program for pedigree records in BASIC. Hortscience 15:439

Nishimura H, Philip RP, Calvin M (1977) Lipids of *Hevea brasiliensis* and *Euphorbia coerulescens*. Phytochemistry 16:1048–1049

Ohayama K, Uchida Y, Misawa N, Komano T, Fujita M, Ueno T (1984) Oil body formation in *Euphorbia tirucalli* L. cell suspension cultures. Plant Cell Rep 3:21–22

Ripley KP, Preece JE (1986) Micropropagation of *Euphorbia lathyris* L. Plant Cell Tissue Organ Cult 5:213–218

Sun BH, Francis FJ (1967) Apple anthocyanins: Identification of cyanidin-7-arabinoside. J Food Sci 32:647–649

Tideman J, Hawker JS (1982) In vitro propagation of latex-producing plants. Ann Bot 49:273–279

Yamakawa T, Kato S, Ishida K, Komada T, Minoda Y (1983) Production of anthocyanins by *Vitis* cells in suspension culture. Agric Biol Chem 47:2185–2191

Yamamoto Y, Mizuguchi R, Yamada Y (1981) Chemical constituents of cultured cells of *Euphorbia tirucalli* and *E. millii*. Plant Cell Rep 1:29–30

Yamamoto Y, Mizuguchi R, Yamada Y (1982) Selection of a high and stable pigment-producing strain in cultured *Euphorbia millii* cells. Theor Appl Genet 61:113–116

Yamamoto Y, Kadota N, Mizuguchi R, Yamada Y (1983) Computer tracing of the pedigree of cultured *Euphorbia millii* cells that produce high levels of anthocyanin. Agric Biol Chem 47:1021–1026

Yamamoto Y, Kinoshita Y, Watanabe S, Yamada Y (1989) Anthocyanin production in suspension cultures of high-producing cells of *Euphorbia millii*. Agric Biol Chem 53:417–423

XVI *Fritillaria* spp. (Fritillary): In Vitro Culture and the Regeneration of Plants

C.S. SUN[1] and D.Y. WANG[2]

1 Introduction

Fritillaria belongs to the family Liliaceae and its bulb is a traditional Chinese medicine ("Beimu" in Chinese). The bulb of the fritillary is divided into two groups according to its medical use: the fritillary bulb of zhebei and the fritillary bulb of chuanbei. The former is the underground bulb of *Fritillaria thunbergii* Miq. and the latter the underground bulb of *F. sungbei* Hsiao et K.C. Hsia, mss, *F. cirrhosa* D. Don, *F. cirrhosa* D. Don var. *paohsinensis* S.C. Chen, *F. delavayi* Franch., *F. pallidiflora* Schrenk., *F. sichuanica* S.C. Chen, and *F. ussuriensis* Maxim.

1.1 Geographic Distribution

F. thunbergii Miq. is a glabrous perennial plant. Its semi-globate bulb is white, 2–6 cm in diameter, and contains two or three thick bulb scales which are fused at one end. It is an erect, cylindrical, single stem with no branches, 30–70 cm high and green or light purple. The leaf is monophyllous and sessile. The leaves are opposite in the lower part of the stem, whorled with three to five leaves in the middle part of the stem, and alternate at the top of the stem. The leaves at the top of stem are shorter than those in the middle, and are lanceolate. The leaves above the middle of the stem and the apex of the leaf-like bract appear cirrhiform. Anthesis occurs in March-April, and usually two to six flowers develop in the stem apex or leaf axil of the top leaves. The pedicel is 1–1.5 cm long. The leaf-like bract is a very narrow strip. The flower is campanulate, overhanging, yellowish or yellow-green in color. The six perianths are obovate to ovate, 2–3 cm long with a light net and a glandular organ at the base, but with no light purple trellis stripe. There are six stamens, which are half the length of the perianth. The pistil contains one ovary with three parts, each of which contains a number of ovules. The fruit is capsulate (Fig. 1A). In nature, *F. thunbergii* Miq. grows in the woodland or on grassy mountains. At present, the medicinal *F. thunbergii* Miq. is cultivated in the Provinces of Zhejiang, Jiangsu, and Hunan, especially in the Ningbo County of Zhejiang Province. It is also distributed in Japan (Anonymous 1976a,b, 1980).

F. pallidiflora Schrenk is also a perennial herb with a glabrous, erect stem. Leaves are opposite or alternate. The leaves at the base of stem are lanceolate to

[1]Institute of Botany, Academia Sinica, Beijing, China
[2]China National Rice Research Institute, Hangzhou, China

Biotechnology in Agriculture and Forestry, Vol. 15
Medicinal and Aromatic Plants III (ed. by Y.P.S. Bajaj)

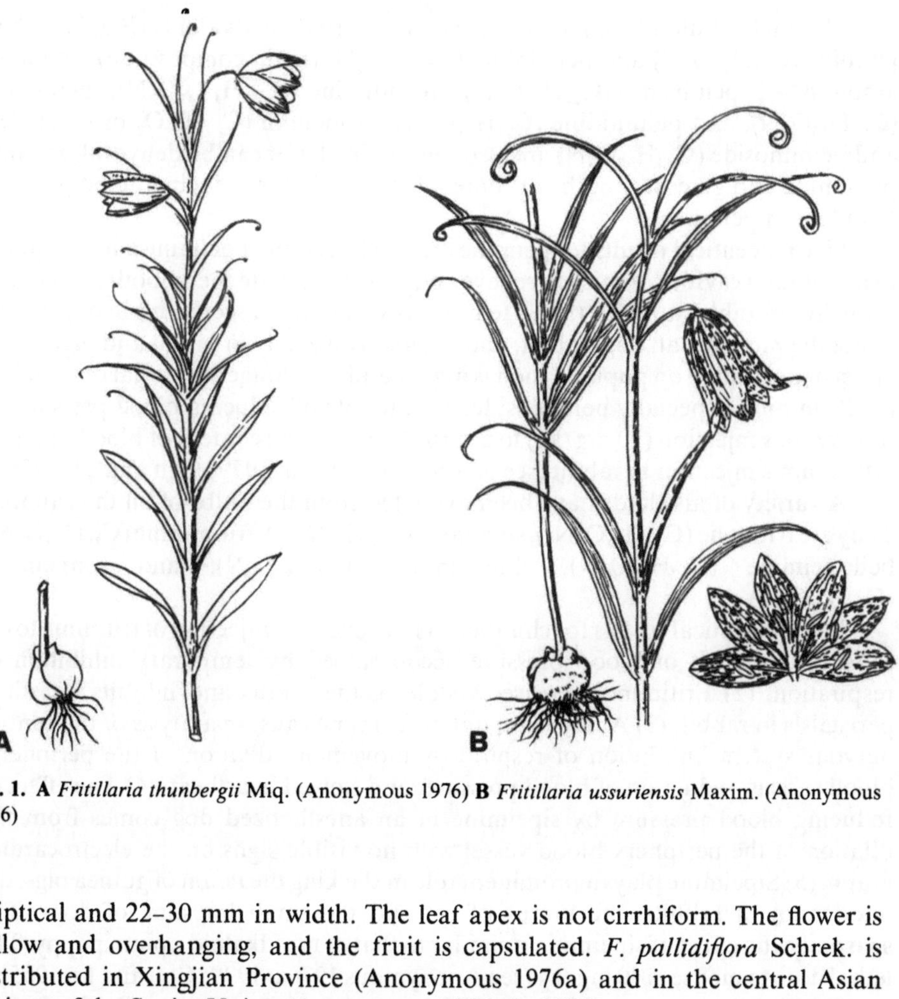

Fig. 1. A *Fritillaria thunbergii* Miq. (Anonymous 1976) **B** *Fritillaria ussuriensis* Maxim. (Anonymous 1976)

elliptical and 22–30 mm in width. The leaf apex is not cirrhiform. The flower is yellow and overhanging, and the fruit is capsulated. *F. pallidiflora* Schrek. is distributed in Xingjian Province (Anonymous 1976a) and in the central Asian regions of the Soviet Union (Anonymous 1976a,b, 1980).

F. ussuriensis Maxim. is a perennial herb, with capsulate fruit (Fig. 1B). It is distributed in the Provinces of Heilongjian, Jilin, and Liaoning, and is also found in Korea and the Far-Eastern area of the Soviet Union (Anonymous 1976a,b, 1980).

The chromosome number of sporophyte of *Fritillaria* is 24 ($2n = 2x = 24$).

1.2 Propagation Methods and Medical Uses

Both the zhebei bulb and the chuanbei bulb are used medicinally to relieve a cough, facilitating expectoration and removing endogenous heat. The medical difference is that chuanbei is applied for chronic cough (such as bronchoblennorrhea), a hacking cough, and the cough caused by tuberculosis, while zhebei is applied to cough and phlegm caused by the common cold. Both of them, especially zhebei, can cure intramammary abscess and scrofula.

The zhebei bulb (*F. thunbergii*) contains a variety of alkaloids (Fig. 2) such as peimine ($C_{27}H_{45}O_3N$) and peiminine ($C_{27}H_{43}O_3N$) (main composition), and trace amounts of peimisine ($C_{27}H_{43}O_4N$), peimiphine ($C_{27}H_{45\text{-}47}O_3N$), peimidine ($C_{27}H_{45}O_2N$), and peimitidine ($C_{27}H_{47}O_3N$). Propeimin ($C_{24}H_{44}O_3$ or $C_{27}H_{46}O_3$) and peiminoside ($C_{33}H_{55}O_3N$) are also found. The latter can be dehydrolyzed into peiminine and glucose, and has a more effective role for reducing blood pressure than that of peimine.

Pharmaceutical results for peimine are: (1) Injection of peimine and peiminine to abdominal cavities of mice to relieve cough. (2) To dilate the smooth muscle and bronchus of rabbit and cat. (3) To temporarily stop saliva secretions in dog. (4) To dilate the pupil in cat, dog, rabbit, and pigeon. Especially in rabbit and pigeon the effect of peiminine on pupil dilation is higher and lasts longer than that of atropine. (5) Peiminine, especially peiminoside, has the role of reducing blood pressure on intravenous injection (10 mg/kg) to cat. (6) Increases the content of blood sugar on intravenous injection to rabbit (see Anonymous 1976a, 1979; Chu and Lu 1955a).

A variety of alkaloids have been extracted from the bulbs of all the chunbei. They are fritimine ($C_{38}H_{62}O_3N_2$), sipeimine ($C_{27}H_{43}O_3N$), fritiminine ($C_{26}H_{39}O_3N$), beilupeimine ($C_{27}H_{43}O_3N$), chinpeimine ($C_{27}H_{43}O_2N$), and sonpeimine ($C_{27}H_{43}O_4N$).

Pharmaceutical results for chunbei: (1) Intravenous injection of fritimine to cat causes depression of blood pressure accompanied by temporary inhibition of respiration. (2) Fritimine increases systole of the uterus and inhibits intestinal peristalsis in rabbit. (3) A large amount of fritimine causes paralysis of the central nervous system, inhibition of respiratory movement, dilation of the peripheral blood vessels, reduction of blood pressure, and slowed heartbeat. (4) The effect of reducing blood pressure by sipeimine in an anesthetized dog comes from the dilation of the periphery blood vessel with no visible signs on the electrocardiograph. (5) Sipeimine plays a prominent role in slacking the ileum of guinea pigs, the duodenum of rabbits, the uterus of rat and the small intestine of dogs. The spasmolysitic effect of sipeimine is similar to that of the alkaloid upium poppy. The lethal dose to mouse is 40 mg/kg (see Anonymous 1976a, 1979; Chu and Lu 1955b).

Fritillary is propagated by bulb and the propagation rate is low. In the case of *F. thunbergii* Miq. one bulb can generate only two bulbs. Considering the loss from insects and disease, an average of only 1.5–1.6 bulbs can be regenerated from one bulb in 1 year. After saving seed bulbs for next year's propagation, only 0.5–0.6 bulbs from one parent bulb can be used for medicine. This low propagation ratio restricts cultivation (Sun et al. 1977). Although seed propagation of *F. thunbergii* Miq. considerably increases the propagation rate, the seedling is weak and the bulbs develop slowly. It takes 5–6 years to develop a bulb to commercial size (Anonymous 1974), so seed propagation has no practical value.

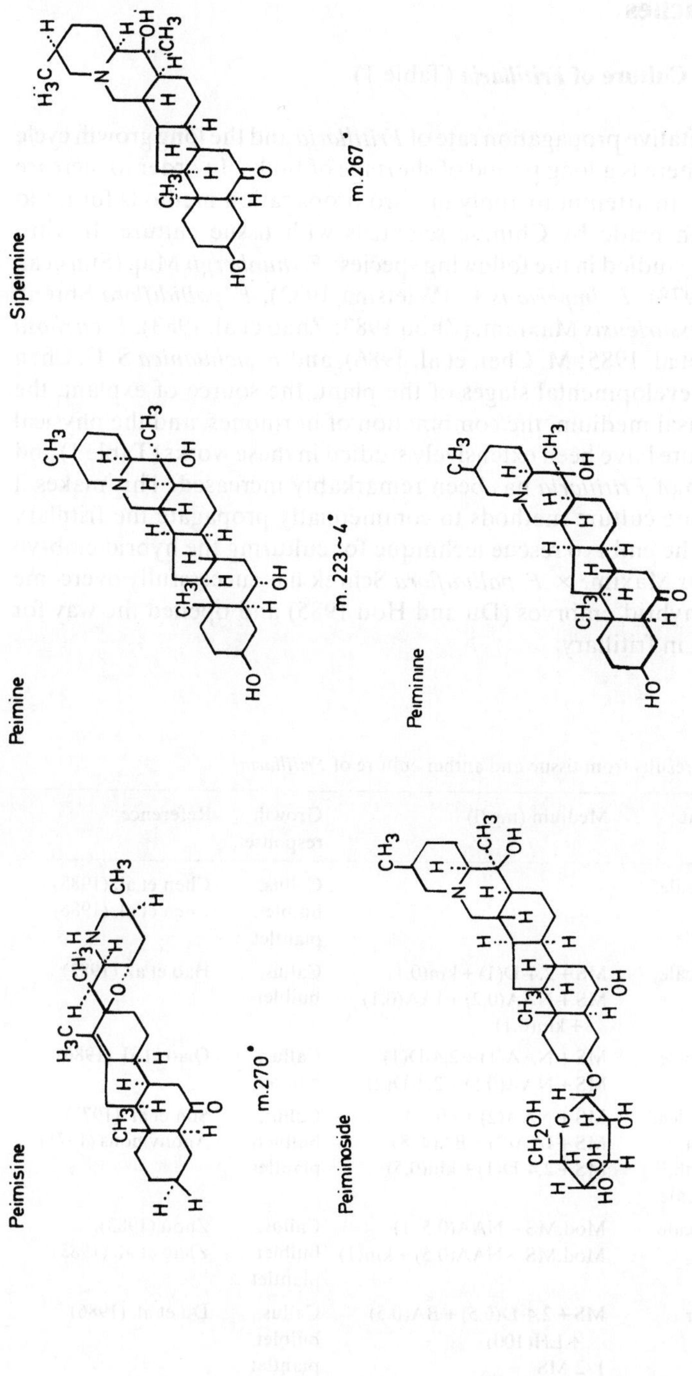

Fig. 2. Alkaloid content of *F. thunbergii*

2 In Vitro Approaches

2.1 Review of Tissue Culture of *Fritillaria* (Table 1)

Owing to the low vegetative propagation rate of *Fritillaria* and the long growth cycle of seed propagation there is a long period of shortage of bulbs. In order to increase the propagation rate, an attempt to apply in vitro propagation methods for rapid propagation has been made by Chinese scientists with tissue culture. In vitro propagation has been studied in the following species: *F. thunbergii* Miq. (Sun et al. 1977; Anonymous 1978), *F. imperialis* L. (Wietsma 1982), *F. pallidiflora* Shrenk (Hao et al. 1982), *F. ussuriensis* Maxium. (Zhou 1983; Zhao et al. 1983), *F. cirrhosa* D. Don (H.R. Chen et al. 1985; M. Chen et al. 1986), and *F. sichuanica* S. C. Chen (Qiao et al. 1986). Developmental stages of the plant, the source of explant, the constitution of the basal medium, the combination of hormones, and the physical conditions of the culture have been extensively studied in these works (Table 1) and the propagation ratio of *Fritillaria* has been remarkably increased. This makes it possible to apply tissue culture methods to commercially propagate the fritillary bulbs. Additionally, the embryo rescue technique for culturing the hybrid embryo between *F. ussuriensis* Maxim. × *F. pallidiflora* Schrek has successfully overcome the degeneration of hybrid embryos (Du and Hou 1985) and opened the way for distant hybridization in fritillary.

Table 1. Summary of the results from tissue and anther culture of *Fritillaria*

Species	Explant	Medium (mg/l)	Growth response	Reference
F. cirrhosa	Bulb scale, leaf	–	Callus, bulblet, plantlet	Chen et al. (1985), Chen et al. (1986)
F. pallidiflora	Bulb scale	MS + 2,4-D(1) + kin(0.1) MS + NAA(0.2) + IAA(0.1) + kin(0.1)	Callus, bulblet	Hao et al. (1982)
F. sichuanica	Bulb scale	MS + NAA(1) + 2,4-D(1) MS + NAA(0.5) + 2,4-D(2)	Callus, bulblet	Qiao et al. (1986)
F. thunbergii	Young leaf, pedicel, Perianth, bulb scale	MS + NAA(2) + kin(1) MS + IAA(2) + BA(4–8) MS + 2,4-D(1) + kin(0.5)	Callus, bulblet, plantlet	Sun et al. (1977) Anonymous (1978)
F. ussuriensis	Bulb scale	Mod.MS + NAA(0.5–1) Mod.MS + NAA(0.5) + kin(1)	Callus, bulblet, plantlet	Zhou (1983), Zhao et al. (1983)
	Anther	MS + 2,4-D(0.5) + BA(0.5) + LH(100) 1/2 MS	Callus, bulblet, plantlet	Du et al. (1986)
F. ussuriensis × *F. pallidiflora*	Seed	MS + IBA(0.5) + BA(2) 1/2 MS	Callus, bulblet, plantlet	Du and Hou (1985)

2.2 Establishment of in Vitro Cultures of *F. thunbergii* Miq.

Both the above ground part of the plant, such as young leaves, pedicel, and perianth (Sun et al. 1977), and the underground parts such as the bulb (Anonymous 1978) can be used as explants for tissue culture of *F. thunbergii* Miq. For the former, sterilization is simple; a 15-min treatment of saturated sodium hypochlorite can sterilize the explant. For the latter, sterilization requires repeated rinsing with tap water, removing the cork, and sterilizing in 0.1% mercuric chloride for 20 min. After sterilization, the bulb is cut into small pieces (5 × 5 × 2 mm) and inoculated onto MS (Murashige and Skoog 1962) medium. The temperature of the culture is 18–23 °C and the photoperiod is 10 h light (1500 lx).

There is a close relationship between callus induction from young leaves, pedicels, and perianths and plant growth regulators in the culture medium. NAA (0.5–2 mg/l) can effectively induce callus formation from explants (Fig. 3A). Callus cannot be induced from medium containing less than 0.1 mg/l NAA or auxin-free medium. One mg/l kinetin or 15% coconut milk favor the growth of callus. White bulblets (Fig. 3B) and roots (Fig. 3C) can be regenerated after the calli have been transferred to regeneration medium containing 2 mg/l IAA and 4–8 mg/l BA. At the time of bulblet regeneration or later, a number of green bud-like protuberances appear on the surface of the callus, which continue to grow and form a linear leaf-like structure (Fig. 3D). Most of the bud-like protuberances cannot stand erect and they spread on the surface of the callus (Fig. 3E).

Bulblets can be directly regenerated from the surface of bulb segments from spring harvested explants on MS medium supplemented with 1 mg/l 2,4-D and 0.15 mg/l kinetin. Roots can form from the base of the bulblets. The size of a 4-month-old bulblet is equal to that of a 2–3-year-old bulb from seed propagation.

Plant regeneration is difficult to induce from the callus-derived bulblets or from the bulb-derived bulblets, unless the bulblets have been treated at 2–15 °C followed by light culture at 18–23 °C (Fig. 3F). The plantlets from these low-temperature pretreated bulblets grow vigorously and have a high survival frequency. The chromosome number of the root tips from regenerated bulblets and plantlets is normal, $2n = 24$.

Callus of *F. thunbergii* Miq. can be subcultured for a long time and still keep the capacity of bulblet regeneration and plant regeneration. Some cell lines have been subcultured for 14 years and still retain the capacity for differentiation (Sun unpubl.).

According to the analytical results, the alkaloid in the tissue cultures is the same as that in the commercial bulb. (Sun et al. 1977).

2.3 Establishment of in Vitro Cultures of *F. pallidiflora* Schrenk

The bulb, harvested in March or April, is repeatedly rinsed with tap water, wiped with 70% alcohol, followed by sterilization in 0.1% mercuric chloride for 20 min. Sterilized bulbs are cut into small pieces of 4 mm and cultured.

Experiments indicate that the basal medium has a prominent affect on callus induction. (Table 2). A light yellowish callus can be induced from bulb explant on

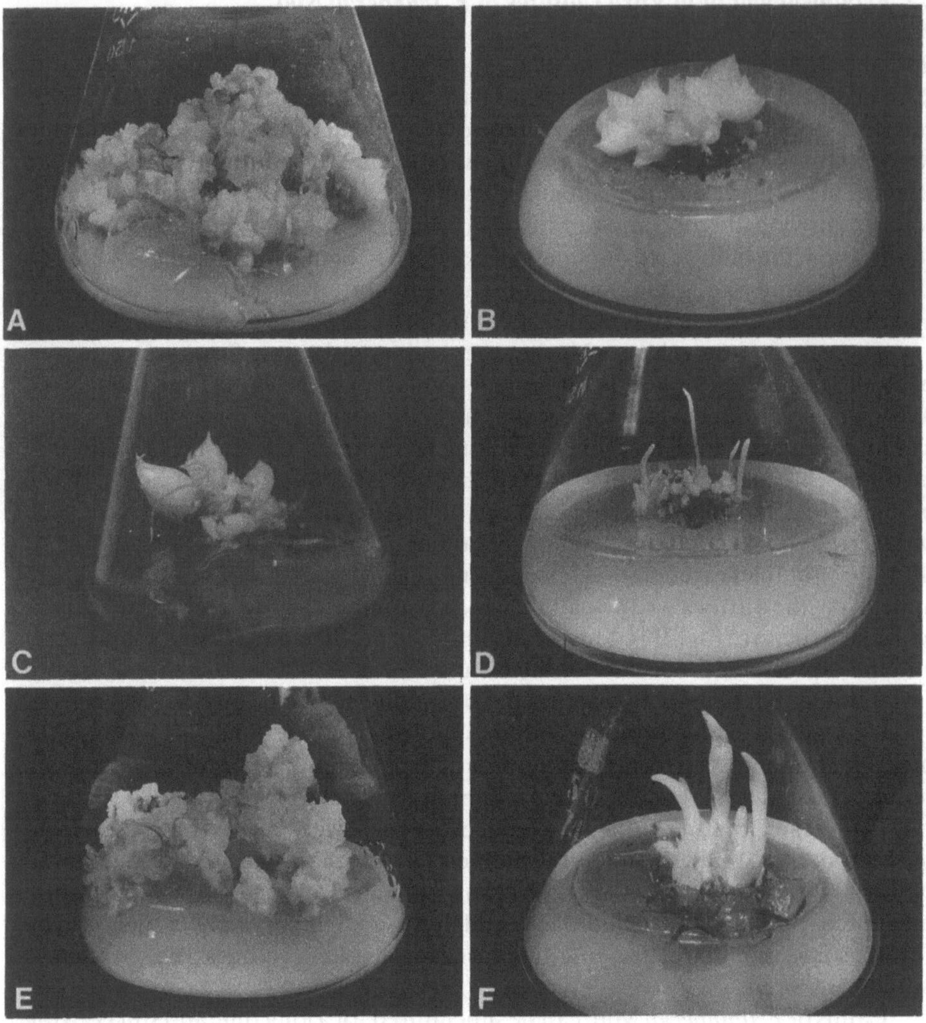

Fig. 3A-F. Callus growth and differentiation of bulblets and plantlets. (Sun et al. 1977). **A** Callus (1 month after inoculation) from perianth of *F. thunbergii* Miq. on MS medium + 1.5 mg/l NAA + 1 mg/l kinetin. **B** Growth and differentiation of bulblets 3 months after callus was transferred onto MS medium + 2 mg/l IAA + 5 mg/l BA. **C** Same as **B**, showing the growth and differentiation of roots. **D** Linear-like leaves differentiated from callus onto the medium which is the same as **B**. **E** Leaves from callus spread onto the medium which is the same as **B**. **F** Plantlet from a bulblet pretreated at low temperature (2–6° C) for 40 days

Table 2. Effect of basal medium on callus induction from bulb of *F. pallidiflora* Schrenk. (Hao et al. 1980)[a]

Medium	No. of explants	15 days after culture		30 days after culture	
		No. of calli	%	No. of calli	%
MS	48	8	16.67	17	35.42
H	16	4	25.00	4	25.00
B5	45	0	0	0	0

[a] Media were supplemented with 1 mg/l 2,4-D and 0.1 mg/l kinetin.

MS medium containing 1 mg/l 2,4-D and 0.1 mg/l kinetin in 2 weeks. The frequency of callus induction can reach as high as 35.42% in 30 days of culture; but the frequency of callus induction is only 25% in the H medium (Nitsch 1972) containing the same hormone composition, and there is no callus induction at all on B5 medium (Gamborg et al. 1968).

The data in Table 3 indicate that callus can be induced from bulb segments of *F. pallidiflora* Schrenk in medium containing 0.5–2 mg/l 2,4-D at an optimum concentration of 1 mg/l 2,4-D. Four mg/l 2,4-D makes the explants blacken, and there is no callus induction.

On MS medium + 1 mg/l 2,4-D + 0.1 mg/l kinetin, callus is yellowish, friable, and translucent. Callus should be subcultured at least once a month otherwise it will turn brown and leads to necrosis.

Besides callus formation from inoculated bulb segments on cytokinin-containing medium, bulblets can be directly regenerated. Experiments indicate that the maximum number of bulblets can be regenerated on MS + 1 mg/l IAA + 0.2 mg/l NAA + 0.1 mg/l kinetin, and 12 bulblets can be induced from one piece of explant.

Callus and regenerated bulblets are stored at 4–10 °C for 40 days, then transferred to the MS + 0.1 mg/l IAA + 0.5 mg/l kinetin and cultured at room temperature. Seedling, roots, and complete plantlets can be regenerated.

Table 3. Effect of the concentration of 2,4-D on callus induction from bulbs of *F. pallidiflora* Schrenk. (Hao et al. 1982)[a]

Conc. of 2,4-D (mg/l)	No. of explants	15 days after culture		30 days after culture	
		No. of calli	%	No. of calli	%
0	20	0	0	1	5
0.5	19	5	26.31	10	52.36
1.0	20	9	40.90	14	63.64
1.5	16	4	25.00	8	50.00
2.0	16	2	12.50	6	37.50
4.0	18	0	0	0	0

[a] MS medium + 0.1 mg/l kinetin.

2.4 Establishment of in Vitro Cultures of *F. ussuriensis* Maxim

2.4.1 Tissue Culture

Bulbs from 2–3-year-old plants of *F. ussuriensis* Maxim. are inoculated onto modified MS medium (250 mg/l KH_2PO_4 and 5 mg/l vitamin B) after sterilization (method as above). Liquid culture is better than solid culture medium. Liquid culture uses a filter paper bridge, i.e., a filter paper roll of 2.5–3 cm high, on the top of which is placed a round filter paper called a "carry bridge", which is placed in a 100 ml flask filled with 30–40 ml liquid medium. The explant is inoculated onto the round filter paper. For callus induction, liquid medium contains 0.5–1 mg/l NAA. For bulblet induction liquid medium contains 0.5 mg/l NAA and 1 mg/l kinetin. The temperature for culturing is 16–21 °C, and the photoperiod is 12 h light (400 lx).

A yellow-green callus is induced from the edge of the explant 3–4 weeks after culture. A freshly harvested bulb is better to culture than one harvested over winter. The frequency of callus induction from the season-harvested bulb is 86%, but only 47% from the over-winter-harvested bulb. Calli were cut from explants and transferred to MS medium supplemented with 0.5 mg/l NAA and 1 mg/l kinetin. White bulblets were induced in the medium, with a frequency of bulblet regeneration of 65–80%. Bulblets and buds can be directly induced from bulb explants on the differentiation medium. Two or three roots from each bulb can be induced on the root-inducing medium: MS + 2 mg/l IBA.

The experiment indicates that liquid medium is better than solid medium in terms of root and shoot growth, bulb growth, and bulb differentiation. The size of a 1-month-old bulblet in the liquid medium is 0.5–0.6 cm in diameter, which is bigger than that of a 3-year-old bulb in nature. The growth rate of bulblets and calli in the ginseng root powder (0.1%)-containing medium is 1.5 times faster than that of the control (based on fresh weight).

2.4.2 Anther Culture

The first report on plant regeneration from anther culture of *F. ussuriensis* Maxim. appeared in 1986 (Du et al. 1986). The anthers, with pollen at the uninucleate stage, were inoculated onto agar medium and cultured at a temperature of 22–28 °C/18–22 °C (day/night). Light intensity was 1600 lx and photoperiod 10 h light. A total of 21,300 anthers were inoculated and calli were induced from 32 anthers. The average callus induction frequency (no. of response/total no. of inoculated anthers) was 0.2%. MS basal medium gave the best result for callus induction. No callus was induced on either Miller (Miller 1963) or White media (White 1943). The frequency of callus induction was less 0.1% on N6 medium (Chu et al. 1975), and 0.39% on MS medium. The maximum callus induction frequency (1.94%) was obtained from MS medium + 0.5 mg/l 2,4-D + 0.5 mg/l BA + 1000 mg/l LH (lactalbumin hydrolysate) + 6% sucrose, about nine times higher than the average callus induction frequency.

At 1 mm diameter calli were transferred to the regeneration medium. Following are the tested regeneration media: (1) MS + 2 mg/l BA (or kin) + 0.5 mg/l IBA + 3%

sucrose; (2) MS + 2 mg/l BA + 0.5 mg/l IBA + 500 mg/l CH + 3% sucrose; (3) 1/2 MS (half strength macroelements of MS medium) + 2 mg/l BA + 0.5 mg/l IBA + 500 mg/l CH + 3% sucrose; and (4) 1/2MS. Callus began proliferating 10 days after transfer onto regeneration medium, then bulblets were differentiated from the surface of callus. An average of three bulblets (maximum seven bulblets) were induced from one callus piece. Green buds from the top of bulbs and roots from the base of bulbs were developed and formed complete plantlets. Most of the regenerated complete plantlets were induced from half-strength MS medium without hormone.

The result of chromosome examination of root tip cells indicated that about one-quarter of the cells had 12 chromosomes which are haploid. The remaining cells were diploid, tetraploid, or aneuploid (Table 4). This heteroploidy may result from spontaneous doubling of the chromosome during culture of callus and bulbs. Sharp et al. (1971) reported that haploidy cannot be maintained for more than four to six subcultures after bulbs formed from haploid callus in *Lilium longilflorum*. This result is similar to that of *F. ussuriensis*.

Table 4. Chromosome number of root tip cells of pollen-derived plants in *F. ussuriensis*. (Du et al. 1986)

No. of cells examined	Cell number of various chromosome number			
	12	24	48	Aneuploid
35	8 (23%)	11 (31%)	4 (11%)	12 (35%)

3 Conclusion and Prospects

An effective tissue culture method for several important medical species of fritillary has been established after 10 years of extensive research in China (Table 1).

Young leaves, pedicels, and perianths can be used as explants in order to induce callus formation from *F. thunbergii* Miq. on the medium MS + 2 mg/l NAA + 1 mg/l kinetin. Bulblet differentiation can be induced after callus is transferred to MS medium with 2 mg/l IAA and 4–8 mg/l BA. Bulblets can also be induced directly from underground bulbs on the medium MS + 1 mg/l 2,4-D + 0.15 mg/l kinetin.

Callus can be induced from bulb segments of *F. pallidiflora* Schrenk on MS + 1 mg/l 2,4-D + 0.1 mg/l kinetin. Root and shoot differentiation can be induced after callus is transferred to MS + 0.1 mg/l IAA + 0.5 mg/l kinetin. Bulblets can also be induced from bulb segments on MS + 1 mg/l IAA + 0.2 mg/l NAA + 0.1 mg/l kinetin.

The optimum condition for callus induction and bulblet regeneration of *F. ussuriensis* Maxim. is filter paper bridge-mediated liquid medium. The basal medium is modified MS medium (250 mg/l KH_2PO_4 and 5 mg/l vitamin B). The medium for callus induction contains 0.5–1 mg/l NAA, and for bulblet differentiation 0.5 mg/l NAA and 1 mg/l kinetin.

Optimum conditions for anther culture of *F. ussuriensis* Maxim. are (1) anthers with uninucleate pollen; (2) MS + 0.5 mg/l 2,4-D + 0.5 mg/l BA + 1000 mg/l LH + 6% sucrose for callus induction; and (3) 1/2 MS medium without hormone for bulb induction and plant regeneration.

Tissue cultures of fritillary need low temperature. Calli and bulblets can normally be induced at 16–23° C supplemented with 10–12 h light (400–1500 lx). In order for the regenerated bulblet to grow further, the bulblet needs a 2–15° C cold treatment for 40 days.

The above in vitro methods can remarkably increase the propagation ratio of *Fritillaria*, and can open a way to apply tissue culture technique to the mass production of stock bulbs of *Fritillaria*. The alkaloid composition of calli and bulblets from tissue culture is the same as that of naturally harvested bulbs, which indicates that there is a potential to apply tissue and cell culture for industrial production of secondary products of the alkaloids of *Fritillaria*, or to utilize the tissue cultures (such as bulblets) directly as crude drugs or dried medicinal herbs.

References

Anon (1974) Seed propagation of *Fritillaria thunbergii* Miq. Commun Chin Med Herbs 1:61–63 (in Chinese)

Anon (1976a) A corpus of traditional Chinese herbs, vol 1. Medical Press, Beijing, pp 127–643 (in Chinese)

Anon (1976b) Iconographia Cormophytorum Scnicorum (tomusV). Science Press, Beijing, pp 461–463 (in Chinese)

Anon (1978) Tissue culture and in vitro propagation of bulb segment of *Fritillaria thunbergii* Miq. Commun Chin Med Herbs 5:39–41 (in Chinese)

Anon (1979) A dictionary of Chinese herbs. Shanghai Scientific and Technological Press, Shanghai, pp 506–546 (in Chinese)

Anon (1980) Florae Reipublicae Popularis Sinica (Tomus 14). Science Press, Beijing, pp 101–112 (in Chinese)

Chen HR, Chen FT, Chen M, Zhong FL (1985) Tissue culture of *Fritillaria cirrhosa*. I.J Tradit Chin Med 10:10 (in Chinese)

Chen M, Chen HR, Zhong FL, Wang BJ (1986) Tissue culture of *Fritillaria cirrhosa*. II. J Tradit Chin Med 11:13 (in Chinese)

Chu CC, Wang CC, Sun CS, Hsu C, Yin KC, Chu CY, Bi FY (1975) Establishment of an efficient medium for anther culture of rice through comparative experiments of the nitrogen sources. Sci Sin 18:559–668

Chu ZC, Lu RR (1955a) A study on alkaloids of *Fritillaria* II. Further study on Peimine A and Peimine B. Acta Chem Sin 21:227–231 (in Chinese with English summary)

Chu ZC, Lu RR (1955b) A study on alkaloids of *Fritillaria* IV. New alkaloids in *F. pallidiflora* (1). Acta Chem Sin 21:241–246 (in Chinese with English summary)

Du LG, Hou YH (1985) Plant regeneration from immature embryo of a hybrid of species cross of *Fritillaria*. Chin Herbs 2:1 (in Chinese)

Du LG, Hou YH, Chang WC, Zhao YH, Li AS, Shao QQ, Yang ZT, Chen ZG (1986) Induction of pollen plant and clone establishment of *Fritillaria ussuriensis* Maxim. Acta Genet Sin 14:262–265 (in Chinese with English summary)

Gamborg OL, Miller RA, Ojima K (1968) Nutrient requirement of suspension culture of soybean root cells. Exp Cell Res 50:151–158

Hao YR, Li MS, Wu YW (1982) Callus induction and plant regeneration in tissue culture of *Fritillaria pallidiflora*. Acta Bot Bor-Occ Sin 2:38–43 (in Chinese with English summary)

Miller CO (1963) Kinetin and kinetin-like compounds. In: Linskens HF, Tracey MV (eds) Moderne Methoden der Pflanzenanalyse, vol 6. Springer, Berlin Heidelberg New York, p 194–202

Murashige T, Skoog F (1962) A revised medium for rapid growth and bioassay, with tobacco cultures. Physiol Plant 15:473–497

Nitsch JP (1972) Haploid plants form pollen. Z Pflanzenzücht 67:3–18

Qiao HL, Ma XF, Li LY (1986) Preliminary report of greenhouse production of bulb of *Fritillaria sichuanica*. J Tradit Chin Med 11:10–11 (in Chinese)

Sharp WR, Raskin RS, Sommer HE (1971) Haploidy in *Lilium*. Phytomorphology 21:334–336

Sun CS, Chu CC, Wang CC (1977) Callus formation and organ regeneration in the tissue culture of *Fritillaria thunbergii* Miq. Acta Bot Sin 19:161–162 (in Chinese with English summary)

White PR (1943) A handbook of plant tissue culture. Cattel, Tempe, Ariz

Wietsma WA (1982) *Fritillaria* in kweekbuize te vermeerderen. Vakbl Bloem 73:35

Zhao GF, Cao Y, Wu Y, Fan F, Zhou LJ, Yang WH (1983) Callus induction and organ regeneration in tissue culture of *Fritillaria ussuriensis* Maxim. Chin Bull Bot 1, 2:40–41 (in Chinese)

Zhou BJ (1983) A study on in vitro culture of bulb of *Fritillaria ussuriensis* Maxim. Plant Physiol Commun 1:30 (in Chinese)

XVII *Glycyrrhiza glabra* L. (Licorice): Cell Culture, Regeneration, and the Production of Glycyrrhizin

M. Henry[1], A.M. Edy[1], P. Desmarest[2], and J. Du Manoir[2]

1 Introduction

1.1 Distribution and Importance of the Plant

The genus *Glycyrrhiza* contains 30 species native to hot temperate or subtropical regions such as the Mediterranean Basin. Among these various kinds of licorice plants, few are of industrial interest, for example, *Glycyrrhiza glabra* L., *G. glabra* L. var. *glandulifera* Rgl. et Herd., *G. uralensis* D.C. and *G. echinata* L. This chapter describes only the results for *G. glabra* L. the licorice material most employed in industry for its aromatic and sweetener properties (Afchar 1981; Boischard 1988).

G. glabra L. is a perennial plant (Fig. 1) belonging to the subfamily Papilionoideae of the family Leguminosae, common around the Mediterranean Sea. It is a herb which can reach 1.50 in height, the most important part being underground. The roots are held by a subterranean stem or rhizome, which are very ramified with long and thick branches. The upper shoot, with a sometimes scabrid and pubescent stem, is erect, with long (5 to 20 cm) imparipinnate leaves, each with 9 to 17 elliptical leaflets.

The hanging, elongate racemes, surrounded by their subtending leaves, contain 20 to 30 flowers, which are relatively short (corolla 8–12 mm with five whitish purple petals) and characterized by a horseshoe-shaped opening of the anthers at the end of the ten stamens. The fruit, up to 30 mm, is compressed, glabrous or glandular-setose, with three to five small (2–3 mm) seeds.

In the species *G. glabra* L., three varieties are generally distinguished:

- *G. glabra* L. var. *typica* called Spanish Licorice after the country where it has long been cultivated over great areas. In 1930, 1648 ha were reported to be cultivated with this variety. It has been found in many neighboring countries in the Mediterranean zone such as Portugal, Italy, Greece, Turkey, and the Caucasus.
- *G. glabra* L. var. *glandulifera* Rgl. et Herd. or Russian Licorice. In this variety, the stolons are missing. It develops best in Hungary, Iran, and Spain.
- *G. glabra* L. var. *violacea*, wrongly called Iran Licorice, is not found in Iran, and is encountered mainly in Iraq.

[1] Laboratoire de Botanique, Faculté des Sciences Pharmaceutiques, 31, Allées Jules Guesde, 31000 Toulouse, France
[2] Centre de Recherches "Pernod-Ricard", 120, Av. du Maréchal Foch, 94003 Creteil, France

Biotechnology in Agriculture and Forestry, Vol. 15
Medicinal and Aromatic Plants III (ed. by Y.P.S. Bajaj)

Fig. 1. Plants of *Glycyrrhiza glabra* var. *typica* 5 years old, obtained by micropropagation, in the Botanical Garden of the Faculty of Pharmacy of Toulouse

1.2 Economical Aspects

Glycyrrhizae Radix is an important commercial product (Molyneux 1975). It is used:

– in the tobacco industry, where licorice powder gives a sweet aromatic perfume, and also protects tobacco from desiccation,
– in the food industry, where licorice is used as an aromatic raw material in sweets or in mixtures in which the sweetening effect needs its aromatic taste,
– in the pharmaceutical industry for the anti-inflammatory and antiulcerative activity of glycyrrhizin, the main component of licorice.

Licorice is mainly known throughout the world as a sweetener, glycyrrhizin being the sweet-tasting component. Glycyrrhizin and its aglycone, glycyrrhetinic acid (Fig. 2), are the most active components of Glycyrrhizae Radix, the root of *Glycyrrhiza* plants. Glycyrrhetinic acid has two stereoisomers, 18α-form (trans) (18α-GA) and 18β-form (cis) (18β-GA), the structure of which was studied particularly by Ruzicka, Jeger and their collaborators (see Beaton and Spring 1955 and references cited therein).

The anti-ulcerative activity of Glycyrrhizae Radix was reported by Landerlan in a patent (1976), and it is well known for its clinical application in the treatment of stomach and duodenal ulcers. The anti-inflammatory action has been attributed essentially to 18α-GA stronger than that of 18β-GA in several experimental models (Amagaya et al. 1984). 18α-GA showed a strong anti-granulomatous action in

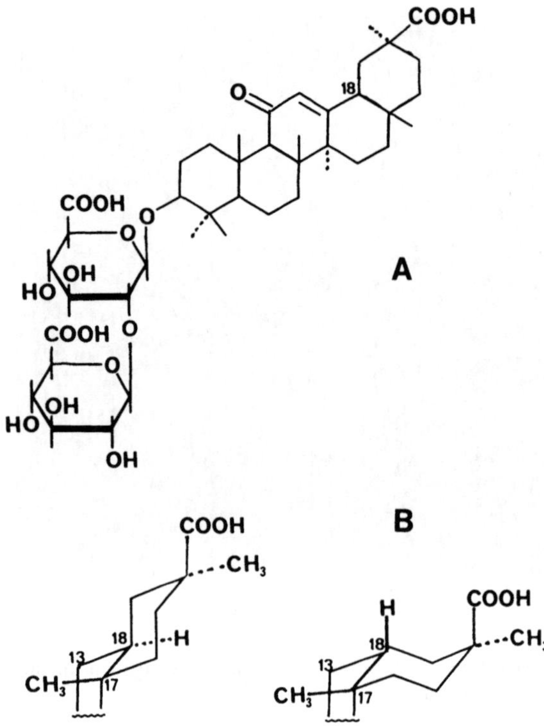

18α-GA 18β-GA

Fig. 2A,B. Glycyrrhizin and its aglycone, glycyrrhetinic acid. **A** Glycyrrhizin: diglucuronide of glycyrrhetinic acid. **B** The two stereoisomers, 18α-form (trans) (18α-GA) and 18β-form (cis) (18β-GA) of Glycyrrhetinic acid

adrenalectomized rats. These results indicate that the action of 18α-GA is similar to that of glucocorticoid.

Glycyrrhizin has also been shown to have anti-allergic and antihepatitis activity (Kiso et al. 1984). Glycyrrhetinic acid possesses an inhibitory action against nonenzymatic lipid peroxidation. The anti-gastric ulcer effect was also attributed to flavonoid constituents of licorice present in Glycyrrhizae Radix (Kobayashi et al. 1985).

1.3 Conventional Practices for Propagation, the Production of Glycyrrhizin, and the Demand in the World Market

Licorice plants can grow only in limited regions in the world around the Mediterranean Sea, and therefore the production of licorice root and industrial reprocessing has necessarily been regionally restricted.

Propagation is performed by cutting. Small pieces of rhizomes, 10 to 15 cm long, cut from the mother plant, contain two to three buds. These buds develop rapidly and later adventitious roots appear on the short rhizome. Licorice plantation is carried out in early spring in deep, light soils.

In France, one of the authors (P. Desmarest) succeeded in cultivation by directly sowing seeds previously treated with concentrated sulfuric acid for 2 h following the technique of Shukurullaev and Khamdamov (1976). The soil had to be perfectly free of weeds for the first 2 years after planting or sowing. Five liter/ha Etazine is spread on the cuttings after planting and 1 l/ha Sinbar on the sowing bed just before drilling.

After 3 years of growth, 26 t/ha of a mixture of roots and rhizomes were extracted from the soil. The glycyrrhizin contents increased with subsequent growth periods: 11.02% were obtained the first year, 11.50% the second, 14.14% the third, and 15.20% the fourth (Yaskonis 1976).

The main countries producing licorice are Turkey, Syria, Iran, Iraq, the U.S.S.R., and China, but Greece, Italy, and Spain come very soon after (Adzet Porredon 1986). The U.S.A. are the main importer, with 20,000 tons of root-rhizome mixture handled by only one company, which imports the roots and rhizomes either directly from Syria, Turkey, and the U.S.S.R., or after a first extraction performed in China (Bozzi M., EVD Company, Marseille, France, Division of MacAndrew and Forbes Company, U.S.A., pers. commun.).

In France, 3500 t are imported each year and transformed for either the pharmaceutical or food industry.

2 In Vitro Culture of Licorice

2.1 Review of Tissue Culture and Biotechnological Studies on Licorice

Only a few studies have been done on tissue culture of licorice. Some of them deal with the production of glycyrrhetinic acid and glycyrrhizin, others have studied the flavonoid production. Most of this research has been done on undifferentiated tissue culture (Table 1).

2.2 Micropropagation

Propagation of a specific genotype of licorice by cutting means immobilizing fields of parental plants to pick out small pieces of rhizomes.

Propagation in vitro can multiply this genotype more rapidly with a great number of plants in a short time to plant in preproduction areas.

Propagation in vitro of licorice was first established by Shah and Dalal (1980) on *G. glabra*. This technique has been applied by the Research Center of the Pernod Ricard Company (Dumanoir and Desmarest unpubl.) to other species: *G. glabra* L. var. *typica* (Fig. 3) and *glandulifera, G. uralensis, G. lepidota,* and *G. pallidiflora*.

Nodal stem segments with an axillary bud excised from well-growing plants were surface sterilized with calcium hypochlorite (70 g/l, 15 min) and washed with sterile distilled water. They were subsequently cultured on a solidified medium containing Shah and Dalal (1980) macroelement mineral salts, Murashige and

Table 1. Summary of in vitro studies on licorice

Plant species	Inoculum	Medium (mg/l)	Growth response	Reference
G. glabra	Nodel stem segments	SD	Plants	Shah and Dalal (1980)
G. glabra typica	Nodal Stem	Macronutrients SD + micronutrients MS	Plants	Du Manoir and
G. g. glandulifera	segments	+ folic acid(0.1) + biotine(0.1)		Desmarest
G. uralensis		+ sucrose(2%) + IAA(0.1) or IBA(0.1)		(unpubl.)
G. lepidota				
G. pallidiflora				
G. glabra	Seedlings	MIL + 2,4-D(0.0221) + kin(2.15)	Differentiated organs (Production of glycyrrhizin)	Fujita et al. (1978)
G. glabra	Seedlings	1) Herligel medium 2) MS with various conditions	Calli	Syrtanova and Mukhitdinova (1984)
		3) SD without hormones	Plants	
G. glabra	Leaves	1) MS + 2,4-D(2) + Ba(0.3)	Cali	Boischard (1988)
	Hypocotyles	2) MS with various conditions		
G. glabra	Seedlings	G-PRL-4C + CM or YE	Single cell suspension	Wu et al. (1974)
G. glabra typica	Inter nodal stem segments	1) NN + aden(2) + 2,4-D(0.1) + sucrose (2%)	Callus induction	Henry et al. (1984) Dorisse et al. (1988)
		2) MS + aden(2) + 2,4-D(0.1) + kin (1.0) + glucose (3%)	Calli Cell suspensions Embryogenesis attempts	Roustan et al. (1988)
G. g. glandulifera	Root fragment	LS + IAA(2) + kin(0.2)	Primary calli and suspensions (Production of glycyrrhizin)	Tamaki et al (1973)
G. echinata	Seedlings	W mod. + 2,4-D(0.1) + YE(0.1%)	Calli Production of flavonoids	Furuya et al. (1971) Furuya et al. (1976) Saitoh et al. (1975) Ayabe et al. (1980a) Ayabe and Furuya (1981)
G. uralensis	Seedlings	1) MS + 2,4-D(3) + kin(0.1) + sucrose 2) MS + 2,4-D(2.5) + BA(0.1) 3) MS + 2,4-D(1) + NAA(2) + BA(0.1)	Calli Production of flavonoids	Kobayashi et al. (1985)

Abbreviations: Basal media: SD: Shah and Dalal (1980), MS: Murashige and Skoog (1962), LS: Linsmaier and Skoog (1965), NN: Nitsch and Nitsch (1969), G-PRL-4C: Gamborg and Eveleigh (1968), Herligel medium: Syrtanova and Mukhitdinova (1984), MIL: Fujita et al. (1978), W mod.: White's medium modified by Ayabe et al. (1980a).

Supplements: IAA: Indoleacetic acid, NAA: Naphthyleneacetic acid, 2,4-D: Dichlorophenoxyacetic acid, kin: kinetin, BA: Benzylamino purine, aden.: adenine, CM: coconut milk, YE: yeast extract.

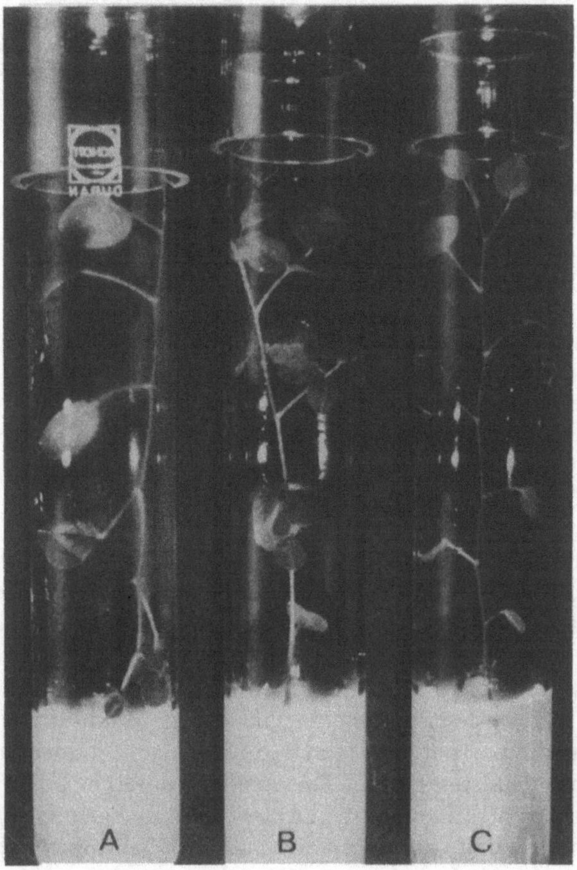

Fig. 3A-C. Micropropagation of *Glycyrrhiza uralensis* (**A**) *Glycyrrhiza glabra* var. *typica* (**B**) and *Glycyrrhiza glabra* var. *glandulifera (***C**) Vitro-plants aged 6 weeks obtained on a solidified medium containing Shah and Dalal (1980) macroelement mineral salts, Murashige and Skoog microelement salts and vitamins (1962), folic acid (0.1 mg/l), biotine (0.1 mg/l), and sucrose (2%), and rooted with IAA (0.1 mg/l) on the same medium

Skoog microelement salts and vitamins (1962), folic acid (0.1 mg/l), biotine (0.1 mg/l), and sucrose (2%).

After 1 month of culture, these explants were transferred on to the same medium supplemented with an auxin to promote shoot growth and root regeneration. Rooting of the explants is necessary before any shoot growth. Most of the varieties studied here could be rooted with 0.1 mg/l IAA, few of them needed 0.1 mg/l IBA. NAA produced unorganizated proliferating growth round the roots.

After 6 weeks, young stems could be cut in any segment as node. This permitted to obtain a multiplying factor of 4 to 7 according to the species or the clones studied (Table 2).

Rooting began during the first week of culture. After 4 weeks, the rooted plants were able to be acclimatizated in an air conditioned room or directly in a greenhouse

Table 2. Multiplication factor and rooting percentage during the micropropagation of licorice

	Glycyrrhiza glabra var *typica*	*Glycyrrhiza glabra* var. *glandulifera*	*Glycyrrhiza uralensis*
Multiplication factor	4.9 ± 0.9	6.8 ± 1.2	3.7 ± 1.7
Rooting percentage	85.0 ± 9.7	96.8 ± 4.8	92.5 ± 5.3

by transplanting them into a compost (Grassland) with a mixture of sand and light and brown peat. Established plants may be transferred to the field 1 month later. The final rate of success (plants in open ground) is usually higher than 80%.

2.3 Callus Culture

The first observations on in vitro culture of licorice tissues were made by Tamaki et al. in 1973. By cultivating root fragments of *G. glabra* var. *glandulifera* on Linsmaier and Skoog's medium (1965) with kinetin 0.2 mg/l and IAA 2.0 mg/l and agar 1%, these researchers obtained white or yellowish white callus in 2–3 weeks. These calli continued to proliferate after three subcultures on the same medium.

It was later proved (Syrtanova and Mukhitdinova 1984; Boischard 1987) that various parts of the vegetative system can produce calli, which can be maintained on a defined medium. These calli of different origin (root, hypocotyl, cotyledon) display no difference in morphology or in regenerating organs, but a relatively high amount of an auxin, differing in nature and in concentration for each part of the plant, is vital to trigger proliferation and maintain it during later subcultures.

No difference could be detected between the two varieties *typica* and *glandulifera*, either in morphological aspect or in the nutritional needs of the calli. Whatever the concentration of 2,4-D and cytokinin, kinetin, or benzyladenin, the calli obtained are always yellowish brown or brown with 2,4-D, and more hydrated with the increase of the concentration of 2,4-D. To induce substantial callusing 0.1 mg/l of 2,4-D is necessary. When the 2,4-D concentration is increased to 5 mg/l, the callus growth and the friability of the aggregates are better, and the calli are less brown. This color is usually attributed to the biosynthesis of polyphenols which accumulate in the calli and inhibit the growth, inducing degeneration of the callus line within a few months. The auxin/cytokinin ratio does not seem to modify either the callus growth or the callus morphology.

When 2,4-D is replaced by NAA, just after the initiation of the preliminary calli, the calli are completely different. They are chlorophyllous with a more compact texture and have a good growth even at 0.01 mg/l of NAA. The callus lines could then be maintained by subculturing for several years, at least 6 years in the laboratory at Toulouse (Henry and Edy unpubl.). The decrease in concentration of auxin (2,4-D or NAA) leads to root formation (Fig. 4), and roots are the only organ that we could obtain whatever the callus line.

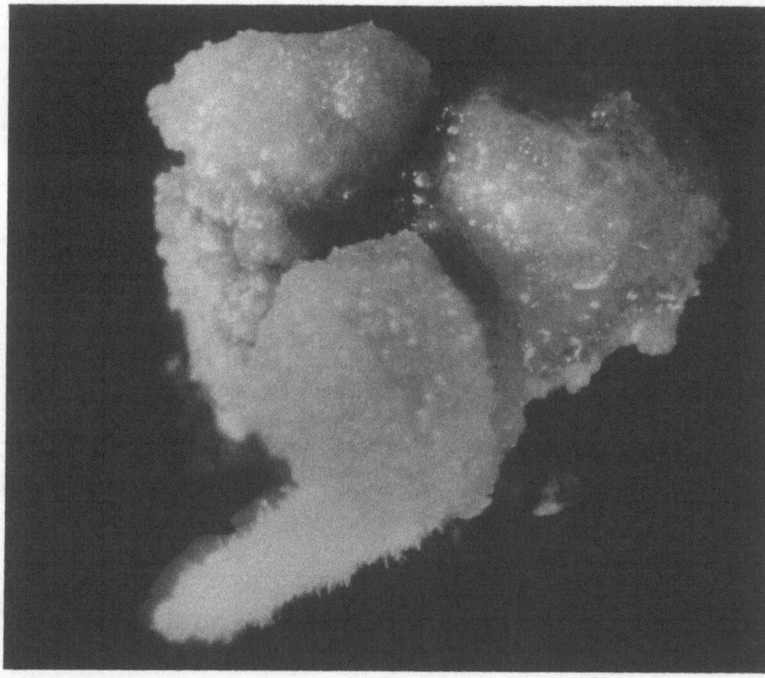

Fig. 4. Root formation in callus culture of *Glycyrrhiza glabra* var. *typica* in MS medium with 2,4-D 0.1 mg/l and kinetin 1 mg/l (x4)

2.4 Cell Suspension Culture

The establishment of cell suspensions from calluses of licorice is very easy, and the medium used for cell suspension culture is generally the same as for callus (Tamaki et al. 1973; Henry et al. 1984; and Dorisse et al. 1988). With a sufficient concentration of 2,4-D, up to 0.1 mg/l, soft and friable colonies produce cell suspension rather quickly. With callus cell lines obtained with NAA, the cell suspension is chlorophyllous with some voluminous aggregates. Once established, these cultures proliferate well and can be maintained for a long time (4 years is our longest). For example, a cell line cultivated on MS medium with a 0.01 mg/l NAA (330 ml in a 1-l Erlenmeyer flask) at 25 ± 1 °C under fluorescent light (2000 lx, 12 h photoperiod) on a gyratory shaker (80 r.p.m.) with an inoculum of 55 fr. wt. g/l (12% packed cell volume) can produce exponentially 400 fr. wt. g/l in 21 days (Dorisse et al. 1988). Single cell suspensions were obtained by Wu et al. (1974) on PRL 4 medium (2 mg/l 2,4-D) (Gamborg and Eveleigh 1968) containing 10% coconut milk.

2.5 Obtaining Protoplasts

A protocol was developed for the isolation of protoplasts from small clusters of a cell suspension established from in vitro plantlets on MS medium supplemented with

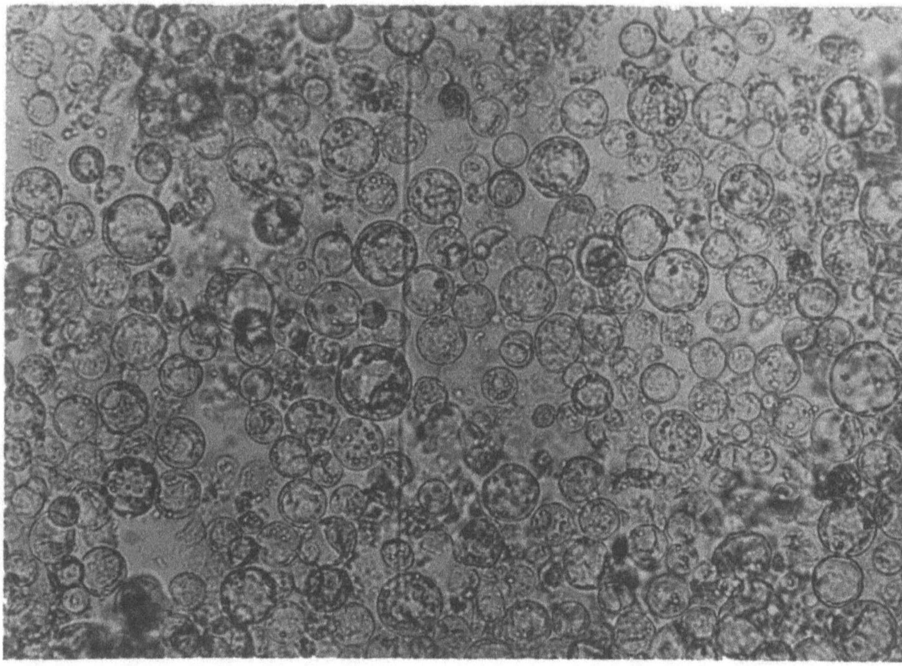

Fig. 5. Protoplasts obtained from cell suspension of *Glycyrrhiza glabra* var. *typica* on the same medium with a mixture of pectinase of *Rhizopus* sp. (2.1%, Sigma), cellulase Onozuka R 10 (4%), mannitol 0.7 M, $CaCl_2$ 1 g/l after 6 h of treatment (x300) (Henry et al. 1984)

0.1 mg/l 2,4-D, 2 mg/l adenine and 1 mg/l kinetin (Fig. 5). Incubation in a mixture consisting of pectinase of *Rhizopus* sp. (2.1%), cellulase Onozuka R 10 (4%), mannitol 0.7 M, $CaCl_2$ 1 g/l yielded viable protoplasts after 6 h of treatment. (Henry et al. 1984).

2.6 Attempts at Somatic Embryogenesis

In spite of many efforts, no general rules have been established for the certain induction of somatic embryos. The only results are, sometimes, embryo-like structures which stop at the torpedo stage (Fig. 6).

An increased arginine decarboxylase (ADC) (EC 4.1.1.19) activity has been reported at an early stage of embryo development in *D. carota* (Montague et al. 1979), which results in elevated levels of putrescine and spermidine (Montague et al. 1978). When licorice cells were placed under the conditions under which carrot cells successfully completed embryogenesis is (dilution of the cell suspension to 2×10^4 cells/ml in a medium without 2,4-D), no increase in ADC activity could be detected (Roustan et al. 1988).

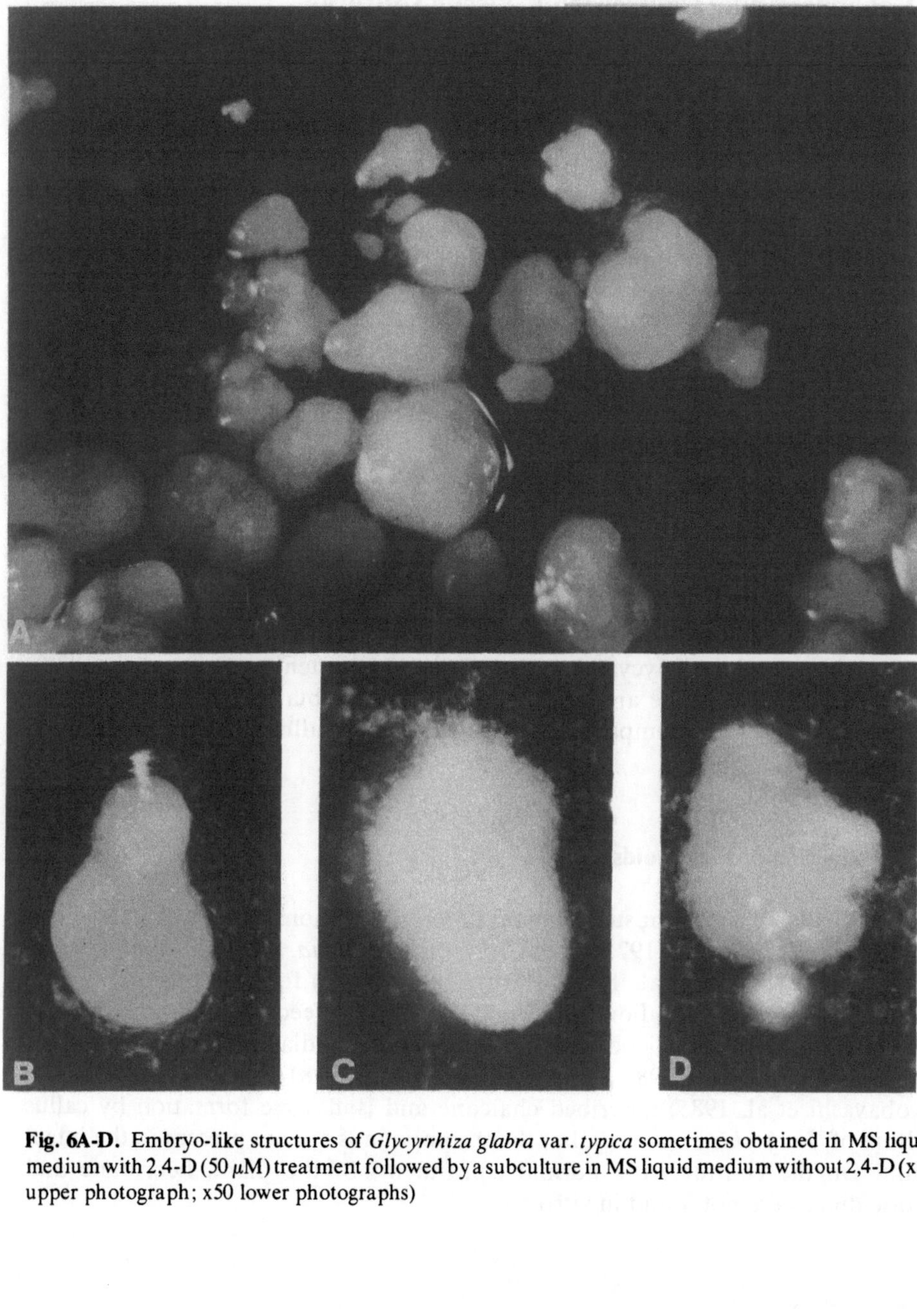

Fig. 6A-D. Embryo-like structures of *Glycyrrhiza glabra* var. *typica* sometimes obtained in MS liquid medium with 2,4-D (50 μM) treatment followed by a subculture in MS liquid medium without 2,4-D (x15 upper photograph; x50 lower photographs)

3 Secondary Metabolites from Tissue Cultures

3.1 Production of Glycyrrhizin

In a patent, Tamaki et al. (1973) gave the procedure for preparing licorice extract-like material for tobacco flavoring from callus and cell suspension of *G. glabra* L., *G. glabra* L. var., *glandulifera* Rgl. et Herd., *G. uralensis* D.C., *G. echinata* L. The aim of this procedure is first to produce biomass containing flavoring products in one growth step from callus initiation to bioreactor, and not to obtain regular metabolite production from an established callus or suspension cell line. The in vitro material was not stabilized and the authors counted on these preliminary cultures being able to retain production. Only a small amount of glycyrrhizin (1.84%) was produced by this technique, but the concentrated extract of licorice had a good aroma, suitable for tobacco flavoring. No information was given on the licorice species from which these results were obtained.

In a single cell suspension culture of *G. glabra* L., Wu et al. (1974) could not detect glycyrrhetinic acid but a volatile apple aroma, which indicated the presence of EtOH and some related esters. Like many undifferentiated cultures, callus and cell suspension cultures of licorice lose their ability to produce glycyrrhizin after some subcultures.

Production of glycyrrhizin is possible with licorice calli which differentiated organs which produced glycyrrhizin as described in a patent by Fujita et al. (1978). A marked increase in the amount of sweetener was obtained with these organ-differentiated calli as compared with conventional callus and cell suspension cultures.

3.2 Production of Flavonoids

A new chalcone with unusual placing of O-substitutions, echinatin, had been isolated by Furuya et al. (1971) from *Glycyrrhiza echinata*, and its biosynthesis has been studied (Saitoh et al. 1975). Furuya et al. (1976) found a new dibenzoylmethane derivative, licodione, and established, by feeding experiments with labeled compounds, that licodione is an obligate intermediate in the biosynthesis of echinatin (Ayabe et al. 1980 a,b; Ayabe and Furuya 1981). Later, the same team (Kobayashi et al. 1985) described chalcone and isoflavone formation by callus culture of *Glycyrrhiza uralensis*, but these compounds are not present in the intact plant. On the contrary, the normal constituents of the plant, licorinone and licoricidine, were not found in vitro.

4 Protocol

1. *Establishment of tissue cultures*
   – A convenient method to obtain callus is to use plantlets grown aseptically as explant source. Dip seeds in concentrated H_2SO_4 for 1 h to allow them to germinate. Rinse seeds and grow them on Knop medium + 2% glucose + 0.7% agar at 25 °C. Maintain germination in complete darkness if hypocotyl fragments are used or, else in light for cotyledons, petiole, stem and roots.

– After a few weeks explants from sterile plantlets are cultured on a medium containing MS salts, LS vitamins, 3% glucose or sucrose, 0.1 to 1 mg/l 2,4-D, 0.1 to 0.3 mg/l kinetin and 0.7% agar. Grow in a 25°C chamber, preferentially in light. Most explants will proliferate and produce yellow calli a few months later. Alternatively, use NAA instead of 2,4-D to obtain chlorophyllous and compact calli.

2. *Maintenance*

– Growth rate and morphology of primary calli often differ greatly from one explant to the other. When necessary (usually 4-6 months later) select the calli you need and transfer pieces onto fresh medium. Subculture at 4–6-week intervals, depending on the growth rate of the lines.

– When cell suspensions are needed, choose a soft, fast-growing callus, transfer the tissue into a small volume of the same medium without agar, and put on a shaker. Subculture the suspension every 2 weeks by adding 20–50 ml stationary phase cells to 330 ml fresh medium in a 1-l Erlenmeyer flask. Transfer more frequently if you need to maintain suspension in the log phase of growth.

3. *Regeneration of plants*

– To establish clones from aseptic plantlets, select those having at least one well-spread leaf, discard their hypocotyl and root, and grow buds on the micropropagation medium as described in Section 2.2 in order to have many nodal segments. Dissect these nodal segments and transfer each bud onto fresh medium. Rooting is necessary prior to the following multiplication step. Allow the plantlet to grow roots and six to seven nodal segments for 6 weeks on the same medium with 0.1 mg/l IAA or 0.1 mg/l IBA as the chosen species needs, and without cytokinins, and then dissect them as described above.

– When complete plantlets are needed, apply the above technique to obtain rooted buds and remove them to small pots containing a peat-vermiculite mixture (1:1). Maintain plantlets for about 1 week in a greenhouse and under mist. Pay great attention to root rot and possibly apply preventive treatments with convenient antifungal products. Established plants may be transferred to the field 1 month later.

5 Conclusion

Micropropagation of licorice is actually the only technique which can be used in industry. Given the difficulty of multiplying this plant by classical methods, it represents important progress. Until now, undifferentiated cultures of licorice failed to regenerate the plant, or to produce metabolites in vitro. Further investigations are necessary to elucidate the induction of the embryogenesis and the regulation of the biosynthesis of secondary products.

References

Adzet Porredon T (1986) La culture des plantes médicinales et aromatiques en Espagne. Coll Trento, pp 37–38

Afchar D (1981) Contribution à l'étude des Régisses d'Iran. *Glycyrrhiza glabra* et var. *glandulifera*, *Glycyrrhiza echinata*. Thèse Doct Pharm, Univ Paris XI

Amagaya S, Sugishita E, Ogihara Y, Ogawa S, Okada K, Aizawa T (1984) Comparative studies of the stereoisomers of glycyrrhetinic acid on anti-inflammatory activities. J Pharm Dyn 17:923–928

Ayabe S, Furuya T (1981) Biosynthesis of a retrochalcone, echinatin: a feeding study with advanced precursors. Tetrahedron Lett 22, 22:2097–2098

Ayabe S, Kobayashi M, Hikichi M, Matsumoto K, Furuya T (1980a) Flavonoids from the cultured cells of *Glycyrrhiza echinata*. Phytochemistry 19:2179–2183

Ayabe S, Yoshikawa T, Kobayashi M, Furuya T (1980b) Biosynthesis of a retrochalcone, echinatin: involvement of O-methyltransferase to licodione. Phytochemistry 19:2331–2336

Beaton JM, Spring FS (1955) Triterpenoids, pt 42. The configuration of the carboxyl group in glycyrrhetic acid. J Chem Soc 3126–3129

Boischard L (1987) Etude de la callogénèse et de la régénération de plantules chez *Glycyrrhiza glabra* L. et *Glycyrrhiza uralensis* Fish. DEA Biol Prod Veg Univ Paris VI, 51 pp

Boischard L (1988) Contribution à la connaissance des Réglisses. Méthodes analytiques d'identification. Thèse Doct Pharm Univ Paris XI, 86 pp

Dorisse P, Gleye J, Loiseau P, Puig P, Edy AM, Henry M (1988) Comparison between two plant cell suspensions towards papaverine biotransformation. J Nat Prod 51, 3:532–536

Fujita Y, Teranishi K, Furukawa T (1978) Glycylglycine. Patent Jpn Kokai 78, 91, 188, Aug 10, 1978

Furuya T, Matsumoto K, Hikichi M (1971) Echinatin, a new chalcone from tissue culture of *Glycyrrhiza echinata*. Tetrahedron Lett 27:2567–2569

Furuya T, Ayabe S, Kobayashi M (1976) Licodione, a new dibenzoylmethane derivative from cultured cells of *Glycyrrhiza echinata*. Tetrahedron Lett 29:2539–2540

Gamborg OL, Eveleigh DE (1968) Culture methods and detection of glucanases in suspension cultures of wheat and barley. Can J Biochem 46:417–421

Henry M, Edy AM, Marty B (1984) Obtention de protoplastes de Réglisse (*Glycyrrhiza glabra* L. var. typica Reg. et Hed.) à partir de suspensions cellulaires dépourvues d'acide glycyrrhétique. C R Acad Sci Ser III 299:899–903

Kiso Y, Tohkin M, Hikino H, Hattori M, Sakamoto T, Namba T (1984) Mechanism of antihepatotoxic activity of glycyrrhizin, I: Effect on free radical generation and lipid peroxidation. Planta Med, pp 298–302

Kobayashi M, Noguchi H, Sankawa U (1985) Formation of chalcones and isoflavones by callus culture of *Glycyrrhiza uralensis* with different production patterns. Chem Pharm Bull 33, 9:3811–3816

Landerlan SA (1976) New triterpene derivatives. Brit Patent 1,447,162, 25 Aug 1976

Linsmaier EM, Skoog F (1965) Organic growth factor requirements of tobacco tissue cultures. Physiol Plant 18:100–127

Molyneux S (1975) Licorice production and processing. Food Technol Aust 27, 6:231–234

Montague MJ, Koppenbrink JW, Javorski EG (1978) Polyamine metabolism in embryogenic cells of *Daucus carota*: I. Changes in intracellular content and rates of synthesis. Plant Physiol 62:430–433

Montague MJ, Armstrong TA, Javorski EG (1979) Polyamine metabolism in embryogenic cells of *Daucus carota*: II. Changes in arginine decarboxylase activity. Plant Physiol 63:341–345

Murashige T, Skoog F (1962) A revised medium for rapid growth and bioassays with tobacco tissue cultures. Physiol Plant 15:473–497

Roustan JP, Henry M, Fallot J (1988) Effect of auxin on arginine decarboxylase activity in cell suspension cultures. CR Acad Sci 307, Série III:781–784

Saitoh T, Shibata S, Sankawa U, Furuya T, Ayabe S (1975) Biosynthesis of echinatin, a new biosynthetical scheme of retrochalcone. Tetrahedron Lett 50:4463–4466

Shah RR, Dalal KC (1980) In vitro multiplication of *Glycyrrhiza*. Curr Sci 49,2:69–71

Shukurullaev PH, Khamdamov IK (1976) Uniform germination of *Glycyrrhiza glabra* seeds. Uzbek Biol Zh 2:64–65

Syrtanova GA, Mukhitdinova ZR (1984) Clonal multiplication attempts of *Glycyrrhiza glabra* L. and of *G. uralensis* Fisch. on culture media. Rastit Resur 1:85–88

Tamaki E, Morishita I, Nishida K, Kato K, Matsumoto T (1973) Process for preparing licorice extract-like material for tobacco flavoring. US Patent 3,710,512, Jan 16, 1973

Wu C-H, Zabawa EM, Townsley PM (1974) Single cell suspension culture of the licorice plant, *Glycyrrhiza glabra*. Can Inst Food Sci Technol J 7,2:105–109

Yaskonis YA (1976) The propagation and growth of licorice and the active principle content in root I. Propagation and growth. Liet TSR Nokslu Akad Darbai Sev B 2674:45–52

XVIII *Lavandula* spp. (Lavender): In Vitro Culture, Regeneration of Plants, and the Formation of Essential Oils and Pigments

J. SEGURA and M.C. CALVO[1]

1 Introduction

1.1 Distribution and Importance

Lavandula species are of great interest due to their content of essential oils, which are important to the perfume, cosmetic, flavoring and pharmaceutical industries. They also have numerous other applications, including their uses as ornamental and melliferous plants.

The genus *Lavandula*, of the family Labiatae (Lamiaceae), encompasses about 20 species almost exclusively distributed in the Mediterranean area, where they are characteristic components of the degradated mediterranean shrublands (Rivas Goday and Rivas-Martinez 1967; Rivas-Martinez 1979). They usually grow in zones with scanty summer rainfall and exhibit marked xerophytic tendencies. *Lavandula* plants are evergreen shrubs up to 100 cm (-200 cm). Their stems have a characteristic square shape and the leaves are almost linear to oblong-lanceolate, and usually tomentose. The upper branches bear terminal spikes consisting of verticillasters six- to ten-flowered, with purplish corollas. Most aboveground plant organs are covered in hairs and glands that emit an aromatic fragrance (Tutin et al. 1972). Their morphology can be seen from Fig. 1, exemplified by *Lavandula latifolia*.

Within the genus *Lavandula*, only three species are considered to be of current economic interest: *Lavandula angustifolia* Miller (Synonym: *L. officinalis* Chaix, *L. vera* C.D., *L. spica* L. nom ambig.), *Lavandula latifolia* Medicus (synonym: *L. spica* auct., non L.), and the hybrid lavandin (*L. angustifolia* × *L. latifolia*). The distribution and importance of these chalk-growing species are summarized below. For a more detailed review, see Paris and Moyse (1971), Font Quer (1978), Belaiche (1979), Reynolds (1982), De Pascual-T et al. (1983), Gennaro (1985), Duke (1985), and Bezanger-Beauquesne et al. (1986).

L. angustifolia is native throughout the western Mediterranean areas, growing wild, at altitudes between 600 and 1000 m, on the dry, barren, sunny mountain slopes of southern France or Italy and the northeast of Spain. It is commonly cultivated in many countries, in addition to those in the Mediterranean Basin, England, U.S.S.R., U.S.A., India etc. *L. latifolia* has a geographical distribution similar to that of *L. angustifolia*, growing wild at low altitudes (200–1000 m). Usually it is not planted, since it grows naturally and abundantly in Spain, the main

[1]Departament de Biologia Vegetal, Facultat de Farmàcia, Universitat de València, Avgda. Blasco Ibàñez, 13, 46010 València, Spain

Biotechnology in Agriculture and Forestry, Vol. 15
Medicinal and Aromatic Plants III (ed. by Y.P.S. Bajaj)

Fig. 1. Flowering *Lavandula latifolia* plants growing in their natural habitats

producing country of this species. The hybrid lavandin grows wild at medium altitudes, in those regions where the parent plants coexist. This plant is cultivated on a large scale, especially in France, Italy, and England.

The essential oils (Fig. 2) of *Lavandula* species are obtained by steam distillation from the fresh flowering spikes. The amount varies according to the variety, season, and method of distillation. The essential oil of *L. angustifolia* (lavender oil) has a weight per ml of 0.878 to 0.892 g. Genuine continental lavender oil normally contains not less than 35% esters, calculated as linalyl acetate, the chief constituent of this oil. Other constituents include: cineol; amyl alcohol; d-borneol; geraniol; lavadulol; linalool; nerol; acetic, butyric, valeric, and caproic acids (as esters); traces of d-pinene, limonene, and the sesquiterpene caryophyllene; ethyl n-amil ketone; an aldehyde (probably valeric aldehyde); and coumarin (Gennaro 1985). Lavender oil is chiefly used in perfumery, although it also has uses as a carminative and as a coloring or flavoring agent. It is sometimes applied externally as an insect repelent and is occasionally used in ointments and other pharmaceutical preparations to cover disagreeable odors.

The essential oil of *L. latifolia* (spike lavender oil) has a wt./ml of 0.894 to 0.915 g. It contains a small amount of esters but a high proportion of free alcohols, especially cineol, camphor, linalool, and borneol (De Pascual-T et al. 1983). Spike oil resembles lavender oil in its properties and is mainly used in perfumery.

The oil from hybrids of *L. latifolia* and *L. angustifolia* (lavandin oil) has a wt./ml of about 0.89 g. It shows characteristics of both lavender and spike oils, containing

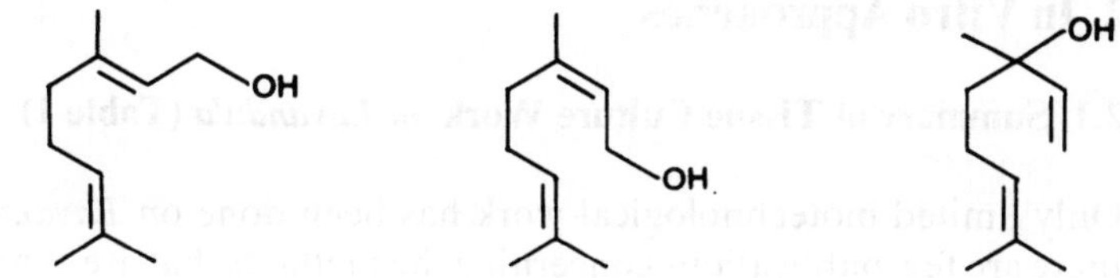

Fig. 2. Some examples of monoterpenoids found in the essential oils of *Lavandula* species

20 to 28% esters, calculated as linalyl acetate (Reynolds 1982). Lavandin oil is a substitute for lavender oil in perfumery, being more fragrant than spike oil. It is used largely in the soap industry as a source of linalool and linalyl acetate.

1.2 Conventional Methods for Propagation

For large-scale plantations, *Lavandula* species are grown either from seeds or stem cuttings. Usually, the seeds are kept in humid sand for some time in order to bring about germination. Alternatively, treatments such as soaking in H_2O_2 or incubation in the presence of gibberellic acid significantly improve seed germination. In our laboratory, best results are currently obtained by soaking the seeds in 0.5% H_2O_2 for 24 h. Propagation by cuttings requires considerable care and is not always successful. This technique is, however, the only one available to propagate the hybrid lavandin. A comprehensive review on *Lavandula* and other medicinal species concerning conventional methods for propagation and cultivation has been recently published (Muñoz 1987).

2 In Vitro Approaches

2.1 Summary of Tissue Culture Work on *Lavandula* (Table 1)

Only limited biotechnological work has been done on *Lavandula* (Table 1), and there are few publications concerning the methods that are commonly reported for other aromatic or medicinal plants (Bajaj et al. 1988; Mulder-Krieger et al. 1988). Most of the in vitro studies on *Lavandula* are related to the conditions for callus establishment and growth, isolation and culture of callus-derived cells, and plant regeneration from primary explants, callus or single cells (Quazi 1980; Gomez et al. 1987a,b; Calvo and Segura 1988, 1989a,b; Calvo et al. 1988; Panizza and Tognoni 1988; Jordan et al. 1988). Protoplasts have recently been isolated and cultured using mesophyll and cotyledon tissues of *L. latifolia* (Fernandez-Feijoo et al. 1988), but plant regeneration has not yet been achieved. Apart from this, some groups have investigated the possibility of secondary product formation or biotransformation in tissue culture as well as cryopreservation of callus tissues (Watanabe and Yamada 1982a,b; Watanabe et al. 1982, 1983, 1985a; Webb et al. 1984; Banthorpe et al. 1985, 1986; Nakajima et al. 1985; Lappin et al. 1987).

This chapter primarily concerns tissue culture studies on *L. stoechas* L. subsp. *stoechas* and *L. latifolia* Medicus. The former is a calcifuge species, abundant in the wild state in Spain, with uses similar to *L. latifolia*. Data without citations of the source in the following sections are from our laboratory.

2.2 Establishment of Tissue Culture

Most of the material used in tissue culture of *Lavandula* came from seedlings obtained by germinating seeds under sterile conditions. *L. stoechas* or *L. latifolia* seeds are surface sterilized with 2% Chloramine T and 0.1% Tween 20 for 1 h, rinsed with sterile distilled water, and germinated in Petri dishes on solid medium (0.6% agar in distilled water) at 26 ± 2° C in darkness (Calvo and Segura 1988). Leaf buds from 1-year-old *L. angustifolia* or *L. latifolia* plants were sterilized with 70% ethanol for 30 s and 0.32% sodium hypochlorite for 20 min, followed by three washes in sterile water (Quazi 1980). Sodium hypochlorite (7% chlore-active) was also used for sterilization of nodal explants from mature lavandin cv. Grosso plants (Panizza and Tognoni 1988).

Although medium is an important factor for successful establishment of tissue culture, information is still sparse on the media used for *Lavandula*. All species under study were cultured almost exclusively on MS medium (Murashige and Skoog 1962), sometimes half-strength. Only callus and cell suspensions of *L. vera* (Watanabe et al. 1982) or explants from lavandin (Panizza and Tognoni 1988) have also been cultured on LS medium (Linsmaier and Skoog 1965). The carbon source usually utilized for studies on *Lavandula* cultures was sucrose, although an investigation used alternative carbon sources in the form of glucose, galactose, and mannitol (Gomez et al. 1987a).

Table 1. Summary of tissue culture work on *Lavandula*

A Culture of *Lavandula* explants in vitro

Lavandula sp.	Inoculum	Medium/growth regulators (mg/l)	Response	Reference
L. angustifolia and	Leaf bud	MS/2,4-D(2) + BA(4)	Callus, embryoids	
L. latifolia	Stem-derived callus	MS/GA_3(5) + BA(0.5) + IAA(5)	Plants	Quazi (1980)
L. angustifolia	Seedling	MS/BA (20)	Shoots	Webb et al. (1984)
L. latifolia		MS/diverse	Callus, roots, shoots, plants	Calvo and Segura (1988)
L. latifolia	Hypocotyl with adventitious buds	MS/CM + BA(2) + IAA(0.1 or 1)	Shoots, plants	Calvo and Segura (1989a)
L. latifolia	Hypocotyl- and cotyledon-derived callus	MS/BA (0.1, 2, or 4) + IAA or NAA (0.01, 0.1, or 1)	Shoots, plants	Jordan et al. (1988)
L. latifolia	Leaf explants	MS/BA (2, 4, 6, or 8) + IAA or NAA (0.01, 0.1, 1 or 2)	Callus, roots, shoots, plants	Calvo and Segura (1989b)
L. latifolia	Single cells from hypocotyl-derived callus	MS/BA (0.1 or 2) + IAA (0.1, 1, or 2)	Callus, shoots, plants	Calvo et al. (1988)
L. latifolia	Leaf and cotyledon explants	Diverse	Protoplasts	Fernandez-Feijoo et al. (1988)
L. stoechas	Hypocotyl explants	MS + several sugars/ NAA(0.1) + BA (0.1, 1, or 2)	Callus	Gomez et al. (1987a)
L. stoechas	Seedling	MS/diverse	Callus, roots	Calvo and Segura (1988)
L. stoechas	Single cells from hypocotyl- and root-derived callus	MS/diverse	Callus	Gomez et al. (1987b)
L. angustifolia × *L. latifolia* (lavandin)	Nodal explants	LS/BA(0.2 or 10) or BA(0.2) + GA_3(0.5)	Plants	Panizza and Tognoni (1988)

(*to be continued*)

Table 1. *(Continued)*

B In vitro formation of secondary products

Lavandula sp.	Explant source	Culture type	Compounds	Reference
L. angustifolia	Stem	Callus Adventitious shoots	Blue pigment monoterpenes	Webb et al. (1984)
L. angustifolia	Stem	Cell suspension	Biotransformation of monoterpenes	Lappin et al. (1987)
L. angustifolia	Stem	Callus	Blue pigment	Banthorpe et al. (1985)
L. vera	–	Callus and suspension	Blue pigment	Watanabe et al. (1985a)
L. vera	–	Calcium alginate-entrapped cells	Blue pigment	Nakajima et al. (1985)
L. vera	–	Synthetic resin prepolymer-entrapped cells	Blue pigment	Nakajima et al. (1986)
L. vera	–	Cell aggregate cloning	Biotin	Watanabe et al. (1982) Watanabe and Yamada (1982a,b)

C Cryopreservation

Lavandula sp.	Culture	Cryoprotectant	Freezing	Thawing	Response	Reference
L. vera	Cell suspension	DMSO (10%) + glue (20%)	1°C/min, –40, –196°C	Rapid	Retention of metabolic and plant regeneration potentials	Watanabe et al. (1983)

2.3 Bud Culture

Bud cultures are initiated from either terminal or axillary buds, usually with the stem segment attached, using either a growing or dormant shoot. This technique can be used to propagate lines from mature selected *Lavandula* plants. To date, however, only one study has been reported on micropropagation of *L. angustifolia* and *L. latifolia* with leaf buds (Quazi 1980). In this study, leaf buds with 5–10-mm stems, excised from 1-year-old plants, were able to proliferate when cultured, on filter paper bridges, in liquid medium with GA_3, BA, and IAA. The nutrient medium consisted of MS salts, B5 amino acids (Gamborg et al. 1968) and 2% sucrose. More than 90% of the buds survived in cultures of the two species, but *L. angustifolia* produced more rooted plants than *L. latifolia* (Table 2). After 4 weeks in culture, young plants with two or three roots were successfully planted in soil and grown in the greenhouse (Quazi 1980). Recently, successful micropropagation of lavandin has been reported (Panizza and Tognoni 1988). Nodal segments from mature plants of this species were cultured on LS medium (Linsmaier and Skoog 1965) supplemented with BA or BA and GA_3. Regenerated shoots were transferred to the same medium with NAA for rooting. Shoot proliferation rates were similar in all hormonal treatments tested, but rooting percentage was negatively influenced by GA_3 in the shoot multiplication medium. Micropropagated plants were easily acclimatized to outdoor conditions.

Table 2. Response of leaf buds from *Lavandula* spp. to medium containing MS inorganic salt mixture and amino acids from B5 medium supplemented with GA_3, BA, and IAA, after 3 weeks in culture. (Quazi 1980)

	L. angustifolia	*L. latifolia*
No. of leaf buds cultured	114	82
Forming shoots only (%)	34.2	54.9
Developing roots and shoots (%)	57.0	35.4
Died (%)	8.8	9.7

2.4 Callus and Cell Suspension Cultures

Callus cultures of *Lavandula* can be initiated from a variety of tissues of seedling and adult plants. Callus cultures of *L. latifolia* and *L. stoechas* were easily obtained from hypocotyl, cotyledon, and root explants cultured on MS medium supplemented with varying concentrations of auxins (NAA, IAA, or 2,4-D), cytokinins (BA or kin), or different combinations of both types of growth regulators (Calvo and Segura 1988). Although both auxins and cytokinins promoted callus induction and growth, the former were more effective. Only in root explants from *L. stoechas* did cytokinins, especially BA, promote a better callus growth than auxins. Nevertheless, the efficiency of the auxins on callus formation and growth was generally increased by adding cytokinins to the media. Note, however, that cytokinins also antagonized the auxin-induced growth in *L. stoechas*; this effect was observed when the explants were cultured in media with auxin concentrations optimal for callus induction and growth. Calli induced by auxins were friable and yellowish in color, while those obtained in media including cytokinins had a compact consistency and were bright green. Leaf explants from both species behaved similarly to cotyledons in their requirements for callus induction and growth.

Callus cultures were established from leaf bud of *L. angustifolia* or *L. latifolia* cultured on MS salts, B5 vitamins, coconut milk (CM), 4 mg/l BA and 1 mg/l NAA (Quazi 1980). *L. angustifolia* callus has also been initiated from stem explants on MS medium supplemented with IAA and kin (Webb et al. 1984) or NAA, kin, asparagine, and CM (Banthorpe et al. 1985). It is worth noting that the stem-derived callus has been maintained, apparently unaltered in its characteristics, for 5 years (Banthorpe et al. 1986).

The lack of information concerning *Lavandula* carbohydrate utilization prompted an investigation to select alternatives to sucrose as carbon source in callus cultures of *L. stoechas* (Gomez et al. 1987a). Hypocotyl explants from 30-day-old seedlings were cultured on MS medium containing: (1) different concentrations of sucrose, glucose, or galactose; (2) varying levels of sucrose or sucrose plus mannitol to give 30 g/l molar equivalent of sucrose at all sucrose levels. The medium was supplemented with three combinations of the growth regulators NAA (0.1 mg/l) and BA (0.1, 1, and 2 mg/l).

Calli of *L. stoechas* were able to use sucrose, glucose, or galactose as carbon and energy sources, although galactose was the most effective sugar maintaining callus growth (Fig. 3).

The effect of sucrose concentration on callus growth from hypocotyl explants of *L. stoechas* depended on the relative auxin/cytokinin ratio in the nutrient medium (Fig. 4). In media containing balanced ratios of NAA and BA, no significant differences were observed in callus growth at sucrose concentrations higher than 5 g/l. In contrast, when the hormone ratios were favorable to BA, callus growth increased as the concentration of sucrose in the medium was increased up to 20 g/l. A reduction in the callus fresh weight was, however, observed on media with sucrose levels higher than 20 g/l (Fig. 4). This inhibition could be caused by a reduction in the water potential of the media supplemented with 30 g/l sucrose. To test this possibility, hypocotyl explants were also cultured on media containing varying levels of sucrose plus mannitol. As is shown in Fig. 5, mannitol was unable to replace sucrose requirement for growth. This was independent of the relative concentrations of NAA and BA. In addition, no significant differences were detected in the growth of calli cultured in the presence of 20 or 30 g/l sucrose, which clearly indicates that the water potential of the medium supplemented with 30 g/l sucrose limits callus induction and growth of *L. stoechas* (Gomez et al. 1987a).

Cell suspensions of *L. angustifolia* were established from callus induced on agar-solidified MS medium supplemented with 2,4-D and kin (Lappin et al. 1987); about 2 g of callus was subcultured into liquid medium (callus medium without agar) and incubated in a reciprocal shaker in the dark. These cultures had been used for biotransformation of monoterpenes (see Sect. 2.8.1). Watanabe et al. (1982) also established suspension cultures of green *L. vera* cells, which were subsequently used to select for high vitamin-producing cells (see Sect. 2.8.2), production of pigments by immobilized cells (see Sect. 2.8.3) and germplasm conservation studies (see Sect. 2.9).

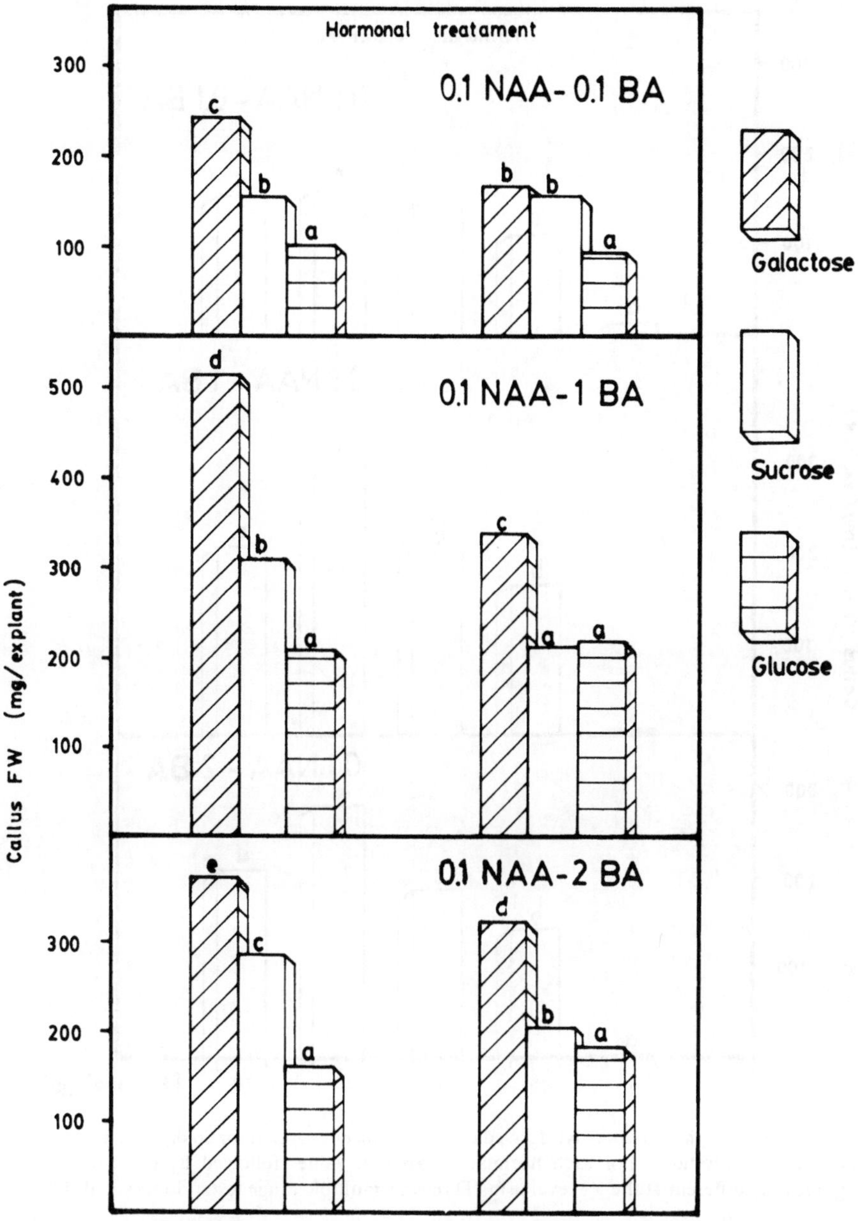

Fig. 3A,B. Effect of carbon sources on callus growth from *L. stoechas* hypocotyl cultures. Sugar concentration: **A** 20 g/l and **B** 30 g/l. For each hormone treatment, values followed by the same letter are not significantly different at the 5% level using Duncan's multiple range test. (Gomez et al. 1987a)

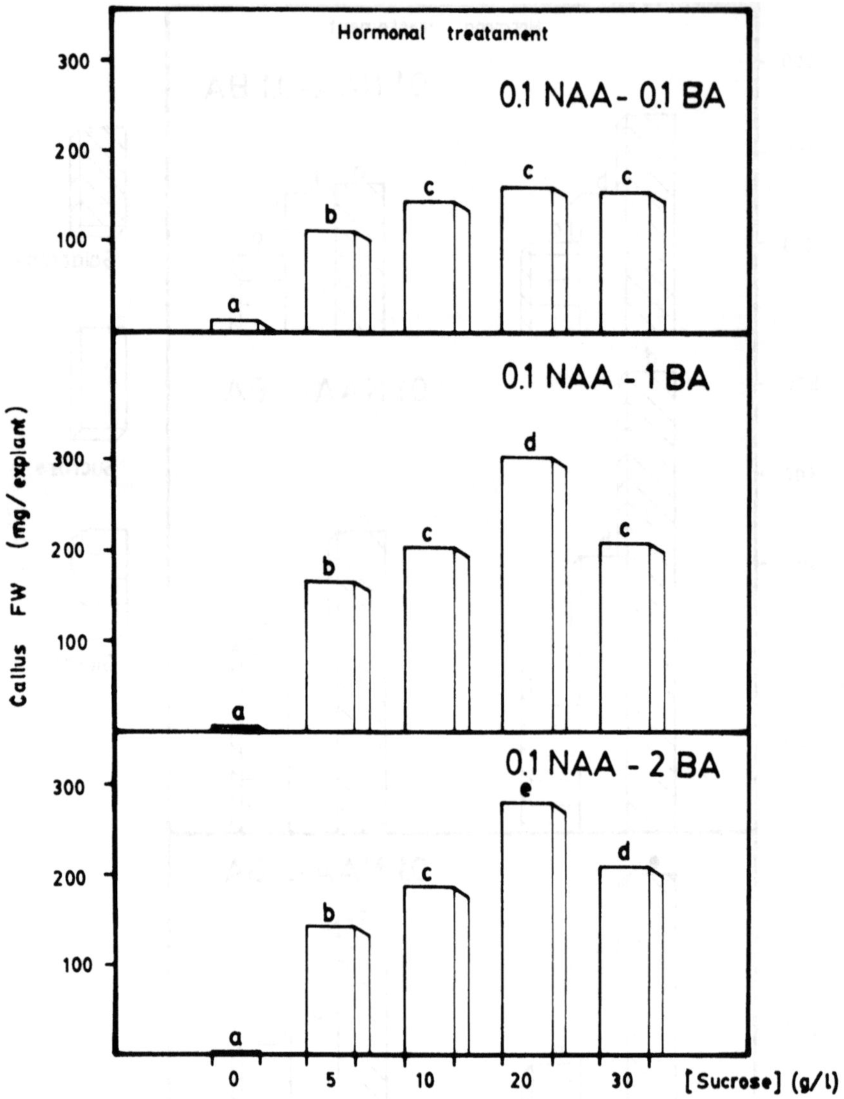

Fig. 4. Fresh weight of hypocotyl-derived calli of *L. stoechas* grown for 30 days on media with various sucrose concentrations. For each hormone treatment, values followed by the same letter are not significantly different at the 5% level using Duncan's multiple range test. (Gomez et al. 1987a)

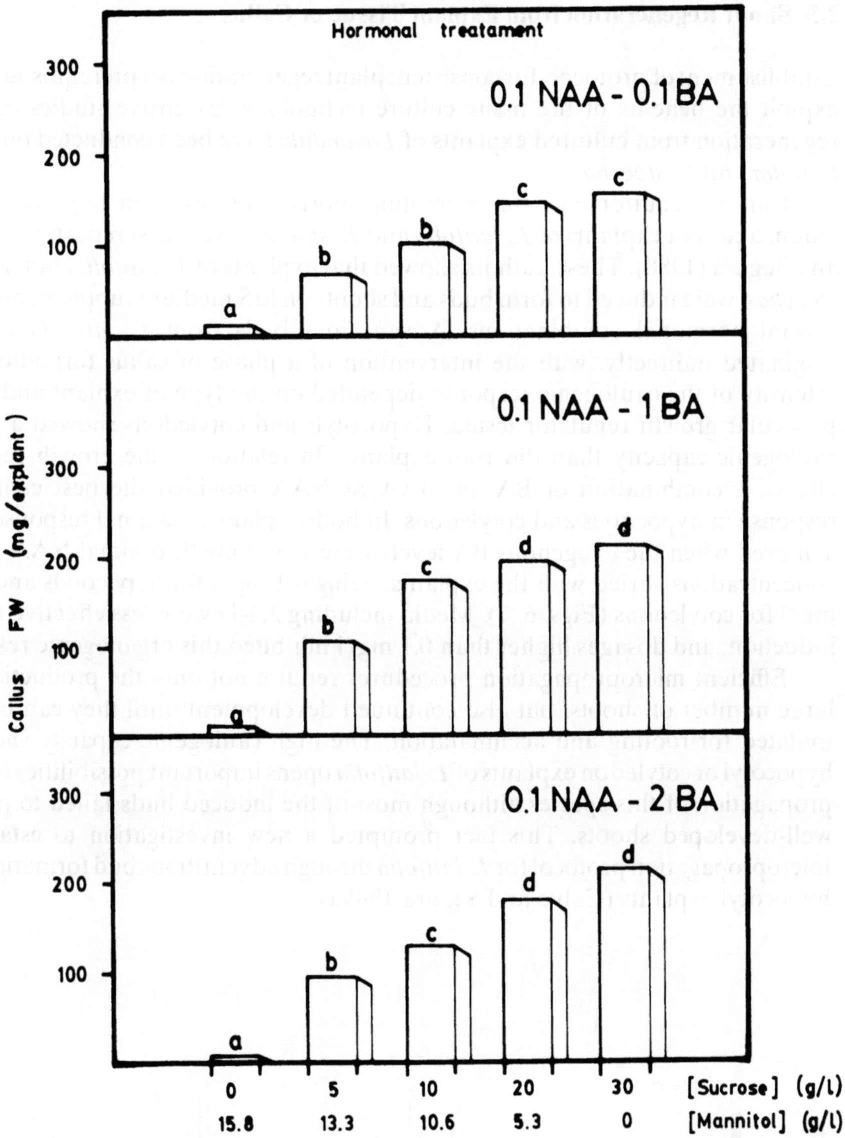

Fig. 5. Fresh weight of hypocotyl-derived calli of *L. stoechas* grown for 30 days on media with varying levels of sucrose plus mannitol. Mannitol was added to give a 3% molar equivalent of sucrose at all sucrose levels. For each hormone treatment, values followed by the same letter are not significantly different at the 5% level using Duncan's multiple range test. (Gomez et al. 1987a)

2.5 Shoot Regeneration from Explant Tissue or Callus

Establishment of protocols for consistent plant regeneration is a prerequisite to fully exploit the benefits of the tissue culture technology. Extensive studies on plant regeneration from cultured explants of *Lavandula* have been conducted only in *L. latifolia* and *L. stoechas*.

Culture requirements for promoting morphogenesis from hypocotyl, cotyledon, and root explants of *L. latifolia* and *L. stoechas* were first reported by Calvo and Segura (1988). These authors showed that explants of *L. latifolia*, but not of *L. stoechas*, were induced to form buds and shoots on MS medium supplemented with several hormonal combinations. Adventitious buds from *L. latifolia* explants originated indirectly, with the intervention of a phase of callus formation. The intensity of the caulogenic response depended on the type of explant and on the particular growth regulator tested. Hypocotyls and cotyledons showed a greater caulogenic capacity than did root explants. In relation to the growth regulator effects, a combination of BA and IAA or NAA provided the best caulogenic response in hypocotyls and cotyledons. In both explants, maximal responses were achieved when the exogenous BA levels were 1 or 2 mg/l; optimal NAA or IAA concentrations varied with the explants, being 0.1 mg/l for hypocotyls and 1 or 2 mg/l for cotyledons (Figs. 6, 7). Media including 2,4-D were less effective for bud induction, and dosages higher than 0.1 mg/l inhibited this organogenic response.

Efficient micropropagation procedures require not only the production of a large number of shoots, but also continued development until they can be manipulated for rooting and acclimatation. The high caulogenic capacity shown by hypocotyl or cotyledon explants of *L. latifolia* opens important possibilities for mass propagation of this species although most of the induced buds failed to produce well-developed shoots. This fact prompted a new investigation to establish a micropropagation protocol for *L. latifolia* through adventitious bud formation from hypocotyl explants (Calvo and Segura 1989a).

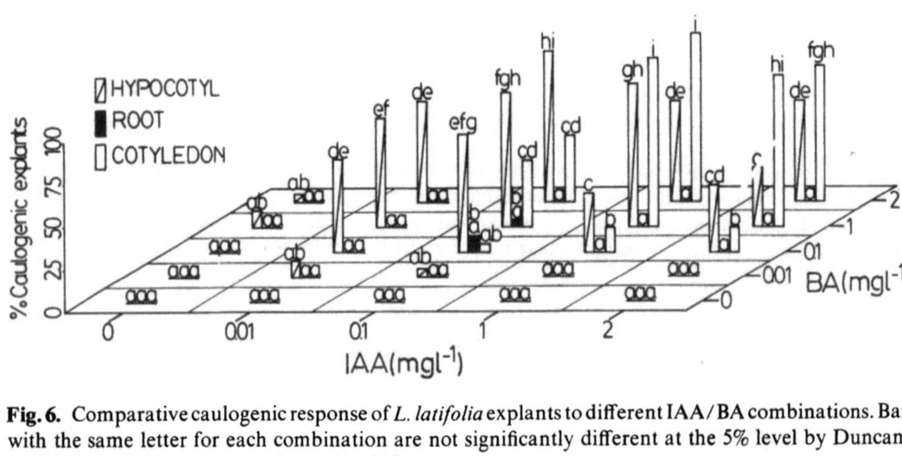

Fig. 6. Comparative caulogenic response of *L. latifolia* explants to different IAA/BA combinations. Bars with the same letter for each combination are not significantly different at the 5% level by Duncan's multiple range test. (Calvo and Segura 1988)

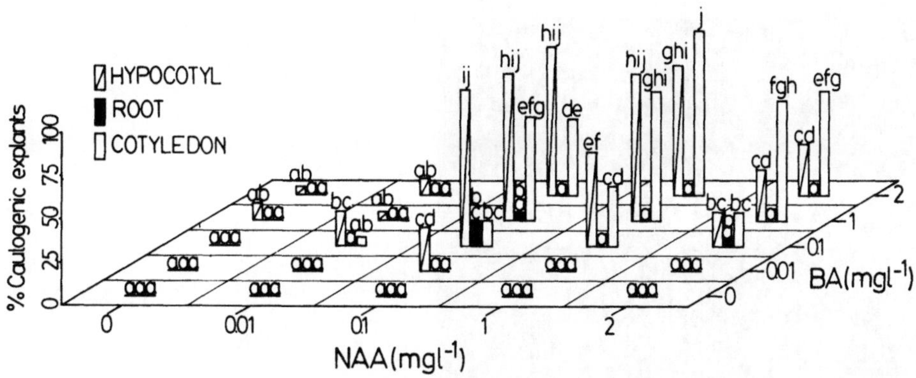

Fig. 7. Comparative caulogenic response of *L. latifolia* explants to different NAA/BA combinations. Bars with the same letter for each combination are not significantly different at the 5% level by Duncan's multiple range test. (Calvo and Segura 1988)

For bud induction, hypocotyl explants from *L. latifolia* seedlings were cultured on MS medium supplemented with 0.1 mg/l IAA and 2 mg/l BA. After 30 days in this medium, explants with 10–15 discernible buds were transferred to several shoot-elongation media (see Table 3). Explant subculture enhanced the elongation of the adventitious buds, and shoots longer than 10 mm were obtained in all treatments tested. Media supplemented with IAA and BA were more effective in promoting shoot development than those without growth regulators or supplemented only with CM or BA. However, when CM was added in combination with IAA or BA, greater numbers of usable shoots were obtained (Table 3, Fig. 8).

Table 3. Shoot development on hypocotyl explants of *Lavandula latifolia* after 30 days. The data show the number of shoots > 10 mm in length per explants. Buds were previously induced on MS medium supplemented with 0.1 mg/l IAA and 2 mg/l BA. (Calvo and Segura 1989a)

Growth regulators (mg/l)		MS macronutrient	Coconut milk conc. (%)			
BA	IAA	strength	0	10	20	Mean[a]
0	0	Half	0.7	1.8	2.9	1.5 a
		Full	0.6	1.1	1.7	
2	0	Half	1.4	1.0	4.1	2.2 a
		Full	1.4	1.5	3.8	
2	0.01	Half	5.8	10.1	16.2	10.7 c
		Full	8.3	8.6	15.4	
2	0.1	Half	3.4	6.8	11.5	6.9 b
		Full	3.5	6.2	10.0	
Mean[a]			3.1 a	4.6 b	8.2 c	

[a] Entries with the same letter within a group are not significantly different at the 5% level, according to Tukey's test.

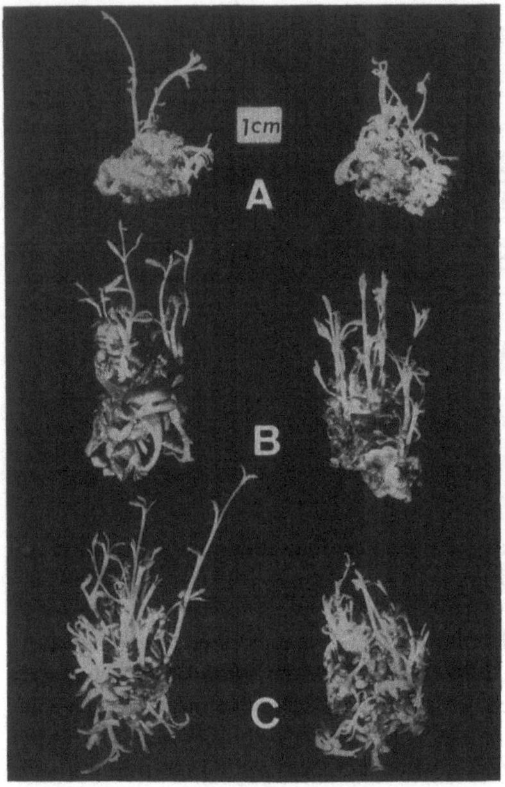

Fig. 8A-C. Effect of medium composition on the elongation of buds previously induced from hypocotyl explants of *L. latifolia* M. Basal media containing MS major salts at full strength (*right*) or at half strength (*left*) were supplemented with: **A** (0.01 mg/l IAA + 2 mg/l BA); **B** (0.01 mg/l IAA + 2 mg/l BA + 10% CM); **C** (0.01 mg/l IAA + 2 mg/l BA + 20% CM)

Leaf explants from *L. latifolia* have also been cultured for shoot regeneration (Calvo and Segura 1989b). The abundance of leaf material makes it the preferable explant for genetic manipulation work, especially transformation via *Agrobacterium* (Klee et al. 1987). Leaves were obtained from 4-week-old seedlings of *L. latifolia* grown in vitro. Leaf explants were then cultured on MS medium supplemented with BA, auxins (IAA or NAA), or several combinations of both types of growth regulators, and maintained at 26 $\pm$ 2°C for 8 weeks under the following three different illumination conditions: darkness, light, and 4 weeks dark plus 4 weeks light. The standard light treatment consisted of a 16-h photoperiod of 20 W m^{-2} from Gro-lux fluorescent tubes.

Irrespective of illumination conditions, bud regeneration from leaf explants of *L. latifolia* was achieved only in media containing BA or BA auxin combinations, with the best results being obtained in the presence of BA and 0.01 or 0.1 mg/l IAA or NAA. However, light conditions during culture had a marked effect on the morphogenic capacity of these explants. A photoperiod of 16 h appeared to yield the best response in terms of bud regeneration percentage, while darkness clearly lessened the shoot-bud differentiation efficiency of the cultured leaves, especially when high auxin concentrations were added to the media (Tables 4 and 5). On the other hand, maximum shoot development was observed under light conditions in

Table 4. Effects of illumination conditions, IAA, and BA on bud induction (% caulogenic explants) from leaves of *Lavandula latifolia* after 8 weeks of culture (Calvo and Segura 1989b)

Illumination conditions	IAA (mg/l)	BA (mg/l) 0.0	2.0	4.0	6.0	8.0	Mean[a]
Light	0.0	0	45	35	25	25	26 abc
	0.01	0	65	60	55	40	44 cdef
	0.1	0	85	80	90	85	68 def
	1.0	0	60	90	90	90	66 def
	2.0	0	35	60	55	45	39 bcd
	Mean[b]	0 a	58 cd	65 d	63 d	57 cd	
Dark + light	0.0	0	75	50	55	60	48 cdef
	0.01	0	95	100	75	55	65 ef
	0.1	0	85	95	95	80	71 f
	1.0	0	20	35	30	60	29 abc
	2.0	0	5	10	10	40	13 ab
	Mean[b]	0 a	56 cd	58 d	53 cd	59 cd	
Dark	0.0	0	25	40	50	60	35 bc
	0.01	0	30	70	55	55	42 bcde
	0.1	0	15	30	40	65	30 abc
	1.0	0	10	20	20	50	20 abc
	2.0	0	0	5	10	25	8 a
	Mean[b]	0 a	16 b	33 bc	35 bcd	51 cd	

[a] Interaction of illumination conditions with levels of IAA.
[b] Interaction of illumination conditions with levels of BA. For each interaction, values followed by the same letter are not significantly different according to the Tukey's test at $p = 0.05$.

the presence of BA and 0.01 to 0.1 mg/l NAA or 0.01 to 1 mg/l IAA, whereas both high auxin concentration and darkness inhibited this response.

Since caulogenesis was also observed in darkness (Fig. 9), it is apparent that light is not directly involved in the process of bud differentiation from cultured leaves of *L. latifolia*. Based on these data, it could be inferred that light may be exerting its influence through hormonal changes or through changes in tissue sensitivity. It is known that light modifies endogenous levels of hormones, and there are several reports showing light-induced reduction of free endogenous IAA (Evans 1984; Rajagopal et al. 1986). Also, tissues grown in darkness are seemingly more sensitive to auxins (Rajagopal et al. 1986). If the endogenous levels of auxin rise in tissues maintained in darkness, the exogenous addition of high auxin concentrations will decrease organ regeneration. This effect and a possible change in tissue sensitivity may explain the inhibitory action of darkness on bud formation from cultured explants of *L. latifolia*.

Regeneration of *Lavandula* from established callus cultures was also found feasible. Calli from hypocotyls and cotyledons of *L. latifolia*, induced on MS

Table 5. Effects of illumination conditions, NAA, and BA on bud induction (% caulogenic explants) from leaves of *Lavandula latifolia* after 8 weeks of culture. (Calvo and Segura 1989b)

Illumination conditions	NAA (mg/l)	BA (mg/l) 0.0	2.0	4.0	6.0	8.0	Mean[a]
Light	0.0	0	45	35	25	25	26 b
	0.01	0	60	75	80	60	55 cd
	0.1	0	90	80	85	70	65 d
	1.0	0	75	80	85	55	59 cd
	2.0	0	45	35	25	25	26 b
	Mean[b]	0 a	63 e	61 e	60 e	47 de	
Dark + light	0.0	0	75	50	55	60	48 bcd
	0.01	0	75	65	80	60	56 cd
	0.1	0	80	60	55	55	50 bcd
	1.0	0	45	35	30	25	27 b
	2.0	0	5	5	10	5	5 a
	Mean[b]	0 a	56 e	43 de	46 de	41 cde	
Dark	0.0	0	25	40	50	60	35 bc
	0.01	0	35	35	60	55	37 bc
	0.1	0	30	40	45	50	33 bc
	1.0	0	0	0	0	0	0 a
	2.0	0	0	0	0	0	0 a
	Mean[b]	0 a	18 b	23 bc	31 bcd	33 bcd	

[a] Interaction of illumination conditions with levels of NAA.
[b] Interaction of illumination conditions with levels of BA. For each interaction, values followed by the same letter are not significantly different according to the Tukey's test at $p = 0.05$.

medium containing 0.08 mg/l 2,4-D, produced adventitious buds when transferred to MS medium with BA alone or in combination with auxins (IAA or NAA). Nevertheless, the cotyledon-derived calli showed a greater caulogenic capacity than those obtained from hypocotyl (Table 6). To promote shoot development, calli bearing buds were subcultured into MS medium containing 2 mg/l BA and 15% CM (Jordan et al. 1988). In *L. angustifolia*, shoot regeneration from stem-derived callus was observed after transfer to MS medium supplemented with 20 mg/l BA. However, success occurred in 10% of the calli and the shoots remained small and did not enlarge after subculture (Webb et al. 1984).

Regeneration via somatic embryogenesis has been reported only from leaf bud-derived calli of *L. angustifolia* and *L. latifolia* (Quazi 1980). These calli were grown on MS medium containing auxins and cytokinins in various concentrations. In a treatment with 2 mg/l 2,4-D and 4 mg/l BA, the calli became deep green in color and formed embryoids that grew into plantlets within 4 weeks of subculturing. The embryoid induction efficiency of these cultures was not reported by the author.

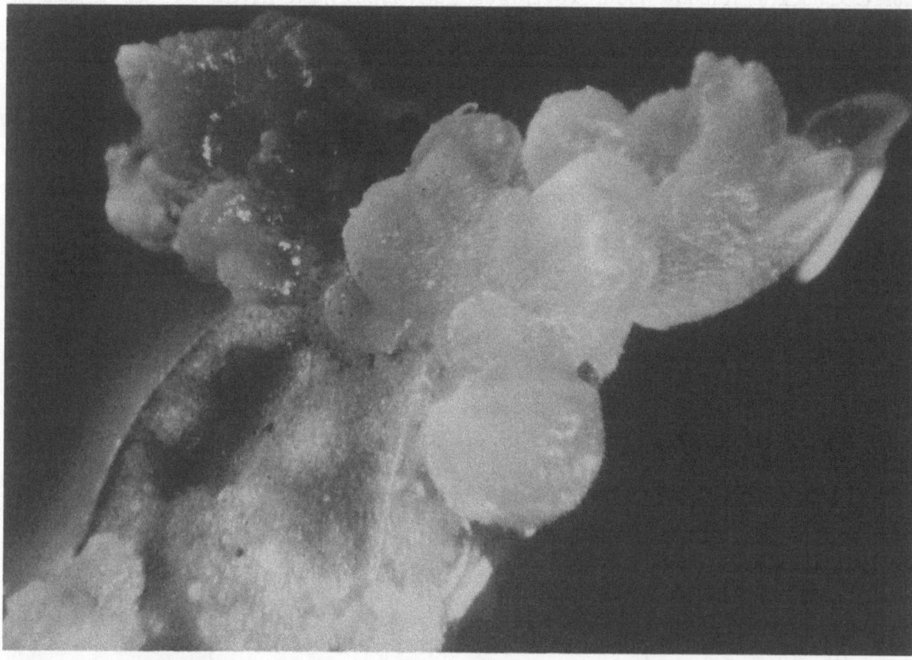

Fig. 9. Buds induced in darkness from cultured leaf explants of *L. latifolia* after 1 month of culture on MS medium containing 0.01 mg/l IAA and 4 mg/l BA

2.6 Isolation and Culture of Cells and Protoplasts

Single cell cultures have been established from callus of *L. stoechas* (Gomez et al. 1987b) and *L. latifolia* (Calvo et al. 1988). For cell isolation, samples of calli (ca. 1 g fr. wt.), chosen from several hypocotyl or root sources, were transferred to 125-ml Erlenmeyer flasks containing 35 ml of sterile liquid MS medium and agitated for 6 to 24 h (130 r.p.m.) at 26 $\pm$ 2°C. The resulting cell suspension was filtered through a stainless steel sieve (60 μM) and centrifuged at 100 *g* for 5 min. Isolated cells were finally resuspended in the medium to be used for culture. This procedure gave cell populations of 95–100% single cells, as determined microcopically (Fig. 10). Yields of cells isolated from 1 g of calli ranged between 1.0×10^5 and 3.0×10^5. Isolated cells were plated, at 10^4 cells/ml, in MS medium supplemented with different hormonal treatments, and kept in darkness at 27 $\pm$ 1°C for 30 to 45 days.

Experiments carried out with *L. stoechas* cells (Gomez et al. 1987b), showed that hormonal treatments including both auxins (IAA or NAA) and cytokinins (BA) allowed continuous cell division and callus formation after 30 days of culture. Best plating efficiency was achieved with cells isolated from hypocotyl-derived calli cultured in the presence of 0.1 mg/l NAA and 1 mg/l BA. To date, all attempts to promote caulogenesis from cell-derived calli of *L. stoechas* have been unsuccessful.

Cells isolated from hypocotyl-derived callus of *L. latifolia* required the addition of both IAA and BA to promote cell proliferation (Table 7). Single cells began to

Table 6. Caulogenic responses induced by growth regulators on hypocotyl (H) and cotyledon (C) derived callus of *L. latifolia* established on MS medium with 0.08 mg/l 2,4-D. Each treatment consisted of 12 replicates and the results were scored after 45 days of culture. (Jordan et al. 1988)

Growth regulators (mg/l)			Callus with buds (%)	
	Auxin	BA	H	C
	0	0	0	0
	0	0.1	0	17
	0	2.0	17	42
	0	4.0	25	25
IAA	0.01	0.1	0	8
	0.01	2.0	0	50
	0.01	4.0	0	17
	0.1	0.1	0	8
	0.1	2.0	0	50
	0.1	4.0	0	17
	1.0	0.1	0	0
	1.0	2.0	0	0
NAA	0.01	0.1	0	17
	0.01	2.0	0	58
	0.01	4.0	25	17
	0.1	0.1	0	17
	0.1	2.0	0	41
	0.1	4.0	17	8
	1.0	0.1	0	25
	1.0	2.0	0	0

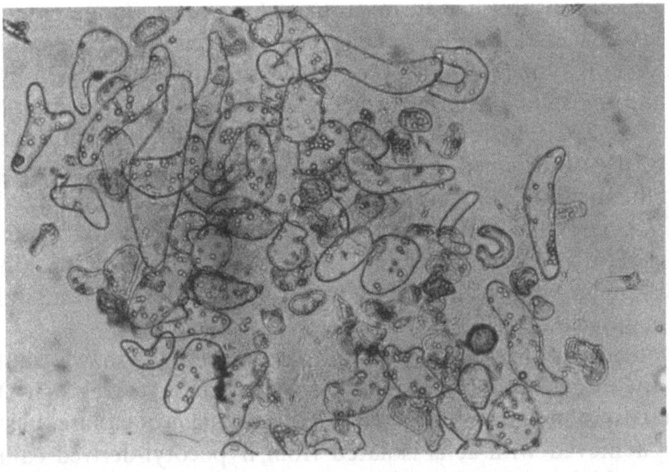

Fig. 10. Freshly isolated cells from hypocotyl-derived callus cultures of *L. latifolia* M

Table 7. Plating efficiencies of isolated cells from hypocotyl-derived callus of *Lavandula latifolia*, cultured at 10^{-4} cells/ml in BM with IAA and BA[a]. (Calvo et al. 1988)

Growth regulators (mg/l)		Plating efficiency (%) after:		
IAA	BA	20 d	30 d	45 d
0	0.1	0.00 a	0.00 a	0.00 a
0	2	0.00 a	0.00 a	0.00 a
0.1	0	0.00 a	0.00 a	0.00 a
1	0	0.00 a	0.00 a	0.00 a
2	0	0.00 a	0.00 a	0.00 a
0.1	0.1	1.34 c	1.34 c	1.34 c
0.1	2	0.85 b	1.18 c	1.24 c
1	0.1	0.00 a	0.00 a	0.00 a
1	2	0.94 b	1.31 c	1.39 c
2	0.1	0.00 a	0.00 a	0.00 a
2	2	0.00 a	0.00 a	0.00 a

[a] Values followed by the same letter are not significantly different at the 5% level using Duncan's Multiple Range Test.

divide within 4–7 days, giving rise to colonies after 10–15 days in culture (Fig. 11). Many of these colonies developed into calli after 20 days (Fig. 12). In order to induce organogenesis, cell-derived calli were transferred to various regeneration media (Table 8). Shoot bud differentiation efficiency depended on the composition of both the callus induction and the shoot regeneration media, best results being obtained when calli grown in 1 mg/l IAA and 2 mg/l BA were subcultured to media containing 2 mg/l BA and 15% CM (Fig. 13). To induce shoot development, calli showing buds were subcultured to two different elongation media. As is shown in Table 7, both media promoted bud elongation, but best results were obtained in the presence of coconut milk.

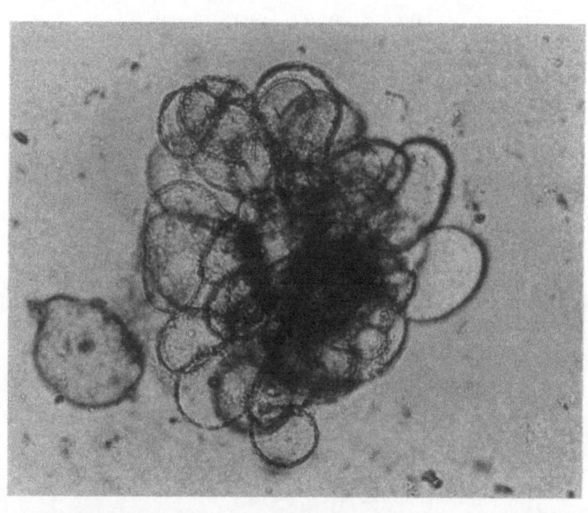

Fig. 11. Colony from single cell of *L. latifolia* after 15 days of culture on MS medium containing 1 mg/l IAA and 2 mg/l BA. (Calvo et al. 1988)

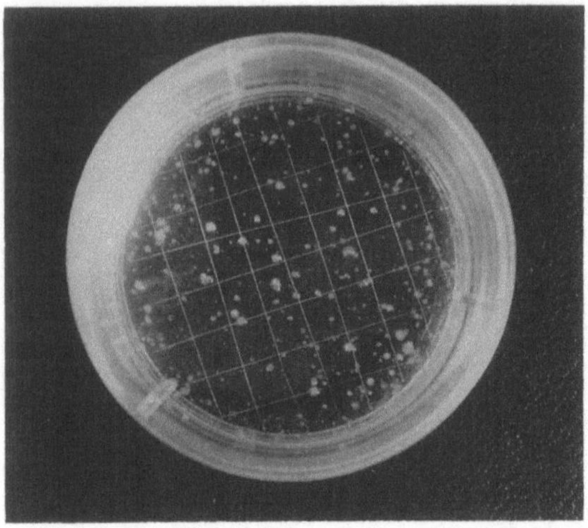

Fig. 12. Calli developed from isolated cells of *L. latifolia* after 30 days of culture on MS medium containing 1 mg/l IAA and 2 mg/l BA. (Calvo et al. 1988)

Table 8. Comparative caulogenic responses of calli derived from isolated cell of *Lavandula latifolia*. (Calvo et al. 1988)

Cell culture medium		Bud induction medium			Shoot-bud formation (%)		Elongation and multiplication medium		No. of shoots > 1 cm
IAA (mg/l)	BA (mg/l)	IAA (mg/l)	BA (mg/l)	CM (%)	After 30 d	After 60 d	BA (mg/l)	CM (%)	After 60 d
0.1	2	0	2	0	0	37.5	2	0	13
							2	15	25
		0.1	2	0	0	6.2	–	–	–
							–	–	–
		0	2	15	0	12.5	2	0	7
							2	15	12
		0.1	2	15	0	0	–	–	–
							–	–	–
		0	0	15	0	0	–	–	–
							–	–	–
1	2	0	2	0	0	0	–	–	–
							–	–	–
		0.1	2	0	0	0	–	–	–
							–	–	–
		0	2	15	25	75	2	0	105
							2	15	362
		0.1	2	15	12.5	62	2	0	83
							2	15	264
		0	0	15	0	0	–	–	–
							–	–	–

Fig. 13. Shoot development on caulogenic callus from isolated cells of *L. latifolia* 60 days after transfer of the calli onto MS medium supplemented with 2 mg/l BA and 15% CM (Calvo et al. 1988)

Cultures of *L. vera* green cells have also been established (Watanabe et al. 1982). These cultures have been used to select for high biotin-producing cells (see Sect. 2.8.2).

Protoplasts have been obtained from *L. latifolia* seedlings, but there is no detailed work on their culture. Protoplasts from leaf and cotyledon explants of *L. latifolia* were isolated using several enzyme mixtures of pectinase (Macerozyme R-10 or Pectolyase Y-23) and cellulase RS (Fernandez-Feijoo et al. 1988). All enzymatic treatments were effective on protoplast release, but yield of protoplasts was affected by pectinase type, best results being obtained when Pectolyase Y-23 was used. The addition of Ca^{2+} and spermine to the isolation medium seems to be important in obtaining highly viable protoplasts. So far, no reports exist on plant regeneration from *Lavandula* protoplasts.

2.7 Plant Regeneration

Shoots derived from primary explants, callus, or isolated cells of *L. latifolia* may be used to obtain complete plants, which can be transferred to soil (Fig. 14). Regenerated shoots are easily rooted on hormone-free MS medium with macronutrients at half strength. The establishment of the in vitro-grown plants in soil is usually achieved as follows: plantlets with well-developed roots are potted in a mixture of vermiculite and peatmoss. Subsequently, the plants are hardened off under inverted plastic cups in a growth chamber (26 ± 2 °C with a 16-h photoperiod) for 15–20 days. Once the acclimatation is accomplished, the plants are transferred to the greenhouse (Calvo and Segura 1988, 1989a,b; Calvo et al. 1988).

Fig. 14. In vitro-regenerated plants of *L. latifolia*

2.8 Secondary Products in Tissue Culture

The major secondary products reported from cultures in vitro of *Lavandula* are essential oils, biotin, and pigments. Work done in this field is summarized below.

2.8.1 Essential Oils

Essential oils constitute an important group of natural products with a high industrial interest (see Sect. 1.1). Consequently, the production of these compounds by plant cell cultures has been the subject of numerous studies (Mulder-Krieger et al. 1988). However, the exciting promise of in vitro technology for the production of volatile oils has been somewhat tempered by the observation that most cultures are unable to produce an adequate yield of essential oils (Mulder-Krieger et al. 1988). It has been postulated that synthesis and accumulation of volatile secondary metabolites may usually require a degree of morphological and cellular differentiation (Collin and Watts 1983). This is particularly true in *Lavandula* cultures. Several authors (Webb et al. 1984; Banthorpe et al. 1986; Lappin et al. 1987) reported that no monoterpenes could be detected in callus or suspension cultures of *L. angustifolia*. The callus was friable and homogeneous with neither apparent differentiation nor pigment formation. Nevertheless, the shoots regenerated from these calli (see Sect. 2.5) accumulated some monoterpenes characteristic of the intact plant. The accumulation of these products, which are components of the lavender oil, was low (12 and 0.1% respectively of that found in 5-week-old seedlings and in leaves from mature plants), but was at least 10^4-fold greater than in the parent callus (Webb et al. 1984).

Although undifferentiated tissues of *L. angustifolia* did not detectably accumulate monoterpenes or secrete them into the medium, these cultures yielded cell-free extracts containing prenyltransferase and an isomerizing system with activities greater than those extracted from seedlings and mature plants. The callus also contained the crucial enzymes for the synthesis of the parents of the lower terpenoids, and in addition the MVA-activating pathway (MVA-kinase, MVAP-kinase, MVAPP-decarboxylase, and IPP-DMAPP-isomerase) was present and could be extracted (Banthorpe et al. 1986).

There appears to be no explanation for the absence of terpenes in undifferentiated tissues of *Lavandula*, despite the presence of the enzymes responsible for monoterpenoid anabolism. Lappin et al. (1987) reported that suspension cultures of *L. angustifolia* reduced monoterpenoid aldehydes and structurally related compounds to their corresponding primary alcohols. Acyclic primary monoterpenoid alcohols were further metabolized to unidentified compounds. Cyclic, secondary, tertiary, and aromatic alcohols were not metabolized. According to Lappin et al. (1987), these results raise the possibility that the absence of monoterpenes in callus cultures may be due to their catabolism immediately following their formation. In the intact plant, the accumulation of essential oils in metabolically remote oil glands and the seasonal appearance of catabolic enzymes may control the rate of turnover and thereby the composition of the oil. Such controls may not exist in undifferentiated cultures (Lappin et al. 1987). The lack of accumulation of monoterpenes in callus cultures may also be due to the absence or low activity of an enzyme for a step before MVA (Banthorpe et al. 1986).

2.8.2 *Biotin*

Cell aggregate cloning has been used successfully to select for high biotin-producing cells of *L. vera* (Watanabe and Yamada 1982a,b; Watanabe et al. 1982). In these studies, green *L. vera* cells, which had large amounts of free biotin, were repeatedly cultured on LS medium containing 10^{-5} M IBA and 10^{-6} M BA, analyzed for free biotin content, and reselected. A combination of pimelic acid and L-alanine (precursors of biotin) was used as a selection agent to screen for cells that could amass large amounts of biotin (Watanabe et al. 1982). Some cell lines were also irradiated with several gamma ray dosages to determine the relation of the dosage to the production of variant cells containing large amounts of free biotin (Watanabe and Yamada 1982a).

L. vera cell lines with high levels of free biotin were obtained from cells which survived on a medium containing pimelic acid and L-alanine or from gamma-irradiated cells (Watanabe et al. 1982). One *L. vera* cell line, obtained from irradiated cells, contained 15 times the amount of free biotin found in the original unselected cultured cells and ten times that found in the leaves. The cell line selected with pimelic acid had an average value for free biotin content that was one and a half times higher than that of the control (Watanabe and Yamada 1982b).

2.8.3 Pigments

Cultured cells of *L. angustifolia* have been found to secrete a blue pigment into the medium under a variety of conditions (Banthorpe et al. 1985 and cited references). The pigment has been identified as the (Z,E)-2-(3,4-dihydroxyphenyl) ethenyl ester of 3-(3,4-dihydroxyphenyl)-2-propenoic acid and its (E,E)-isomer. The excreted product complexes with Fe^{2+} in the medium, producing the blue coloration. These unusual enol esters could not be detected in the foliage, stem, and flower-heads of *L. angustifolia*, although the (Z,E)-isomer occurred in shoots and foliage of *Plectranthus caninus* Roth (Banthorpe et al. 1985).

Blue pigment accumulation seems to be an almost exclusive property of cultured *Lavandula* cells, since it did not occur in cultures of some 40 species of higher plants, many of these being Labiatae (Banthorpe et al. 1985). Furthermore, we have also observed blue pigment accumulation in cultures of *L. latifolia*.

Several factors may affect blue pigment formation by callus of *L. angustifolia*. Banthorpe et al. (1985) reported that pigmentation was increased by darkness, high sucrose levels, addition of asparagine, and accidental bacterial or fungal infection of cultures. Pigment formation was reduced by addition of CM, and was completely suppressed by casein hydrolysate or by removal of Fe^{2+} from the culture medium. In cultured green *L. vera* cells, blue pigment formation did not occur unless cystein was added to the medium (Watanabe et al. 1985a). No blue pigment was seen in suspension cultures established from stem-derived callus of *L. angustifolia*, although this callus did produce the pigment when cultured on solid medium (Lappin et al. 1987). In contrast, Banthorpe et al. (1985) reported that the liquid medium became intensely blue within 5 days of initial transfer of the *L. angustifolia* callus, but after this the color disappeared within days. There appears to be no available explanation for the differences in blue pigment production based on the plant variety, explant source, medium composition, or culture technique used, although the reported data are insufficient.

So far, the major interest in immobilized plant cell techniques has been centered upon their use for the production of phytochemicals (Hall et al. 1988). A significant finding in this area was published by Nakajima et al. (1985), who reported a prolonged excretion of the blue pigment by immobilized green *L. vera* cells. These results clearly demonstrated that entrapment in proper gels is very effective to stabilize viability and metabolic activities of plant cells, thus permitting the repeated production of metabolites, which is hardly possible with the free counterparts. *L. vera* strain L_{10}4–2154, which was selected as a potent biotin-producing strain (see Sect. 2.8.2), was used by Nakajima et al. (1985) in their expriments. Finely suspended cells of the strain were obtained by cultivating the cells in the presence of 50 mM $CaCl_2$. Cultured cells were immobilized by entrapment in gels formed from natural polysaccharides, agar, alginate, and k-carrageenan. The gel-entrapped cells grew well inside the gel matrices, the growth being confirmed by the increases in oxygen uptake, cell number, and chlorophyll content. Blue pigment was synthesized de novo in the presence of 1 mM L-cysteine. This response occurred in both free and immobilized cultures, although synthesis in the latter began earlier. Calcium alginate gel-entrapped cells were successfully employed for the repeated production of the pigment for over 7 months by alternating growth (-cysteine) and production (+cysteine) phases every 14 days. In contrast, the free counterparts died after only one subculture.

L. vera cells were also entrapped with photosensitive synthetic resin prepolymers (Nakajima et al. 1986). It is noteworthy that these entrapped cells were superior to calcium alginate-entrapped cells judging from cell growth and total blue pigment productivity. Release of the pigment from the gels was markedly enhanced by the increase in hydrophilicity of the cell-entrapping gels. The entrapped cells could also be used repeatedly for the pigment production.

2.9 Germplasm Conservation

The evidence available so far (Kartha 1987; Bajaj 1988) indicates that cryopreservation is the most reliable approach for the long-term storage of cell cultures with capacity for the synthesis and accumulation of secondary metabolites. Cryopreservation techniques have been developed for the biotin-forming cell strain of *L. vera* (Watanabe et al. 1983). Green *L. vera* cell suspensions were established in liquid LS medium supplemented with 10^{-5} M IBA and 10^{-6} M BA. The cell suspension was cooled at 1°C/min to -40°C, then transferred into LN for 6 to 21 days. After a rapid thawing in a 60°C water bath, the cell suspension was diluted with fresh liquid medium and centrifuged. The resulting cells were plated on solidified LS medium and incubated for 3 to 5 months.

Watanabe et al. (1983) found that *L. vera* calli, grown from cells recovered after cryopreservation, retained not only the biosynthetic capability for biotin, but also the plant regeneration potential, as did the original green callus cultures. The average values of free biotin content in calli before and after freezing were 0.07 ± 0.039 μg/g fr. wt. and 0.065 ± 0.054 μg/g fr. wt. respectively. In further experiments, Watanabe's group reported changes in freezing resistance for cryopreserved *L. vera* cells and demonstrated that the increase in the unsaturation of membrane phospholipid fatty acids plays an important role in the freezing resistance of cultured *Lavandula* cells (Watanabe et al. 1985b, 1986).

3 Conclusions and Prospects

Clonal propagation of *Lavandula* by means of tissue culture technology appears feasible. A review of the present literature clearly demonstrates that in vitro regeneration of *Lavandula* can be easily achieved through organogenesis from a variety of explant sources, including isolated cells. These results suggest potential applications for selection experiments and somaclonal variation studies, specially for fast multiplication and high yield of desired products. Nevertheless, in order to fully exploit the benefits of the tissue culture technology, more work is required to establish the conditions for plant regeneration from *Lavandula* protoplasts. Also, more and detailed studies on factors controlling micropropagation of *Lavandula* via somatic embryogenesis must be undertaken. All these in vitro techniques will certainly improve the breeding and productivity of this important aromatic crop in coming years.

The major secondary products reported from in vitro cultures of *Lavandula* are essential oils, biotin, and blue pigments. One great success has been the induction and selection of high biotin-producing cell strains. Another important achievement was the immobilization of *Lavandula* cells and the prolonged production of the blue pigment by gel-entrapped cells. In contrast, progress on essential oil production by cultured *Lavandula* cells is still lacking. Callus and suspension cultures of *Lavandula* are unable to accumulate monoterpenes, but contain the enzymes responsible for terpenoid anabolism and may biotransform some monoterpenoid aldehydes to their corresponding primary alcohols. Up till now, our knowledge is not sufficient to overcome the absence of monoterpenes in cultured cells of *Lavandula*. Consequently, more studies are needed to elucidate the suppression mechanism(s) of terpenoid accumulation to raise its productivity in callus or suspension cultures.

Cryopreservation techniques have been used successfully in suspension cultures of *Lavandula*. The retention of metabolic and plant regeneration potentials of the retrieved cultures suggests the use of this technology for the storage of germplasm from *Lavandula* plants with high essential oil content.

4 Protocol

The following protocols are used in our laboratory for micropropagation and establishment of single cell cultures of *L. latifolia*

1. Micropropagation

1.1 Hypocotyl explants from 30-day-old seedlings are cultured on agar-solidified medium containing BA (2mg/l) and IAA (0.1 mg/l) for bud induction. Buds appear after 2–3 weeks of incubation at 26 ± 2°C under 16-h photoperiod.
1.2 Cultures bearing buds are transferred to MS medium supplemented with BA (2 mg/l), IAA (0.01 mg/l) and CM (20%) for shoot development.
1.3 The obtained shoots are transplanted to rooting medium (hormone-free MS medium with macronutrients at half strength). Rooted plantlets are subsequently acclimated to greenhouse conditions after gradual hardening off.

2. Single cell culture

2.1 Hypocotyl explants from 30-day-old seedlings are cultured on MS medium supplemented with 0.1 mg/l 2,4-D for callus proliferation. Cultures are incubated for 30–40 days at 26 ± 2°C and 16-h photoperiod.
2.2 The obtained calli are transferred to liquid MS medium and agitated for 24 h (130 r.p.m.) at 26 ± 2°C. The resulting cell suspension is filtered throught a stainless steel sieve (60 μM) and centrifuged (100 g for 5 min).
2.3 Isolated cells are plated, at 10^4 cells/ml, in MS medium supplemented with BA (2 mg/l) and IAA (1 mg/l), and kept in darkness at 27 ± 1°C for 30–45 days.
2.4 To induce organogenesis, cell-derived calli are transferred to MS medium supplemented with BA (2 mg/l) and CM (15%). Buds appear after 4-8 weeks of incubation at 26 ± 2°C under 16 h photoperiod.
2.5 For shoot development, rooting and acclimation proceed as indicated in (1.2) and (1.3).

Acknowledgments. The financial support of the Dirección General de Investigación Cientifica y Técnica, Madrid (Spain), project N° PB86-0066 is gratefully acknowledged.

References

Bajaj YPS (ed) (1988) Cryopreservation and the retention of biosynthetic potential in cell cultures of medicinal and alkaloid-producing plants. In: Biotechnology in agriculture and forestry, vol 4: Medicinal and aromatic plants I. Springer, Berlin Heidelberg New York Tokyo, pp 169–187

Bajaj YPS, Furmanowa M, Olszowska O (1988) Biotechnology of the micropropagation of medicinal and aromatic plants. In: Bajaj YPS (ed) Biotechnology in agriculture and forestry, vol 4: Medicinal and aromatic plants I. Springer, Berlin Heidelberg New York Tokyo, pp 60–103

Banthorpe DV, Bilyard HJ, Watson DG (1985) Pigment formation by callus of *Lavandula angustifolia*. Phytochemistry 24:2677–2680

Banthorpe DV, Branch SA, Njar VCO, Osborne MG, Watson DG (1986) Ability of plant callus cultures to synthesize and accumulate lower terpenoids. Phytochemistry 25:629–636

Belaiche P (1979) Traité de phytothérapie et d'aromathérapie, vol 1: L'aromatogramme. Maloine, Paris

Bezanger-Beauquesne L, Pinkas M, Torck M (1986) Les plantes dans la therapeutique moderne. Maloine, Paris

Calvo MC, Segura J (1988) In vitro morphogenesis from explants of *Lavandula latifolia* and *Lavandula stoechas* seedlings. Sci Hortic 36:131–137

Calvo MC, Segura J (1989a) In vitro propagation of lavander. HortScience 24:375–376

Calvo MC, Segura J (1989b) Plant regeneration from cultured leaves of *Lavandula latifolia* Medicus: influence of growth regulators and illumination conditions. Plant Cell Tissue Org Cult 19:33–42

Calvo MC, Jordan A, Segura J (1988) Plant regeneration from isolated cells of *Lavandula latifolia* Medicus. In Vitro 24:943–946

Collin HA, Watts M (1983) Flavor production in culture. In: Evans DA, Sharp WR, Ammirato PV, Yamada Y (eds) Handbook of plant cell culture vol 1. Macmillan, New York, pp 729–747

De-Pascual-T J, Caballero C, Machin G (1983) Constituents of the essential oil of *Lavandula latifolia*. Phytochemistry 22:1033–1034

Duke JA (1985) Handbook of medicinal herbs. CRC, Boca Raton, Fla

Evans ML (1984) Functions of hormones at the cellular level of organization. In: Scott TK (ed) Hormonal regulation of development II. Encyclopedia of plant physiology, N S, vol 10. Springer, Berlin Heidelberg New York, pp 23–79

Fernandez-Feijoo J, Brisa MC, Segura J (1988) Isolation of cotyledon and leaf protoplasts from *Lavandula latifolia*: Preliminary report. In: 6th Congr FESPP, Split, Abstr, pp 14–09

Font Quer P (1978) Plantas medicinales. Labor, Barcelona

Gamborg OL, Miller RA, Ojima K (1968) Nutrient requirements of suspension cultures of soybean root cells. Exp Cell Res 50:151–158

Gennaro AR (ed) (1985) Remington's pharmaceutical sciences, 17th edn. Mack, Easton, Penn

Gomez MP, Calvo MC, Segura J (1987a) Callus formation from hypocotyls of *Lavandula stoechas*: Interaction between nutritional and hormonal factors. Gartenbauwissenschaft 52:6–9

Gomez MP, Calvo MC, Cornejo MJ, Segura J (1987b) Isolation and culture of *Lavandula stoechas* cells. Plant Physiol Life Sci Adv 6:159–164

Hall RD, Holden MA, Yeoman MM (1988) Immobilization of higher plant cells. In: Bajaj YPS (ed) Biotechnology in agriculture and forestry, vol 4: Medicinal and aromatic plants I. Springer, Berlin Heidelberg New York Tokyo, pp 136–156

Jordan A, Calvo MC, Segura J (1988) Morphogenesis in callus and isolated cell cultures of *Lavandula latifolia*. In: 6th Congr FESPP, Split, Abstr, pp 14–14

Kartha KK (1987) Cryopreservation of secondary metabolite-producing plant cell cultures. In: Constabel F, Vasil IK (eds) Cell culture and somatic cell genetics of plants, vol 4: Cell culture in phytochemistry. Academic Press, New York London San Diego, pp 217–227

Klee H, Horsch R, Rogers S (1987) *Agrobacterium*-mediated plant transformation and its further applications to plant biology. Annu Rev Plant Physiol 38:467–486

Lappin GJ, Stride JD, Tampion J (1987) Biotransformation of monoterpenoids by suspension cultures of *Lavandula angustifolia*. Phytochemistry 26:995–997

Linsmaier EM, Skoog F (1965) Organic growth factors requirements of tobacco tissue culture. Physiol Plant 18:100–127

Mulder-Krieger TH, Verpoorte R, Baerheim Svendsen A, Scheffer JJC (1988) Production of essential oils and flavours in plant cell and tissue cultures, a review. Plant Cell Tissue Org Cult 13:85–154

Muñoz F (1987) Plantas medicinales y aromáticas: estudio, cultivo y procesado. Mundi-Prensa, Barcelona
Murashige T, Skoog F (1962) A revised medium for rapid growth and bioassays with tobacco tissue cultures. Physiol Plant 15:473–497
Nakajima H, Sonomoto K, Usui N, Sato F, Yamada Y, Tanaka A, Fukui S (1985) Entrapment of *Lavandula vera* cells and production of pigments by entrapped cells. J Biotechnol 2:107–117
Nakajima H, Sonomoto K, Morikawa H, Sato F, Ichimura K, Yama DA, Tanaka A (1986) Entrapment of *Lavandula vera* cells with synthetic resin prepolymers and its application to pigment production. Appl Microbiol Biotechnol 24:266–270
Panizza M, Tognoni F (1988) Clonal propagation, callus formation and plant regeneration of Lavandin. Sci Hortic 37:157–163
Paris RR, Moyse H (1971) Précis de matière médicale, vol 3: Pharmacognosie spéciale. Masson, Paris
Quazi MH (1980) In vitro multiplication of *Lavandula* spp. Ann Bot (London) 45:361–363
Rajagopal R, Uluskov P, Marcussen J (1986) Effects of light on endogenous hormone levels in plants. In: Purohit SS (ed) Hormonal regulation of plant growth and development, vol 3. Agrobotanical Publishers, Bikaner, India, pp 29–90
Reynolds JEF (ed) (1982) Martindale, The extra pharmacopoeia. Pharmaceutical Press, London
Rivas Goday S, Rivas-Martinez S (1967) Matorrales y tomillares de la Peninsula Ibérica comprendidos en la clase Ononido-Rosmarinetea. An Inst Bot AJ Cavanilles 25:1–20
Rivas-Martinez S (1979) Brezales y jarales de Europa occidental (Revisión Fitosociológica de las clases Calluno-ulicetea y Cisto-Lavanduletea). Lazaroa 1:5–127
Tutin TG, Heywood VH, Burges NA, Moore DM, Valentine DH, Walters SM, Webb DA (eds) (1972) Flora Europea, vol 3. Univ Press, Cambridge
Watanabe K, Yamada Y (1982a) Selection of variants with high levels of biotin from cultured green *Lavandula vera* cells irradiated with gamma rays. Plant Cell Physiol 23:1453–1456
Watanabe K, Yamada Y (1982b) Selection of high biotin producing strains of cultured *Lavandula vera* cells. In: Fujiwara A (ed) Plant tissue culture 1982. Maruzen, Tokyo, pp 357–358
Watanabe K, Yano S, Yamada Y (1982) The selection of cultured plant cell lines producing high levels of biotin. Phytochemistry 21:513–516
Watanabe K, Mitsuda H, Yamada Y (1983) Retention of metabolic and differentiation potentials of green *Lavandula vera* callus after freeze-preservation. Plant Cell Physiol 24:119–122
Watanabe K, Sato F, Furuta M, Yamada Y (1985a) Induction of pigment production by S-containing compounds in cultured *Lavandula vera* cells. Agric Biol Chem 49:533–534
Watanabe K, Yamada Y, Ueno S, Mitsuda H (1985b) Change of freezing resistance and retention of metabolic and differentiation potentials in cultured green *Lavandula vera* cells which survived repeated freeze-thaw procedures. Agric Biol Chem 49:1727–1731
Watanabe K, Kito M, Ueno S, Mitsuda H (1986) Relationships of membrane phospholipids to freezing resistance of cultured green *Lavandula vera* cells. Agric Biol Chem 50:1791–1796
Webb JK, Banthorpe DV, Watson DG (1984) Monoterpene synthesis in shoots regenerated from callus cultures. Phytochemistry 23:903–904
Whitaker RJ, Hashimoto T (1986) Production of secondary metabolites. In: Evans DA, Sharp WR, Ammirato PV (eds) Handbook of plant tissue culture, vol 4. Macmillan, New York, pp 264–286

XIX *Nigella* spp.: In Vitro Culture, Regeneration, and the Formation of Secondary Metabolites

H.-P. SCHMAUDER[1] and P. DOEBEL[2]

1 Characteristics of the Whole Plants and Their Secondary Metabolites

The genus *Nigella* belongs to the family Ranunculaceae and includes about 20 species. Assignation to the genus *Nigella* changed repeatedly during the last century; some selected species are listed in Table 1. *Nigella* species are located mainly in the Middle East and the Mediterranean region. A characteristic description of the *Nigella* species can be found in Tutin et al. 1964, Rechinger 1964, Davis et al. 1965.

The plants are annuals and often have spirally arranged bi- or tripimatisect leaves with linear or capillary lobes and solitary, terminal, hermaphrodite, actinomorphic, bluish flowers. Some subspecies or varieties also have white flowers or flowers of other pigmentation (mostly breeding forms).

Only a small number of *Nigella* species are of practical interest as ornamental plants or as plants used in traditional folk medicine (Table 1). Most of them are only known as weeds or wild plants. The habitats of the *Nigella* species are, for instance, south + east Europe, south + western and central Asia and Asia Minor.

In *Nigella* species many biologically active compounds, mostly secondary plant products, are found, for example, enzymes, other proteins and peptides, fatty acids, fats, oils, essential oils, and alkaloids, as well as saponins. A selection of such substances is summarized in Table 2 and formulas of some of them are given in Fig. 3. Detailed tests have shown a clear pharmacological action for only a small part of these compounds or mixtures.

Analyses for a chemical characterization of *Nigella sativa* seeds were done by Babayan et al. (1978) and Bose et al. (1981). Esters of unsaturated fatty acids (e.g., dehydrostearinic or linolenic acid) with C_{15} or higher terpenoid or aliphatic alcohols, and α,β-unsaturated hydroxyketone ($C_{18}H_{28}O_4$), as well as a trace of an alkaloid of the pyridine group, were detected (Bose et al. 1981). Babayan et al. (1978) found the following chemical composition: protein 21%, fat 5.5%, ash 3.7%, the remaining percent being carbohydrates. The fatty acids are a mixture of linoleic acid (56%), oleic acid (24.6%), palmitic acid (12%), stearic acid (3%), eicosadienoic acid (2.5%), linolenic acid (0.7%), and myristic acid (0.16%). By hydrolysis of the proteins, 15 amino acids, nine of them essential, could be detected.

[1] Friedrich-Schiller-Universität Jena, Biologisch-Pharmazeutische Fakultät, Institut für Mikrobiologie, Neugasse 23, O-6900 Jena, FRG
[2] Akademie der Wissenschaften, Institut für Biotechnologie, Permoserstraße 15, O-7050 Leipzig, FRG

Biotechnology in Agriculture and Forestry, Vol. 15
Medicinal and Aromatic Plants III (ed. by Y.P.S. Bajaj)

Fig. 1. *Nigella damascena*, flowers. (Schmauder and Doebel unpubl.)

Two different kinds of proteins of *Nigella* have been studied in detail. The homologies of seed storage proteins within the genus *Nigella* compared with proteins obtained from the agriculturally important families Fabaceae and Poaceae, as well as the properties of coding nucleic acids, were studied; they can be useful markers for taxonomic studies (e.g., Jensen and Grumpe 1983; Jensen 1984).

Rudyuk's group, Higginson (1981) Gmoshinskii et al. (1982) characterized lipases from *Nigella damascena* seeds as surfactants, possibly therapeutics, or for use in technical systems. Main lines for isolation, purification, stabilization, and some forms for application were elaborated (Korchagina et al. 1973; Korchagina and Rudyuk 1975, 1979; Rudyuk and Korchagina 1975, 1977, 1981; Gorodetzkii et al. 1976; Potekhina et al. 1982; Rudyuk et al. 1984).

Fig. 2. *Nigella arvensis*, plant. (Schmauder and Doebel unpubl.)

Datta et al. (1987) analyzed proteins extracted from normal and mutant lines of *Nigella sativa* by use of PAA electrophoresis. The different normal and mutant lines studied showed great heterogenities.

Some workers have analyzed *Nigella* species for their alkaloid and terpenoid content. Protoanemonine-forming compounds and berberine could not be detected. Only magnoflorine (*8*) was detectable in these experiments (Frencel 1965; Hegnauer 1973). The alkaloid-like substance damascenine (Fig. 3, *6*) formed by *N. damascena* was a model for the study of alkaloid formation in the whole plant because it has a simple structure. Nevertheless, the biosynthetic pathway for the formation of damascenine is not as yet completely clear. The formation of this compound is bound to a differentiation process in the development of the plants. Only seeds contain this compound. The biosynthesis probably starts from anthranilic acid, an intermediate of the tryptophan branch in the biosynthesis of aromatic amino acids. Some methylations and a hydroxylation step convert anthranilic acid into damascenine and related compounds (see, for example,

Table 1. Some selected of *Nigella* species (see also: Rechinger 1964; Tutin et al. 1964; Davis et al. 1965)

Species	Main habitat	Remarks[a]
N. arvensis L.	Europe, Asia	More than eight ssp./var., use in folk medicine and as a spice. Older names are *N. latifolia* Mill., *N. tenuiflora* Gilib. [a]Acker-Schwarzkuemmel, Wilder Schwarzkuemmel, field *Nigella*, field fennelflower, nigelle des champs, estrella de campo, aranuela menor neguila, tschernuschka polevaja
N. assyrica Boiss.	Asia minor, Turkey, Asia	synonym for *N. arvensis* varieties
N. cretica Miller	Crete	
N. damascena L.	Europe, Asia	Ornamental plant [a]Jungfer im Gruenen, love-in-the-mist, tschernuschka damasskaja
N. degenii Vierh.	Europe, Asia	
N. doerfleri Vierh.	S-E. Europe, Asia	Perhaps closely related to *N. stellaris* Boiss. from Asia minor
N. elata Boiss	Europe, Asia	
N. fumariifolia Kotschy	S-E. Europe, Asia, Turkey	
N. gallica Jordan	Europe, Asia	
N. glauca Schkuhr	Asia minor, Asia	
N. hispanica L.	Europe, Asia	Two or more ssp./var.
N. latisecta P.H. Davis	Turkey, Asia	
N. nigellastrum (L.) Wilk.	Europe, Asia	Now with a new taxonomic name: *Garidella nigellastrum* L.
N. orientalis L.	S-E. Europe, Asia	
N. oxypetala Boiss.	Turkey, Asia	
N. sativa L.	Europe, Asia	Aromatic seeds, use in traditional medicine [a]Echter Schwarzkuemmel, black cumin, tschernuschka posevnaja
N. segetalis Bieb.	Europe, Asia, Turkey	[a]Tschernuschka paschennaja
N. stellaris Boiss.	Turkey, Asia	
N. unguicularis (Lam.) Spenner	Turkey, Asia	

[a] Selected common names.

Table 2. Selected products isolated from *Nigella* species

Class of Compound	Compound(s)	*Nigella* sp.	Remarks	Reference
Enzymes/ proteins/ peptides	Storage proteins (legumin-, vicilin-like)	*N. damascena* *N. sativa* *N. arvensis* *N. hispanica* *N. nigellastrum*	Seeds	Jensen and Grumpe (1983); Jensen (1984)
	Lipases	*N. damascena*	Seeds	Vishin (1961); Rudyuk and Korchagina (1976)
	Cytochrome C	*N. damascena*	Sequence	Brown and Boulter (1973)
	β-Fructofuranosidase	*N. damascena*	Therap. use?	Kabachnyi et al. (1986)
	Amylolytic enzymes	*N.* sp.		Kabachnyi et al. (1987)
Fats/fatty acids	Olefinic fatty acids (*1*)	*N. damascena*		Takagi et al. (1983)
	Fatty acids mostly unsaturated (e.g., petroselinic acid (2), unsat. C_{18}-acids (*3*)/isoseries	*N. sativa*		Kartha and Khan (1969)
	Fatty acids oleic acid (*4*) linolic acid (*5*)	*N. damascena* *N. sativa*	Seeds	Vishin (1961); Gad et al. (1963)
Essential oils	Essential oil (0.45%)	*N. sativa*	Seeds	Jukneviciene et al. (1977)
	Essential oil	*N. sativa*		Canonica et al. (1963)
	Sesquiterpenes (β-elemene, *23*, o-selinene, *24*)	*N. damascena*		Tillequin et al. (1976)
Alkaloids	Damascenine (*6*)	*N. damascena*	Seeds	Keller (1908)
	Damascenine/ damascinine (*7*)	*N. damascena*	Seeds	Doepke and Fritsch (1970)
	Magnoflorine (*8*)	*N.* sp.		Frencel (1965); Nijland and Uffelie (1965)
	Nigellicine (*9*)/ nigellimine-N-oxide-(*10*)	*N. sativa*	Seeds	Atta-Ur-Rahman et al. (1985a,b)
Saponines	Melanthigenine (syn. hederagenine, *11*)	*N. sativa*	Seeds	Mustafa and Soliman (1943)
	Saponins (melanthine)	*N. arvensis*	Seeds	Hampel (1957)
Phenols	Kaempferol (*12*), quercetine (*13*)	*N. arvensis*	Green parts	Kirichenko et al. (1971)
Polyols	Acyclic poly-6 diff. ols, mainly sorbitol (*14*) accumulation	*N.* sp.	Taxonomic markers	Wallaart (1981)

(*to be continued*)

Table 2. (*continued*)

Class of Compound	Compound(s)	*Nigella* sp.	Remarks	Reference
Ascorbic acid (*15*)	1119 mg%/dry weight	*N. arvensis*		Istratescuguti and Forstner (1974)
Sterols	Cholesterol (*16*), campesterol, stigmasterol (*17*), β-sitosterol(*18*), α-β-spinasterol	*N. sativa*	Seed oil	Salama (1973)
Quinones	Thymoquinone (*19*) and di-/polymers	*N. sativa*		Canonica (1963); El-Dakhakhny (1963, 1965)

Munsche 1964; Vishin et al. 1960, 1964). Some minor alkaloids were isolated by Atta-Ur-Rahman et al. (1985a,b; see Table 3).

Much experience in the use of *Nigella* species, plant extracts or isolated compounds or their mixtures, can be found in the area of traditional medicine, for example in Sri Lanka, India, Arabia, or Saudi Arabia. Namba et al. (1985) analyzed the use of these drugs in Sri Lanka. Among other things, extracts of *N. sativa* seeds (in 50% methanol) also have the effect of inhibiting viable cells of *Streptococcus mutans* (IC 10–30 μg/ml). Besides other drugs, Al-Yahya (1986) also qualitatively analyzed the *Nigella* species used in traditional medicine. Sayed (1980) discussed some plant uses in folk medicine. In this report, *N. sativa* is termed a drug plant which is already industrialized and known in Arabia as Habbett et Barakh. In Egypt this drug is used as a diuretic and carminative. The oil shows activity in treating asthma, respiratory congestion, and coughs. Sayed named nigellone as an active principle. The nigellone-containing fraction of the volatile oil of *N. sativa* is described as a useful agent in the treatment of bronchial asthma. Canonica et al. (1963) showed that nigellone is probably thymoquinone or a polymer of it. Going further, El-Dakhakhny (1963, 1965) isolated thymoquinone as a yellow compound from seeds of *Nigella sativa*. He also termed this compound the main pharmacologically active principle used in folk medicine in the orient. Hampel (1957) made reference to the use of *N. arvensis* seeds as a spice and in folk medicine as a diuretic. The unripe seeds are described as a source for homeopathic mixtures.

Vohora et al. (1973) studied the influence of different plant extracts on the fertility of animals. In their experiments *N. sativa* seeds did not show any anti-ovular activity in rabbits. Deshpande et al. (1974) described pest control through *N. sativa* extracts. Oleic acid- and linoleic acid-containing extracts have shown insecticidal activity against insects which attack stored grains.

In a German patent a pharmaceutical preparation containing five components, e.g. fluid extracts, are given:

1. extracts of *Carum carvi* (caraway);
2. extracts of *Foeniculum vulgare*;

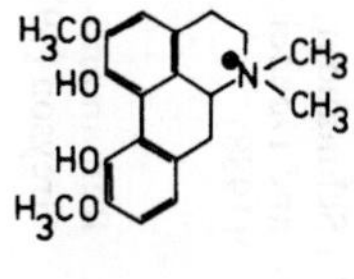

8

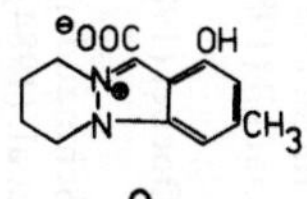

9

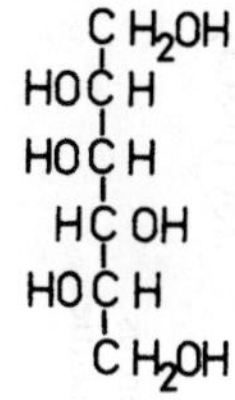

14

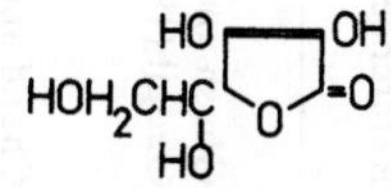

15

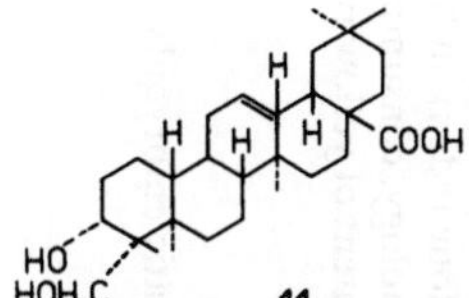

11

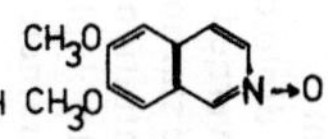

10

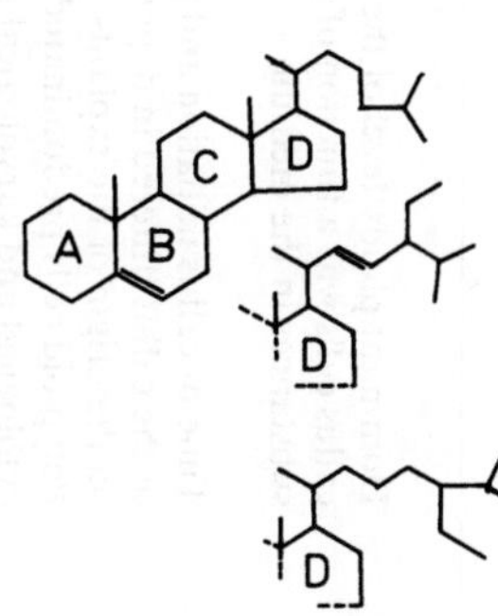

16

17

18

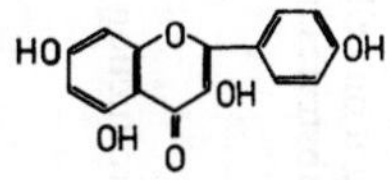

12

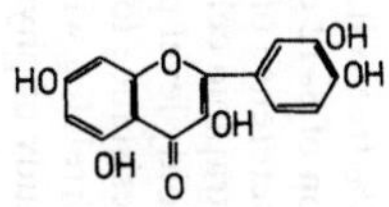

13

$R\,CH=CH(CH_2)_3COOH$ (trans) 1

$CH_3(CH_2)_nCH=CH(CH_2)_mCOOH$

n	m	
10	4	2 (cis)
9	5	3
7	7	4 (cis)

$CH_3(CH_2CH{=}CH)_3(CH_2)_7COOH$ 5

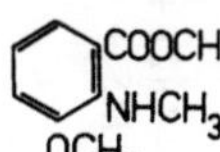

6

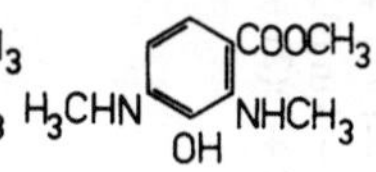

7

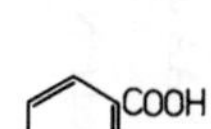

20

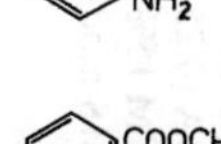

21

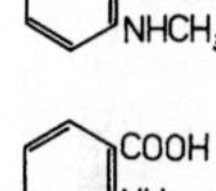

22

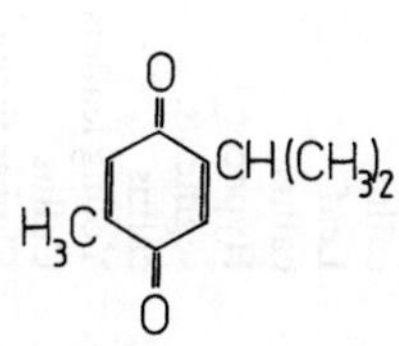

19

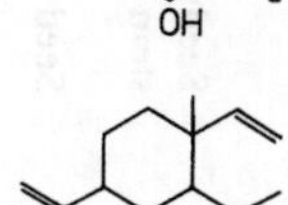

23

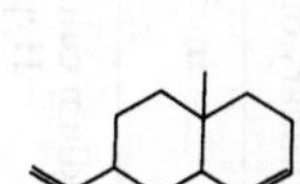

24

Fig. 3. Overview of plant products isolated from different *Nigella* species

Table 3. Summary of some in vitro studies on *Nigella*

Species	Inoculum	Medium (mg/l)	Culture conditions	Response	Reference
A. Solid medium culture					
N. sativa	Hypocotyl segments	MS + NAA,IBA,IAA, 2,4-D:(1); 2,4-D or IAA or NAA:(1) + kin (0.1); IBA (1) + BA (0.25); NAA or 2,4-D: (1) + CM 15%	Continuous darkness	Dominant ploidy level of the dividing callus cells was a function of growth substances in the medium	Ghosh and Gadgil (1979)
N. sativa	Seeds, leaf stem	MS modified ("N-1")	16 h light	Time of callus initiation and properties of the callus differed in dependence on the origin of the explants	Chand and Roy (1980a)
N. sativa	Seeds	MS + 2,4-D(2) + kin (0.3) + CM 10% + YE (1000)	16 h light	Polyploid cells predominated; numerous cytological and karyological abnormalities	
B. Suspension culture					
N. sativa	Stem callus	HI + NAA(0.05) + CM 15%	?	Frequency of diploid cells decreased after 42 months from 78% to 47%	Gupta (1972)
N. sativa	Leaf callus	MS + NAA (0.5) + CM 15%; W + CH (100) + inositol (50) + adenine sulphate (2)	Continuous darkness	91% free cell in suspension; formation of 8–12 embryoids/ml	Banerjee and Gupta (1975c)
N. sativa	Hypocotyl segments, callus	MS + NAA(2); MS + NAA (0.25) or MS + 2,4-D (0.25)	Continuous darkness	Preference of divisions in diploid and tetraploid cells largely governed by the ploidy level pattern of the starting inoculum	Ghosh and Gadgil (1980)
N. damascena	Young leaflets Callus	MS + 2,4-D(2) + kin (0.2); MS + 2,4-D (5–100) + kin (0.2)	12 h light	Resistance to high concentrations of 2,4-D correlated with decreased polyploidy	Doebel (1983)
N. damascena	Flower tissue Callus	MS + 2,4-D(2) + kin (0.2)	Airlift fermentator; 12 h light vs. continuous darkness	Study of physiological parameters (enzymes, growth, mitotic index), influence of light on cell morphology, consumption of substrates, and content of yellow pigments	Schmauder et al. (1984); Schmauder and Doebel (1988)
C. Differentiation and regeneration					
N. damascena	Pedicels + floral buds Callus Suspension culture	MS + 2,4-D(2) + CM 10%; MS + 2,4-D (2)	Continuous light	Production of somatic embryoids	Raman and Greyson (1974)

N. sativa	Root, stem, leaf segments Callus	W + NAA (0.5) + CM 15% W + NAA (0.5) W	8 h Light	Root callus: formation of roots; stem callus and leaf callus: formation of roots and shoots	Banerjee and Gupta (1975a)
N. sativa	Root cultures	MS + IAA(0.5) + CH (100); MS + IAA (0.5) + CH (500)	16 h Light	Formation of somatic embryoids from the distal ends of the cultivated roots cultivated roots	Banerjee and Gupta (1975b)
N. damascena	Germinating seeds, callus	MS + 2,4-D(10^{-5}M)	Continuous darkness – continuous light	Formation of filamentous embryoids from endosperm callus	Sethi and Rangaswamy (1976)
N. sativa	Leaf segments Callus	W + NAA(0.5) + CM 15%; MS + IAA (0.5) + CM 15%; MS + IAA (0.5) + CH (500)	16 h Light	Continuous production of somatic embryos	Banerjee and Gupta (1976)
N. sativa	Leaf segments Callus	MS + 2,4-D(2) + kin (0.1) + CM 10%; MS + IAA(0.5) + kin (1) + CM 10%; MS	16 h Light	Formation of shoot buds and plantlets	Chand and Roy (1981)
D. Protoplast culture					
N. sativa	Seed callus Suspension culture	MS + CH(1)	Continuous low illumination	Rhizogenesis from protoplasts	Iha and Roy (1979)
N. arvensis	Apex and leaves	VKM or NT: + BA(2.5 μM) + NAA (10 μM) or MS + BA (2.5 μM)	Continuous light	Shoots regenerated and plants by rooting of the shoots	Binding et al. (1981)
N. damascena	Leaf from shoot culture	VKM + BA(2.5 μM) + NAA (5 μM) + 2,4-D(0.5 μM)	Continuous light	Roots formed at callus	Binding et al. (1981)

Abbreviations:
Basal media: HI = Hildebrandt et al. (1946); MS = Murashige and Skoog (1962); NT = Nagata and Takebe (1971); VKM = Binding and Nehls (1977); W = White (1943).
Supplements: BA = 6-benzylaminopurine; CH = casein hydrolysate; CM = coconut milk; IAA = indole-3-acetic acid; IBA = 3-indolebutyric acid; kin = kinetin; NAA = α-naphthalene acetic acid; YE = yeast extract; 2,4-D = 2,4-dichlorophenoxyacetic acid.
Concentrations of supplements are given in mg/l, if not otherwise stated.

3. plant extracts containing terpenoid ketones, and/or terpenoid ethers, and/or terpenoid alcohols, and/or phenolic derivatives of terpenes;
4. ethanol;
5. water.

For preparation of the third component, *N. sativa* was used among other plants. The agents for extraction and the parts of the plants used for this patent cover a wide range. The described mixtures were tested for an indirect parasympathomimetic action and can be applied in the therapy of dystonic (atonic and/or spastic) and dysbacterial illness as well as flatulence in stomach or intestine (Gracza 1986).

Data for the cultivation of *Nigella* species in the field are rare. Duke (1982) named the following conditions for *N. sativa*:

Precipitation (decimeters)	mean	7.9
	min.	4.3
	max.	15.3
Temperature (°C)	mean	13.0
	min.	7.8
	max.	21.0
pH of the soil	mean	6.9
	min.	5.6
	max.	8.2

2 Callus Cultures and Regeneration of Plants

The most important works on in vitro cultivation of *Nigella* spp. are listed in Table 3.

2.1 Initiation

Nigella has proven to be an excellent subject for cell, tissue, and organ culture. Proliferating callus cultures may be obtained from different organs, preferably from tissue explants of young seedlings. Leaf segments are better than the other parts of the seedling for quick callusing (Chand and Roy 1980a). Another very suitable source for callus culture initiation is flower buds. The stamen filaments react especially easily with callusing under appropriate culture conditions. This is presumably due to the persistence of a diffuse meristem in the stamens up to an advanced developmental stage of the flower bud (Greyson and Tepfer 1966). For initiating and establishing callus cultures of *Nigella*, primarily Murashige and Skoog or White basal media were used. The callusing rate of the explants may be enhanced by supplementing the basal medium with a relatively high concentration of auxin (2–4 ppm 2,4-D or NAA).

2.2 Characteristics

The consistency of established callus cultures of *Nigella* ranges from friable to compact, and the degree of organization varies from undifferentiated parenchymatous cells to organized cell masses with organoid character. The explant source may influence the type of organization evolved.

Considerable attention has been paid to the nuclear character of *Nigella* tissue cultures. This is due to the excellent karyological properties of the genus *Nigella*. The two *Nigella* species commonly used for tissue culturing (*N. damascena*, *N. sativa*) have a low chromosome number (2n = 12) and relatively large chromosomes. Their karyotypes consist of five metacentric and one acrocentric chromosome. The chromosomes are characterized by a relatively small amount of heterochromatin, which is demonstrable by the C-banding technique. The C-bands are localized in the primary and secondary constrictions and in some intercalary regions of the chromosomes (Gill and Srivastava 1974; Klasterska and Natarajan 1975; Gilot-Delhalle et al. 1976, Fig. 4).

N. sativa and *N. damascena* have frequently been used for investigations on the induction of chromosomal aberrations. Root tip cells of these species are very suitable short-term test subjects for eventual mutagenic effects of pharmaceutics, agrochemicals, etc. (Moutschen et al. 1965, 1969, 1987; Degraeve et al. 1972; Bhaskar et al. 1983). Nevertheless, investigations concerning the genetic aspects of *Nigella* are very scarce (Toxopeus 1927; Raman and Greyson 1977, 1978; Datta and Biswas 1984). This is not because the object is unsuitable; on the contrary it is ideal for genetic studies because of its short life cycle, small plant size, self-compatibility, and excellent karyology. If more attention is paid to genetic basis, *Nigella* cultures might represent a very attractive model system for further research on the cytogenetics of plant tissue cultures.

The results obtained until now in studies on the nuclear cytology of *Nigella* callus cultures are concerned primarily with variability in chromosome number and chromosome structure. Some evidence suggests that media components, the type of the explant tissue, and the culture age affect the occurrence of chromosomal instabilities. Ghosh and Gadgil (1979) found a shift from diploidy to higher ploidy

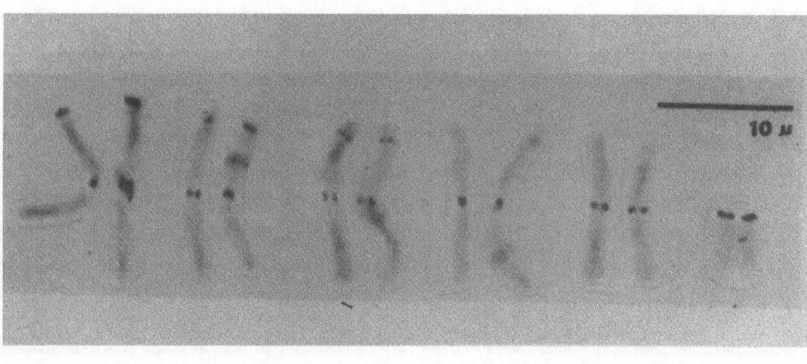

Fig. 4. Chromosome set of *N. damascena* with C-bands in the centromeric regions and in the secondary constrictions. (Klasterska and Natarajan 1975)

levels during subsequent passages of a callus culture of *N. sativa*. The level of ploidy dominant among the dividing cells of the culture was found to be dependent on the phytohormones used in the nutrient medium. The rate of the shift reached a maximum value when coconut milk was used in combination with 2,4-D. Chand and Roy (1980b) found that in the callus of *N. sativa* seed polyploid cells predominated from the first stages of the culture. In this case the frequency of diploid cells also decreased as the age of the culture increased. Numerous chromosome aberrations such as stickness, fragmentation, laggards, ring chromosomes, multipolarity, and unequal separation of chromosomes were found. Some abnormalities (nuclear budding, chromatin extrusion, and laggards) were interpreted as symptoms of the elimination of nuclear material from the high polyploid cells. In this way, new aneuploid cells may originate.

2.3 Differentiation and Regeneration

Cell and tissue cultures of *Nigella* generally have a high differentiation and regeneration potential. The differentiation of organs and plantlets from callus and suspension cultures may be induced regularly by changes in the phytohormone balance of the nutrient medium. Several authors have used *Nigella* cultures for studying the general influence of a variety of factors on the differentiation and regeneration ability.

The role of the nutrient mediums' phytohormone balance in organ differentiation on callus cultures of *N. sativa* has been investigated by Chand and Roy (1981). They showed that differentiation of roots and shoots occurred after removing auxin and kinetin from the medium (Fig. 5). Exposure to auxin for a few days before putting the callus into an auxin-free medium was, however, essential for subsequent organ differentiation. Plantlets could be obtained from the differentiated shoot buds. They were diploid, which was in contrast to the undifferentiated callus tissue, which showed a wide variation in chromosome number. Banerjee and Gupta (1975a) were able to demonstrate the persistence of

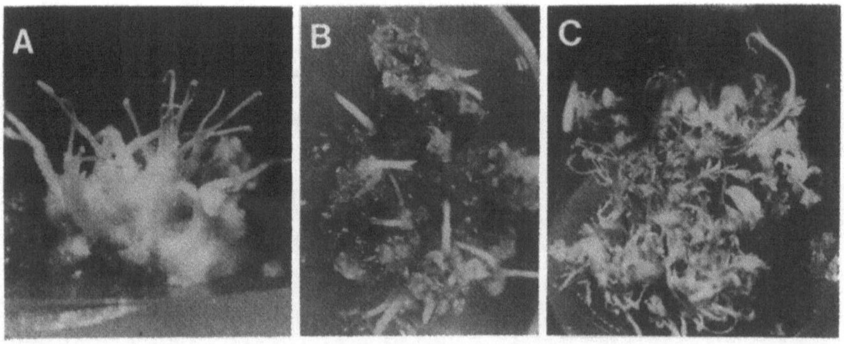

Fig. 5A-C. Organ differentiation in callus cultures of *N. sativa*. **A** Formation of shoot buds. **B** Formation of roots. **C** Formation of plantlets. (Chand and Roy 1981)

explant tissue-specific states in respect to the morphogenetic capacity. The authors investigated callus cultures isolated from different vegetative parts of *N. sativa* plants. The cultures obtained from roots, stems, or leaves formed roots in response to the withdrawl of auxin and CM from the medium. Shoots, however, were formed only in stem- and leaf-derived cultures. The reasons for this difference in the organogenic potential are not clear.

A number of investigations are concerned with the formation of somatic embryos from callus cultures of *Nigella*. Banerjee and Gupta (1975b) found CH necessary for the induction of embryogenesis in fresh callus of roots of *N. sativa*. The authors presume that a high nitrogen level caused by CH in the nutrient medium may alter the auxin balance and facilitate embryogenesis. The same authors found formation of proembryo tissue in leaf callus of *N. sativa* grown on MS medium supplemented with IAA and coconut milk. Omission of CM was necessary for the further development of the embryoids, which grew to maturation only on a medium enriched by organic nitrogen in the form of casein hydrolysate. Embryogenesis and differentiation were inhibited by the exogenous application of kinetin in the presence of auxin (Banerjee and Gupta 1976).

There are indications that explants of floral tissues are particularly prone to produce embryogenic cultures in *Nigella*. Raman and Greyson (1974) initiated callus formation from floral primordia of *N. damascena* on MS medium supplemented with CM and 2,4-D. During subculturing on this medium the callus mainly consisted of parenchymatous cells and tracheary elements. In response to the transfer in liquid MS medium devoid of CM, however, globular structures were organized which proved to be somatic embryoids. Sethi and Rangaswamy (1976) obtained abundant formations of early, filamentous embryoid stages in a *N. damascena* endosperm culture. In the aging endosperm tissue some embryoids differentiated up to the globular stage.

3 Cell Suspension Cultures of *Nigella*

3.1 Establishment, Characterization, and Differentiation

Suspension cultures of *Nigella* may be obtained from freshly initiated callus tissues as well as from established callus cultures. Compact callus tissue should eventually be subcultured on a medium with a high concentration of auxin in order to obtain a friable callus. Friable callus of *Nigella* will, in almost all cases, grow as a well-proliferating suspension culture in the MS basal medium supplemented with 2 ppm of 2,4-D and 0.2 ppm kinetin. The subculturing of *Nigella* suspension cultures has been completed at intervals varying from 3 1/2 days (Doebel 1983) to 30 days (Banerjee and Gupta 1975c).

The structural features of suspension cultures of *Nigella* have been found to be somewhat different. According to our experience, most cultures consist of loose aggregates of 3–50 cells (Fig. 6) which are interspersed with a few free single or double cells. Large oblong parenchymatous cells occur most frequently. A minority of small meristematic cells is also present. In contrast, Banerjee and Gupta (1975c)

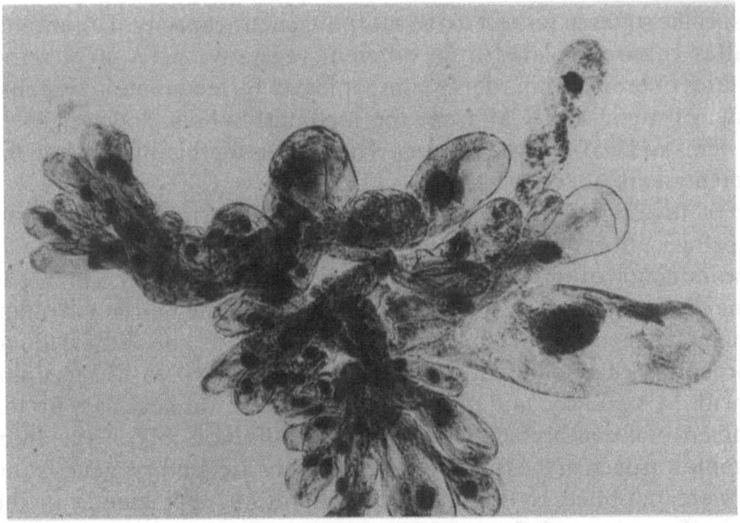

Fig. 6. Typical cluster of parenchymatous cells, representing the main component of a *N. damascena* suspension culture. (Doebel unpubl.)

found a remarkably high number (more than 90%) of free single cells in a suspension culture of *N. sativa* grown in White's medium supplemented with CH, inositol, and adenine sulfate.

In addition to the normal cell plate formation, Banerjee and Gupta (1975c) also found some cell divisions by budding and internal septation in the above-mentioned suspension culture of *N. sativa*. The ploidy level variation in the suspension cultures is similar to that of the callus cultures. A suspension culture of *N. sativa* investigated by Gupta (1972) initially showed 78% diploid cells and 4% aneuploids. After 42 months of culture, the frequency of diploid cells had decreased to 47%, while that of aneuploids had increased to 23%. Numerous abnormalities of the mitosis were also detected. Endomitotic doubling of the number of chromatids occurred in 1.5%–2% of the cells. Ghosh and Gadgil (1980) found no comparable shifts in the ploidy level of a suspension culture of the same *Nigella* species. Instead, the initial diploid cell population of the explant remained dominant. In the case of a dominant tetraploid inoculum, however, the tetraploid cells remained dominant. The preference of division in diploid or tetraploid cells in this culture was largely governed by the ploidy level pattern of the starting inoculum.

In a suspension culture of *N. damascena* obtained from young leaflets, tetraploid and octaploid mitoses predominated. Furthermore, a sizable number of C-mitoses and decondensing chromosomes were found. This indicates an arrest of mitoses followed by the formation of restitution nuclei. Restitution nucleus formation as a consequence of spindle failure or chromosome lagging at anaphase may be an essential reason for polyploidization in this culture (Doebel 1983). As another, obviously infrequent mechanism of polyploidization in *Nigella* cultures, endoreduplication as indicated by metaphases with diplochromosomes, has been found in suspension cultures of *N. sativa* (Gupta 1972) and *N. damascena* (Doebel unpubl.).

In general, the results of the karyological investigation of *Nigella* suspension cultures show that very different factors may contribute to the chromosomal variability in plant tissue cultures. Moreover, they demonstrate some possibilities for influencing this variability (change of subculture intervals, variation of phytohormone concentration, and use of explant tissue with defined ploidy level).

The differentiation of organs and plantlets in suspension cultures of *Nigella* may be induced in the same manner, and with the same results, as in callus cultures. To quote one example: Banerjee and Gupta obtained embryo formation in a suspension culture (1975c), as in a callus culture (1976) of *N. sativa*, through the withdrawl of auxin and the addition of CH.

Some authors have proved the regeneration ability of protoplasts obtained from *Nigella* and reported on the easy accessibility of *Nigella* for this technique as well (Iha and Roy 1979; Binding et al. 1981). Protoplasts isolated from a suspension culture of *N. sativa* regenerated cell walls and started to proliferate within 48 h. Subsequently, friable callus masses were formed. Approximately 73% of these cells were diploid, 25% hyperdiploid, and 2% hypodiploid. On a modified MS medium lacking phytohormones but containing CH, the formation of a high number of roots could be induced (Fig. 7). These roots were invariably diploid. The authors suppose that the roots were regenerated only from the diploid fraction of the mixoploid callus masses (Iha and Roy 1979). This is in agreement with similar observations in many other regenerating plant tissue cultures. Thus, the process of regeneration may act as a screen for gross chromosomal changes (Torrey 1967; Orton 1980; Bayliss 1980). This may also be true in the case of an embryogenic long-term suspension culture of *N. damascena* which showed a high numerical and structural chromosome variability. Somatic embryoids in the globular stage (Fig. 8), which appeared spontaneously in this culture, showed a considerably smaller karyotype heterogeneity than the parenchymatous tissue (Fig. 9; Doebel unpubl.). Therefore, in this culture, selection seems to act against higher ploidy levels and higher karyotype heterogeneity which may be incompatible even with the relatively low degree of organization realized in the globular embryoids.

3.2 Metabolic Activities

Metabolic activities of *Nigella* have been studied in vitro. There has also been interest in the status of cultures in relation to illumination and the status of phytohormones, above all the IAA content. This topic was mainly studied by Indian workers (Banerjee and Gupta, 1975c, Chattopadhyay et al. 1980, Mukherjee et al. 1981), who found close correlation between the IAA content, exposure, and growth rate.

Lehmann et al. (1983a,b) studied the metabolism of abscisic acid in suspension cultures of *Nigella damascena*. They isolated a new metabolite of abscisic acid, (S)-3-methyl-5-(l′-hydroxy-2′-hydroxymethyl-6′dimethyl-4′-oxo-cyclohex-2′-phenyl)-penta-2Z,4E-dienoic acid.

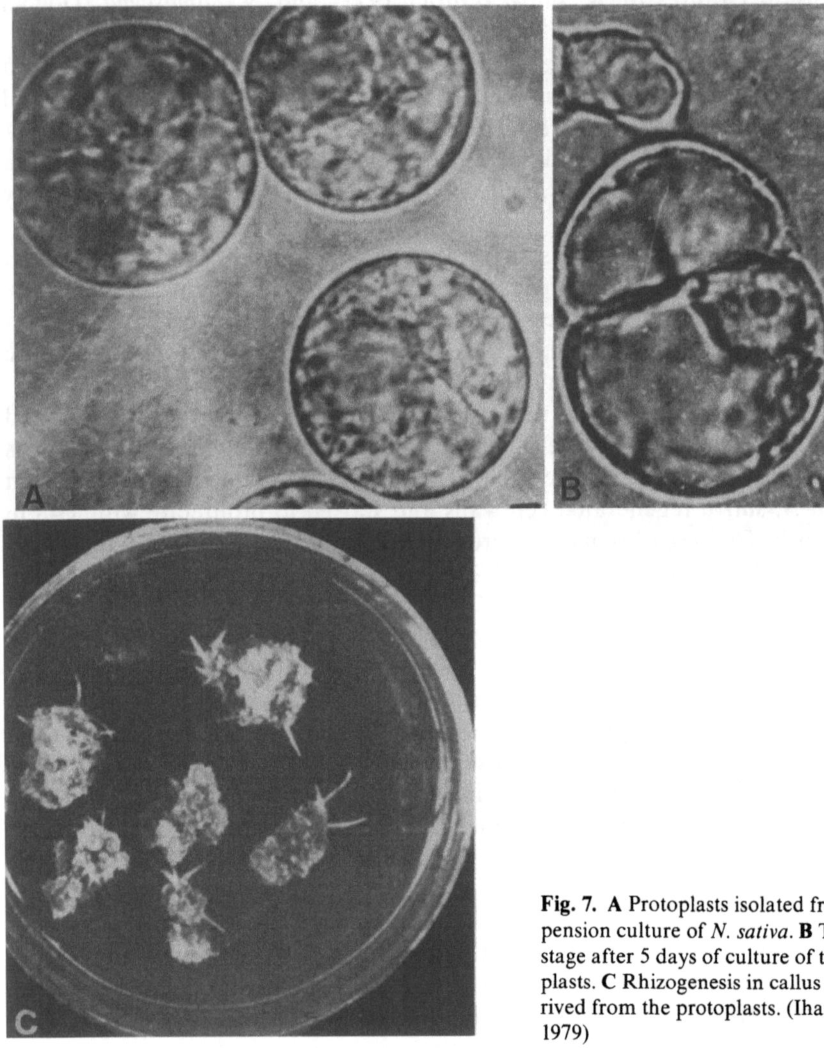

Fig. 7. A Protoplasts isolated from a suspension culture of *N. sativa*. **B** Two-celled stage after 5 days of culture of the protoplasts. **C** Rhizogenesis in callus tissues derived from the protoplasts. (Iha and Roy 1979)

Baumert et al. (1983) used suspension cultures of *Nigella damascena* to study a special metabolic reaction, the N-methylation of anthranilic acid, important for the biosynthesis of acridone alkaloids in *Ruta graveolens* cell cultures. This methylation is the starting reaction of alkaloid formation and the authors tested the specificity of the enzyme of *Ruta graveolens*. Since the formation of damascenine in *Nigella* plants also starts from anthranilic acid, and methylation steps are necessary in the formation of this compound, the suspension of *Nigella damascena* was an important control of this enzyme specificity. The cultures of *Nigella damascena* used were not able to form damascenine and an unspecific N-methyltransferase was not detectable in cell-free extracts.

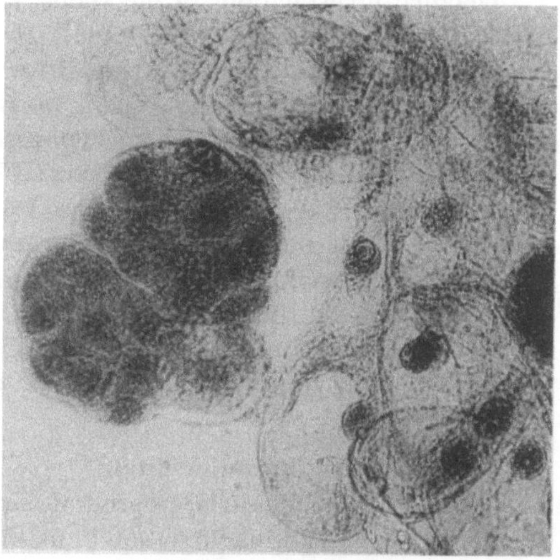

Fig. 8. Somatic embryoid in the globular state developing spontaneously in a suspension culture of *N. damascena*. (Doebel unpubl.)

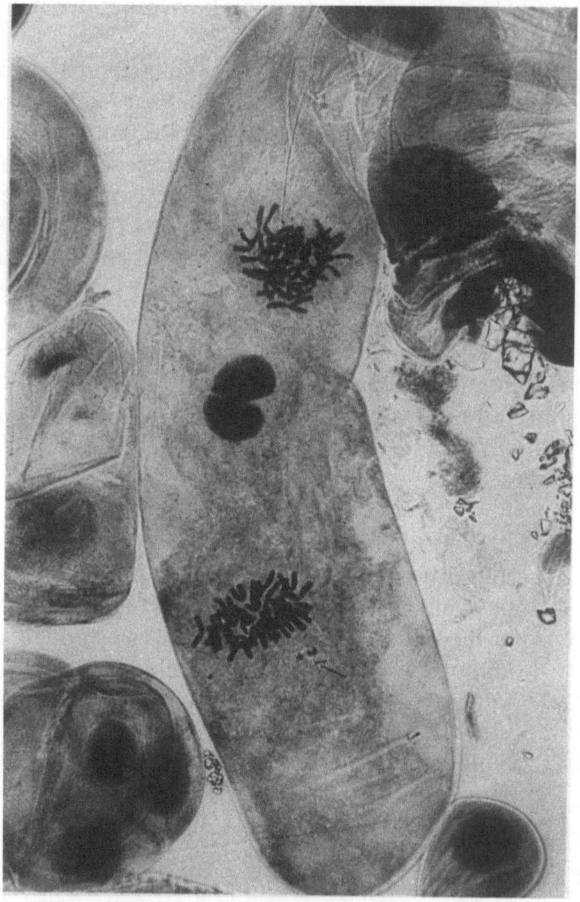

Fig. 9. Karyological abnormal parenchymatous cell of an embryogenic suspension culture of *N. damascena* showing multinucleate character, polyploid metaphase plates, and chromosome fragments (*arrow*). (Doebel unpubl.)

Detailed investigations of metabolic activities in cell suspension cultures of *Nigella* species were done mostly with *N. damascena*. These were aimed mostly at formation of damascenine as a model for simple alkaloid-like structures or at finding possible intermediates of this biosynthesis. There are no data for such damascenine formation in current literature. A trace of damascenine or a compound with similar properties was detectable on a dc chromatogram in only one cultured strain. However, this formation was not stable (Schmauder and Doebel, unpubl.). Because the damscenine biosynthesis is limited to ripened seeds of *N. damascena* (Munsche 1966; Mothes 1966), a complete de novo synthesis in undifferentiated or differentiated cell cultures cannot be expected (Boehm 1988). On the one hand, Munsche (1966) and Mothes (1966) have hypothesized that in damascenine biosynthesis the methylation of the precursor anthranilic acid (*20*) results through:

a) methylation of the amino group,
b) methylation of the carboxyl group to form *21*,
c) hydroxylation and methylation or methoxylation to damascenine (*6*).

On the other hand, Boehm (1988) postulated that 3-hydroxyanthranilic acid (*22*) could be an intermediate in this biosynthetic pathway. The details of this process are still unclear.

The compound *21* as a possible intermediate of the first hypothetical pathway was identified in suspension cultures of *N. damascena* by Schmauder et al. (1984). These findings are in contrast to Boehm's postulation that a transformation of anthranilic acid in damascenine is impossible in cell cultures of *N. damascena* because on the one hand, necessary enzyme(s) are absent, and on the other hand, there is a deficiency of anthranilic acid (Boehm 1988).

Many experiments were done by Schmauder et al. (1984, 1991), Boehm (1988), and Schmauder and Doebel (unpubl.) to study the relations between the biosynthesis of the main precursor of damascenine formation (anthranilic acid, *20*), the influence of some antimetabolites of tryptophan, the number of chromosomes, and the differentiation pattern of suspension cultures of *N. damascena* and damascenine biosynthesis. None of these experiments has shown the path to stable damascenine formation in suspension cultures. The results of these investigation are summarized below:

1. In experiments for studying the role of desirable genotypes, Boehm established four inbreeding lines from seedlings of plants with unusual morphological or physiological characteristics. For better solubility of the lipophilic damascenine in the culture liquid paraffin oil was added to the cultures. Traces or larger amounts of damascenine were not detectable at any cultivation stage. Differences in the level of auxins in the nutrient broth did not affect minimal biosynthesis (Boehm 1988).
2. Differences in morphogenesis, for example embryogenesis, which occurs spontaneously, were often found parallel to a yellow pigmentation (Schmauder and Doebel 1988). These pigments are similar to the class of quinones, for example thymoquinone or its polymers, which are identified as the active principle(s) of the essential oil of *Nigella* species. These biosynthetically active

cultures, however, were also unable to form damascenine (Schmauder and Doebel 1988 and unpubl.; Boehm 1988).

3. Experiments with substances which are able to influence the genetic material, for example 5-azacytidine, showed a negative result in relation to damascenine formation. The same results were obtained in works for the induction of haploids, which were more useful in mutation experiments (Boehm 1988).
4. A high intensity of research was aimed at the formation of anthranilic acid as the main precursor of damascenine biosynthesis. This acid is an intermediate of the tryptophan branch in the system of biosynthesis of aromatic amino acids. Most of the important enzymes of this pathway were detectable in cell-free extracts obtained from suspension cultures of *N. damascena* (e.g., DAHP synthase, shikimic acid dehydrogenase, shikimate kinase, chorismate mutase, prephenate dehydratase, prephenate dehydrogenase, anthranilate synthase, anthranilate-phosphoribosyl transferase, tryptophan synthase; Schmauder unpubl.). The most important enzymes for regulation of the biosynthesis of aromatic amino acids are characterized briefly in Table 4 (Fig. 10). The "bottle-necked" enzyme in this complex system is also the anthranilate synthase, which shows a strong inhibition through tryptophan in a μmolar range in this plant. Nevertheless, possible ways for attaining higher cellular amounts of anthranilic acid were studied. The long-range adaptation of cell cultures to 5-methyltryptophan (MT) or 5-fluorotryptophan (FT) showed:

Table 4. Some properties of key enzymes of aromatic amino acid biosynthesis in suspension cultures of *N. damascena*. (Schmauder et al. 1991)

Enzyme	K_m-Value[a]	Regulatory properties[b]
DAHP synthase (EC 4.1.2.15)	E4P: 0.4–0.65 PEP: 0.45	Inhibition by trp 0–82% tyr 0–9% phe 0–37%
chorismate mutase (EC 5.4.99.5)	CA: 0.8–1.15	Inhibition by phe 9–100% tyr 25–90% activation by trp 10–66%
anthranilate synthase (glu-utilising, EC 4.1.3.27)	CA: 0.2–0.25 glu: 1.25–1.45 Mg++: 5–6.5	K_i (trp) between 1–5 μmol/l
tryptophan synthase (indole-utilising, EC 4.2.1.20)	PLP: 0.02 ind: 0.2–0.35 ser: 1.3–1.5	No regulatory influences

[a] DAHP:3-deoxy-D-arabino-heptulosonic acid 7-phosphate, E4P: erythrose-4-phosphate, PEP: phosphoenolpyruvate, CA: chorismic acid, glu: glutamine, trp: tryptophane, tyr: tyrosine, phe: phenylalanine, ser: serine, ind: indole. Concentrations are given in mmol/l.

[b] Given is the range of inhibition/activation in the in vitro test with cell-free extracts from cultures of different lines of *Nigella damascena* (controls and antimetabolite treated suspensions). The large group of effects found shows the difficulty in working with this heterogenous (and not haploid) material. A scheme of the complex regulatory system is given for one of the used lines in Fig. 10. (If not otherwise noted the concentrations of effectors are 1 mmol/l).

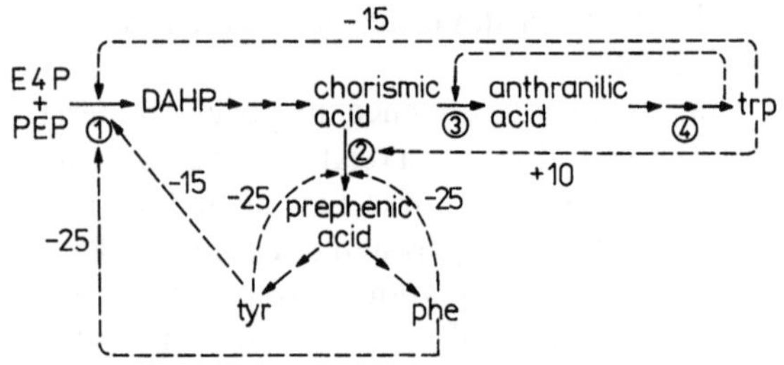

Fig. 10. Common regulation pattern of the key enzymes of biosynthesis of aromatic amino acids from 5-day-old *N. damascena* cells cultivated in an airlift fermentor in the dark. The values are given in % of inhibition or activation in comparison with untreated controls. The concentration used in vitro was 1 mmol/l. For anthranilate synthase see Table 4. The activities in the control were: *1* DAHP synthase, 6.12 pkat/mg protein; *2* chorismate mutase, 822 pkat/mg protein; *3* anthranilate synthase, 2.7 pkat/mg protein; *4* tryptophan synthase, 5.2 pkat/mg protein. (Schmauder et al. 1984)

- a dramatic decrease of the tryptophan content of the cells (from 5.3 μg trp/mg dry wt. of the control to 1.25 or 1.8 in cells treated with 10 or 20 ppm FT, respectively),
- an increase of the cellular chorismic acid level through the use of increasing FT concentrations (up to 60 ppm FT),
- a partial decrease in the activity of selected enzymes of primary metabolism independent of the biosynthesis of aromatic amino acids and in close connection with growth reduction,
- a much lower level of activity of anthranilate synthase and a lower degree of DAHP synthase, both combined with greater or lesser changes in some parameters of feedback inhibition,
- uncharacteristic changes in the content of free amino acids in the cells,
- no cases of significantly higher activity of anthranilate synthase in cell-free extracts (Schmauder et al. 1991).

Activity of key enzymes of aromatic amino acid biosynthesis measured in vitro is given in Table 5. The summarized results show, in the case of line 65, that activities of chorismate mutase are distinctly higher after cultivation with 5-fluorotryptophan than in the control which was cultivated without this antimetabolite. The content of anthranilate synthase is decreased after the influence of fluorotryptophane in amounts higher than 10 ppm.

5. Feeding experiments with anthranilic acid were unsuccessful with respect to damascenine synthesis. This cytotoxic compound is difficult to apply in such feedings. The anthranilic acid was transported quickly into the cells and converted into other compounds, obvious mainly in glycosides (Boehm 1988). With cell-free extracts of *N. damascena* cell cultures a conversion of anthranilic acid into another fluorescent compound was detectable. A nearer characterization was impossible, but it was not damascenine or another intermediate in the latter's hypothetical pathway (Schmauder unpublished).

Table 5. Influence of cultivation with and without antimetabolites of tryptophan on activities of four key enzymes of biosynthesis of aromatic amino acids. The cultivation took place in the following way: Precultivation in a suspension culture – plating on agar – suspension culture – preservation (with antimetabolite in the culture medium) – suspension culture for testing of enzyme activities. The enzyme activity is given in pkat/mg protein

Line	Age (d)	Supplements in the culture medium (ppm)	DAHP synthase	Chorismate mutase	Anthranilate synthase	trp synthase
Control	2	–/–/–/–	0.001	0.51	0.0004	0.015
43	2	20MT/40MT/10FT/5FT	0.002	0.57	0.005	0.024
152/3	2	5FT/10FT/–/–	0.0023	0.39	0.0004	0.015
65	4	20MT/40MT/–/–, suspension for testing without FT	0.003	0.19	0.0006	0.011
65	4	As above, suspension with 10 ppm FT	0.004	0.52	0.0007	0.007
65	4	As above, suspension with 20 ppm FT	0.007	0.46	0.0003	0.01
65	4	As above, suspension with 80 ppm FT	0.005	0.41	0.0002	0.015

Abbreviations: MT = 5-methyltryptophan, FT = 5-fluorotryptophan. (Schmauder et al. 1991).

3.3 Cultivation in Fermentors

Schmauder et al. (1984) and Schmauder and Doebel (1988) investigated the behavior of *N. damascena* suspension cultures in fermentors. They rebuilt stirred tank reactors into airlift fermentors with low shearing stress. The working volumes were in the range between 3.5 and 9 l. The different strains used in these experiments were grown in the fermentors under illumination and also as a control in the dark. Both kinds of cultivated cells, an almost undifferentiated line, and a strain which tends to differentiation (meristemoids as a first stage of embryoid formation) showed sensitivity to illumination. In the first case, differences in growth and rate of mitosis were observed. In the illuminated fermentor, the mitosis was strongly reduced (Table 6) and the growth was stopped on the 7th day of cultivation. Parallel to this phenomenon, the activity of three key enzymes of biosynthesis of aromatic amino acids, DAHP synthase, chorismate mutase, and anthranilate synthase, is decreased (Fig. 11).

In the second example of fermentor cultivation, a yellow pigment-forming strain was used. In these experiments, besides the degree of differentiation, the growth, pH of culture liquid and the "specific pigmentation" (OD_{365}/mg dry wt.) were followed up (Fig. 12). The aim of the supplementary measuring was to judge the utilization of nutrients (Fig. 13). In all these cases the cultures were also inhibited by exposure to light and on the 10th day of cultivation under illumination it was found that all activity stopped. The yellow pigment was thymoquinone-like and showed a spectrum similar to extracts from plant parts grown in the dark, for example testa or root tips (Fig. 14). These yellow pigment(s) were only synthesized de novo in the dark.

Table 6. Mitosis rates of cultures of *N. damascena* grown in airlift fermentors in the dark and under illumination. (12h dark/light cycle 250–300 lx; Schmauder et al. 1984)

Age of culture (d)	Cells in mitosis (%)	
	Dark culture	Illuminated fermentor
0	2.92	2.92
2	4.86	0.70
4	6.27	0.86
7	6.60	1.54
9	5.51	1.43
11	4.72	1.20
13	4.36	0.92
15	3.98	0.72

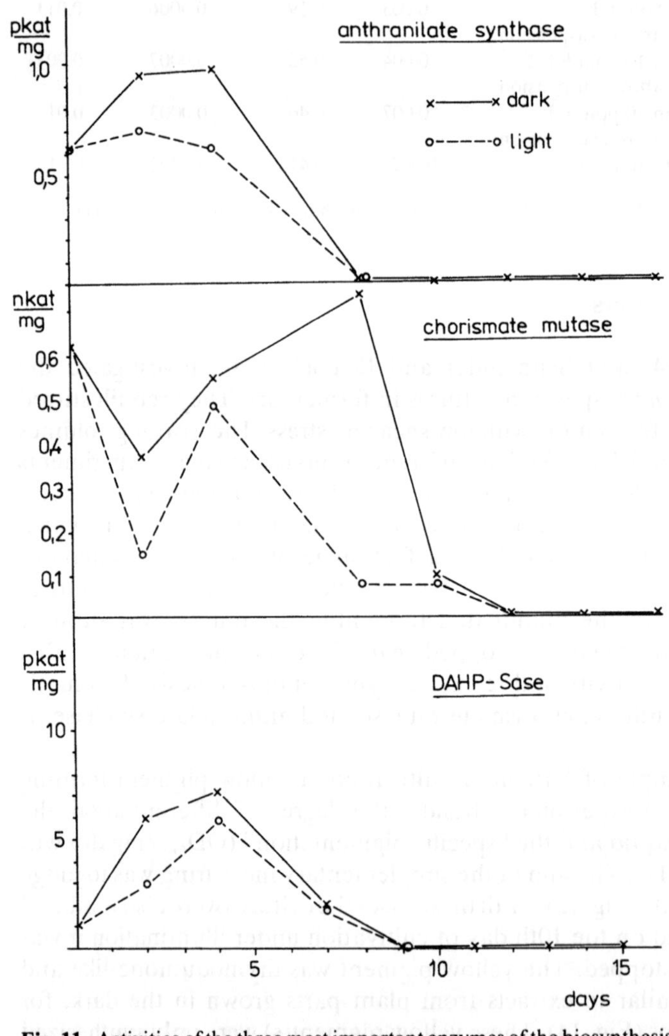

Fig. 11. Activities of the three most important enzymes of the biosynthesis of aromatic amino acids from *N. damascena* cells cultivated in airlift fermentors in relation to the age of culture. ×——× dark cultures, ○----○ illuminated cultures. (Schmauder et al. 1984)

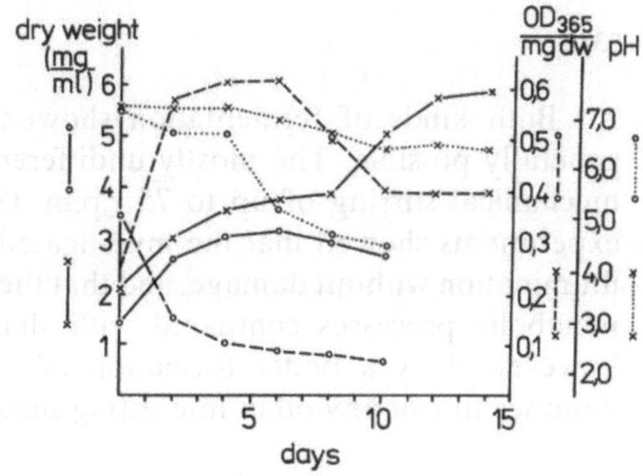

Fig. 12. Growth, pH values, and specific pigmentation (OD_{365}/mg dry wt.) of *N. damascena* fermentor cultures. ○----○ OD_{365}/mg dry wt., illuminated cultures; ×----× –″–, cultivation in the dark; ○—○ dry weight (mg/ml), illuminated cultures; ×—× –″–, cultivation in the dark; ○····○ pH, illuminated cultures; ×····× pH, cultivation in the dark. (Schmauder and Doebel 1988)

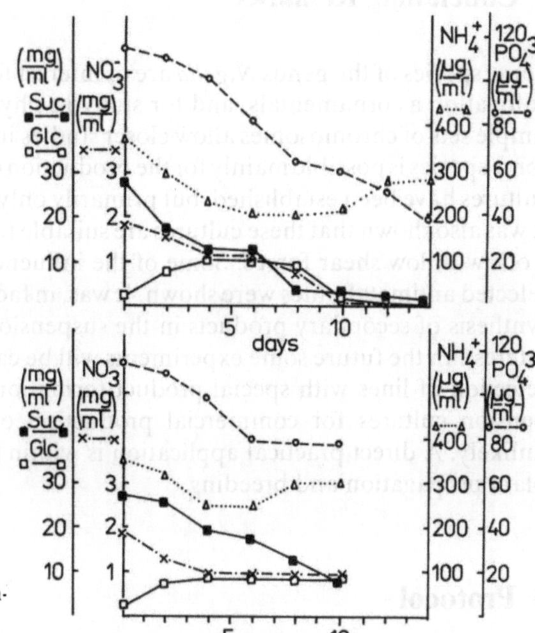

Fig. 13. Nutrient consumption in fermentor cultures of *N. damascena*. *Upper part* dark cultures; *below* illuminated cultures. (Schmauder and Doebel 1988)

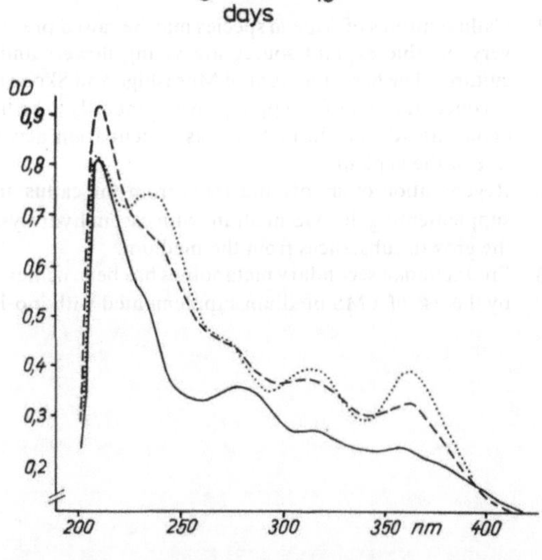

Fig. 14. Absorption curves of the pigment mixtures isolated from yellow fermentor cultures (——), the testa (····) or root tips (----) of *N. damascena* plants. (Schmauder and Doebel 1988)

Both kinds of fermentation showed that the cultivation in fermentors is generally possible. The mostly undifferentiated cells also tolerated a maximum mechanical stirring of up to 75 r.p.m. (Schmauder unpubl.). In general, these experiments showed that the investigated *N. damascena* cultures do not tolerate illumination without damage, and that the higher oxygen level also activates some metabolic processes compared with shaken cultures. This activation did not, however, show a better formation of secondary products or the synthesis of damascenine or any other interesting substances.

4 Concluding Remarks

Some species of the genus *Nigella* are of interest for use in traditional medicine, for cultivation as ornamentals, and for study in physiology and genetics because the simple sets of chromosomes allow closer studies in this range. In vitro cultivation of some species is possible mainly for the production of regenerated plants. Suspension cultures have been established, but primarily only with *N. damascena* and *N. sativa*. It was also shown that these cultures are suitable for cultivation in fermentors which work with low shear forces. Some of the influences of light, phytohormones, and selected antimetabolites were shown. It was, in fact, not possible to induce a specific synthesis of secondary products in the suspension cultures which are of practical interest. In the future some experiments will be carried out for the isolation and/or selection of lines with special product formation. A practical use of *Nigella* suspension cultures for commercial production of such compounds is, however, unlikely. A direct practical application is within sight for the in vitro methods for plant propagation and breeding.

5 Protocol

1. Callus cultures of *Nigella* species may be raised preferably from tissues of young seedlings. Another very suitable explant source are young flowers and flower buds, if obtained from axenic shoot cultures. The basic medium of Murashige and Skoog (1962) supplemented with 2,4-D, IAA or NAA in concentrations of 2–4 ppm proved generally to be the most suitable for callus induction. Dim light or darkness during the first 3 weeks of incubation, according to our experience, increased the callusing rate of the explants.
2. Regeneration of shoots and roots from the callus and suspension cultures has been achieved by supplementing the MS medium with casein hydrolysate during some passages, and later removing the growth substances from the medium.
3. Production of secondary metabolites has been achieved in suspension cultures of *Nigella damascena* by the use of a MS medium supplemented with inositol, casein hydrolysate, 2,4-D, and kinetin.

References

Al-Yahya MA (1986) Phytochemical studies of the plants used in traditional medicine of Saudi Arabia. Fitoterapia 57:179–182

Ansari AA, Osman SM, Subbaram MR (1975) Component acids of minor seeds oils. I. J Oil Technol Assoc 7:26–27

Atta-Ur-Rahman, Malik S, He C, Clardy J (1985a) Isolation and structure determination of nigellicine, a novel alkaloid from the seeds of *Nigella sativa*. Tetrahedron Lett 26:2759–2762

Atta-Ur-Rahman, Malik S, Ahmad S, Chaudhary I, Habib-Ur-Rahman (1985b) Nigellimine-N-oxide – a new isoquinoline alkaloid from the seeds of *Nigella sativa*. Heterocycles 23:953–955

Babayan VK, Kootungal D, Halaby GA (1978) Proximate analysis, fatty acid and amino acid composition of *Nigella sativa* L. seeds. J Food Technol 43:1314–1315

Banerjee S, Gupta S (1975a) Morphogenesis in tissue cultures of different organs of *Nigella sativa*. Physiol Plant 33:185–187

Banerjee S, Gupta S (1975b) Embryoid and plantlet formation from stock cultures of *Nigella* tissues. Physiol Plant 34:243–245

Banerjee S, Gupta S (1975c) Suspension culture of *Nigella sativa*. Experientia 31:793–795

Banerjee S, Gupta S (1976) Embryogenesis and differentiation in *Nigella sativa* leaf callus in vitro. Physiol Plant 38:115–120

Baumert A, Hieke M, Groeger D (1983) N-Methylation of anthranilic acid to N-methylanthranilic acid by cell-free extracts of *Ruta graveolens* tissue culture. Planta Med 48:258–262

Bayliss MW (1980) Chromosomal variation in plant tissues in culture. Int Rev Cytol (Suppl) 11A:113–144

Bhaskar KV, Bahadur RB (1983) Effect of semithion on the dividing root tip cells of *Nigella sativa* L. Indian J Bot 6:103–107

Binding H, Nehls R, Koch R, Finger J, Mordhorst G (1981) Comparative studies on protoplast regeneration in herbaceous species of the dicotyledoneae class. Z Pflanzenphysiol 101:119–130

Boehm H (1988) Qualitative Aspekte der Bildung von Sekundärstoffen durch pflanzliche Zellkulturen. In: Boehm H (ed) Pflanzliche In-vitro-Systeme für Züchtung und Stoffproduktion. Biol Ges DDR/AG Pflanzl Zell Gewebekult, Halle, pp 71–83

Bose B, Ghosh O, Singh RP (1981) Study on the chemical constituents of seeds of *Nigella sativa* (Kalajira) – a preliminary note. J Inst Chem (India) 53:273

Brown RH, Boulter D (1973) Amino acid sequence of cytochrome c from *Nigella damascena* (Love-in-a-mist). Biochem J 133:251–254

Canonica L, Jommi G, Scolastico C, Bonati A (1963) Sul principio farmacologicamente attivo della *Nigella sativa* L. Gazz Chim Ital 93:1404–1407

Chand S, Roy SC (1980a) Study of callus tissues from different parts of *Nigella sativa* (Ranunculaceae). Experientia 36:305–306

Chand S, Roy SC (1980b) Cytological abnormalities during culture of *Nigella sativa*. Protoplasma 104:353–357

Chand S, Roy C (1981) Induction of organogenesis in callus cultures of *Nigella sativa*. Ann Bot (London) 48:1–4

Chattopadhyay S, Bandyopadhyay MN, Mukherjee BB (1980) Growth and IAA balance in *Nigella sativa* L. callus in vitro influenced by light and culture media. Indian J Exp Biol 18:93–95

Datta AK, Biswas AK (1984) Cytomixis and a trisomic in *Nigella sativa* L. Cytologia 49:437–445

Datta AK, Das JL, Biswas AK (1987) Electrophoretic characterization and evaluation of proteins in control and mutant lines of *Nigella sativa* L. Cytologia 52:317–322

Davis PH, Cullen J, Coode MJ (1965) Flora of Turkey, vol 1. Univ Press, Edinburgh

Degraeve N, Gilot-Delhalle J (1972) Cytological effects of vinblastine in plants. Experientia 28:581–582

Deshpande RS, Adlukary PR, Tipnis HP (1974) Stored grain pest control agents from *Nigella sativa* and *Pogostemon heyneanus*. Bull Grain Technol 12:232–234

Doebel P (1983) Verringerung des Ploidiegrades und der Restitutionskernbildung in einer gegen hohe 2,4-D-Konzentrationen resistenten Suspensionskultur von *Nigella damascena*. Biol Zentralbl 102:431–444

Döpke W, Fritsch G (1970) Der Alkaloidgehalt von *Nigella damascena* L. Pharmazie 25:69–70

Duke JA (1982) Ecosystematic data on medicinal plants. In: Atal CK, Kapur BM (eds) Cultivation and utilization of medicinal plants. Reg Res Lab, Jammu-Tawi, India, pp 13–23
El-Dakhakhny M (1963) Studies on the chemical constitution of Egyptian *Nigella sativa* L. seeds. II. The essential oil. Planta Med 11:465–470
El-Dakhakhny M (1965) Studies on the Egyptian *Nigella sativa* L. Arzneimittelforschung 15:1227–1229
Frencel I (1965) Diss Pharm Pol 17:577–584
Gad AM, El-Dakhakhny M, Hassan MM (1963) Studies on the chemical constitution of Egyptian *Nigella sativa* L. oil. Planta Med 11:134–138
Ghosh A, Gadgil VN (1979) Shift in ploidy level of callus tissue: a function of growth substances. Indian J Exp Biol 17:562–564
Ghosh A, Gadgil VN (1980) Selection pressure theory and predominance of diploidy in suspension culture. Indian J Exp Biol 18:958–961
Gill BS, Srivastava PK (1974) A simple technique for demonstrating heterochromatin in *Nigella*. Experientia 30:1483
Gilot-Delhalle J, Degraeve N, Moutschen J (1976) Cytotaxonomic investigation of the genus *Nigella* (Helleboreae) with C-banding techniques. Caryologia 29:139–154
Gmoshinskii IV, Yampol'Skaya GP, Rotanova TV, Ginodman LM (1982) Topographical study of the active center of lipase from *Nigella damascena* L. (Damask love-in-a-mist). Bioorg Khim 8:822–829
Gorodetzkii IP, Chernov NE, Korchagina LN, Rudyuk VF, Chernobai VT (1976) Study of the process of sublimation drying of lipase from *Nigella damascena* L. seeds. Farm Zh (Kiev) 4:55–57
Gracza L (1986) Pharmazeutische Zubereitung. Patent Ger Offen 3 511 862, Int Cl A 61 K 35/78, Appl 1.4.1985, Publ 9.10.1986
Greyson RI, Tepfer SS (1966) An analysis of stamen filament growth of *Nigella hispanica*. Am J Bot 53:485–490
Gupta S (1972) Tissue culture of *Nigella sativa*. I. The behaviour of nucleus. Experientia 28:441–443
Hampel G (1957) Über die anatomischen Merkmale von Getreide-Unkrautsamen. II. Mitt. Acker-Schwarzkümmel und Acker-Wachtelweizen. Z Lebensmittelunters forsch 105:373–378
Hegnauer R (1973) Chemotaxonomie der Pflanzen, vol 6. Birkhäuser, Basel, pp 14, 19, 28, 33, 43, 44, 274, 594
Higginson D (1981) Lipases as natural high-molecular-weight surfactants. I. Isolation and properties of lipase from *Nigella damascena* L. Depos Doc 1981: VINITI, 575–582, 828–832
Hildebrandt AC, Riker AJ, Duggar BM (1946) The influence of the composition of the medium on growth in vitro of excised tobacco and sunflower tissue cultures. Am J Bot 33:591–597
Iha TB, Roy SC (1979) Rhizogenesis from *Nigella sativa* protoplasts. Protoplasma 101:139–142
Istratescu-Guti L, Forstner S (1974) Ascorbic acid content of *Batrachiophyta* plants. Farmacia (Bucharest) 22:489–492
Jensen U (1984) Legumin-like and vicilin-like storage proteins in *Nigella damascena* (Ranunculaceae) and six other dicotyledonous species. J Plant Physiol 115:161–170
Jensen U, Grumpe B (1983) Seed storage proteins. In: Jensen U, Fairbrothers DE (eds) Proteins and nucleic acids in plant systematics. Springer, Berlin Heidelberg New York, pp 238–254
Jukneviciene G, Dagyte S, Stankeviciene N (1977) Biological properties and essential oils of some spice plants grown at the Kaunas Botanical Garden. 2. Plants, the seeds of which are used as a raw material for spicery. Liet TSR Mokslu Akad Darb Ser C 3:9–16
Kabachnyi PI, Gorodetskaya ZV, Chernobai VT (1986) β-Fructofuranosidase activity of the seed of some cultivated plants. Farm Zh (Kiev) 1986(3):59–62
Kabachnyi PI, Kortunova TV, Chernobai VT (1987) Activity of amylolytic enzymes in extracts of agricultural plant seeds. Farm Zh (Kiev) 1987:45–48
Kartha ARS, Khan RA (1969) Proportions of $\Delta^{7,8}$-octadecanoic acids in seed fats from ten umbelliferae species. Chem and Ind 1969:1869–1872
Keller O (1908) Studien über die Alkaloide der *Nigella*-Arten. Arch Pharm 246:1–50
Kirichenko BM, Koreshchuk KE, Drozd GA, Kremzer AA (1971) Kaempferol and quercetine from *Nigella arvensis*. Khim Prir Soed 7:371–372
Klasterska I, Natarajan AT (1975) Distribution of heterochromatin in the chromosomes of *Nigella damascena* and *Vicia faba*. Hereditas 79:154–156
Korchagina LN, Rudyuk VF (1975) Conditions for isolation of the lipase enzyme from love-in-a-mist (*Nigella damascena*) seeds. Prikl Biokhim Microbiol 11:443–446

Korchagina LN, Rudyuk VF (1979) Effect of fatty and bile acids and thiol reagents on lipase activity in *Nigella damascena* L. seeds. Prikl Biokhim Mikrobiol 15:112–116
Korchagina LN, Rudyuk VP, Chernobai VT (1973) Lipase activity of seeds and vegetative organs of several plants. Rast Resur 9:577–581
Lehmann H, Preiss A, Schmidt J (1983a) A novel abscisic acid metabolite from cell suspension cultures of *Nigella damascena*. Phytochemistry 22:1277–1278
Lehmann H, Boehm H, Schuette HR (1983b) The metabolism of abscisic acid in cell cultures of various plant species. Z Pflanzenphysiol 109:423–428
Mothes K (1966) Diskussion zu Munsche. In: Mothes K, Gross D, Liebisch H-W, Schuette HR (eds) Abh Dtsch Akad Wiss Berlin, Kl Chem Geol Biol, 13:615
Moutschen J, Moutschen-Dahmen M (1965) A suitable material for chromosome breakage studies: *Nigella damascena* L. Naturwissenschaften 20:566–567
Moutschen J, Moutschen-Dahmen M, Gilot-Delhalle J, Woodley R (1969) The relative biological effectiveness of different kinds of radiations on chromosome aberrations in *Nigella damascena* seed. Int J Radiat Biol 15:525–540
Moutschen J, Gilot-Delhalle J, Moutschen-Dahmen M (1987) Clastogenic effects of benzodiazepines in a *Nigella damascena* seed test. Environ Exp Bot 27:227–231
Mukherjee BB, Chattopadhyay S, Ganguli SN (1981) Spectrofluorimetric determination of endogenous level of IAA from tissues of *Nigella sativa* L. grown in vitro. Indian J Exp Biol 19:295–297
Munsche D (1964) Zytologische Untersuchungen zur Biosynthese des Damascenins. 5. Mitteilung über das Damascenin. Flora 154:317–324
Munsche D (1966) Biosynthese des Damascenins. In: Mothes K, Gross D, Liebisch H-W, Schuette HR (eds) Abh Dtsch Akad Wiss Berlin, Kl Chem Geol Biol 3:611–614
Murashige T, Skoog F (1962) A revised medium for rapid growth and bioassays with tobacco tissue cultures. Physiol Plant 15:473–497
Mustafa Z, Soliman G (1943) Melanthigenin and its identity with hederagenin. J Chem Soc 1943:70–71
Nagata PR, Takebe I (1971) Plating of isolated tobacco mesophyll protoplasts on agar medium. Planta 99:12–20
Namba T, Tsunezuka M, Dissanayake DMRB, Pilapitiya U, Saito K, Kakiuchi N, Hattori M (1985) Studies on dental cares prevention by traditional medicines, pt 7. Screening of ayurvedic medicines for anti-plaque action. Shoyakugaku Zasshi 39:146–153
Nijland MM, Uffelie OF (1965) Voorkomen van Magnoflorine in de Ranunuculaceae. Pharm Weekbl 100:49–51
Orton TJ (1980) Chromosomal variability in tissue culture and regenerated plants of *Hordeum*. Theor Appl Genet 56:101–112
Potekhina TS, Shataeva LK, Rudyuk VF, Samsonova GV (1982) Purification of plant lipase on a carbocyclic cation exchange resin. Khim Farm Zh 16:462–465
Raman K, Greyson RI (1974) In vitro induction of embryoids in tissue cultures of *Nigella damascena*. Can J Bot 52:1988–1989
Raman K, Greyson RI (1977) Changes during development in the compartmentalization patterns of extractable gibberellin-like substances in "single" and "double" genotypes of *Nigella damascena*. Can J Bot 55:2115–2121
Raman K, Greyson RI (1978) Further observations on the differential sensitivities to plant growth regulators by cultured "single" and "double" flower buds of *Nigella damascena* L. (*Ranunculaceae*). Am J Bot 65:180–191
Rechinger KH (ed) (1964) Flora of lowland Iraq. Cramer, Weinheim
Rudyuk VF, Korchagina LN (1975) Properties of lipase from *Nigella damascena* seeds. Khim Prir Soed 5:645–648
Rudyuk VF, Korchagina LN (1976) Isolation and partial purification of lipase from *Nigella damascena* seeds. Prikl Biokhim Mikrobiol 12:45–48
Rudyuk VF, Korchagina LN (1977) Effect of metal ions on the activity of lipase from *Nigella damascena* L. seeds. Prikl Biokhim Mikrobiol 13:319–323
Rudyuk VF, Korchagina LN (1981) Effect of digestive apparatus secretions on the lipase activity of *Nigella damascena* L. seeds. Khim Farm Zh 15:59–62
Rudyuk VF, Korchagina LN, Chernobai VT (1984) Nigedase enzymatic preparations. USSR Patent 1 118 673, Int Cl C 12 N 9/18, Appl. 3.12.82, Publ 15.10.84

Salama RB (1973) Sterols in the seed oil of *Nigella sativa*. Planta Med 24:375–377

Sayed MD (1980) Traditional medicine in health care. J Ethnopharmacol 2:19–22

Schmauder H-P, Doebel P (1988) Zum Lichteinfluß auf Fermentorkulturen von *Nigella damascena* L. Acta Biotechnol 8:151–159

Schmauder H-P, Doebel P, Gröger D (1984) Charakterisierung von *Nigella damascena*-Zellkulturen in Airlift-Fermentoren. Biochem Physiol Pflanzen 179:611–622

Schmauder H-P, Schröder W, Böhm H, Klossek P (1991) Einfluß von 5-Fluor-D,L-tryptophan auf Zellkulturen von *Nigella damascena* L. Biochem Physiol Pflanzen (in preparation)

Schröder W (1985) Einsatz genetisch-züchterischer Methoden zur Verbesserung der Sekundärstoffbildung in pflanzlichen Zellkulturen. Diss A, Akad Wiss DDR

Sethi M, Rangaswamy NS (1976) Endosperm embryoids in cultures of *Nigella damascena*. Curr Sci 45:109–111

Takagi T, Itabashi Y, Kaneniwa M, Mizukami M (1983) Trans-5-olefinic unusual fatty acids in seed lipids of *Aquilegia*. Yukagaku 32:367–374

Tillequin F, Leconte C, Paris M (1976) Carbures sesquiterpenes des graines de *Nigella damascena*. Planta Med 30:59–61

Torrey JG (1967) Morphogenesis in relation to chromosomal constitution in long-term plant tissue cultures. Physiol Plant 20:265–275

Toxopeus HJ (1927) Erblichkeitsuntersuchungen an *Nigella damascena* L. Genetica 9:341–440

Tutin TG, Heywood VH, Burges NA, Valentine DH, Walters SM, Webb DA (eds) (1964) Flora europaea, vol 1. Univ Press, Cambridge

Vishin ML (1961) A note on the fatty oil obtained from *Nigella damascena* seeds. Curr Sci 30:55–56

Vishin ML, Mothes K, Engelbrecht L, Schröter HB (1960) Biosynthesis of damascenine in *Nigella damascena* L. Nature (London) 188:61–62

Vishin ML, Munsche D, Schröter HB (1964) Untersuchungen über den Bildungsort des Damascenins in *Nigella damascena*. 4. Mitteilung über das Damascenin. Flora 154:299–316

Vohora SB, Khan MSY, Afaq SH (1973) Antifertility studies on Unani herbs. Part II, Antiovulatory effects of 'Hanzal', 'Halun', 'Kalonij' and 'Sambhalu'. Indian J Pharm 35:100–102

Wallaart RAM (1981) Acyclic polyols as taxonomic characters 4. Distribution of polyols in Ranunculaceae. Proc Koeninkl Ned Akad Wet, Ser C 84:465–477

White PR (1943) A handbook of plant tissue culture. Cattel, Tempe, Ar

XX *Pelargonium* spp. (Geranium): In Vitro Culture and the Production of Aromatic Compounds

B.V. CHARLWOOD[1] and K.A. CHARLWOOD[2]

1 Introduction

The genus *Pelargonium* is a member of the family Geraniaceae in which grouping are also included the genera *Geranium*, *Erodium*, *Monsonia* and *Sarcocaulon*. The vast majority of the 250 or so natural species of *Pelargonium* derive from South Africa, although a few species are native to Australia, East Africa and Syria. The genus is subject to large morphological diversity and, for descriptive purposes, has been subdivided into 15, or sometimes 16, sections (or subgenera) based on leaf and flower characteristics and on habitat. The leaves of many of the species and numerous artificial hybrids are scented, and members of the subgenera *Pelargonium*, *Cortusina* and *Polyactium* appear to be especially rich in essential oil (Webb 1984).

"Geranium oil" is the commercial name given to the product obtained by the steam distillation of the green parts of several variants of *Pelargonium* – namely *P. graveolens*, *P. capitatum*, *P. odoratissimum* and *P. radula* (otherwise known as *P. radens* or *P. roseum*). The highest quality oil possesses a delicate rose-like fragrance and is used in perfumes and toilet waters. Lower quality product has been widely used as a general purpose perfume for hand creams, soaps and other toilet requisites, although alternative, synthetic products are now being employed more frequently for this purpose.

World production of geranium oil is estimated at between 250 and 300 tonnes per year, with perhaps 200 tonnes being traded on an international basis and the balance being retained for home consumption (Robbins 1985). China is now the major producer of medium quality geranium oil, exporting around 60–70 tonnes per year. Good quality oil is produced in North Africa, notably in Egypt, Algeria and Morocco, although only small amounts of oil are regularly exported from the latter two countries. Up until 1983, Egypt was the major supplier of geranium oil, with production of over 100 tonnes per year in the middle 1970's. During the last decade, however, production has declined significantly from this level and was estimated at around only 40–50 tonnes in 1985. The highest quality oil (so-called Bourbon oil) is from Reunion Island, but here too production has fallen from around 60 tonnes in 1978 to nearer 35 tonnes per year at the present time. A major portion of the oil from

[1] Plant Cell and Molecular Sciences Group, Division of Biosphere Sciences, King's College London, Campden Hill Road, London W8 7AH, United Kingdom
[2] School of Chemistry, Thames Polytechnic, Wellington Street, Woolwich, London SE18 6PF, United Kingdom

Biotechnology in Agriculture and Forestry, Vol. 15
Medicinal and Aromatic Plants III (ed. by Y.P.S. Bajaj)

this source is exported to southern France to supplement the small quantity of very high quality oil produced in the region of Grasse for use by the exclusive perfumery houses in the area.

Geranium oil is also produced on a relatively small scale in a number of other countries, for example Spain, Hungary, USSR, Yugoslavia, Israel, India, Haiti, Congo, Kenya, Tanzania, Zaire, Madagascar and the Ivory Coast (Robbins 1985). Production is mostly for internal consumption, but commercial quantities of oil from several of these countries are sometimes available in the market.

Optimal yields of geranium oil are obtainable when plants are grown under a subtropical climate in a light, fertile and well-drained soil. Reasonable rainfall and a humid atmosphere are advantageous to oil accumulation and temperatures must remain above 5 °C throughout the year. Plants are generally propagated by taking cuttings from healthy plants (Fig. 1) and rooting them in nursery beds prior to transfer into the field. Young plants are planted in rows 60–80 cm apart, individual plants being spaced 30 to 40 cm apart. After 6 months growth, plants may be cropped by removing all the aerial material except for one stalk and associated leaves. Two, and sometimes three, crops are possible per year from a mature plantation, and such a plantation can remain economic for up to 6 years (Tropical Products Institute 1961). Yields of fresh plant material vary between 10 and 40 tonnes per year per

Fig. 1. Preparation of geranium cuttings in Morocco. (Crown Copyright: by kind permission of Dr. Clinton Green)

hectare, but may be as high as 70 tonnes per year per hectare (Fleisher and Fleisher 1985).

The essential oil is obtained by steam distillation of the green material either directly after harvesting, or after 24 h storage in order to reduce the volume. Any further storage of harvested material leads to the onset of fermentation and a deterioration in oil quality. Best quality oil is obtained through distillation from stainless steel vessels with steam supplied from an external source. Yields of oil vary according to the time of cropping and, of course, environmental factors. Maximal yields of up to 0.4% may be obtained from herbage harvested just before and during flowering. Typical overall yield is within the range 15–80 l of oil per hectare per year. However, yields of up to 185 l per hectare per year have been obtained, at least on an experimental scale, by selection and subsequent propagation of high-yielding plants (Fleisher and Fleisher 1985).

Commercial geranium oil is a complex mixture of over 120 mono- and sesquiterpenes and low molecular weight aroma compounds (Vernin et al. 1983). The main components (Fig. 2) are citronellol (*1*), geraniol (*2*), linalol (*3*) and their methanoyl (and other) esters, and together these components constitute some 60–70% of the total oil. The other significant components in all oil types are isomenthone (*4*), menthone (*5*), nerol (*6*), *cis* and *trans*-rose oxides (*7*, *8*), α-terpineol (*9*), α-pinene (*10*), myrcene (*11*) and β-phellandrene (*12*). Bourbon and Egyptian oils have relatively low levels of citronellol and its esters and much higher levels of geraniol, geranyl esters and linalol when compared with the Chinese oil. The main difference between North African and Bourbon oils is that the former uniquely contain 10-epi-γ-eudesmol (*13*) whilst the latter, and incidentally the Chinese oil, contain the sesquiterpene hydrocarbon 6,9-guaiadiene (*14*) (Teisseire 1987).

Apart from the species of *Pelargonium* that are cultivated for geranium oil, there are many other scented-leaved variants within the genus whose essential oils display wide chemical diversity. For example, the oil of *P. tomentosum* consists almost entirely (96%) of isomenthone (*4*) and menthone (*5*) whereas *Pelargonium* Mabel Grey produces an oil containing 88% of the citral isomers, geranial (*15*) and neral (*16*). The leaves of these variants are used for culinary purposes to impart, respectively, a delicate peppermint and lemon flavour to food. On the other hand, the oil from *P. fragrans* is an extremely complex mixture containing many different classes of lower isoprenoids such as fenchone (*17*), camphene (*18*), sabinene (*19*), 1,8-cineole (*20*), limonene (*21*), p-cymene (*22*) and farnesene (*23*). The leaves of some scented pelargoniums have even been used as medicinal remedies, for example, the camphoraceous vapour from steamed leaves of *P. betulinum* is inhaled for coughs and chest problems, whilst the powdered leaves of *P. inquinans* are said to alleviate the symptoms of headaches and colds (van der Walt 1977).

2 Biotechnological Approaches

The wide range of monoterpene components found in the essential oil of *Pelargonium* variants suggests that the genus might provide a good model system in which to investigate the production of this class of natural product in vitro.

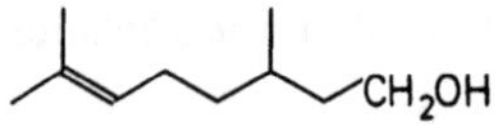

(1) citronellol

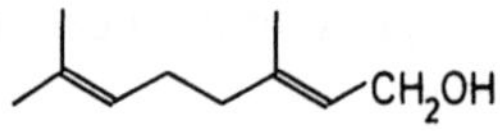

(2) geraniol

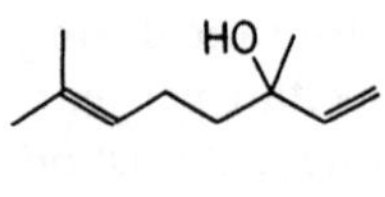

(3) linalol

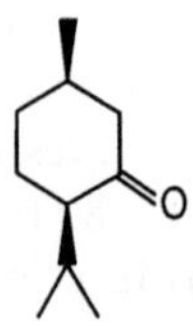

(4) isomenthone

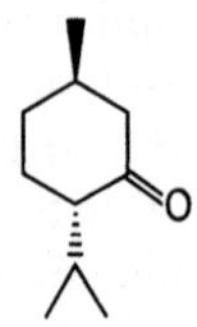

(5) menthone

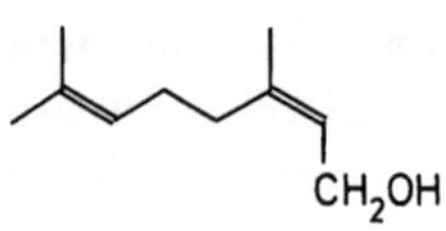

(6) nerol

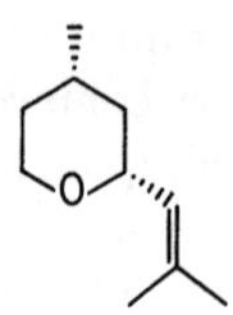

(7) *cis*-rose oxide

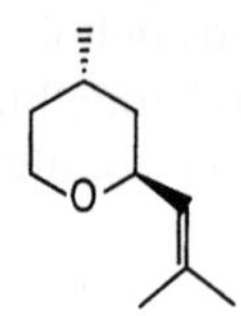

(8) *trans*-rose oxide

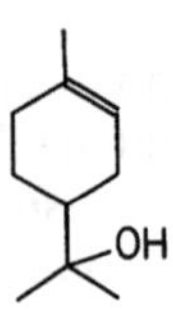

(9) α-terpineol

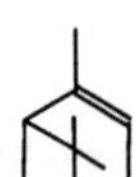

(10) α-pinene

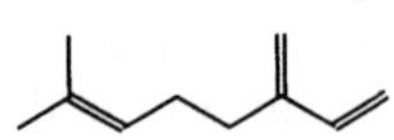

(11) myrcene

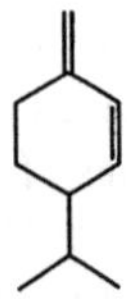

(12) α-phellandrene

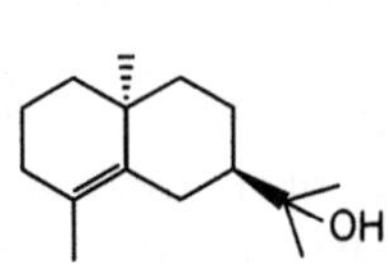

(13) 10-epi-γ-eudesmol

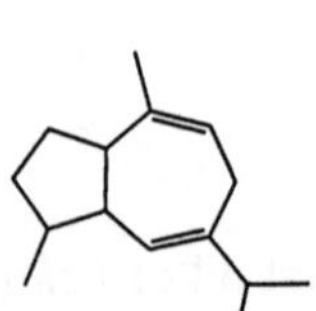

(14) 6,9-guaiadiene

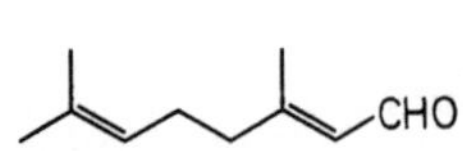

(15) geranial

Fig. 2. Structures of various secondary metabolites of *Pelargonium*

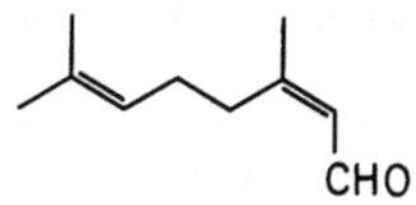
(16) neral

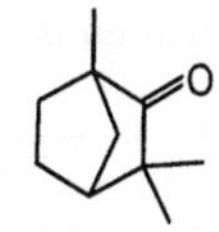
(17) fenchone

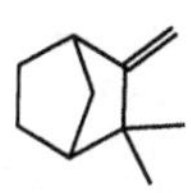
(18) camphene

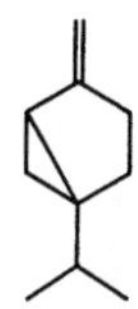
(19) sabinene

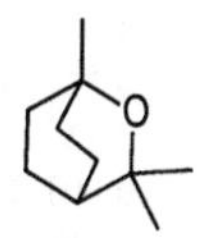
(20) 1,8-cineole

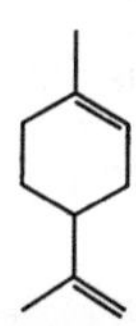
(21) limonene

(22) p-cymene

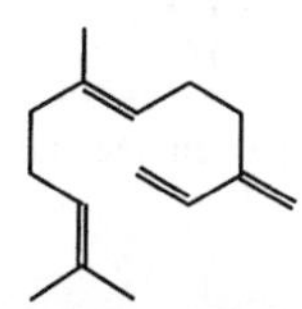
(23) farnesene

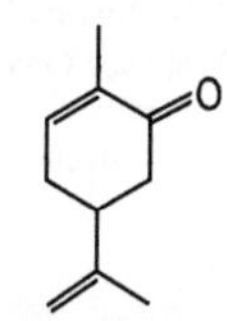
(24) carvone

Fig. 2. cont'd.

2.1 Induction of Unorganized Cultures

Chen and Galston (1965) showed that callus cultures of *P. hortorum* could be initiated readily by incubating sections of pith tissue on White's medium (supplemented with thiamine, pyridoxin, nicotinic acid, glycine, myo-inositol and sucrose) containing kinetin (0.22 mg/l) and 2,4-D (0.22 mg/l). Rapidly growing, white, friable calli were also obtained from stem tips, internodal portions and petioles, as well as pith, which had been cultured on medium containing mineral salts, sucrose and coconut milk both in the presence and absence of 2,4-D (Pillai and Hildebrandt 1969).

In the case of scented-leaved pelargoniums, efficient callus initiation was obtained from stem explants of some 30 variants incubated on a specialized, defined medium (Charlwood and Charlwood 1983) supplemented with kinetin (0.2 mg/l) and 2,4-D (1 mg/l) and maintained at 27°C under continuous subdued light (3 $\mu E/m^2/s$) or in a 16-h light / 8-h dark photoperiod (Table 1). Successful callus induction was obtained within 5 to 22 days for over 90% of all explants plated. Immediately after initiation, callus tissue was transferred to Murashige and Skoog medium supplemented with BAP (5 mg/l) and NAA (1 mg/l) and maintained by subculture at 21-day intervals under the above conditions. Calli have remained stable over periods in excess of 5 years and no significant morphological differentiation has occurred under such maintenance conditions. Callus morphology appears to be under genetic control for each variant: most calli are white and friable, but some variants (especially Mabel Grey and Duke of York) produce hard, highly pigmented cell masses (Charlwood et al. 1988b).

Table 1. The accumulation of essential oil in callus tissue form *Pelargonium* variants

Pelargonium variant	Accumulation of oil[a] (%)	Callus type
P. australe Willd.	2.18	White, friable
P. citriodorum Mart.	0.03	Pigmented, friable
P. fragrans Willd.	0.01	Pigmented, friable
P. glutinosum L'Her	0.07	Pigmented, hard
P. graveolens L'Her	0.001–0.03	White, friable
P. nervosum Sweet	3.5	White, friable
P. quercifolium	0.35–1.42	Pigmented, friable
P. radula L'Her	0.13	Pigmented, hard
P. tomentosum Jacq.	0.001–0.67	Pigmented, hard
P. Clorinda	< 0.01	White, friable
P. Duke of York	0.31	Pigmented, hard
P. Lillian Pottinger	1.48	Pigmented, friable
P. Mabel Grey	2.55	Pigmented, hard
P. Miss Australia	5.0	Pigmented, friable
P. Rollisons Unique	0.58	White, friable
P. Royal Oak	0.01	White, friable

[a] Oil content expressed as percentage of the oil accumulated by the parent plant.

2.2 Monoterpene Accumulation in Unorganized Cultures

Monoterpene accumulation in the majority of these callus cultures is very low (Table 1); for example, white, friable calli derived from *P. fragrans* and Royal Oak accumulated only 0.01% of the oil stored by the parent. In only six cell lines out of 34 tested was the oil level greater than 1% of that of the parent plant: thus callus from Mabel Grey accumulated 2.6% of the oil of the plant, and a maximal level of 5% was detected in a white, friable cell line from Miss Australia. There is no apparent correlation between level of oil accumulation and the morphology of the callus, suggesting that the regulation of these two characters is not linked. Typically, the composition of the oil accumulated by callus tissue is very similar to that of the parent plant.

Finely divided suspension cultures can easily be produced from friable callus tissue by gentle shaking in liquid MS medium supplemented with BAP (5 mg/l) and NAA (1 mg/l) and maintained under continuous subdued light at 27°C. Such cultures accumulate amounts of essential oil similar to those found in the original callus. However, it is sometimes possible to increase monoterpene accumulation by altering the light regime under which the cells are incubated. Thus, when finely divided suspension cultures of *P. fragrans* were subjected to a photoperiod regime of 18-h light/6-h dark, oil accumulation increased by typically 165-fold: however, even for the best cell line this still represents only 3% of the yield of the intact plant (Brown and Charlwood 1986a). Other techniques aimed at stimulating oil accumulation in suspension cultures, such as alteration of the nutrient composition of the growth medium or addition of "elicitors" (Table 2), lead to increases of the same order of magnitude (Charlwood et al. 1988b).

Table 2. The accumulation of essential oil in various culture systems of *Pelargonium fragrans*

Culture type	Accumulation of oil (μg/g fresh weight)
[Intact leaf tissue (average)	620]
Callus culture – unorganized	0.06–0.09
Callus culture – partial shoot differentiation (0.5 mg/l BAP: 0.5 mg/l NAA)	1.3
Callus culture – considerable shoot differentiation (0.5 mg/l BAP: 0.25 mg/l NAA)	19.3
Suspension culture – light-grown	0.13
Suspension culture – dark-grown	2.6
Suspension culture – 16 h light photoperiod	11.5
Aggregate culture – 16 h light photoperiod (0.5 mg/l BAP: 0.1–0.4 mg/l NAA)	35.0
Immobilized culture – 16 h light photoperiod	110
Supplementation with organic acids (suspension cells – light-grown)	up to 14.5
Supplementation with fungal cell-wall extracts (suspension cells – light grown)	0.12–0.16
Two-phase culture system with Miglycol (suspension cells – 16 h light photoperiod)	6300

[a] All cultures maintained in MS medium containing 5 mg/l BAP and 1 mg/l NAA except where otherwise stated.

2.3 In Vitro Regeneration

The findings indicated above are in line with the generally accepted view that monoterpenes are not accumulated in significant concentrations in unorganized callus or finely dispersed suspension cultures (Charlwood et al. 1986). Whereas organogenesis and differentiation of distinct tissue types are not pre-requisites of the synthesis of monoterpenes, it is likely that a degree of organization is necessary for the accumulation of product. However, unlike the situation for solanaceous plants, *Pelargonium* callus will not readily differentiate organs without the application of the correct sequence of stimuli. Callus cultures of *P. hortorum* developed vascular nodules when maintained on White's medium supplemented with BAP (0.23 mg/l) and IAA (0.88 mg/l), but would only undergo shoot differentiation after passage through medium containing no plant growth regulators (Chen and Galston 1967). On the other hand, Pillai and Hildebrandt (1969) report that calli from a variety of cultivars of *P.hortorum* would undergo shoot organogenesis readily only if transferred to MS medium containing kinetin (10 mg/l) and NAA (0.1 to 1 mg/l) and incubated under a 16-h light/8-h dark photoperiod immediately after initiation. Calli that had undergone three passages on initiation medium differentiated with much greater difficulty, and none would differentiate after six subcultures under initiation conditions. More recent studies have indicated that not all genotypes may be capable of regeneration (Grunewaldt 1981).

Differentiation in calli derived from scented-leaved pelargoniums may be controlled by exogenous growth regulators provided that the stock cultures have been maintained on MS medium supplemented with BAP (5 mg/l) and NAA (1 mg/l) since initiation (Brown and Charlwood 1986b). Although each variant responds in a specific manner to such treatment, the general developmental pattern observed is shown in Fig. 3. Continued formation of callus tissue occurred when auxin and cytokinin concentrations were in excess of 1 mg/l, and there was a tendency for calli to become hard and pigmented at auxin:cytokinin ratios greater than 1:1. When the cytokinin level was less than 0.25 1 mg/l, increasing root formation was observed at higher auxin levels. Shoot production was detected only when both auxin and cytokinin levels were lower than 0.5 mg/l, although at cytokinin concentrations of less than 0.05 mg/l the response was somewhat unpredictable. Regenerated plantlets of most scented-leaved pelargoniums could be obtained after ca. 4–14 weeks' incubation of the calli on medium containing BAP (0.5 mg/l) and NAA (0.05 mg/l). Calli did not lose their ability to regenerate even after continued subculture on maintenance medium over a period in excess of 4 years. Some variants regenerated only after prolonged culture on this medium (up to 51 weeks for Rober's Lemon Rose). Some variants, for example *P. echinatum*, failed to regenerate under any of the experimental regimes applied.

2.4 Essential Oil Accumulation in Differentiated Tissue

The essential oils accumulated by most calliclones that had been grown under greenhouse conditions for 3 months after rooting were similar both qualitatively and quantitatively to those of the parent plants cultivated under identical conditions

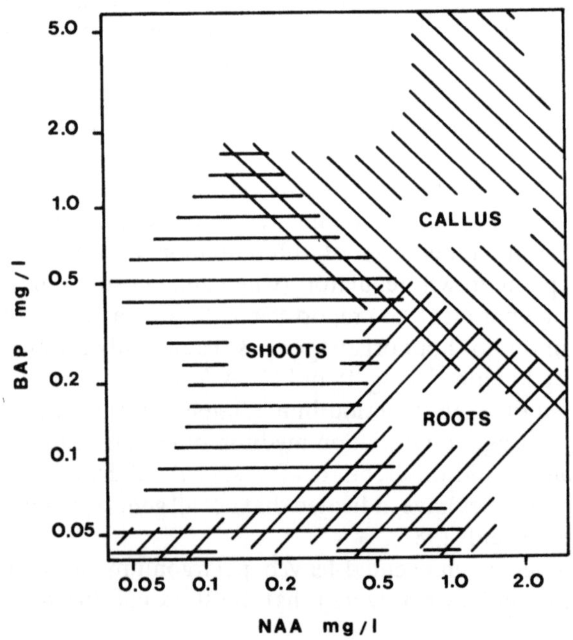

Fig. 3. Effects of growth regulator levels on the growth and differentiation of *Pelargonium* variants. (After Brown and Charlwood 1986b)

(Brown and Charlwood 1986c). It would thus appear that tissue culture-induced variation in essential oil composition is relatively uncommon. This view is supported by an early study of 166 first-generation calliclones of the cultivar. Rober's Lemon Rose by Skirvin and Janick (1976), who demonstrated that in only one plant, with an aberrant flat-leaf morphology, was the essential oil significantly different from that of the parent. The alteration in essential oil composition involved loss of a single, major (unidentified) component, but such loss could not be correlated unambiguously with change in chromosome number. Much greater variability in oil composition was reported in calliclones derived from anther cultures of a tetraploid of *P. roseum* (Tokumasu and Kato 1979). In this study, derivatives obtained from the tetraploid (which had been induced by colchicine treatment of a diploid plant) had diploid number and were thought to have originated from pollen. The essential oils of these derivatives showed much wider variation in content of citronellol (*1*), geraniol (*2*) and isomenthone (*4*) when compared to the oil from the diploid calliclones derived from anther culture of diploid plants.

Whilst unorganized callus of *P. fragrans* accumulated insignificant levels of essential oil, the induction of a low level of shoot initiation by transfer of callus material to solid medium supplemented with BAP (0.5 mg/l) and NAA (0.5 mg/l) increased monoterpene accumulation 20-fold (Table 2). Maximal frequency of shoot induction occurred on solid medium containing BAP (0.5 mg/l) and NAA (0.25 mg/l) and this was accompanied by a 300-fold increase in oil accumulation (Charlwood et al. 1988b). This oil was, however, markedly different from that of the intact plant and consisted almost entirely of the monoterpene hydrocarbons [α-pinene (*10*) being the major component] with few oxygenated derivatives being present even in trace amounts.

Shoot morphogenesis could also be induced in submerged callus cultures of *P. fragrans* and *P. tomentosum* by transfer of greening aggregates to MS medium supplemented with BAP (0.5 mg/l) and NAA (0.05 mg/l) and incubated under a 16-h light/8-h dark regime. Shoot-proliferation cultures (Fig. 4) formed after 4–18 weeks under these conditions and have remained stable for over 3 years (Charlwood and Moustou 1988). Shoot-proliferation cultures grown in 5-1 air-stirred bio-reactors (Fig. 5) show an eight to ten fold increase in biomass with a concomitant increase in essential oil accumulation during a 14-day incubation cycle. The final yield of oil is dependent upon the culture environment and can vary between 10 and 50% of that of the intact plant. In contrast to the situation for most culture systems, the oil derived from shoot-proliferation cultures is qualitatively comparable with that of the whole plant. Thus the oil from both the plant and the shoot-proliferation culture of *P. tomentosum* consisted almost entirely of the two menthone isomers (*4*) and (*5*). The spectrum of mono- and sesquiterpene hydrocarbons in the plant and in the shoot proliferation culture of *P. fragrans* is also very similar but there is a significant reduction in the accumulation of oxygenated monoterpenes, particularly fenchone (*17*), in the latter. Katagi et al. (1986) have produced submerged, shoot-proliferation cultures directly from adventitious shoots of *P. graveolens* pre-cultured on agar medium. Such cultures, grown in Linsmaier Skoog medium with BAP (0.23 mg/l) and IAA (1.8 mg/l), showed an increase in biomass of 14.5-fold within a 21-day culture period, although the yield could be doubled by using indolebutanoic acid as auxin and by reducing the

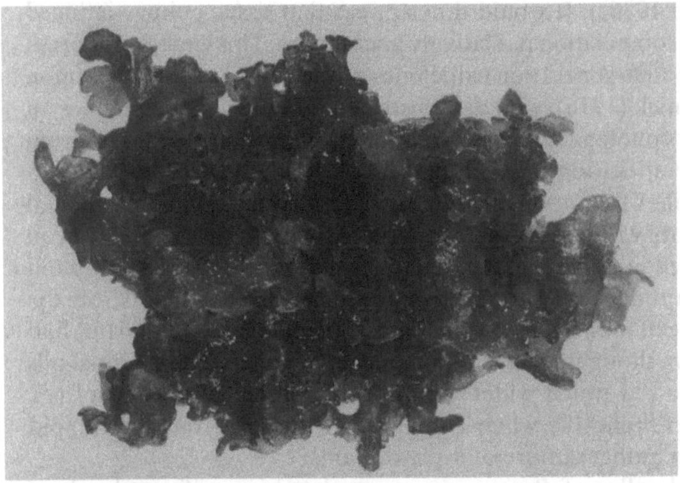

Fig. 4. Shoot-proliferation culture of *Pelargonium fragrans* incubated in MS medium supplemented with BAP (0.5 mg/l) and NAA (0.05 mg/l) under a 16-h light/8-h dark photoperiod

ammonium concentration in the medium. In each case the essential oil produced in vitro was qualitatively similar to that of the plant but reduced with respect to citronellol (*1*) content.

In *Pelargonium* plants, essential oil is stored in specialized secretory cells contained within glandular hairs (Fig. 6). Shoot-forming cultures possess both peltate and capitate glandular hairs on the surfaces of the leaflet and the petiole and these are the exclusive sites of oil storage. It has been demonstrated that many monoterpenes (hydrocarbons, alcohols, ketones and aldehydes) are highly toxic to cultured *Pelargonium* cells, giving rise to significant reductions in viability and cell growth at concentrations in the order of 0.4 mg/ml of culture medium (Brown et al. 1987). Furthermore, cells of individual variants show no specific tolerance to the monoterpenes accumulated in the oils of their parent plants. It is presumed that end-product toxicity is alleviated in whole plants and in shoot-forming cultures by sequestration of the essential oil in specialized storage sites. Indeed, there is increasing evidence to suggest a significant correlation between accumulation of essential oil and the density of the storage sites in a number of culture systems. Thus when shoot-proliferation cultures of *P. tomentosum* were treated with bleaching herbicides such as AMO 1618, metflurazon, desmethyl-norflurazon and 5C cycocel, progressive de-differentiation of the tissue was accompanied by decreasing glandular hair density and a reduction in essential oil accumulation (Charlwood and Moustou 1988). Continued treatment with metflurazon and desmethyl-norflurazon gave rise eventually to nonaccumulating aggregates.

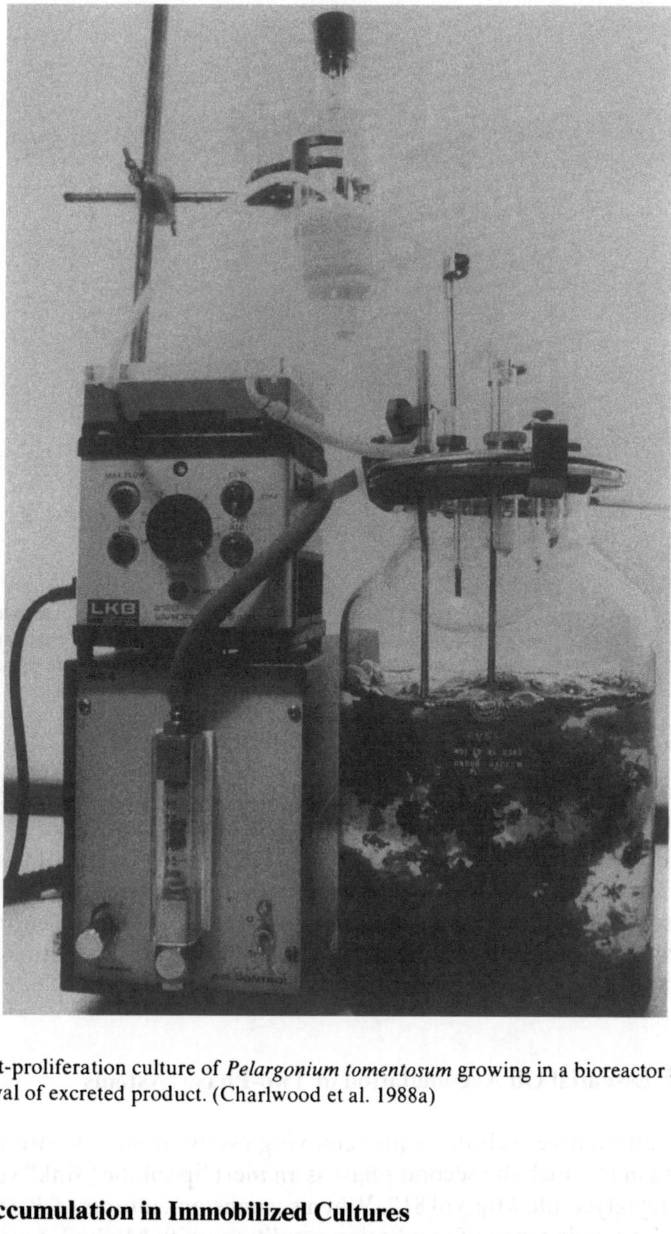

Fig. 5. Submerged shoot-proliferation culture of *Pelargonium tomentosum* growing in a bioreactor set up for continuous removal of excreted product. (Charlwood et al. 1988a)

2.5 Essential Oil Accumulation in Immobilized Cultures

It is not unreasonable to suggest that end-product toxicity might determine, in part, the upper limit of monoterpene accumulation in unorganized cultures. In such systems it would clearly be necessary to remove the product as soon as it is synthesized in order to reduce feed-back inhibition. One technique by which this might be achieved is to immobilize cells on an inert matrix and to flush the product away with a continuous flow of medium. Finely divided *Pelargonium* suspension cultures may be readily immobilized by entrapment within the pores of reticulate

Fig. 6. Glandular hair stained with Nile Blue Sulphate and Oil Red O to show storage site of essential oil. (Charlwood et al. 1988a)

polyurethane foam. Such immobilized cells, incubated in medium supplemented with BAP (5 mg/l) and NAA (1 mg/l) and grown under a 16-h light/8-h dark photoperiod, produce around 110 μg of oil/g fresh weight of tissue (18% of the yield of the plant), most of which is excreted into the medium (Table 2). As is often the case for nondifferentiated cultures, the oil consisted almost entirely of monoterpene hydrocarbons with limonene (*21*) being the major constituent (Charlwood et al. 1988b).

2.6 Essential Oil Accumulation in Two-Phase Systems

An alternative technique for removing essential oil is to use a two-phase culture system in which the second phase is an inert lipophilic "sink" such as hexane or the triacylglyceride Miglyol 812. When suspension cultures of *P. fragrans* were treated under a wide range of incubation conditions with Miglyol, a minimum of a ten fold increase in essential oil accumulation (compared to suspension cells grown under identical conditions but in the absence of Miglyol) was observed (Charlwood and Brown 1988). A maximum stimulation of oil production of some 500-fold was achieved after six passages through a two-phase system in which cultures were grown in medium supplemented with BAP (0.5 mg/l) and NAA (0.05 mg/l) and exposed to a 16-h light/8-h dark photoperiod (Table 2). A pronounced alteration in the constitution of the oil occurred after each subculture. Initially cells grown in the presence of Miglyol accumulated limonene (*21*) together with the oxygenated monoterpenes, 1,8-cineole (*20*) and carvone (*24*). After four or five passages the amount of accumulated limonene decreased and the major components were p-cymene (*22*) and sesquiterpene hydrocarbons [especially farnesene (*23*)]. Miglyol is clearly not an ideal lipophilic phase, however, since the growth rate of the cultures decreased 16-fold during the series of seven subcultures. These results do demonstrate, however, that under certain conditions, oil yields from in vitro systems can attain the levels found in the most productive tissues of the whole plant.

3 Future Prospects

It is quite clear that in vitro techniques are unlikely to provide a commercially attractive, alternative source for natural geranium oil. On the other hand, tissue culture techniques can provide an element of variation in essential oil composition (both in vivo and in vitro) that could be exploited in a number of ways. In this context, recent experiments (Moustou, C. unpubl.) have shown that when suspension cells of *P. fragrans* are treated over a 5-day period with colchicine, accumulation of essential oil is considerably enhanced and the product profile of the oil is quite variable and very different from that of the untreated cells. Since perception of aroma of an oil does not necessarily depend upon the presence or concentration of a single component but upon the relative amounts of a large number of, perhaps minor, constituents, techniques to induce such variation in oil product could be used to provide novel fragrances or flavours. If this is the case, then the technologies required to produce commercial quantities of these oils are already in place.

On the other hand, formation of calliclones from cell lines with changed oil composition could lead to new varieties of pelargoniums for use in a traditional cropping regime. A further possibility for the production of such new hybrids is provided by the recent demonstration (Yarrow et al. 1987) of plant regeneration from culture-derived protoplasts of a number of *Pelargonium* species. If the reported methods are generally applicable to scented-leaved pelargoniums, then any number of novel genotypes could be produced via somatic hybridization of culture-induced or colchicine-induced aberrant lines.

Acknowledgments. The authors wish to thank Dr. Clinton Green of the Overseas Development Natural Resources Institute, London for Fig. 1 and for information about the commercial aspects of geranium oil production. The ownership of all registered trade names is acknowledged.

References

Brown JT, Charlwood BV (1986a) The accumulation of essential oils by tissue cultures of *Pelargonium fragrans* Willd. FEBS Lett 204:117–120

Brown JT, Charlwood BV (1986b) The control of callus formation and differentiation in scented pelargoniums. J Plant Physiol 123:409–417

Brown JT, Charlwood BV (1986c) Differentiation and monoterpene biosynthesis in plant cell cultures.

In: Morris P, Scragg AH, Stafford A, Fowler MW (eds) Secondary metabolism in plant cell cultures. Univ Press, Cambridge, pp 68–74
Brown JT, Hegarty PK, Charlwood BV (1987) The toxicity of monoterpenes to plant cell cultures. Plant Sci 48:195–201
Charlwood BV, Brown JT (1988) The accumulation of mono- and sesquiterpenes in plant cell cultures. Biochem Soc Trans 16:61–63
Charlwood BV, Charlwood KA (1983) The biosynthesis of mono- and sesquiterpenes in tissue culture. Biochem Soc Trans 11:592–593
Charlwood BV, Moustou C (1988) Essential oil accumulation in shoot-proliferation cultures of *Pelargonium* spp. In: Robins RJ, Rhodes MJC (eds) Manipulating secondary metabolism in culture. Univ Press, Cambridge, pp 187–194
Charlwood BV, Hegarty PK, Charlwood KA (1986) The synthesis and biotransformation of monoterpenes by plant cells in culture. In: Morris P, Scragg AH, Stafford A, Fowler MW (eds) Secondary metabolism in plant cell cultures. Univ Press, Cambridge, pp 15–34
Charlwood BV, Brown JT, Moustou C, Charlwood KA (1988a) Pelargoniums: flavours, fragrances and the new technology. Plants Today 1:42–46
Charlwood BV, Brown JT, Moustou C, Morris GS, Charlwood KA (1986b) The accumulation of isoprenoid flavour compounds in plant cell cultures. In: Schreier P (ed) Bioflavour '87. De Gruyter, Berlin, pp 303–314
Chen H-R, Galston AW (1965) Growth and development of *Pelargonium* pith cells in vitro. I. Induction of cell division. Physiol Plant 18:454–461
Chen H-R, Galston AW (1967) Growth and development of *Pelargonium* pith cells in vitro. II. Initiation of organised development. Physiol Plant 20:533–539
Fleisher A, Fleisher Z (1985) Yield and quality of essential oil from *Pelargonium graveolens* cultivated in Israel. J Sci Food Agric 36:1047–1050
Grunewaldt J (1981) Züchtungsforschung an *Pelargonium-zonale-* und *Pelargonium-peltatum*-Hybriden. I. Regeneration von *Pelargonium-peltatum*-Hybriden in vitro. Gartenbauwissenschaft 46:250–252
Katagi H, Takahashi E, Nakao K, Inui M (1986) Shoot-forming cultures of *Pelargonium graveolens* by jar fermentation. Nippon Nogeikagaku Kaishi 60:15–17
Pillai SK, Hildebrandt AC (1969) Induced differentiation of geranium plants from undifferentiated callus in vitro. Am J Bot 56:52–58
Robbins SRJ (1985) Geranium oil: market trends and prospects. Trop Sci 25:189–196
Skirvin RM, Janick J (1976) Tissue culture-induced variation in scented *Pelargonium* spp. J Am Soc Hortic Sci 101:281–290
Teisseire P (1987) Industrial quality control of essential oils by capillary GC. In: Sandra P, Bicchi C (eds) Capillary gas chromatography in essential oil analysis. Huethig, Heidelberg, pp 215–258
Tokumasu S, Kato M (1979) Variation of chromosome numbers and essential oil components of plants derived from anther culture of the diploid and the tetraploid *Pelargonium roseum*. Euphytica 28:329–338
Tropical Products Institute (ed) (1961) Memorandum on geranium oil. Trop Prod Inst Rep 31
van der Walt JJA (1977) In: Pelargoniums of southern Africa, vol 1. Purnell, Cape Town
Vernin G, Metzger J, Fraisse D, Scharf C (1983) Etude des huiles essentielles par GC-SM-banque specma: essences de geranium. Parf Cosm Arom 52:51–61
Yarrow SA, Cocking EC, Power JB (1987) Plant regeneration from cultured cell-derived protoplasts of *Pelargonium aridum*, *P.* × *hortorum* and *P. peltatum*. Plant Cell Rep 6:102–104
Webb WJ (1984) The *Pelargonium* family. Croom Helm, London

XXI *Perilla frutescens* (L.) Britton: In Vitro Culture and the Production of Caffeic Acid

N. ISHIKURA[1]

1 Introduction

1.1 Distribution and Importance of *Perilla* Plants

Perilla frutescens (L.) Britton var. *crispa* (Thunb.) Decaisne, belonging to the family Labiatae, is distributed widely in Japan, China, and southeast Asia. It is an annual herb about 1 m in height, and includes several forms such as f. *purpurea* Makino (Chirimen-aka-shiso or Aka-shiso) (Fig. 1) and f. *viridis* Makino (Chirimen-ao-shiso or Ao-shiso) having deep red-purple and green leaves, respectively. *Perilla* plants contain essential oil at about 0.5% of fresh leaf weight and give out a fragrance, principles of which are *Perilla*-aldehyde (55%), d-limonene (20–30%), and α-pinene (Yoshiki 1911); linoleic, stearic, and palmitic acids are also contained as the major aliphatic acids. According to the extensive study on pigments of Chirimen-aka-shiso (Ishikura 1981), 16 kinds of flavonoids including five anthocyanins, two flavones, and nine flavone glycosides are present in the mature dark-red leaves and seeds. Among these flavonoids, the 3-p-coumaroylglucoside-5-glucoside of cyanidin (shisonin) and the 7-caffeoylglucosides of apigenin and luteolin are the major component in the leaves. In seeds, apigenin and luteolin are present in a ratio of about 1:1. With other phenolics, a large amount of caffeic acid derivatives are present in the leaves.

Perilla plants are used for various purposes as both food stuff itself and garnish in Japanese cooking. Red-purple or green *Perilla* leaves, spikes, and seeds are also used as spices for buckwheat noodle, baked fish, fried food (tempura), etc. (Shibata 1949). The red-purple leaves of Aka-shiso have also been used for coloring *umeboshi* which is a pickled ume (Japanese apricot, *Prunus mume* Sieb. et Zucc.), but recently there has been increased demand for the red-purple leaves of Chirimen-aka-shiso because of the presence of a large amount of the anthocyanin, shisonin.

1.2 Conventional Practices for Propagation, Production of Medicinal Components, and Demand on the Market

Perilla seedling grows moderately at about 25 °C and is a short-day plant. Seed is about 1.5 mm in diameter and the weight of 100 seeds is about 0.2 g. Seeds produced

[1]Department of Biological Science, Faculty of Science, Kumamoto University, 2-Chome, Kurokami, Kumamoto-Shi, 860, Japan

Biotechnology in Agriculture and Forestry, Vol. 15
Medicinal and Aromatic Plants III (ed. by Y.P.S. Bajaj)

Fig. 1. Mature plant (4 months old) of *Perilla frutescens* var. *crispa* Chirimen-aka-shiso, growing on the campus of Kumamoto University

in autumn are dormant until spring and their life span is usually 2 years when stored in a refrigerator. Germination rate of the first-year seeds is about 60–80%.

Perilla plants are known as a herbal medicine, being effective as diaphoretic, anodyne, sedative, diuretic, and cough remedy, and they have powerful antiphlogistic activity due to the presence of rosmarinic acid (Gracza et al. 1985). Essential oil of *Perilla* plant functions as a strong antiseptic (Furukawa and Tomisawa 1920). Fresh, and some salted leaves (*sisoyoo*), seeds (*sisosi*), and the whole mature plants are marketed throughout the seasons. Although their prices on the market depend on the season, it is usually about 200 yen per kilogram of fresh plants. Essential oil of the *Perilla* plant (siso-oil) is sold by the druggist.

2 In Vitro Approaches

2.1 Review of Tissue Culture Work on *Perilla*

Sugisawa and Ohnishi (1976) induced callus from young leaf sections (5 mm) of Ao-shiso on Murashige and Skoog's (1962) basal medium (MS medium) containing 1 mg/l 2,4-D, 5 mg/l kinetin, 30 g/l sucrose, and 0.9% w/v agar. The callus was subcultured every 3 weeks at 25°C in dim light. On the other hand, callus tissue of *P. ocymoides* cv. *crispa* (= *P. frutescens* var. *crispa*) was initiated from 7-day-old cotyledons and was brought into suspension in MS medium containing 1 mg/l 1-naphthaleneacetic acid (NAA), 0.1 mg/l kinetin, and 30 g/l sucrose (Ibrahim and Edgar 1976). In the author's experiments (Ishikura et al. 1983), MS medium (MS-I medium) containing 1 mg/l 2,4-D, 0.1 mg/l kinetin, 30 g/l sucrose, and 0.9% w/v agar was the most effective for inducing Chirimen-aka-shiso callus, compared with several media supplemented with kinetin and 2,4-D or NAA at various concentrations. In addition, the lowering of 2,4-D concentration and the high kinetin concentration in the medium gave rise to increased caffeic acid formation.

2.2 Establishment of Tissue Culture/Cell Suspensions

Chirimen-aka-shiso seeds were soaked in water for 2 days, sterilized with 0.01% $HgCl_2$ aqueous solution for 10 min and washed with distilled water. They were then sown on solidified agar (0.5%) in Petri dishes, and germinated in the dark at 25°C. The callus was induced from 6–7-day-old seedlings (height 10 mm) with MS-I medium, and it was subcultured by using Gamborg's (1966) basal medium (G medium) supplemented with 1 mg/l 2,4-D, 30 g/l sucrose and 0.9% (w/v) agar every 3 weeks at 25°C under continuous 1.6 klx illumination with cool white-fluorescent lamps. Callus tissues in the G medium were white to faint yellow and grew vigorously, whereas in the MS-I medium they were watery and their aged inner parts were necrotic and colored brownish black.

The cell suspension culture derived from the callus at the 8th generation was maintained in modified Schenk and Hildebrandt medium (SH-M medium) (Mitchell and Gildow 1975) supplemented with 0.5 mg/l 2,4-D and 0.01 mg/l kinetin. Cells were cultured in 50 ml of the liquid medium contained in Erlenmeyer flasks (200 ml). The flasks were agitated on a reciprocal shaker at 62 osillations per min with a horizontal excursion of 5.5×2 cm in a constant temperature room of 25°C under continuous illumination (350 lx) with cool white-fluorescent lamps. Aliquots of the culture (50 ml each) were transferred every 10 days, assuring logarithmic growth conditions. The suspension culture in SH-M medium contained free cells and quite a number of aggregates of less than about 40 cells. Most of the cells in suspension were nearly spherical, but some divided into four daughter units with intersecting walls. Among several media examined, SH-M medium was the most adequate for the suspension culture of *Perilla* cells and the G medium was satisfactory as the agar medium (Ishikura et al. 1983). Figure 2 shows the growth of *Perilla* cells in the SH-M medium. Fresh weight of the cells increased exponentially for about 11 days after transfer to a fresh medium. As estimated from the increase

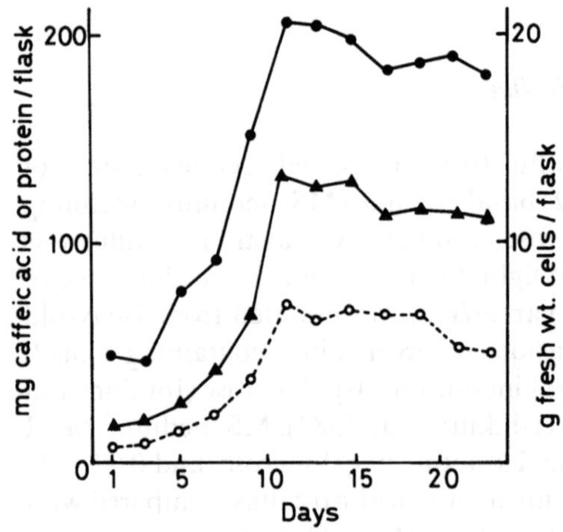

Fig. 2. The growth (▲——▲) of *Perilla* cells cultured in SH-M liquid medium, and changes in the contents of protein (●——●) and caffeic acid (○---○) of the cells. (Ishikura et al. 1983)

in fresh weight, the doubling time of the cells in the logarithmic growth phase was about 60 h. The protein content of the culture also reached a maximum on the 11th day and decreased gradually after entering the stationary growth phase.

2.3 Extraction and Structure of Caffeic Acid

Perilla callus tissues were found to contain almost exclusively caffeic acid derivatives as the phenolic extractives. Therefore, the relative amount of phenolic compounds in methanol could be determined as caffeic acid equivalent spectrophotometrically at 325 nm, because the compounds in the methanol cell extract very closely resembled authentic caffeic acid in their UV-absorption spectra (Fig. 3). As shown in Fig. 2, changes in the caffeic acid content proceeded nearly in parallel with

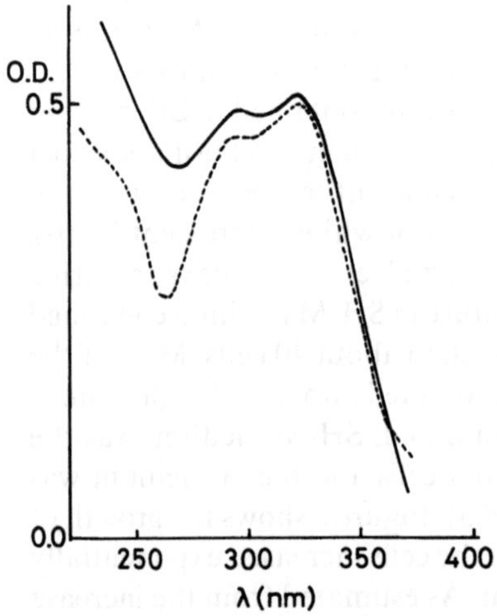

Fig. 3. Absorption spectra of the methanolic solutions of *Perilla* cell extract (*solid line*) and authentic caffeic acid (*broken line*). (Ishikura et al. 1983)

those of the growth parameters such as fresh weight and protein content. The content of caffeic acid in a culture flask as transferred to a fresh medium was ca. 3 mg, corresponding to 0.3 and 6% of fresh and dry weights of the cells, respectively. The amounts of caffeic acid in the stationary growth phase (e.g., the 15th day) became ca. 66 mg, corresponding to 0.5 and 9.5% of fresh and dry weights of the cells, respectively. Thus, the caffeic acid content increased about 22 times on a culture flask basis and about two times on g dry weight basis.

For the detection of caffeic acid, fresh cells were thoroughly extracted with 70% ethanol under reflux for 1 h, and this extraction was repeated. The combined extracts were concentrated in vacuo to a small volume at 45 °C. Aliquots of the concentrate were hydrolyzed by aqueous 10% NaOH solution under N_2 gas for 5 h followed by acidification with dilute HCl. After extraction with ethyl ether, the main phenolic component in the extracts was identified as caffeic acid by direct paper chromatographic (PC) comparison with a authentic sample as described previously (Ishikura 1972). The content (average) of caffeic acid obtained from the cells from the 10-day culture (the stationary growth phase) by alkali-hydrolysis was determined to be 0.073% of the fresh weight. The major component of caffeic acid derivatives in the cells was rosmarinic acid (caffeoyl-3,4-dihydroxyphenyllactic acid) (Fig. 4), consisting of about 50% of the phenolics in the cell extract, and also caffeoylglucose (20–30%) and caffeic acid (trace) (Fig. 4) were present (unpubl.). PC Rf values of caffeic and rosmarinic acids were 0.86 and 0.84 in n-BuOH/HOAc/H_2O (4:1:5), 0.82 and 0.61 in n-BuOH/EtOH/H_2O (4:1:2.2) and 0.42 and 0.57 in 15% HOAc, respectively. Both acids gave blue or blue-green fluorescence under UV light with or without NH_3 vapor. Caffeic and rosmarinic acids were also identified by co-chromatography with authentic samples on a HPLC column (Finepak SIL C_{18} 4.6×250 mm) in a Twincle HPLC (Japan Spectroscopic Co., Ltd.) using the isocratic solvent system methanol-0.02 N acetic acid (1:1) at a flow rate of 1 ml/min. t_R values (min) of caffeic and rosmarinic acids were 5.40 and 6.35, respectively.

The effects of the broad-spectrum herbicide glyphosate (N-[phosphonomethyl]glycine) on the metabolism of phenolic compounds in several plants have been studied (Holländer and Amrhein 1980; Rubin et al. 1982). When 1 mM glyphosate was administered to the cell culture in the logarithmic growth phase (the 8th culture day), marked inhibition of caffeic acid formation was observed (Ishikura and Takeshima 1984), and the level of caffeic acid was almost stable during the following 5 days (Table 1). In the stationary growth phase, the level

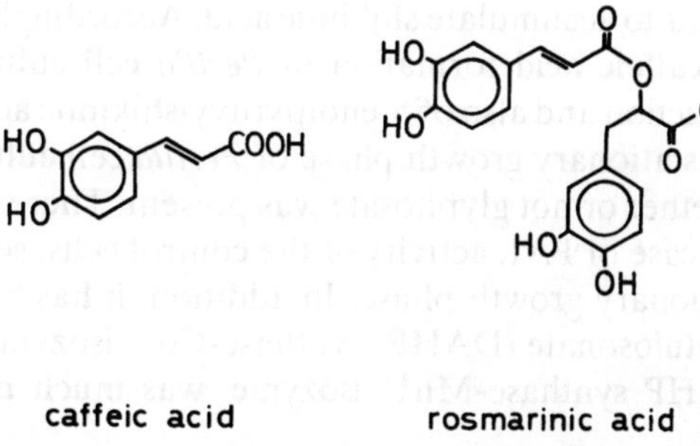

Fig. 4. Chemical structure of caffeic and rosmarinic acids

Table 1. Time courses for changes in fresh weight and caffeic acid content of *Perilla* cell suspension cultures in the absence and presence of 1 mM glyphosate. (Ishikura and Takeshima 1984)

Culture days	Days after dosage of glyphosate	Fr. cell wt. (g) per flask		Caffeic acid content			
				mg/flask		mg/g fr. wt.	
		Control	+ Glyphosate	Control	+ Glyphosate	Control	+ Glyphosate
8	0	5.39	–	8.40	–	1.55	–
9	1	8.78	7.89	13.87	12.57	1.58	1.59
10	2	12.97	8.98	19.98	10.92	1.54	1.22
11	3	13.70	7.89	24.39	10.35	1.78	1.31
12	4	14.94	7.89	28.98	10.57	1.94	1.34
13	5	14.02	7.99	35.89	10.16	2.56	1.27
14	0	13.38	–	37.41	–	2.80	–
16	2	11.32	11.58	28.45	30.05	2.51	2.59
18	4	10.39	10.52	26.02	29.07	2.51	2.76
20	6	11.87	10.85	28.70	25.74	2.50	2.37
22	8	10.58	10.47	26.45	22.77	2.50	2.17

of caffeic acid in the cells hardly decreased, like that of the control cells, during 8 days after the dosage of glyphosate on the 14th culture day. Accordingly, caffeic acid synthesis most likely ceased in the stationary phase, and this became more evident by a decline of phenylalanine ammonia-lyase (PAL) activity during the cell culture. PAL catalyzes the conversion of phenylalanine to t-cinnamic acid as the first committed step in the biosynthesis of phenolic substances such as caffeic acid.

PAL activity in the cells, as shown in Fig. 5, fluctuated with the day of culture, that is, the enzyme activity increased immediately after transfer of the cells to a fresh medium, reached maximum in the early logarithmic growth phase and decreased rapidly in the midst of the same phase. In the early stationary growth phase, PAL activity became almost undetectable (Ishikura and Takeshima 1984). Glyphosate at 1 mM inhibited the PAL activity and cell growth about 82% and 28%, respectively, during 48 h after its administration in the midst of the logarithmic growth phase (the 8th culture day). However, a marked inhibition of the PAL activity and the growth of *Perilla* cells was observed when glyphosate was administered to the cells in the early logarithmic growth phase. Glyphosate also caused reduction of the phenylalanine content of *Perilla* cells (Table 2) (Ishikura et al. 1986). On the other hand, the shikimic acid content of cells grown in the presence of 1 mM glyphosate increased up to 74.9 μg per g fresh weight during 6-day culture, while that of the control cells was nil (Table 3). Amrhein et al. (1980) have reported that glyphosate is an inhibitor of 5-enolpyruvylshikimic acid-3-phosphate synthase and causes the plants to accumulate shikimic acid. Accordingly, glyphosate seems to exert its effect on caffeic acid formation in *Perilla* cell culture through the inhibition of PAL induction and also of 5-enolpyruvylshikimic acid-3-phosphate synthase activity. In the stationary growth phase of *Perilla* cell culture, shikimic acid was not detected whether or not glyphosate was present. Therefore, the shikimic acid pathway, like the case of PAL activity of the control cells, seems to be almost inoperative in the stationary growth phase. In addition, it has been found that 3-deoxy-D-arabino-heptulosonate (DAHP) synthase-Co^{2+} isozyme from *Perilla* cells, in contrast with DAHP synthase-Mn^{2+} isozyme, was much more sensitive to inhibition by gly-

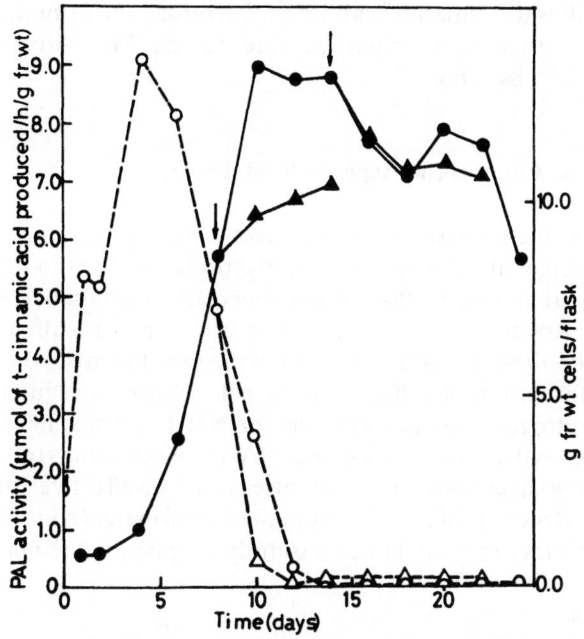

Fig. 5. Time course for changes in fresh weight (*solid line*) and PAL activity (*broken line*) of *Perilla* cell suspension cultures in the late logarithmic and stationary growth phases in the absence (●, ○) or presence (▲, △) of 1 mM glyphosate. Glyphosate was added on the day indicated by the *arrow*. (Ishikura and Takeshima 1984)

Table 2. Changes in phenylalanine content of *Perilla* cell suspension cultures in the absence and presence of 1 mM glyphosate. (Ishikura et al. 1986)

Culture days	Days after dosage of glyphosate	Phe content (µg/g fr. wt.)	
		Control	+ Glyphosate
6	0	35.5	–
8	2	131.4	31.2
10	4	82.5	15.7
12	6	81.9	43.8

Table 3. Changes in shikimic acid content of *Perilla* cell suspension cultures in the absence and presence of 1 mM glyphosate. (Ishikura and Takeshima 1984)

Culture days	Days after dosage of glyphosate	Shikimic acid content (µg/g fr. wt.)	
		Control	+ Glyphosate
10	2	nil	17.6
12	4	nil	21.8
14	6	nil	74.9
16	2	nil	nil
18	4	nil	nil
20	6	nil	nil

phosate (Ishikura et al. 1986). Therefore, the formation of shikimate in the presence of glyphosate might be due to the Mn^{2+}-isozyme activity rather than the Co^{2+}-isozyme.

2.4 Commercial Aspects/Exploitation

As mentioned above, the intact *Perilla* plants contain a large amount of flavones, apigenin, luteolin and their glycosides, and anthocyanins in addition to caffeic acid derivatives, but the callus tissues contain only caffeic acid derivatives as the phenolic extractives. Therefore, it is very easy to isolate caffeic and rosmarinic acids from the *Perilla* cell culture rather than the intact plants. It could be possible to promote the formation of caffeates in the cell culture by regulating the applicable amount of nitrogen sources (NO_3 and/or NH_4) or phosphate and also the combination of growth regulators, 2,4-D or NAA and cytokinins (cf. Ishikura et al. 1983). Thus, the *Perilla* cell suspension cultures could be effectively used for medicinal purposes as a dry cell powder or as an alcoholic cell extract containing caffeate at a high rate, and further they could be constantly supplied to the market throughout the seasons.

3 Conclusions and Prospects

Caffeic and rosmarinic acids are useful secondary metabolites as an antiphlogistic agent and for other remedies. The cell suspension culture of *Perilla frutescens* var. *crispa* produces caffeic and rosmarinic acids and caffeoylglucose as the major phenolic extractives. *Perilla* callus was induced from 6–7-day-old seedlings with Murashige and Skoog's (1962) basal medium supplemented with 2,4-D (1 mg/l) and kinetin (0.1 mg), which was the most effective for inducing the callus. The cell suspension culture was maintained in the modified Schenk and Hildebrandt medium supplemented with 2,4-D and kinetin. The doubling time of the cells in the logarithmic growth phase was about 60 h. Total caffeic acid content of the cells in the stationary growth phase corresponds to 0.5% and 9.5% of fresh and dry weights, respectively. The lowering of 2,4-D concentration and the high kinetin concentration in the medium increased caffeic acid formation. The best medium for the formation of caffeic acid was Murashige and Skoog's basal medium supplemented with 2,4-D (1 mg/l), kinetin (5 mg/l), sucrose (30 g/l) and agar (0.9% w/v). It is easier to isolate caffeic and rosmarinic acids from the cell suspension cultures than from the intact plants. It is also possible that the *Perilla* cell suspensions having a high caffeate content are constantly supplied to the market as a medicinal source throughout the seasons.

References

Amrhein N, Schab J, Steinrücken HC (1980) The mode of action of the herbicide glyphosate. Naturwissenschaften 67:356

Furukawa S, Tomisawa Z (1920) Distribution, yields, physiology and chemical components of shiso-oil from *Perilla nankinensis*. Kogyokagaku Kaishi 2 3:342-362 (in Japanese)

Gamborg OL (1966) Aromatic metabolism in plants II. Enzymes of the shikimate pathway in suspension cultures of plant cells. Can J Biochem 44:791-799

Gracza L, Koch H, Löffler E (1985) Über biochemisch-pharmakologische Untersuchungen pflanzlicher Arzneistoffe, 1. Mitteilung. Isolierung von Rosmarinsäure aus *Symphytum officinale* und ihre antiinflammatorische Wirksamkeit in einem In-vitro-Modell. Arch Pharm (Weinheim) 318:1090-1095 318:1090-1095

Holländer H, Amrhein N (1980) The site of the inhibition of the shikimate pathway by glyphosate I. Inhibition by glyphosate of phenylpropanoid synthesis in buckwheat (*Fagopyrum esculentum* Moench). Plant Physiol 66:823-829

Ibrahim RK, Edgar D (1976) Phenolic synthesis in *Perilla* cell suspension cultures. Phytochemistry 15:129-131

Ishikura N (1972) Anthocyanins and other phenolics in autumn leaves. Phytochemistry 11:2555-2558

Ishikura N (1981) Anthocyanins and flavones in leaves and seeds of *Perilla* plant. Agr Biol Chem 45:1855-1860

Ishikura N, Takeshima Y (1984) Effects of glyphosate on caffeic acid metabolism in *Perilla* cell suspension cultures. Plant Cell Physiol 25:185-189

Ishikura N, Iwata M, Mitsui S (1983) The influence of some inhibitors on the formation of caffeic acid in cultures of *Perilla* cell suspensions. Bot Mag (Tokyo) 96:111-120

Ishikura N, Teramoto S, Takeshima Y, Mitsui S (1986) Effects of glyphosate on the shikimate pathway and regulation of phenylalanine ammonia-lyase in *Cryptomeria* and *Perilla* cell suspension cultures. Plant Cell Physiol 27:677-684

Mitchell JP, Gildow FE (1975) The initiation and maintenance of *Vicia faba* tissue cultures. Physiol Plant 34:250-253

Murashige T, Skoog F (1962) A revised medium for rapid growth and bioassays with tobacco tissue cultures. Physiol Plant 15:473-497

Rubin JL, Gaines CG, Jensen RA (1982) Enzymological basis for herbicidal action of glyphosate. Plant Physiol 70:833-839

Shibata K (ed) (1949) A cyclopedia of useful plants and plant products. Hokuryukan, Tokyo, p 312 (in Japanese)

Sugisawa H, Ohnishi Y (1976) Isolation and identification of monoterpenes from cultured cells of *Perilla* plant. Agr Biol Chem 40:231-232

Yoshiki Y (1911) Studies on the components of shiso-oil. Yakugaku Zasshi 31:1-6

XXII *Podophyllum* spp.: In Vitro Regeneration and the Production of Podophyllotoxins

Yasuo Fujii[1]

1 Introduction

Lignans, as natural products, are distributed widely in the plant kingdom. More than 200 compounds in this general class have been identified. Lignans have aroused considerable interest because some of them display antitumor activities. This is particularly true of the podophyllotoxin group of lignans, which are constituents of the medical resin extracted from *Podophyllum* species. Podophyllotoxins are a particularly instructive class of natural products for consideration in the design and synthesis of potential anticancer agents based upon natural product prototypes.

1.1 History

The medical use of *Podophyllum* species dates back over 1000 years. At that time the roots of wild chervil were used in a salve for treating cancer in England. About 400–600 years ago, the natives of the Himalayas and the American Indians independently discovered that the aqueous extracts of the roots (podophyllin) from *Podophyllum* species was a canthartic and poison. After the American Indians introduced the use of podophyllin to the American colonists, it became such a popular drug that it was included in the US Pharmacopoeia in 1820 as a canthartic and cholagog and remained until 1942, when it was removed because of its severe toxicity.

However, Kaplan (1942) revived interest in podophyllin when he cured the venereal wart (condyloma acuminatum) with topical application of podophyllin oil. This led to studies of the action of podophyllin on tumor tissue and to intensive chemical examinations of the constituents of podophyllin. King and Sullivan (1946, 1947) demonstrated that podophyllin produced this therapeutic effect by acting as a mitotic poison, and Hartwell and Shear (1947) first reported that podophyllin inhibited the growth of experimental cancer cells in animals.

At the same time, the chemical structure of podophyllotoxin was established by Hartwell and Schrecker (1951). In the 1960's, Stahelin and his coworkers at Sandoz tested many semi-synthetic podophyllotoxin derivatives as anticancer agents. They (Stahelin 1970, 1973) found two derivatives, VP-16–213 and VM-26, which had potential anticancer activity with minimal toxicity. VP-16–213 (Etoposide) has

[1]Takasaki Research Laboratory, 219 Iwahana, Takasaki, Gunma, 370–12, Japan

Biotechnology in Agriculture and Forestry, Vol. 15
Medicinal and Aromatic Plants III (ed. by Y.P.S. Bajaj)

passed governmental approval for use in lung cancer and testicular cancer in several European countries and approved by the FDA in the USA in 1983 and in Japan in 1986.

1.2 Distribution

Podophyllum species are the perennial plants in the family Berberidaceae, and they have the chromosome number 2n = 12. One of the most common species is *Podophyllum peltatum*, (May apple), which grows in gregarious groups in the oak-hickory forests all over the eastern United States and southern Canada. It has whitish nodding flowers with parts in whorls of three between palmately dissected peltate leaves. Plants flower from March to May. Meijer (1974) surveyed the distribution of *P. peltatum* in eastern North America and identified the optimum conditions for its cultivation on a commercial scale.

A second species, *P. hexandrum* (*P. emodi*), distributed from the Himalayas to southwest China (India, Bhutan, Pakistan, China, and Formosa), has properties and uses similar to the May apple. It has white or pale red rotate flowers with palmately parted peltate leaves. The flowering time is May and the ripe fruits are red, ellipsoidal, and edible.

A third species, *Podophyllum pleianthum* (*Dysosma pleiantha*), grown in the southern part of China and Formosa, has two palmately lobate peltate leaves and has five to eight dark red nodding flowers at the forked point of the petioles.

Furthermore, *Dipphylleia sinensis* (China) and *D. cymosa* (Japan) and *Podophyllum japonica (Ranzania japonica,* Japan) are known to contain some podophyllotoxins.

In addition to the *Podophyllum* species, *Polygala polygama* (Hokanson 1979), *Juniperus virginiana* (Kupchan et al. 1965), *Linum album* (Weiss et al. 1975), and *Linum flavum* (Berlin et al. 1986) contain some podophyllotoxins.

1.3 Extraction and Structure of Podophyllotoxins

Analysis of nonglucosidic lignans present in an alcoholic extract of the dried root/rhizome of *Podophyllum* species showed that the range of lignans observed in *P. peltatum*, *P. pleianthum*, and *P. hexandrum* was the same except that no peltatins were detected in *P. pleianthum* (Jackson 1984, 1985). The lignan contents of *Podophyllum* species and their chemical structures are listed in Table 1 and in Fig. 1A.

1.4 Medicinal Components and Demand on the World Market

Etoposide, VP-16-213, 4′-demethyl epipodophyllotoxin ethylidene β-D-glucose (Fig. 1B), a semi-synthesitic derivative of podophyllotoxin, has been shown to be an effective antitumor agent in the treatment of lung cancer, a variety of leukemias, and other solid tumors. Etoposide is administered by injection or orally as a capsule. Etoposide is used mainly in the U.S.A. and Europe by the trade name of Vepeside.

Table 1. Lignan content of *Podophyllum* species (Jackson and Dewick 1985)

Lignan	*P. pleianthum*	*P. hexandrum*	*P. peltatum*
Podophyllotoxin (1)	13.5	128	7.5
4′-Demethylpodophyllotoxin (2)	4.1	13.5	0.2
Desoxypodophyllotoxin (3)	1.0	0.5	0.7
4′-Demethyldesoxypodophyllotoxin (4)	0.3	0.3	0.2
β-Peltatin (5)	–	0.3	10.0
α-Peltatin (6)	–	0.2	7.5
Podophyllotoxone (7)	3.7	1.7	0.6
4′-Demethylpodophyllotoxone (8)	1.3	0.4	0.2
Isopicropodophyllone (9)	2.0	1.0	0.2
4′-Demethylisopicropodophyllone (10)	0.5	0.2	0.1

Lignan content expressed as mg/10 g dried root/rhizome material.

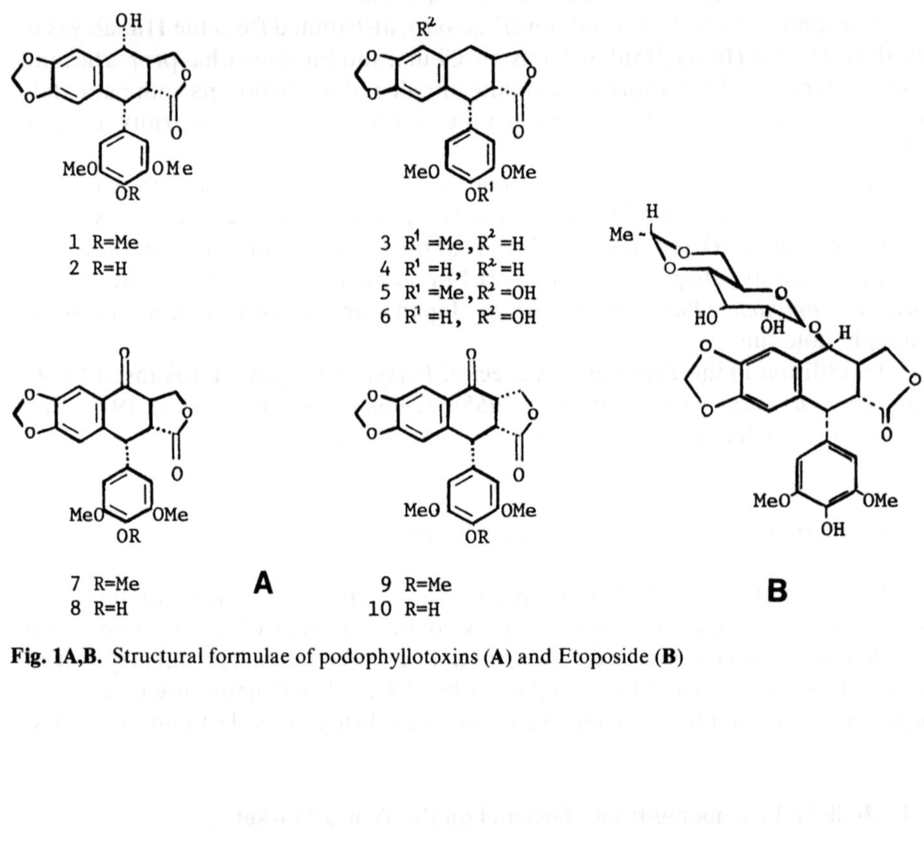

Fig. 1A,B. Structural formulae of podophyllotoxins (**A**) and Etoposide (**B**)

The estimated sold units (1 unit = 100 mg) in the U.S.A. and Europe were more than 830,000 units in 1986 and the gross sale of Vepeside in the U.S.A. and Lastet (trade name in Japan) were estimated to be 100 million $ and 3.8 billion yen in 1989, respectively.

2 In Vitro Approaches

2.1 Review of Tissue Culture

Published data on the tissue culture of *Podophyllum* species are very few. Kadkade (1981) first reported that rhizome segments of *P. peltatum* formed callus on MS medium and the 8-week-old callus produced α,β-peltatin and podophyllotoxin. Later he (Kadkade 1982) studied the effect of growth regulators, natural factors, carbon sources, and monochromatic light irradiation on the growth and podophyllotoxin production of *P. peltatum* callus tissues. Tissue cultures were established from root, rhizome, leaf, and stem segments on MS medium (Murashige and Skoog 1962) supplemented with 2,4-D, IAA, NAA, kinetin, and BA solidified with 0.5% agar. Of the three auxins, 1 ppm of NAA and IAA supported maximum growth of callus. Increasing levels of auxins were inhibitory to growth as well as podophyllotoxin biogenesis, while 2,4-D caused a marked increase in podophyllotoxin production. With respect to kinetin and 2,4-D, the optimum concentration for callus growth and podophyllotoxin formation was 0.2 and 1 ppm, respectively. Addition of casamino acid 0.05% was the most favorable for growth and podophyllotoxin biosynthesis. The podophyllotoxin contents of callus tissue derived from root and rhizome were higher than those from the other plant parts. Only sucrose was an effective carbon source for podophyllotoxin production. Callus production was promoted by 590 and 660 nm light, and the total podophyllotoxin levels were maximally affected by red light at 660 nm with the intensity of 750 $\mu W/cm^2$. The maximum podophyllotoxin level in callus was 0.74% on a dry weight basis.

Kashiwagi et al. (1987) reported that callus was induced from the root segments of *P. peltatum* on White's medium (1963) supplemented with 2,4-D 10^{-4} M, BA 10^{-4} M, and 3% sucrose. This callus regenerated unorganized roots on the same medium with the increased ratio of 8.2. Thirteen mg of podophyllotoxin were detected in 375 g of the unorganized roots. The podophyllotoxin content in unorganized roots was equal to that in original plant roots.

Formation of 5-methoxypodophyllotoxin and its glucoside by root cultures of *Linum flavum* was reported by Berlin et al. (1986, 1988). Extracts of root cultures contained 0.7–1.3% podophyllotoxins on a dry mass basis. Growth and product accumulation varied greatly with the 2,4-D concentration in the medium.

Recently, Chuang and Chang (1987a) reported the embryoids production and their development into plantlets from calli of somatic tissues of *P. pleianthum*, and furthermore, they (1987b) reported the somatic embryogenesis and plantlet regeneration in callus cultures derived from immature seeds' mature zygotic embryos of *P. pleianthum*.

Uden et al. (1989) obtained callus cultures from the roots of *P. hexandrum* Royle in B5 (Gamborg et al. 1968) medium supplemented with 2% coconut milk, 4 mg/l NAA, and 4% sucrose. Roots of in vitro plantlets were used as starting material for the induction of callus cultures. Suspension-grown cells in the above medium produced podophyllotoxin. The highest cellular accumulation of podophyllotoxin was measured at about 0.09% dry weight basis at day 15 in the dark.

2.2 Callus Induction of *P. pleianthum*, *P. peltatum*, and *P. hexandrum*

Plants that were purchased commercially and preserved in our greenhouse were used (Fig. 2). Newly developed leaves and stems were sterilized by immersing them in 70% ethanol for 10 s followed by 5% hypochlorite for 20 min. Callus cultures were initiated by incubating the sterile leaf and stem segments on 1% agar MS, B5 and White's medium containing 25 different concentrations of phytohormones consisting of BA, kinetin, NAA, 2,4-D, and IAA.

In *P. pleianthum*, B5 medium containing BA(0.02–10 ppm) and 2,4-D(0.2–10 ppm) gave the best results for callus induction (Table 2; Fig. 3A). The callus was transferred to MS and B5 agar and MS liquid medium and subcultured every month and every week, respectively.

In the case of *P. peltatum*, callus was readily formed on MS agar medium supplemented with 0.2 ppm kinetin and 1 ppm 2,4-D, as shown by Kadkade (1982). On the other hand, *P. hexandrum* formed hardly any callus.

Fig. 2. *P. pleianthum* more than 4 years old; grown in a greenhouse

Table 2. Effect of hormones on callus induction in *P. pleianthum* on B5 agar medium. (Fujii et al. 1986)

	2,4-D (mg/l)									
	0		0.02		0.2		2.0		10.0	
BA (mg/l)	Stem	Leaf	Stem	Leaf	Stem	Leaf	Stem	Leaf	Stem	Leaf
0	–	–	–	–	+	+	+	+	++	+
0.02	–	–	–	–	+++	+	+	++	++	+
0.2	–	–	+	–	+++	+	+++++	+++	++++	+
2.0	–	–	±	–	+++	+++	+	+	–	+
10.0	–	–	–	–	+	+++	+++	+	–	±

The callus of *P. pleianthum* and *P. peltatum* was suspended in 20 ml of MS liquid medium supplemented with 0.2 ppm kinetin and 1 ppm 2,4-D or IAA in a 100-ml flask shaken at 90 r.p.m. at 24 °C in 16 h light and 8 h dark. The tissues of *P. peltatum* and *P. pleianthum* showed steady growth from about the 3rd and 7th day of culture with the doubling time of about 5 and 10 days, respectively. Lag time depended largely on the amount of inoculated cells (Fig. 4).

Podophyllotoxins were analyzed with HPLC (Hitachi 655, Zorbax sil:25 cm × 4.6 mm) equipped with a UV detector. Elution solution was composed of 98 vol. chloroform, 2 vol. ethanol and 0.1 vol. water; the flow rate was 1 ml/min throughout analysis, and the effluent was monitored by measuring the absorbance at 254 nm. Podophyllotoxins were identified with mass spectrometry-Direct-EIMS and FAB-MS (JEOL JMS D-300/JMA-200 computer system).

During the suspension culture, no podophyllotoxins were detected in the cells and medium. In contrast, 0.001% to 0.04% (dry weight base) of podophyllotoxin was detected in *P. pleianthum* callus cultured on B5 agar medium supplemented with 0.2 ppm BA.

2.3 Plant Regeneration from Callus Tissue of *P. pleianthum*

Subcultured callus was transferred to shoot-forming medium and kept in 16 h light and 8 h dark at 25 °C. The surface of the transferred calli turned hard, and some of them regenerated colorless shoots and others produced green shoots directly (Fig. 3B,C). No further growth and greening were observed in colorless shoots after 3-month culture in the shoot-forming medium. The morphology of green shoots resembled those of the original plants. The green shoots were then transferred to agar medium or filter paper medium for rooting. Rooting medium contained no phytohormones. After root regeneration, the plantlets were planted in a glass pot of vermiculite (Fig. 3D) and were grown in a chamber with 16 h light and 8 h dark at 25 °C.

Shoot regeneration was observed in wide ranges of hormone contents. Table 3 shows that shoot regeneration from the callus required or was accelerated by the addition of a higher amount of cytokinin than that of auxin.

Multiple shoot formation was observed during callus culture. Table 4 shows that two to 20 shoots were regenerated from each callus. As shown in Table 5, podo-

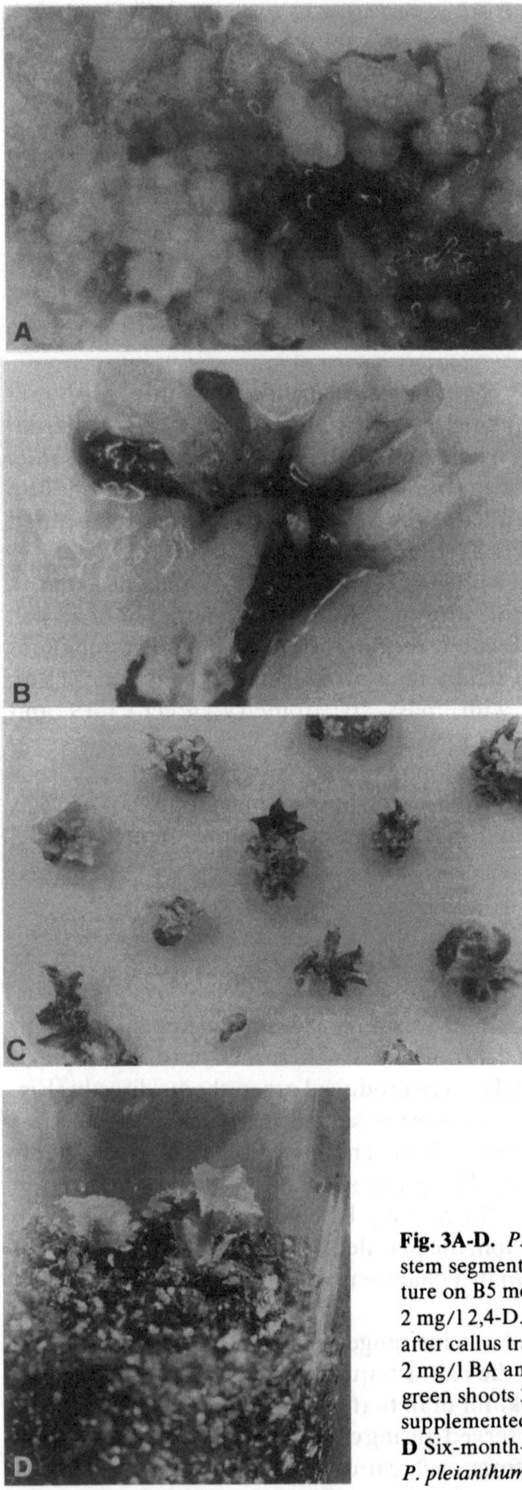

Fig. 3A-D. *P. pleianthum.* **A** Callus formation from a stem segments of *P. pleianthum* 1 month after explant culture on B5 medium supplemented with 0.2 mg/l BA and 2 mg/l 2,4-D. **B** Regeneration of colorless shoots 1 month after callus transfer to B5 medium supplemented with 2 mg/l BA and 0.02 mg/l NAA. **C** Regeneration of green shoots 3 months after callus transfer to B5 medium supplemented with 2 mg/l BA and 0.02 mg/l NAA. **D** Six-month-old regenerated plants from callus of *P. pleianthum*

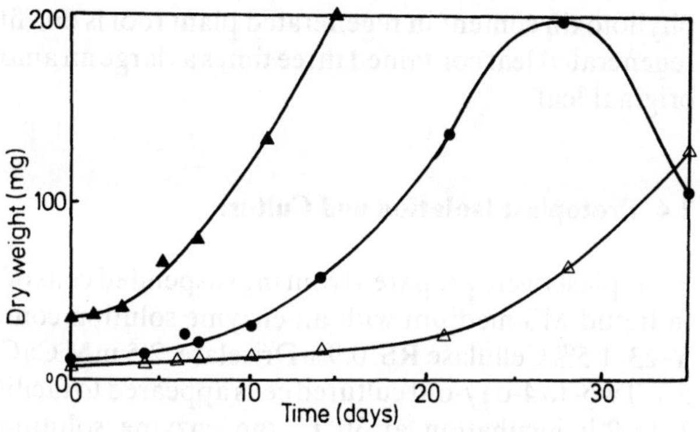

Fig. 4. Growth curves of *Podophyllum pleianthum* (●) and *Podophyllum peltatum* (▲, △) cell suspensions cultured on MS medium with 0.2 ppm kinetin and 1 ppm 2,4-D. Inoculated *P. peltatum* cells in curve (▲) were four times as much as in curve (△)

Table 3. Effect of hormones on shoot regeneration in *P. pleianthum* on B5 agar medium. (Fujii et al. 1986)

BA (mg/l)	NAA (mg/l) 0	0.02	0.2	2.0	10.0
0	–	++++	++++	+++S	+++
0.02	++	+S	+S	++++	++++
0.2	++++S	++S	+++S	+++++	+++++
2.0	++++S	+++S	+++S	+++S	++
10.0	+	+	+	–	+

+: callus growth, S: shoot formation.

Table 4. Shoot regeneration frequency of *P. pleianthum* callus on MS agar medium. (Fujii et al. 1986)

Phytohormones (mg/l)	No. of shoot forming callus/total callus	Shoot number/callus
2.0 BA + 0.2 NAA	18/26	4–20
2.0 BA + 0.2 2,4-D	10/28	2–15
3.0 BA + 0.05 NAA(MS/2)	3/11	6–12
0.3 BA + 0.01 IAA	4/5	3–11
2.0 Kn + 0.01 NAA	5/23	4–7

Table 5. Podophyllotoxin contents of original plants and regenerated plants of *P. pleianthum*. (Fujii et al. 1986)

Plant part	Podophyllotoxin (mg/g dry wt.) Min.	Max.	Mean
Original plant			
Root	4.83	17.69	12.17
Leaf	0.04	2.52	1.09
Regenerated plant			
Root	1.20	3.98	2.61
Leaf	1.07	6.88	3.14

phyllotoxin content in regenerated plant root is one fifth of that in original root, but regenerated leaf contained three times as large an amount of podophyllotoxin as the original leaf.

2.4 Protoplast Isolation and Culture

Protoplasts were prepared from the suspended cells of *P. peltatum* and *P. pleianthum* in liquid MS medium with an enzyme solution containing 0.1% (w/v) pectolyase Y-23, 1.5% Cellulase RS, 0.9% Driselase, 2.5 mM $CaCl_2$, and 0.45M Mannitol at pH 5.7. Two- to 4-day-old cultured cells appeared to facilitate protoplast isolation. After 1 to 2 h incubation at 30°C, the enzyme solution was pipetted. The isolated protoplasts were filtered through a 64-μm nylon net and washed three times by centrifugation. The protoplasts were cultured at a density of 1×10^4/ml in a 2.5-cm well containing 1 ml of a liquid medium. The culture medium was composed of MS medium without NH_4NO_3, supplemented with 0.2 ppm BA, 1 ppm 2,4-D, 1% glucose and 0.4M mannitol at pH 5.7. After 3 to 7 days of culture, the medium was replaced with a half volume of fresh medium or solidified by adding a half volume of 0.2% Gellan gum solution containing 1% glucose and 0.4 M mannitol. From the 2nd week, the fresh medium with reduced mannitol content to 0.2 M was replaced every week. Protoplast-derived calli were transferred to MS agar medium for further growth and plant regeneration.

The yield of protoplasts was $2\text{–}4 \times 10^6$ per 1 g fresh weight of cultured cell in both species (Fig. 5A,B).

First protoplast division was observed at 3 to 10 days after culture (Fig. 6A,B) in both species. Cell wall regeneration was confirmed with staining for 5 min with a 0.1% solution of Calcofluor White ST.

In *P. peltatum*, the divided cells grew to the cluster during 1 to 2 months culture (Fig. 7). Two months after, the developing calli were transferred to MS agar medium for further growth and plant regeneration (Fig. 8). Protoplasts exhibited the best survival and colony formation when they had been plated at densities of 1×10^4 cells/ml.

The necrotic behavior of the cells in MS medium originated from the high concentration of ammonium. In modified MS medium, in which no $NH_4\,NO_3$ was added, sustained cell divisions led to colony formation. Another critical factor affecting sustained cell division was the kind of sugar. When sucrose was used, first cell divisions occurred, but daughter cells finally ceased to divide. In contrast, glucose induced sustained cell divisions to form colonies.

Both auxin and cytokinin were required for protoplast division. 2,4-D appeared to give more reproducible results than NAA. The optimal concentrations of 2,4-D and BA were 1 and 0.2 ppm, respectively.

Fig. 6A,B. First cell division of *P. pleianthum* (**A**) and *P. peltatum* (**B**) protoplast after 10 days in culture. × 200

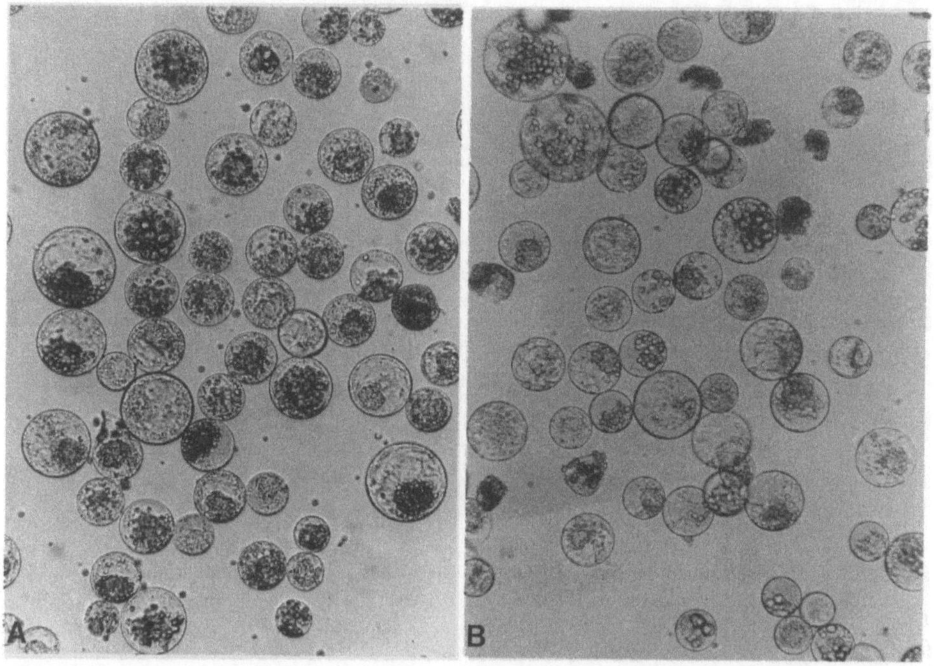

Fig. 5A,B. Freshly isolated protoplasts from suspension culture of *P. pleianthum* (**A**) and *P. peltatum* (**B**). ×200

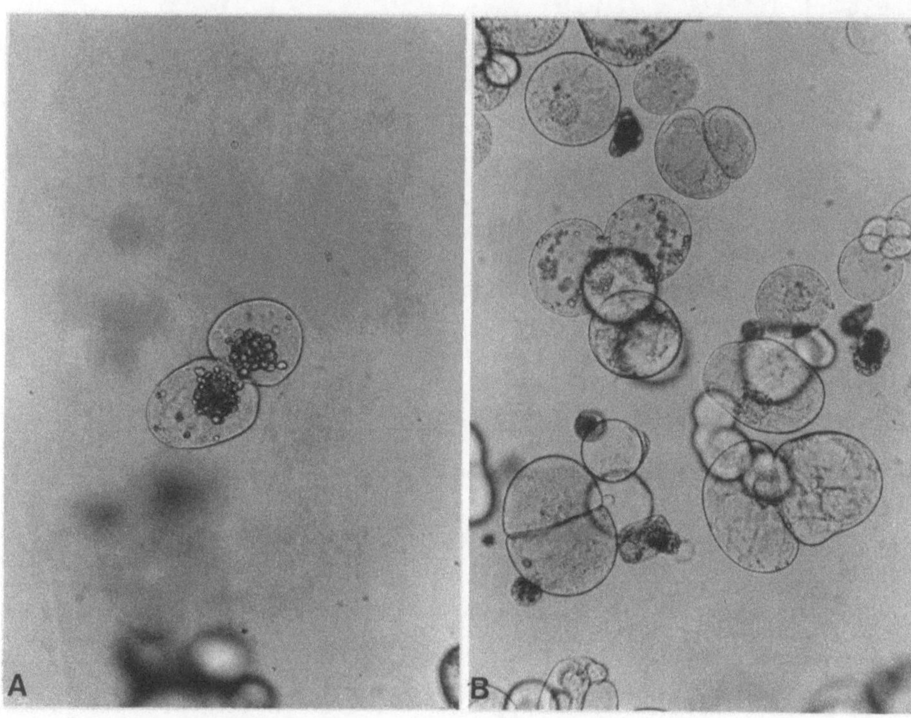

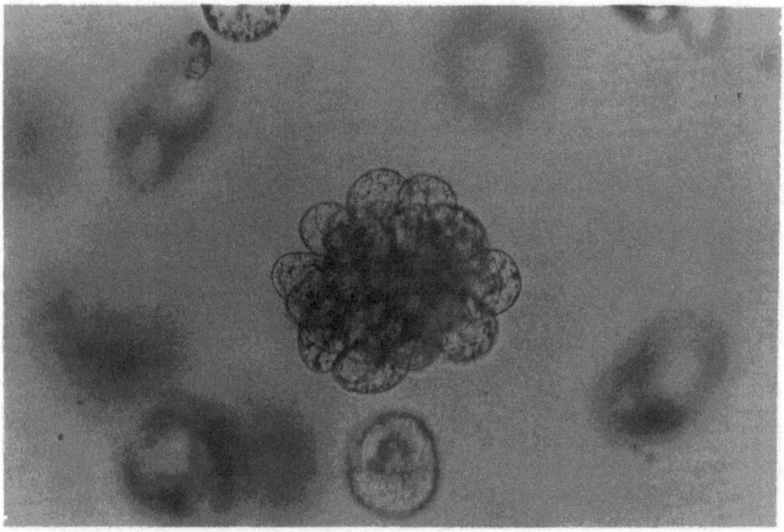

Fig. 7. Cluster derived from *P. peltatum* protoplasts at 50 days in Gellan gum medium. ×100

Fig. 8. Growing calli from *P. peltatum* protoplasts on MS agar medium supplemented with kinetin (0.2 ppm) and 2,4-D (1 ppm)

3 Conclusions and Prospects

Callus induction, suspension culture, plant regeneration from callus and protoplast culture of three *Podophyllum* species (*P. pleianthum*, *P. peltatum*, and *P. hexandrum*) were examined under various culture conditions. 2,4-D or NAA in the presence of BA promoted callus induction and callus growth. The best medium for callus culture was B5 medium supplemented with 0.2 ppm BA and 2 ppm 2,4-D or NAA. Suspension culture was established from both *P. pleianthum* and *P. peltatum* callus in MS liquid medium supplemented with 0.2 ppm kinetin and 1 ppm 2,4-D. Plant regeneration from callus tissue of *P. pleianthum* was accomplished by culturing the callus on a cytokinin-rich medium. The best medium for plant regeneration was B5 medium containing 2 ppm BA and 0.2 ppm NAA. Podophyllotoxin contents of the regenerated plant leaves was three times higher than that of original plant leaves, but the regenerated plant roots contained 1/4 podophyllotoxin compared to that in original plant roots.

Etoposide is one of the most successful medicines derived from plant material. A major direction for future studies of podophyllotoxins should focus on the mechanism of action and on the development of new podophyllotoxin analogs which not only have different pharmacological potencies but also possess unique biological properties. There are several studies demonstrating that Etoposide acts in acute nonlymphoblastic leukemia and lymphoma (Bennett 1982), and furthermore Etoposide has therapeutic value in the treatment of AIDS-associated Kaposis sarcoma (Laubenstein LJ et al. 1983).

It is estimated that the world market of chemicals produced by plants is over 30 billion $, and about 20 of these chemicals will be produced by tissue culture or by genetically improved plants, by 1995 (Misawa 1987). For industrial production, highly valuable chemicals (> 1000 $/kg) such as antitumor substances must be chosen, largely due to the low and/or unstable productivity of many undifferentiated cultures. To some degree, this deficiency can be overcome by screening for cell clones derived from individual high-producing cells, as with the production of ajmalicine by *Catharanthus roseus* cultures (Zenk et al. 1977) and shikonin by cultures of *Lithospermum erythrorhizon* (Fujita et al. 1984).

A development that may revolutionize the role that in vitro culture plays in fine chemical synthesis is the discovery of rapidly growing, productive, and stable "hairy root" cultures obtained by the genetic transformation of plant tissue by the pathogenic soil bacterium, *Agrobacterium rhizogenes* (Tepfer 1984). The induction of hairy root and production of secondary metabolites were established in the formation of pigments and alkaloids such as shikonin (Shimomura et al. 1986), betacyanin (Hamill et al. 1986), atropine, scopolamine, and hyoscyamine (Mano et al. 1986). These hairy roots produced as many alkaloids as normal roots and the alkaloid pattern was similar to that of the original plants. Thus, hairy root cultures may be a useful system for large-scale production of secondary metabolites, yet they are amenable to the techniques of selection and genetic manipulation that have been developed with dispersed cell systems. The infection of *Agrobacterium rhizogenes* to *Podophyllum* species was tried in our laboratory, but no hairy roots have been formed yet.

The large-scale production of podophyllotoxins by tissue culture has not succeeded at present; however, plant regeneration from callus, suspension culture, and protoplast culture were partially successful.

Regarding the protection of natural resources, large-scale propagation of plants using a technique of plant regeneration which can produce multiple clonal plants in vitro is strongly desired.

References

Bennett JM (1982) A phase II trial of Acridinyl anisidine/M-AMSA/ and VP-16-213 in refractory acute mycloid leukemia in adults. Proc Am Soc Clin Oncol 1:135

Berlin J, Wray V, Mollenshott C, Sasse F (1986) Formation of β-peltatin-A-methyl ether and coniferin by root cultures of *Linum flavum*. J Nat Prod 49:435–439

Berlin J, Bedorf N, Mollenschott C, Wray V, Sasse F, Hofle G (1988) On the podophyllotoxins of root cultures of *Linum flavum*. Planta Med 54:204–206

Chuang MJ, Chang WC (1987a) Embryoid formation and plant regeneration in callus cultures derived from vegetative tissues of *Dysosma pleiantha* (Hance) Woodson. J Plant Physiol 128:279–284

Chuang MJ, Chang WC (1987b) Somatic embryogenesis and plant regeneration in callus culture derived from immature seeds and mature zygotic embryos of *Dysosma pleiantha* (Hance) Woodson. Plant Cell Rep 6:484–485

Fujii Y, Ishida K, Kawashima H (1986) Plant regeneration from callus of *Podophyllum pleianthum* Hance. In: Proc 6th Int Congr Plant tissue and cell culture, Univ Minnesota, Minneapolis, p 37

Fujita Y, Takahashi S, Yamada Y (1984) Selection of cell lines with high productivity of shikonin derivatives through protoplast of *L. erythrorhizon*. Eur. Congr. Biotechnology, 1:161–166

Gamborg OL, Miller RA, Ojima K (1968) Nutrient requirements of suspension cultures of soybean root cells. Exp Cell Res 50:151–158

Hamill JD, Parr AJ, Robins RJ, Rhodes MJC (1986) Secondary product formation by cultures of *Beta vulgaris* and *Nicotiana rustica* transformed with *Agrobacterium rhizogenes* Plant Cell Rep 5:111–114

Hartwell JL, Schrecker AW (1951) Components of podophyllin. 5. The constitution of podophyllotoxin. J Am Chem Soc 73:2909–2916

Hartwell JL, Shear MJ (1947) Chemotherapy of cancer. Classes of compounds under investigation and active components of podophyllin. Cancer Res 7:716–717

Hokanson GC (1979) Podophyllotoxin and 4′-demethylpodophyllotoxin from *Polygala polygama* (Polygalaceae). Lloydia 41:497–498

Jackson DE, Dewick PM (1984) Aryltetralin lignans from *Podophyllum hexandrum* and *Podophyllum pleianthum*. Phytochemistry 23:1147–1152

Jackson DE, Dewick PM (1985) Tumour-inhibitory aryltetralin lignans from *Podophyllum pleianthum*. Phytochemistry 24:2407–2409

Kadkade PG (1981) Formation of podophyllotoxins by *Podophyllum peltatum* tissue cultures. Naturwissenschaften 68:481–482

Kadkade PG (1982) Growth and podophyllotoxin production in callus tissues of *Podophyllum peltatum*. Plant Sci Lett 25:107–115

Kaplan IW (1942) *Condylomata acuminata*. New Orleans Med J 94:388–390

Kashiwagi H, Takahashi E, Takemoto T (1987) Production of anti-tumor substances. JP 62-96088

King L, Sullivan M (1946) The similarity of the effect of podophyllin and colchicine and their use in the treatment of condylomata acuminata. Science 104:244–245

King L, Sullivan M (1947) Effect of podophyllin and of colchicine on normal skin, on condyloma acuminatum and on verruca vulgalis. AMA Arch Pathol 43:374–386

Kupchan SM, Hemingway JC, Knox JR (1965) Tumor inhibitors 7. J Pharm Sci 54:659–660

Laubenstein LJ, Krigel RL, Hymes KB, Muggia FM (1983) Treatment of epidemic Kaposi's Sarcoma with VP-16-213 (Etoposide) and a combination of Doxorubicin. Bleomycin, and Vinblastine (ABV). Proc Am Soc Clin Oncol 2:228

Mano Y, Nabeshima S, Ohkawa H (1986) Production of tropane-alkaloid from roots of tropane-alkaloid-producing plants infected with *Agrobacterium rhizogenes*. JP 61:254195

Meijer W (1974) *Podophyllum peltatum*-May apple a potential new cash-crop plant of eastern North America. Econ Bot 28:68–72

Misawa M (1987) Production of tumor inhibitors by cell cultures (in Japanese) Hakko to Kogyo 45, 12:1195–1203

Murashige T, Skoog F (1962) A revised medium for rapid growth and bioassays with tobacco tissue culture. Physiol Plant 15:473–497

Shimomura K, Satake M, Kamada H (1986) Production of useful secondary metabolites by hairy roots transformed with Ri plasmid. In: Proc 6th Int Congr Plant tissue and cell culture, Univ Minnesota, Mineapolis, p 50

Stahelin H (1970) 4′-Demethyl-epipodophyllotoxin thenylidine glucoside (VM-26), a podophyllum compound with a new mechanism of action. Eur J Cancer 6:303–311

Stahelin H (1973) Activity of a new glycosidic lignan derivative (VP-16-213) related to podophyllotoxin in experimental tumors. Eur J Cancer 9:215–221

Tepfer D (1984) Transformation of several species of higher plants by *Agrobacterium rhizogenes*. Cell 37:959–967

Uden van W, Pras N, Visser JF, Malingré M (1989) Detection and identification of Podophyllotoxin produced by cell cultures derived from *Podophyllum hexandrum* royle. Plant Cell Rep 8:165–168

Weiss SG, Tin-wa M, Perdue RE, Fransworth NR (1975) Potential antitumor agents 2. J Pharm Sci 64:95–98

White PR (1963) The cultivation of animal and plant cells. Ronald, New York

Zenk MH, El-Shagi H, Arens H, Stockigt J, Weiler EW, Deus B (1977) Formation of the indole alkaloids serpentine and ajmalicine in cell suspension cultures of *C. roseus*. In: Barz W, Reinhard E, Zenk MH (eds) Plant tissue culture and its biotechnological application. Springer, Berlin Heidelberg New York, pp 27–43

XXIII *Rosa* spp. (Roses): In Vitro Culture, Micropropagation, and the Production of Secondary Products

K.C. SHORT[1] and A.V. ROBERTS[2]

1 Introduction

The genus *Rosa* includes over 100 species which are distributed throughout the temperate and subtropical regions of the Northern Hemisphere (Rehder 1960). Chromosome numbers range from 2n = 2x = 14, to 2n = 8x = 56 (Darlington and Wylie 1955). The DNA content of the rose genome is amongst the lowest recorded in the Angiospermae, the 4C value of *R. wichuraiana* (2n = 14) measuring only 0.45–0.48 pg (Lloyd 1986). The chromosomes are correspondingly small (Fig. 1).

Cultivars which are now most commonly grown in gardens as ornamental plants are classified as Hybrid Tea (large flowered), Floribunda (cluster flowered) and miniature roses. Hybrid Tea and miniature roses are also grown under glass for the sale of cut flowers and pot plants, respectively. Each of these classes of rose has the ability to produce flowers throughout the growing season. This "perpetual flowering" characteristic is determined by a recessive gene (Hurst 1941). The complex origins of modern roses from crosses between perpetual flowering roses which were introduced from China ca. 1800 A.D. and European cultivars with short flowering seasons, have been reconstructed by Hurst (1941) and Wylie (1954, 1955a,b). Wylie (1954) drew attention to the narrow genetic base of the modern garden roses, only eight species having contributed significantly to their gene pool. Attempts to introduce novel genes into modern garden roses by introgressive hybridization are constrained by F_1 sterility, arising either through differences in ploidy level of the parents, or genomic incompatibility.

Cultivars of *R. rugosa* are widely used for amenity horticulture, particularly for roadside verges and urban landscapes. Although they do not possess the gene for perpetual flowering, they do have a longer flowering season than other "species" roses.

Most cultivars are highly heterozygous and do not breed true to type. They are, therefore, propagated vegetatively. Miniatures are often propagated from cuttings, but other classes are usually propagated by budding, or grafting onto root stocks of species which include *R. canina* 'Inermis', *R. multiflora* 'Simplex' and *R. dumentorum* 'Laxa'. The advantages and disadvantages of "own-rooted" plants are important considerations in assessing the potential market of micropropagated roses.

[1]Faculty of Science, Nottingham Polytechnic, Nottingham NG1 4BU, United Kingdom
[2]Plant Biotechnology Research Unit, Polytechnic of East London, London E15 4LZ, United Kingdom

Biotechnology in Agriculture and Forestry, Vol. 15
Medicinal and Aromatic Plants III (ed. by Y.P.S. Bajaj)

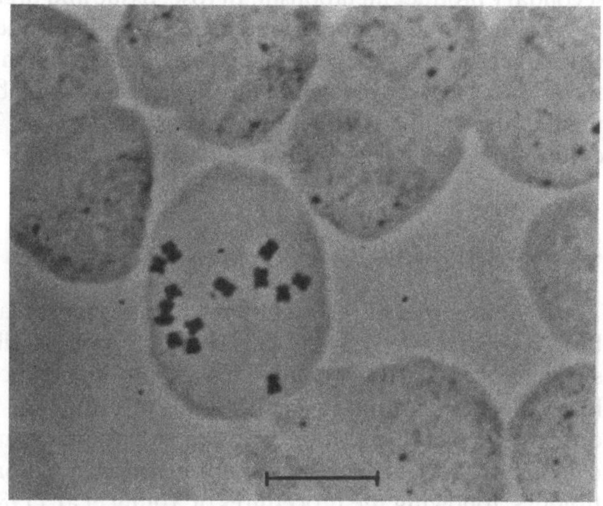

Fig. 1. Chromosomes of *R. wichuraiana* (2n = 14). (Root tips were pretreated with 0.05% colchicine solution for 4 h and stained by the Feulgen procedure. After squashes had been prepared, the slide was frozen to permit the removal of the coverslip, dipped in ethanol and flamed. Supplementary staining was provided by 0.05% toluidene blue in citrate buffer pH 4.0). Bar = 10 μm

Ornamental roses are grown for a variety of purposes, e.g., as cut flowers, pot plants, anemity and garden plants, and as many popular cultivars are not protected by patent, it is difficult to reliably assess production. It is quite clear, however, that throughout the world the rose is associated with beauty and special occasions and in many countries it is the premier ornamental plant. More than 200 million rose bushes are planted each year in gardens the world over, representing a retail market of about US$720 million. The importance of the rose as a cut flower is indicated by sales of more than 4 billion blooms with an appropriate annual retail value of US$11 billion. In the U.K., approximately 30 million field-grown plants and 0.5 million cut flowers are sold annually. In continental Europe, there is a greater demand for cut flowers and 900 million are sold annually in just one market at Aalsmere (Holland).

R. damascena is cultivated in Bulgaria, France, Italy, Turkey, Iran, the U.S.S.R, Morocco and the U.S.A. for the production of attar (otto) of rose or oil of roses (Krussmann 1981). Attar is a highly regarded perfume which is widely used in the cosmetic industry and is one of the most highly prized essential oils (Lata and Gupta 1971). The estimated wholesale price of attar is US$3,000 per kg and the world retail market is US$12 million per annum (Brodelius 1985). The oil of roses is extracted from the blooms of *R. damascena* and *R. centifolia* and about 1 kg can be distilled from 3000 kg of blooms. There are many chemical components in attar which include rhodinol, geraniol, nerol, linalool, citral, phenyl ethyl-alcohol, eugenol and carvone (Krussmann 1981). All of these can be extracted from other flowers or manufactured synthetically; however, the demand for pure attar oil remains.

2 Culture in Vitro

Rose tissue cultures have been exploited for a variety of purposes including studies on the growth in culture of cells and protoplasts, their nutrition, physiology and secondary product metabolism. In rose, tissue and organ cultures are used both for the production of pathogen-free materials and for micropropagation. The latter is extensively used for the commercial production of many cultivars. The impact of the techniques of plant biotechnology have not, to date, been utilized in the development of new cultivars and improved varieties of rose. The literature on the in vitro culture work is summarized in Table 1.

2.1 Micropropagation

The establishment of shoot cultures of rose is most readily achieved from terminal buds taken from actively growing shoots. Surface sterilization is reliable if buds are stripped of all partly expanded leaves and then exposed to a solution of sodium hypochlorite containing a small volume of surfactant. They can then be inoculated directly onto multiplication medium. Any contamination which is subsequently evident is likely to have arisen from endogenous bacteria. Treatment with antibiotics, following the procedures of Young et al. (1984), eliminates bacteria without detriment to the plant.

Multiplication of roses in vitro (Fig. 2A) is based on the subdivision of plantlets into branches, terminal buds and nodal sections. Multiplication rates of between two- and ninefold, over a 4-week period, as exemplified by Fig. 3, are obtained on multiplication media (Table 2) containing Murashige and Skoog (1962) (MS) salts, vitamins and amino acids with or without modification (Table 3); 30–40 g/l sucrose; 0.5–3.0 mg/l benzylaminopurine (BAP); 0.004–0.1 mg/l naphthaleneacetic acid (NAA) or 0.1 mg/l indoleacetic acid (IAA) or no auxin; 0.1 mg/l gibberellic acid (GA_3) or more usually no GA_3; pH 5.6–5.8; 7–8 g/l agar; sterilized by autoclaving at 121 °C for 15 min. Hasegawa (1980) found that multiplication was less satisfactory if BAP was replaced by 2-isopentyl adenine (2-iP) or kinetin. Bressan et al. (1982) found that a temperature of 21 °C and a photon flux density (400–700 nm) of 17 $\mu E/m^2/s$ for 12 to 24 h daily was optimal for shoot multiplication.

The number of plantlets which can be induced to root on transfer to rooting medium (Table 1) varies with genotype. The percentage of plants which can be successfully transplanted to compost varies, typically within the range 20–100%, as exemplified by Table 4. After successful transplantation, growth proceeds rapidly (Fig. 4). Several authors have found that rooting is enhanced if the concentration of salts is reduced to 1/4–1/2 strength (Table 2). Hyndman et al. (1982) found that these beneficial effects could be attributed to reduced levels of nitrogen. All authors (Table 1) excluded BAP from the medium, most used higher levels of NAA or IAA than in multiplication media, but some excluded all growth regulators. Khosh-Khui and Sink (1982a) found that indolebutyric acid (IBA) was less satisfactory than either NAA or IAA, but that best rooting occurred when NAA and IBA were used in combination. Bressan et al. (1982) found that a temperature of 21 °C and a photon flux density of 66 $\mu E/m^2/s$ was optimal for root initiation. Mortality on transplantation to compost can be significantly reduced if plantlets are rooted in vitro in Sorbarod plugs (Baumgartner Papiers SA, Lausanne, Switzerland) (Lloyd et al. 1988; Douglas et al. 1989; Fig. 2B). These plugs are cylindrical rods loosely packed with a cellulose matrix and wrapped in a paper sleeve. When shoots are inserted into

Table 1. Roses which have been cultured in vitro

Species/hybrid	Inoculum	Medium[a] (mg/l)	Growth response	Reference
R. arvensis and R. cooperi	Lateral buds	MS + BAP (0.2)	Multiplication	Short et al. (1981)
	In vitro shoots	MS + NAA(0.1) or IAA(1)	Rooting	
R. canina	Embryos and achenes	K	Germination	Graifenberg (1973)
	Shoot tips/meristems	Mod MS + BAP(1) + NAA(0.5)	Multiplication	Martin et al. (1981);
	In vitro shoots	Mod MS × 1/2 + NAA(0.1)	Rooting	Khosh-Khui and Sink (1982b)
R. damascena	Anthers	MS + kin(0.4) + IAA(2)	Callus	Tabaeezadeh and Khosh-Khui (1981)
	Shoot tips	Mod MS + BAP(1) + NAA(0.1)	Multiplication	Khosh-Khui and Sink (1982b)
	In vitro shoots	Mod MS × 1/2 + NAA(0.1)	Rooting	
R. hybrida	Heat-treated buds	MS + BAP(0.5–1)	Virus-free shoots	Bjarnason et al. (1985)
R. hydria				
Alexander, Athena, Champanger, Flamingo, Gabriella, Golden Times, Kardinal, Lilly Marleen, Lorena, Mercedes, New Dawn, Pasadena, Red Gold, Queen Elizabeth	Lateral buds	MS + BAP(2) + NAA(0.004) + GA(0.1)	Initiation Multiplication Rooting	Horn et al. (1988)
Baccara, Espoir, Lily Marlene,	Shoot tips/meristems	Mod MS + BAP(0.1–2) + NAA(0.1) + GA(0.1–1)	Multiplication	Martin et al. (1981)
Orange Sensation and Sonia	In vitro shoots	Mod MS + NAA(0.1–1) + GA(0.1–1)	Rooting	
Baccara	Anthers	MS + kin(0.8) + IAA(7.5)	Callus	Tabaeezadeh and Khosh-Khui (1981)
Bellona, Bingo, Candia,	Lateral buds	Syn med + activated charcoal(4g)	Initiation	Curir et al. (1986)
Cocktail 80 and Sonia	In vitro shoots	Syn med + BAP(1)	Multiplication	
	In vitro shoots	Syn med + IAA(0.8)	Rooting	
Bluesette	In vitro shoots	Mod MS × 1/2 + BAP(0.5) + IAA(0.1) + GA(10)	Multiplication	de Proft et al. (1985)
Bridal Pink	Shoot tips and lateral buds	Mod MS + BAP(2) + NAA(0.1)	Multiplication	Khosh-Khui and Sink (1982a,b)
	In vitro shoots	Mod MS × 1/2 + NAA(0.1)	Rooting	
Duftwolke, Iseta	Lateral buds	MS + BAP(2) + NAA(0.1)	Multiplication	Sauer et al. (1985);
Lorena and Mercedes	In vitro shoots	MS × 1/3 + IAA(2)	Rooting	Walther and Sauer (1986)
Escapade	Lateral buds	MS + BAP(2)	Multiplication	Short et al. (1981)

Table 1. *Continued*

Species/hybrid	Inoculum	Medium[a] (mg/l)	Growth response	Reference
and Scarlet Gem	In vitro shoots	MS + NAA(0.1) or IAA(1)	Rooting	
Fiona, Sunburst Red and Toy Clown	Shoot tips and Lateral buds	MS + BAP(1) + NAA(0.1) + GA(0.1)	Multiplication	Douglas et al. (1989)
Fiona	In vitro shoots	MS × 1/4 + IAA(0.1)	Rooting	Douglas et al. (1989)
Forever Yours	Shoot tips	Mod MS + BAP(2) + NAA(0.1)	Multiplication	Skirvin and Chu (1979);
	In vitro shoots	Mod MS × 1/4	Rooting	Wilkowske (1981)
Gold Glow, Improved Blaze and Mr Lincoln	Lateral branches	Mod MS + BAP(3)	Multiplication	Bressan et al. (1982)
	In vitro shoots	Mod MS + IAA(1)	Rooting	
Improved Blaze	Shoot tips, lat. buds	Mod MS + BAP(3) + IAA(0.3)	Multiplication	Hyndman et al. (1982)
	In vitro shoots	Mod MS × 1/4–1/2 + NAA(0.03–1) or IAA(1)	Rooting	Hasegawa (1979, 1980)
Kings Ransom, Parade and Plentiful	Lateral buds	MS + BAP(2) + NAA(0.004) + GA(0.1)	Multiplication	Davies (1980)
	In vitro shoots	MS + NAA(0.005–0.1)	Rooting	
Landora	Lateral buds	MS + BAP(0.5) + GA(0.25)		Rout et al. (1989)
Pauls Scarlet	Protoplasts	Mod MS + 2,4-D(1.1) + kin(0.5)	Callus	Strauss and Potrykus (1980)
	Stem explants	Syn med	Callus, cell suspensions	Nash and Davies (1972)
Mercedes	Lateral buds	MS + BAP(2) + NAA(0.1)	Multiplication	Martin et al. (1981);
		MS × 1/3 + IAA(2)	Rooting	Sauer et al. (1985)
Sam McGredy	Meristems	MS + BAP(1) + NAA(0.004) + GA(0.1)	Virus-free shoots	Casey (1985); Douglas et al. (1989)
and Queen Elizabeth	In vitro shoots	MS + BAP(1) + NAA(0.1) + GA(0.1)	Multiplication	
	In vitro shoots	MS × 1/4 + IAA(0.1)	Rooting	
Soraya	Flower buds	Mod W + 2,4-D(0.6)	Callus	Krishnamurthy et al. (1979)
	Callus	Mod MS + NAA(4)	Callus	
	Protoplasts	Mod MS + NAA(4) + mannitol(12%)	Cells	
Super Star	Shoot tips	Syn med + kin(8)	Leaf growth	Jacobs et al. (1969; 1970a,b)
	Shoot tips	Syn med + NAA(0.5–2)	Rooting	
	Shoot tips	Syn med + NAA(8) + GA(16)	Callus	

The Doctor	Stem	Syn med + kin(2) + NAA(0.2)	Callus	Hill (1957)
	Callus	Syn med + kin(0.2) + NAA(0.5) + GA(20)	Shoot primordia	
Tropicana	Shoot tips, lat. buds	Mod MS + BAP(2) + NAA(0.05)	Multiplication	Khosh-Khui and Sink (1982b)
		Mod MS × 1/2 + NAA(0.1)	Rooting	
10 cultivars	Heat-treated buds	MS + BAP(0.5–1)	Virus-free shoots	Bjarnason et al. (1985)
R. indica	Shoot tips, lat. buds	Mod MS + BAP(2)	Multiplication	Graifenberg et al. (1975)
	In vitro shoots	Mod MS + NAA(0.1)	Rooting	Martin et al. (1981)
				Avramis et al. (1982)
R. manetti	Stem, leaf	MS + kin(0.25) + 2,4-D(2) + CH(0.2%)	Callus	Khosh-Khui and Sink (1982c)
	Callus	MS + BAP(0.5) + NAA(2) + CH(0.2%)	Callus	
R. multiflora	Shoot tips	Mod MS + BAP (1–11 mg/l)	Shoot growth and rooting	Elliott (1970); Graifenberg et al. (1975)
R. persica × *xanthina*	Shoot tips	MS + BAP(0.5) + NAA(0.005)	Multiplication	Lloyd (1986);
	In vitro shoots	MS × 1/4 + IAA(0.1)	Rooting	Lloyd et al. (1988)
	Callus, roots leaves	MS + BAP(0.5–3) + NAA(0–0.005)	Shoots	
R. rugosa	Shoot apices	MS + BAP(1) + NAA(0.005)	Multiplication	Tweddle et al. (1984);
	Fruits	Syn med	Callus, cell suspensions	Wegg and Townsley (1988)
R. wichuraiana	In vitro shoots	MS	Rooting	Lloyd (1986)
Rosa spp.	Embryos	Syn med	Seedlings	Asen (1948)

[a] Different media have been used for the same species/hybrid by various workers for the same species/hybrid, however, in Table 1 only the simplest or most clearly defined medium is given.

Abbreviations: Basal media: K: Knudson (1946), MS: Murashige and Skoog (1962), W: White (1934), syn med: synthetic medium defined in cited paper, or undefined.

Supplements: BAP: benzylaminopurine, CH: casein hydrolysate, 2,4-D: dichlorophenoxyacetic acid, GA: gibberellic acid, IAA: indoleacetic acid, kin: kinetin, NAA: naphthaleneacetic acid.

Fig. 2A,B. *Rosa hybrida* 'Frensham' cultured in vitro. **A** Shoot proliferation on MS medium supplemented with 2 mg/l BAP and 0.01 mg/l NAA, culture age 7 weeks. **B** Shoots on sorbarod plugs moistened with rooting medium (half strength MS medium containing 0.1 mg/l NAA, culture age 4 weeks. Shoots are cultured in small sterile plastic containers (Guest Medical Ltd, Sevenoaks, Kent, U.K.). The base of the culture vessel is 20 mm

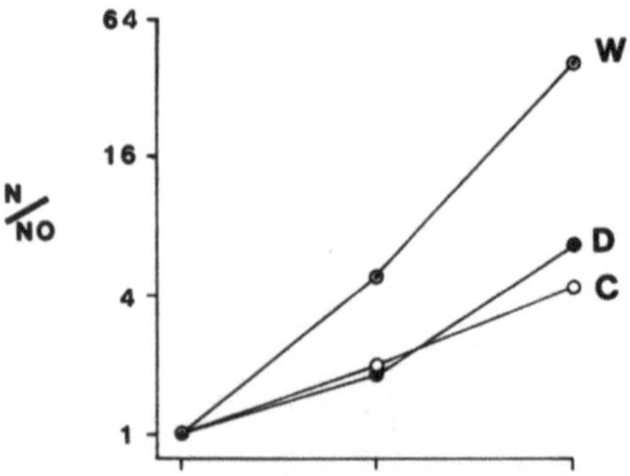

Fig. 3. In vitro multiplication of three rose cultivars. Multiplication (log scale) over two successive propagation cycles, each of 4 weeks. (After Tweddle et al. 1984). *No* and *N* numbers of propagules at start and end of each cycle. *No* = 20 throughout. *W R. wichuraiana* on 1 mg/l BAP, multiplied from nodal sections; *C R. hybrida* Clarissa on 2 mg/l BAP, multiplied from axillary shoots. *D R. hybrida* Dame of Sark on 1 mg/l BAP, multiplied from axillary shoots. All cultures included MS salts and vitamins, 3% sucrose and 0.005 mg/l NAA

Table 2. Multiplication and rooting media

Growth regulators (mg/l)									
Multiplication medium			Rooting medium					Modified strengths	Reference
BAP	NAA	IAA	GA_3	NAA	IAA	IBA	GA_3	of rooting medium	
2				0.1					Avramis et al. (1982)[a]
1		0.3			1				Bressan et al. (1982)[a]
2	0.004		0.1	0.05–0.1					Davies (1980)
3		0.3			0.3				Hasegawa (1979)[a]
1–10		0.3		Either 0.03–0.1 or 1.0				1/4–1/2 strength salts	Hasegawa (1980)[a]
2	0.1			0.1		0.05		1/2 strength salts	Khosh-Khui and Sink (1982a)[a]
1–2	0.05–0.15			0.1				1/2 strength salts	(1982b)[a]
0.1–2		0.1–1			0.1–1		0.1–1		Martin et al. (1981)[a]
2	0.1				2			1/3 strength salts	Sauer et al. (1985)
2	0.1							1/4 strength salts, vits, and acids	Skirvin and Chu (1979)[a]
1–2	0.005			0–0.3					Tweddle et al. (1984)

[a] Modified MS formulation listed in Table 3.
All media including modified[a] or unmodified MS salts, vitamins and amino acids, 30–40 g/l sucrose, pH adjusted to 5.6–5.8 before addition of 7–8 g/l agar, then autoclaved at 121 °C for 15 min. Where authors have varied concentrations of growth regulators, the most successful levels/ranges are indicated.

Table 3. Components of MS salts, vitamins and amino acids modified by some authors in multiplication and rooting media. Modifications of MS formulation (mg/l)

Salts, vitamins and amino acids	Relevant components of MS formulation (mg/l)	Avramis et al. (1982)	Bressan et al. (1982) Hasegawa (1979) Hasegawa (1980) Khosh-Khui and Sink (1982a,b)	Martin et al. (1981)	Skirvin and Chu (1979)
NH_4NO_3	1650.0	1155.0			
$Na_2EDTA \cdot 2H_2O$	37.3	46.62			
$FeSO_4 \cdot 7H_2O$	27.8	34.75			
Nicotinic acid	0.5	1.0		1	
Pyridoxine HCl	0.5	1.0		1	2
Thiamine HCl	0.1	1.0	0.5	0	1
Ca D-pantoghenate	0				1
Biotin	0				1
Folic acid	0				0.5
Choline chloride	0				1
p-Amino benzoic acid	0				0.5
Riboflavin	0				0.5
Nicotinamide	0				2
Vit B12	0				0.0015
Glycine	2.0				0
L-glutamine	0			200	

Table 4. Percentage of plants surviving transplantation to soil

Species	NAA (mg/l) in rooting medium	% Survivors after 4 months (± SE)
R. hybrida Clarissa	0	85 (± 3.4)
R. hybrida Dame of Sark	0.3	20 (± 5.6)
R. rugosa Scabrosa	0	46 (± 6.6)
R. wichuraiana	0	99 (± 0.9)

Rooting medium contained MS salts and vitamins × 1 strength, 3% sucrose, 0.8% agar and 0 or 0.3 mg/l NAA. Plants transplanted to Levington's compost in a mist propagator for 4 weeks. (After Tweddle et al. 1984).

the uppermost end of the plugs, which have been moistened with rooting medium without agar, the roots readily penetrate the cellulose matrix. When plantlets are transferred in the plug to compost, they are protected from physical damage by the cellulose padding. Furthermore, as the cellulose plug is rapidly bio-degraded, it does not hamper the establishment of the root system in soil. The plants can be grown to maturity in a greenhouse or under field conditions (Fig. 4).

The most significant feature about vegetative propagation in vitro is its enormous productive capacity. Martin (1985) has demonstrated that, using this tech-

Fig. 4. *R. hybrida* Clarissa 3 months after transplanting, scale line in cm

nology, up to 400,000 plants can be cloned, on an annual basis, from a single rose. Consequently, such a method has considerable implications for the rose propagator and is of value to the rose breeder as it allows rapid multiplication of new varieties and greater responsiveness to market requirements.

Shoots of Hybrid Tea and Floribunda roses often grow more vigorously on special rootstocks after budding or grafting than on their own roots after propagation from cuttings. Caution is needed in extrapolating from this experience to predict the growth characteristics of micropropagated roses. It should be appreciated that the "juvenile" vigour of micropropagated roses is more akin to that of a seedling than that of a cutting. This phenomenon was interpreted by Martin (1985) in terms of the way in which roots are initiated. He pointed out that roots are formed, in vitro without intervening callus, so that there is a perfect connection between the vascular systems of the roots and stem.

An evaluation of published and unpublished data of rose plants produced by tissue culture and those propagated by in vivo methods, including stenting (van de Pol 1986), grafting, budding and single node cuttings, reveals some consistency of difference. Micropropagated plants are well suited for cut-flower production as they are more compact (Onesto et al. 1985), branch better and sometimes yield more flowers (Reist 1985b). In addition, tissue culture-derived dwarf roses which are used for pot plant production have a faster rate of growth, flower earlier, and exhibit shorter shoots and more laterals than conventionally produced plants (Dubois et al. 1988). Many of these differences favour micropropagation, but advantages are partly offset by higher production costs. An advantage of micropropagated roses which is frequently cited is that, as they grow their own roots, suckers need not be removed. This is an important consideration in garden and amenity horticulture.

2.2 Elimination of Viruses

Two approaches have been taken in the elimination of virus by culture in vitro. Bjarnason et al. (1985) eliminated prunus necrotic ringspot virus (PNRV) from ten cut flower cultivars and one rootstock cultivar by heat treatment (38 °C) followed by culture of axillary buds in vitro. Casey (1985) eliminated PNRV and arabis mosaic virus from 'Queen Elizabeth' rose and rose mosaic virus from Sam McGredy rose by meristem culture of plantlets grown in vitro. After treatment, rose plants may be tested for virus using indicator plants, serologically specific electron microscopy, or enzyme-linked immunosorbent assay (Horst 1983).

2.3 Breeding

Ionizing radiation can be used to induce changes in the colour, size and shape of flowers, and in habit. A review of literature on mutagenesis in vivo was provided by Broertjes and van Harten (1988). After exposure to mutagens, roses exhibit mericlinal and periclinal chimerism. To retrieve pure mutants and stable periclinal chimeras, the technique of pruning to force growth through a succession of lateral buds (Broertjes and van Harten 1988) is generally applied. The procedure of segregating mixed tissues is easier to organize and monitor if irradiated material is multiplied in vitro from nodal sections. It seems likely that this approach will now be preferred.

Sterility of F_1 hybrids is a major obstacle in rose breeding. It is predictable that sterility arising from genomic differences could be overcome by doubling the chromosome number with colchicine (Heslop-Harrison 1958). There is, however, only one published example of the successful use of colchicine in this way (Semenuik and Arisumi 1968). Practical difficulties arise in the controlled application of colchicine to terminal buds, the penetration of colchicine to the meristem, and the subsequent segregation of polyploid plants from heteroploid chimeras. The anticipated advantages of using in vitro-cultured material include: the opportunity to immerse small shoots or terminal buds in liquid medium containing colchicine, more rapid penetration of colchicine to the meristem, and the greater convenience of in vitro culture in the segregation of heteroploid chimeras.

The induction of callus from protoplasts by Strauss and Potrykus (1980) leads to optimism that protoplast technology might be employed in the production of somaclonal variants and somatic hybrids. Likewise, the induction of haploid callus from anther cultures by Tabaeezadeh and Khosh-Khui (1981) suggests the possibility of obtaining dihaploids that could be used in the development of seed-propagated varieties. Exploitation of novel genotypes by either of these approaches depends on the subsequent generation of plantlets from callus. Lloyd et al. (1988) found that adventitious shoots could be induced from leaves and roots (Fig. 5A) (direct organogenesis) and callus (Fig. 5B) (indirect organogenesis) in *R. persica* ×

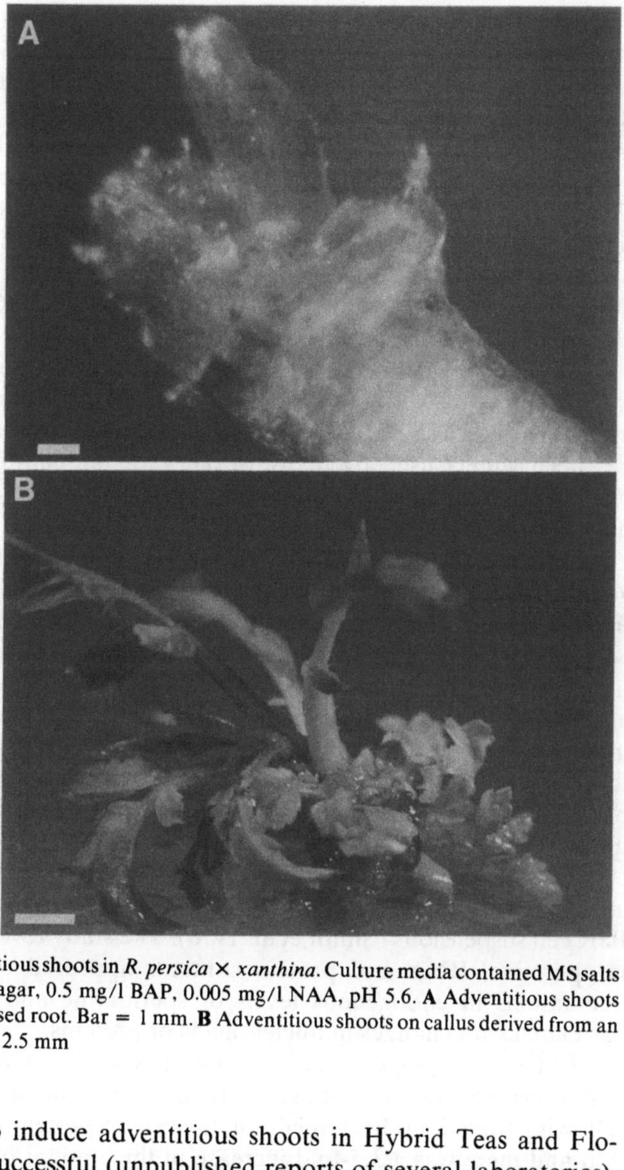

Fig. 5A,B. Formation of adventitious shoots in *R. persica* × *xanthina*. Culture media contained MS salts and vitamins, 3% sucrose, 0.8% agar, 0.5 mg/l BAP, 0.005 mg/l NAA, pH 5.6. **A** Adventitious shoots forming on the cut end of an excised root. Bar = 1 mm. **B** Adventitious shoots on callus derived from an internodal stem segment. Bar = 2.5 mm

xanthina, but attempts to induce adventitious shoots in Hybrid Teas and Floribundas have been less successful (unpublished reports of several laboratories). Direct and indirect organogenesis both have potential value in the generation of somaclonal variants and in the rapid segregation of chimeras following treatment with colchicine or other mutagens.

2.4 Callus and Cell Culture, and the Production of Secondary Products

For over 25 years, callus and cell suspension cultures of rose have been used as tools to investigate the dynamics of plant cell growth and for physiological and biochemical studies. Nickell and Tulecke (1959) initiated callus cultures, on a complex medium, from young stem explants of Paul's Scarlet rose (PSR) and Weinstein et al. (1962) related biochemical changes in amino acids, organic acids, sugars and nucleic acids to the growth characteristics of the cultures. The PSR callus system is ideal for such studies as it has a rapid rate of growth and as a result has been used for biosynthetic studies on chlorophyll (Hilderbrandt et al. 1963), ubiquinone and plastoquinone (Threlfall and Goodwin 1963) and proteins (Dougall 1965). Furthermore, because of its friable nature, PSR callus is easily suspended in liquid medium to produce a homogeneous and undifferentiated suspension culture which was used in the early studies on the large-scale serial culture in 9 l (Nickell and Tulecke 1960) and 134 l volumes (Tulecke and Nickell 1960) and for the continuous culture of plant cells (Tulecke et al. 1965).

A chemically defined organic medium for the growth of PSR cell suspensions was developed by Nesius et al. (1972) and also by Nash and Davies (1972). The latter medium produces a 15-fold increase in cell biomass over a 12-day growth cycle and has been widely used by many workers for the culture of cell suspensions for physiological and biochemical studies, e.g., histidine biosynthesis (Davies 1971) and polyphenol metabolism (Davies 1972).

Nash and Davies (1972) characterized the metabolic events associated with growth and cell division activity in PSR cell suspension cultures. The study showed that respiratory activity, protein accumulation, DNA and RNA content, together with the incorporation of labelled amino acids develops immediately after stationary phase cells are inoculated into fresh medium. These metabolic events reach a peak early in the log phase of growth and sustain cell division activity until factors in the culture medium become limiting. However, Nash and Davies (1972) found that the accumulation of secondary, phenolic metabolites, does not follow this pattern but peaks after the cells have ceased their division and have passed into a quiescent state.

The influence of changes in polyamines, which are implicated in the control of growth in various biological systems, has been followed during the growth cycle of PSR cell suspensions (Smith et al. 1978). The study revealed that cellular changes in spermine, RNA and protein contents appeared to be correlated, whereas fluctuations in endogenous levels of spermidine and putrescine could not be correlated with changes in nucleic acids or proteins.

Fosket (1982) used PSR cell cultures to study protein synthesis in the transition of stationary phase cells to those in an actively dividing state. The study revealed that following transfer to fresh medium, the rate of protein synthesis increased 10–15-fold and there was a 3.5-fold increase in the amount of ribosomal RNA per cell during the log phase of growth. Analysis indicated that there were few differences in the spectrum of polypeptides synthesized during the three phases of growth – stationary phase, lag phase and logarithmic growth. The various phases of growth could be characterized by fluctuations in the rate of synthesis of a few specific polypeptides.

Cell suspension cultures of PSR have been used to demonstrate that the synthesis of certain amino acids is subject to end-product inhibition (Fletcher and Beevers 1971; Fletcher 1975). However, Bradford and Fletcher (1982) have used this system to show that the early steps of nitrogen assimilation (nitrate reductase and nitrite reductase) are not regulated by a similar end-product control mechanism.

Plant tissue cultures have been regarded for many years as an attractive alternative to whole plant-derived, natural secondary products (Bajaj 1988). Cultured cells have the potential of providing a steady supply of commercially important natural products, providing techniques can be developed to grow high-yielding strains on inexpensive media (Fowler 1986). Unfortunately, some of the most highly sought-after components are not produced in cultured cells (Bohm 1982) and Table 5 shows the range of products which have been detected in rose tissue cultures. The much sought-after oil of roses (attar) is produced in quantity only in the petals of roses and there are no reports of its detection in tissue cultures.

PSR suspension cultures have been used as a model system to study the regulation of secondary, phenolic compounds. Amorim et al. (1977) confirmed the report of Nash and Davies (1972) that phenolic production is greatest in the stationary phase of the growth cycle, but these workers found that the phenolic content could be increased by adding more glucose to the culture medium.

More recently, Mubitch and Fletcher (1985) found that supplementing stationary phase cultures of PSR with sucrose and spermidine resulted in an increased yield and wider range of phenols. Therefore it would seem that extending the life

Table 5. Secondary products detected in rose tissue cultures

Compound	Reference
Carotenoids[a]	
Violaxanthin	Williams and Goodwin (1965)
Zeaxanthin	
Neoxanthin	
Auroxanthin	
Steroids and Triterpenes[a]	
β-Sitosterol	Williams and Goodwin (1965)
Stigmasterol	
Campesterol	
β-Amyrin	
Tannins and tannin precursors[b]	
Catechin	Davies (1972)
Epicatechin	
D-Glucogallin	
Epicatechincatechin	Mubitch and Fletcher (1984, 1985)
Gallic acid	
Epigalocatechin gallate	Mubitch and Fletcher (1984, 1985)
Epicatechin gallate	Mubitch and Flecther (1984)
Ferulic acid	
Leucoanthocyanins	Amorim et al. (1977)
Gallotannins	Mubitch and Fletcher (1984)
Vitamins[b]	
Ascorbic acid	Wegg and Townsley (1983)

[a] In callus tissues.
[b] In suspension cultures.

span of mature nondividing cell cultures offers a means of increasing the yield of secondary products.

The addition of precursors of secondary metabolites to cell cultures may also bring about an increase in productivity. Wegg and Townsley (1983) found that suspension cultures of the briar rose, *R. rugosa*, accumulated small quantities of ascorbic acid and the yield could be increased up to five fold by supplementing the culture medium with the precursors L-galactono-1,4, lactone or L-gulono-1,4, lactone (Fig. 6). Unfortunately, this increased yield was much below that found in the fruit of the rose hip.

Several studies have shown that cell suspension cultures are capable of metabolizing a range of organic compounds. Scheel and Sandermann (1977) reported the metabolism of DDT and kelthane in parsley and soybean suspension cultures. Langebartels and Harms (1984) showed that cultures of soybean and wheat metabolized pentachlorophenol, and Fletcher et al. (1987) demonstrated that axenic cell cultures of PSR were able to degrade a chlorinated biphenyl. It would seem, therefore, that cell cultures have potential for the biodegradation of relatively stable organics.

2.5 Protoplast Culture

Suspension cultures of PSR have proved to be a convenient source material for the isolation of protoplasts. Pearce and Cocking (1973) isolated large numbers of protoplasts from actively dividing cell cultures of rose and studied their behaviour and growth in culture. These workers found that rose protoplasts could be readily isolated but their development in culture was limited to cell wall regeneration and occasional cell division. Strauss and Potrykus (1980) devised a procedure for the successful cultivation of rose protoplasts which resulted in sustained cell division and callus formation. A key factor in the in vitro culture of these protoplasts was the rapid serial pre-culture of rose cell suspensions prior to protoplast isolation. The culture of rose protoplasts, with a plating efficiency of the order of 20%, which developed to the callus stage, has been reported by Krishnamurthy et al. (1979). There have been no reports of the regeneration of shoot primordia from in vitro cultured rose protoplasts.

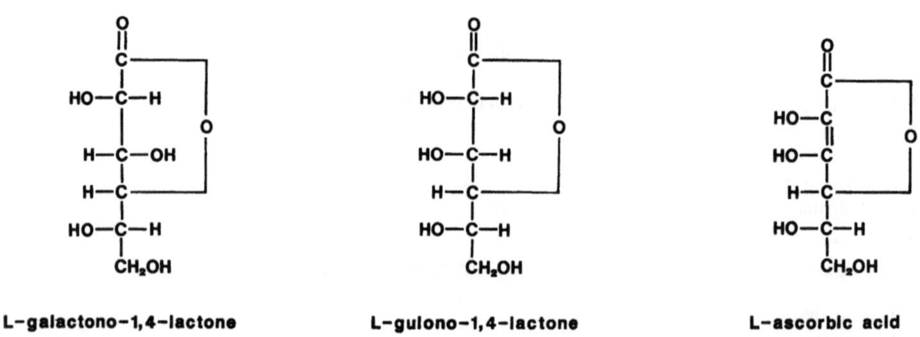

Fig. 6. Structure of L-galactono-1,4-lactone, L-gulono-1,4-lactone and L-ascorbic acid

3 Protocols

1. *Source of explants*

Explants may be taken from field roses in early summer, or at any time from glasshouse-grown plants. Mother plants should be in a state of vigorous growth and without symptoms of disease. Internode stem explants are suitable for the initiation of callus tissue, whereas for the initiation of shoot cultures axillary buds are less suitable than terminal buds because they are in a dormant state and are more likely to be contaminated with endogenous bacteria. All opened and partly opened scale leaves should be removed as these may be a source of microbial contamination.

2. *Surface sterilization*

This may be achieved by immersing up to ten terminal buds for 20 min in 100 ml sodium hypochlorite solution (1–1.5% available chlorine) containing surfactant 7X (0.1%) to ensure that buds are fully wetted. Two washes for 20 min in 100 ml of sterile distilled water are required to remove all traces of sterilant. Buds may then be inoculated either directly onto multiplication medium (Table 6), or if endogenous bacteria resist the above treatment, onto a multiplication medium containing a combination of antibiotics (Table 7).

3. *Shoot multiplication*

A general purpose multiplication medium for rose is provided in Table 6. The optimal combination of growth regulators is likely to vary between species and a range of media should, ideally, be tested.

Depending upon the natural habit of the species and the level of BAP employed, multiplication may be based upon (a) multiple-shoot culture, where the unit of multiplication is a shoot, (b) single-shoot culture, where the unit of multiplication is a terminal bud/nodal explant, or (c) a combination of (a) and (b). The addition of GA_3 (0.01–0.1 mg/l) may be helpful in extending the length of internodes and thereby facilitating the excision of nodal sections.

Table 6. Rose multiplication medium

Murashige and Skoog (1962) salts, amino acids and vitamins	
BAP	2 mg/l
NAA	0.01 mg/l
Sucrose	30 g/l
agar	8 g/l

pH 5.6; optimum culture temperature 21 °C, optimum light intensity, 17 μE/m²/s; photoperiod, 12–24 h per day.

Table 7. Antibiotic solution. (After Young et al. 1984)

Compound	mg
Rifampicin	120
Polymyxin B(sulphate)	120
Cefatoxime (sodium salt)	500
Tetracycline (hydrochloride)	500
DMSO	5 ml

The antibiotics are dissolved sequentially in DMSO and can be deep frozen until required. 0.25 ml of the above solution is added to 1 l of molten, autoclaved, multiplication medium.

4. *Rooting*

Roots can be induced within 2 weeks of the subculture of shoots on a rooting medium containing ½ strength Murashige and Skoog (1962) salts (Table 8). Nodal stem explants, containing axillary buds, are unresponsive to this treatment, as the medium does not alter their quiescent state.

The use of Sorbarod plugs in combination with liquid medium is recommended for rooting. Sorbarod plugs absorb virtually their own volume of liquid medium and, after they have been saturated, can be transferred to the culture vessel (Fig. 2) without further addition of medium. Under these conditions the developing root system is well aerated, grows considerably more rapidly than in agar medium and exhibits precocious root hair development.

5. *Transplantation*

Plants taken from agar-based medium need to be washed in running water to remove residual agar. Following planting in sterile compost they should be carefully weaned for about 2 weeks in an atmosphere of high relative humidity, which is maintained by fogging or misting devices. Plants which have been rooted in Sorbarod plugs do not need to be washed, and can be transferred directly to compost. Furthermore, the period of acclimatization of the plantlets is less demanding and a relative humidity of only 40–60% needs to be maintained for a 10-day period before they are subjected to ambient conditions.

6. *Callus cultures*

Callus cultures can be initiated from soil-grown plants by incubating internodal stem sections on the Murashige and Skoog (1962) medium supplemented with NAA and BAP (Table 9). In vitro-cultured plants may also be used for the initiation of callus cultures and explants from any organ readily produce a friable callus on rose proliferation medium (Table 9). The NAA content of the media may need to be adjusted for each species and concentrations below 0.1 mg/l may lead to the darkening of the explant due to the accumulation of polyphenols. Routinely a four- to sixfold increase in biomass is supported by this medium every 4 weeks.

Table 8. Rose rooting medium

½ strength Murashige and Skoog (1962) salts, amino acids and vitamins	
NAA	0.1 mg/l
Sucrose	30 g/l
Agar	8 g/l

ph 5.6; optimum culture temperature 21 °C, optimum light intensity, 66 μE /m^2/s; photoperiod, 12–24 h per day.

Table 9. Rose callus proliferation medium

Murashige and Skoog (1962) salts, vitamins and amino acids	
BAP	0.5 mg/l
NAA	0.1–1.0 mg/l
Sucrose	30 g/l
Agar	8 g/l

ph 5.6; optimum culture temperature 21–23 °C, optimum light intensity, 17 μE /m^2/s; photoperiod, 12–24 h per day.

7. *Suspension cultures*

Stock suspension cultures can be initiated from small pieces of callus placed in 50 ml liquid medium of Nash and Davies (1972) in a 250-ml Erlenmeyer flask and incubated on a horizontal rotary shaker (80 r.p.m.). Cultures should be maintained on a 14-day cycle by transfer of 6 ml of inoculum to 60 ml of fresh medium in a 250-ml flask. At each subculture an approximate 15-fold increase in biomass is found.

4 Conclusions and Prospects

Rose cell suspension cultures have proved to be an excellent experimental system for investigating the mechanisms which regulate growth, cell division and the physiological and biochemical processes of cultured plant cells. It is likely that future studies will focus very much at the molecular level and this culture system has much potential for elucidating the molecular events which control protein synthesis and cell division activities in plant cells. This information may facilitate the manipulation and growth in culture of isolated plant cells and protoplasts.

The fundamental data obtained from rose cell cultures on the nutritional and hormonal requirements for plant growth are being exploited in the large-scale continuous culture of many plant cells. This is becoming increasingly important with the move away from synthetic food additives, products and fragrances towards nature-derived materials. There is much speculation as to whether cultured plant cells can be utilized as a controllable source of "natural products". This approach has not fulfilled the early promise, and has been limited by the lack of appropriate source material and our inability to modify biochemical pathways to increase product yield. In the case of rose cell suspensions, future studies are likely to be directed to transformation studies, in which target genes from other species will be incorporated to produce, e.g., flavour, fragrances and enzymes.

Micropropagation of roses is proving to be commercially profitable and is now practised on a large scale in France, the U.K., the Netherlands and several other countries (George and Sherrington 1984). A recent survey by the present authors, for example, showed that six companies in the U.K. are currently producing one million roses by micropropagation and that demand is unsaturated. Present outlets include specialist rose growers and garden centres. As evidence of quality and acceptability of micropropagated roses, such as that provided by Dubois et al. (1988), Martin et al. (1981) and Reist (1985a,b), is assimilated by the horticultural industry, demand for micropropagated roses may increase rapidly. Micropropagation may be further enhanced by the transformation of roses by "shooty mutants" of *Agrobacterium tumefaciens* or by the root-inducing strains of *Agrobacterium rhizogenes*. Transformation by the former may result in precocious shoot production (Steffen et al. 1986), whereas the latter could result in roses which produce more vigorous roots. The attractiveness of in vitro technology will be further enhanced by the prospect of establishing elite stocks of virus-free plants and the advantages associated with health certification of in vitro plants for export.

One major area of commercial importance which has not been exploited is that of marketing garden roses by mail order. If customers were able to transplant in vitro plantlets to soil, with acceptable chances of success, distribution of these plantlets by post or carrier could be achieved at a low cost per plant. Rooting of plants in agar does not lend itself to this concept, because of the predictably high levels of mortality. The rooting of plantlets in Sorbarod plugs results in a transplantation unit of plantlet plus plug, which is sufficiently resistant to damage and desiccation to justify careful evaluation of prospects for this material. It is anticipated that the in vitro acclimatization of plantlets will be further enhanced by chemical treatments, some of which are presently being investigated in the authors' laboratories.

The use of in vitro methods in rose breeding offers both immediate and longer-term benefits. It is likely that existing methods could be used to great advantage for chromosome doubling and isolation of somaclones. Both processes are likely to benefit from a capacity to generate adventitious shoots, when this becomes applicable to a wider selection of roses. The potential for the production of new improved varieties of roses by genetic engineering techniques is now a real possibility. By protoplast fusion it may be possible to introduce drought resistance and a central red spot, which is characteristic of the flower of the Persian desert rose, *R. persica*, into important rose cultivars.

Possibly the most significant improvement which could be made to the rose relates to flower colour, and in the near future the techniques of molecular biology are likely to provide an exciting new range of colours. Meyer et al. (1987) have pointed the way by producing a new brick-red flowered petunia by transformation of a colorless petunia with a maize gene. Thus, a new flower pigmentation pathway has been established in petunia. By means of this approach it would seem possible that the delphinidin pigment could be introduced into the rose and so the search for the elusive blue rose may nearly be over.

References

Amorim HV, Dougall DK, Sharp WR (1977) The effect of carbohydrate and nitrogen concentration on phenol synthesis in Paul's Scarlet rose cells grown in tissue culture. Physiol Plant 39:91–95

Asen S (1948) Embryo culture of rose seeds. Am Rose Ann 1948:119–120

Avramis T, Hugard J, Jonard R (1982) La multiplication in vitro du Rosier porte-greffe *Rosa indica major*. Physiol Veg CR Acad Sci Paris Ser 3, 294:63–68

Bajaj YPS (ed) (1988) Biotechnology in agriculture and forestry, vol 4: Medicinal and aromatic plants I. Springer, Berlin Heidelberg New York Tokyo

Bjarnason EN, Hanger BC, Moran JR, Cooper JA (1985) Production of Prunus necrotic ringspot virus-free roses by heat treatment and tissue culture. NZJ Agric Res 28:151–156

Bohm H (1982) The inability of plant cells to produce secondary substances. In: Fujiwara A (ed) Plant tissue culture 1982. Maruzen, Tokyo, pp 325–328

Bradford JA, Fletcher JS (1982) Influence of protein synthesis on NO_3^- reduction, NH_4^+ accumulation, and amide synthesis in suspension cultures of Paul's Scarlet rose. Plant Physiol 69:63–66

Bressan PH, Kim Y-J, Hyndman SE, Hasegawa PM, Bressan RA (1982) Factors affecting in vitro propagation of rose. J Am Soc Hortic Sci 107:979–990

Brodelius P (1985) Utilization of plant cell cultures for production of biochemicals. Hereditas Suppl 3:73–81

Broertjes C, van Harten AM (1988) Applied mutation breeding for vegetatively propagated crops. Elsevier, Amsterdam

Casey A (1985) Virus elimination from rose and popular using meristem tissue culture. M Sc Thesis, Univ Dublin

Curir P, Damiano, Cosmi T (1986) In vitro propagation of some rose cultivars. Acta Hortic 189:221–224

Darlington CD, Wylie AP (1955) Chromosome atlas of flowering plants, 2nd edn. Allen & Unwin, London
Davies DR (1980) Rapid propagation of roses in vitro. Sci Hortic 13:385–389
Davies ME (1971) Regulation of histidine biosynthesis in cultured plant cells: evidence from studies on amitrole toxicity. Phytochemistry 10:783–788
Davies ME (1972) Polyphenol synthesis in cell suspension cultures of Paul's Scarlet rose. Planta 104:50–65
de Proft MP, van den Broeck G, van Dijck R (1985) Implications of the container-atmosphere during micropropagation of plants. Med Fac Landbouww Rijksuniv Gent 50:129–132
Dougall DK (1965) The biosynthesis of protein amino acids in plant tissue culture. Isotope competition experiments using glucose -U-^{14}C and the protein amino acids. Plant Physiol 40:891–897
Douglas GC, Rutledge CB, Casey A, Richardson DHS (1989) Micropropagation of floribunda, ground cover and miniature roses. Plant Cell Tissue Organ Cult 19:55–64
Dubois LAM, Roggermans J, Soyeurt G, de Vries DP (1988) Comparison of the growth and development of dwarf rose cultivars propagated in vitro and in vivo by softwood cuttings. Sci Hortic 35:293–299
Elliott RF (1970) Axenic culture of meristem tips of *Rosa multiflora*. Planta 95:183–186
Fletcher JS (1975) Control of amino acid synthesis in tissue culture cells. Plant Physiol 56:405–451
Fletcher JS, Beevers H (1971) Influence of cycloheximide on the synthesis and utilization of amino acids in suspension cultures. Plant Physiol 48:261–264
Fletcher JS, Groeger AW, Mcfarlane JC (1987) Metabolism of 2-chlorobiphenyl by suspension cultures of Paul's Scarlet rose. Bull Environ Contamin Toxicol 39:960–965
Fosket DE (1982) Protein synthesis during the transition from the resting to the growing state in suspension cultures of Paul's Scarlet rose cells. Physiol Plant 53:146–152
Fowler MW (1986) Process strategies for plant cell cultures. Trends Biotechnol 4:214–219
George EF, Sherrington PD (1984) Plant propagation by tissue culture. Exegetics, Basingstoke, UK
Graifenberg A (1973) Colturo in vitro di embrioni e di parti de seme in *Rosa canina*. Riv Ortoflorofrutticolt Ital 57:374–380
Graifenberg A, Giustiniani L, Papandreou A (1975) In vitro culture of shoot apices of *Rosa multiflora* Thunb. and *R. indica major*. Agric Ital 1/3:119–128
Hasegawa PM (1979) In vitro propagation of rose. Hortic Sci 14:610–612
Hasegawa PM (1980) Factors affecting shoot and root initiation from cultured rose shoot tips. J Am Hortic Sci 105:216–220
Heslop-Harrison JW (1958) Roses and their chromosomes. In: Harkness J (ed) The rose annual RNRS, St Albans, UK, pp 53–61
Hilderbrandt AC, Wilmar JC, Jones N, Riker AJ (1963) Growth of edible chlorophyllous plant tissues in vitro. Am J Bot 50:248–254
Hill G (1957) Morphogenesis of shoot primordia in cultured stem tissue of a garden rose. Nature (London) 216:596–597
Horn W, Schlegel G, John K (1988) Micropropagation of roses. Acta Hortic 226:623–626
Horst RK (1983) Compendium of rose diseases. Phytopathol Soc, St Paul, Minn
Hurst CC (1941) Notes on the origin and evolution of our garden roses. J R Hortic Soc (London) 66:73–82, 242–250, 282–289
Hyndman SE, Hasegawa PM, Bressan RA (1982) Stimulation of root initiation from cultured rose shoots through the use of reduced concentrations of mineral salts. Hortic Sci 17:82–83
Jacobs G, Allan P, Bornman CH (1969) Tissue culture studies on rose: Use of shoot tip explants. I Auxin: cytokinin effects. Agroplantae 1:179–187
Jacobs G, Allan P, Bornman CH (1970a) Tissue culture studies on rose: Use of shoot tip explants. II Cytokinin: gibberellin effects. Agroplantae 2:25–28
Jacobs G, Allan P, Bornman CH (1970b) Tissue culture studies on rose: Use of shoot tip explants. III Auxin: gibberellin effects. Agroplantae 2:45–50
Khosh-Khui M, Sink KC (1982a) Rooting-enhancement of *Rosa hybrida* for tissue culture propagation. Sci Hortic 17:371–376
Khosh-Khui M, Sink KC (1982b) Micropropagation of new and old world *Rosa* species. J Hortic Sci 57:315–319
Khosh-Khui M, Sink KC (1982c) Callus induction and culture of *Rosa*. Sci Hortic 17:361–370
Knudson L (1946) A new solution for germination of orchid seed. Am Orchid Soc Bull 15:214–217
Krishnamurthy KV, Hendre RR, Godbole DA, Kulkarni VM, Mascarenhas AF, Jagannathan V (1979)

Isolation and regeneration of rose bud callus protoplasts (*Rosa* sp. cv. Soraya). Plant Sci Lett 15:135–137
Krussman G (1981) The complete book of roses. Timber Press, Portland, OR
Langebartels C, Harms H (1984) Metabolism of pentachlorophenol in cell suspension cultures of soybean and wheat: pentachlorophenol glycoside formation. Z pflazenphysiol 113:201–211
Lata P, Gupta MN (1971) Effects of gamma rays on stem cuttings of essential oil bearing *Rosa* spp. Flavour Ind 2:421–425
Lloyd D (1986) The induction, in vitro, of chromosomal variation in *Rosa*. PhD Thesis, NE London Polytech
Lloyd D, Roberts AV, Short KC, (1988) The induction in vitro of adventitious shoots in *Rosa*. Euphytica 37:31–36
Martin C (1985) Plant breeding in vitro. Endeavour 9:81–86
Martin C, Carre M, Vernoy R (1981) La multiplication végétative in vitro des végétaux ligneux cultivés: cas des Rosiers. CR Acad Sci Paris Ser 3, 293:175–177
Meyer P, Heidmann I, Forkmann G, Saedler H (1987) A new petunia flower colour generated by transformation of a mutant with a maize gene. Nature (London) 330:677–678
Mubitch MJ, Fletcher JS (1984) Isolation and identification of the phenols of Paul's Scarlet rose stems and stem-derived suspension cultures. Plant Physiol 75:592–595
Mubitch MJ, Fletcher JS (1985) Influence of culture age and spermidine treatment on the accumulation of phenolic compounds in suspension cultures. Plant Physiol 78:25–28
Murashige T, Skoog F (1962) A revised medium for rapid growth and bioassays with tobacco cultures. Physiol Plant 15:473–497
Nash DT, Davies ME (1972) Some aspects of growth and metabolism of Paul's Scarlet rose cell suspensions. J Exp Bot 23:75–91
Nesius KK, Uchytil LE, Fletcher JS (1972) Minimal organic medium for suspension cultures of Paul's Scarlet rose. Planta 106:173–176
Nickell LG, Tulecke W (1959) Responses of plant tissue cultures to gibberellin. Bot Gaz 120:245–250
Nickell LG, Tulecke W (1960) Submerged growth of cells of higher plants. J Biochem Microbiol Tech Eng 3:287–297
Onesto J-P, Poupet R, Julien P (1985) Production de potées fleuries de rosier à partir de plantules obtenus par multiplication in vitro conforme automme 1983 – printemps 1984. Hortic 176:3–10
Pearce RS, Cocking EC (1973) Behaviour in culture of isolated protoplasts from Paul's Scarlet rose suspension culture cells. Protoplasma 77:165–180
Rehder A (1960) Manual of cultivated trees and shrubs hardy in North America, 2nd edn. Springer, Berlin Heidelberg New York
Reist A (1985a) Culture in vitro (civ) en pepinière de rosiers: premiers résultats d'exploitation de plants civ pour la coupe. Rev Suisse Vitic Arboric Hortic 17:173–176
Reist A (1985b) Culture in vitro (civ) en pepinière de rosiers: une alternative au bouturage ou au greffage des variétès? Rev Suisse Vitic Arboric Hortic 17:361–364
Rout GR, Debata BK, Das P (1989) In vitro mass-scale propagation of *Rosa hybrida* cv. Landora. Current Science 58:876–878
Sauer A, Walther F, Preil W (1985) Different suitability for in vitro propagation of rose cultivars. Gartenbauwissenschaft 50:133–138
Scheel D, Sandermann H (1977) Metabolism of DDT and kelthane in cell suspension cultures of parsley (*Petroselinum hortense*, Hoffm.) and soybean (*Glycine max*. L.). Planta 133:315–320
Semenuik P, Arisumi T (1968) Colchicine-induced tetraploid and cytochimeral roses. Bot Gaz 129:190–193
Short KC, Price L, Roberts AV (1981) Micropropagation of roses. In Harkness J (ed) The rose annual. RNRS, St Albans, UK, pp 138–144
Skirvin RM, Chu MC (1979) In vitro propagation of Forever Yours rose. HortScience 14:608–610
Smith TA, Best GR, Abbott AJ, Clements ED (1978) Polyamines in Paul's Scarlet rose suspension cultures. Planta 144:63–68
Steffen A, Eriksson T, Schieder O (1986) Shoot regeneration of mesophyll protoplasts transformed by *Agrobacterium tumefaciens* not achievable with untransformed protoplasts. Theor Appl Genet 72:135–140
Strauss A, Potrykus I (1980) Callus formation from protoplasts of cell suspension cultures of *Rosa* 'Paul's Scarlet'. Physiol Plant 48:15–20

Tabaeezadeh Z, Khosh-Khui M (1981) Anther culture of *Rosa*. Sci Hortic 15:61–66

Threlfall DR, Goodwin TW (1963) Ubiquinone-50 and plastoquinone-45 in plant tissue cultures of Paul's Scarlet rose. Biochem Biophys Acta 78:532–534

Tulecke W, Nickell LG (1960) Methods, problems and results of growing plant cells under submerged conditions. Trans N Y Acad Sci 22:196–206

Tulecke W, Taggart R, Colavito L (1965) Continuous cultures of higher plant cells in liquid media. Contrib Boyce Thompson Inst 23:33–46

Tweddle D, Roberts AV, Short KC (1984) In vitro culture of roses. In: Novak FJ, Havel L, Dolezel J (eds) Plant tissue and cell culture application to crop improvement. Czech Acad Sci, Prague, pp 529–530

van de Pol PA, Joosten MHAJ, Keizer H (1986) Stenting of roses, starch depletion and accumulation during early development. Acta Hortic 189:51–59

Wegg SM, Townsley PM (1983) Ascorbic acid in cultured tissue of briar rose, *Rosa rugosa*. Thunb Plant Cell Rep 2:78–81

Weinstein LH, Tulecke W, Nickell LG, Laurencot HJ (1962) Biochemical and physiological studies of tissue cultures and the plant parts from which they are derived. III. Paul's Scarlet rose. Contrib Boyce Thompson Inst 21:371–386

White PR (1934) Potentially unlimited growth of excised tomato root tips in a liquid medium. Plant Physiol 9:585–600

Wilkowske DR (1981) Propagation in vitro of rose. Hortic Sci 16:464

Williams BL, Goodwin TW (1965) The terpenoids of tissue cultures of Paul's Scarlet rose. Phytochemistry 4:81–88

Wylie AP (1954) The history of garden roses, pt 1. J R Hortic Soc (London) 79:555–571

Wylie AP (1955a) The history of garden roses, pt 2. J R Hortic Soc (London) 80:1–24

Wylie AP (1955b) The history of garden roses, pt 3. J R Hortic Soc (London) 80:77–87

Young PM, Hutchins AS, Cranfield ML (1984) Use of antibiotics to control bacteria in shoot cultures of woody plants. Plant Sci Lett 34:203–209

XXIV *Scutellaria baicalensis* Georgi: In Vitro Culture and the Production of Flavonoids

H. Yamamoto[1]

1 Introduction

1.1 Distribution of *Scutellaria*

The genus *Scutellaria* belongs to the family Labiatae and subfam. Scutellarideae. *Scutellaria* is widely distributed all over the world except for South Africa, and there are about 300 species. The calyx of this genus is remarkably specialized to become two-lip-shaped, and characterized by the upper lip having a flat or dish-shaped upper surface on which a small swelling forms; the dish-shaped part peels off at fruit maturation to allow a seed to fall.

The 15 species, such as *S. maekawa* Hara, *S. brachyspica* Nakai et Hara, *S. laeteviolace* Koizumi, *S. iyoensis* Nakai and others are distributed only in Japan (Kitamura et al. 1975; Satake et al. 1981). Further *S. indica* L., *S. indica* var. *parviflora* Makino, *S. sterigillosa* Hensl, *S. dependens* Maxim. and others are distributed over wide areas in Japan, the Korean Peninsula, the northeastern section of China and the Indonesian Peninsula (Kitamura et al. 1975; Satake et al. 1981). Furthermore, *S. baicalensis* Georgi is native to the region from the northern section of China to Siberia. It was introduced into Japan from the Korean Peninsula in the middle period of the Edo era and has been cultured in various parts of Japan for the medicinal uses of the root. In Japanese Pharmacopoeia (undecima ed. Niphon-koteisho Kyokai 1986) the root of *S. baicalensis* Georgi, excepting the exodermis, is used as medicinal Wogon (Scutellariae Radix).

In China[2], many plants belonging to the *Scutellaria* genus are distributed and the root of *S. baicalensis* Georgi is mainly used as medicinal Wogon but the following seven species are also used as a source for medicinal Wogon, *S. viscidula* Bye, *S. amoena* Wright, *S. rehderiana* Diels, *S. likiangesis* Diels, *S. ikonnikovii* Juz., *S. hypericifola* Levl., and *S. rivularis* Wall (Ban Zhi Lian). Further, in the Alpine districts of Europe, *S. alpina* L. and *S. costericana* H. Wendel are distributed and cultured as garden species for their attractive flowers.

Studies on the constituents of *Scutellaria* species have been made, i.e., *S. amoena* C.H. Wright (Liu et al. 1980), *S. viscidula* Bunge (Yu et al. 1984), *S. rivularis*

[1] School of Pharmacy, Hokuriku University, Kanagawa-Machi, Kanazawa 920-11, Japan

[2] Material for Medicine (Nanking Pharmaceutical Institute 1959); Dictionary of Chinese Medicine (Koso New Medical Institute 1978); Chinese Pharmacopoeia (Institute of Pharmacy, Chinese Ministry of Sciences 1979)

Biotechnology in Agriculture and Forestry, Vol. 15
Medicinal and Aromatic Plants III (ed. by Y.P.S. Bajaj)

Wall (Tomimori et al. 1984c, 1986c) *S. discolor* Colebr (Tomimori et al. 1986a), *S. indica* L. (Miyaichi et al. 1987), *S. scandens* Buch, Ham ex D. Don (Miyaichi et al. 1988), and others.

1.2 Importance of *S. baicalensis*

Wogon is one of the crude drugs frequently used as an important medicine in Chinese clinical practice. The origin of this crude drug is the plant *S. baicalensis* Georgi, and its root, excepting the exodermis, is prescribed in the Japanese Pharmacopoeia JPXI and compounded in many Chinese medicines including Sannosan and others, for treatment of diarrhea, abdominalgia, anorexia, hot flash, stiffness of the shoulders, agrypnia, or the like. *S. baicalensis* is a perennial plant native to the region from the north section of China to Siberia, as mentioned above, and grows to a height of 30–60 cm, spreading in a semi-spherical shape, with a lanceolar leaf having many hairs. From June to July, the leading end of each branch produces purple ear-shaped flowers (Fig. 1) and the root is long and conical in shape, the interior showing a golden color.

The study of components of Wogon prepared from plants originated from *S. baicalensis* has concentrated on flavonoids, and about 40 kinds of flavonoids (Shibata and Hattori 1931; Takido 1973; Takido et al. 1975, 1979; Takagi et al. 1980, 1981a,b; Tomimori et al. 1982, 1983, 1984a,b) ranging from flavonoids relatively

Fig. 1. *Scutellaria baicalensis* Georgi

high in content such as baicalin, baicalein, wogonin or the like to minor flavonoids have been isolated and reported up to now. Other components are β-sitosterol, camphesterol, and stigmasterol, two kinds of sugars (glucose and sucrose), three kinds of amino acids structurally unknown (Takido 1973), and tannin (Izawa 1966).

Many pharmacological functions, e.g. to eliminate bile acid or to act as a diuretic laxative (Kumazaki 1958), anti-inflammatory (Hirai et al. 1983; Koda et al. 1982; Koda 1987), or anti-allergic factor (Koda 1973, 1987; Koda et al. 1970a,b,c,d, 1982; Kubo et al. 1981, 1984, 1985; Nishiyori et al. 1982), lipid metabolism (Aonuma et al. 1957; Kimura et al. 1981a,b, 1982a,b, 1985; Yamahara et al. 1981) and arachidonate metabolism (Kimura et al. 1985), and similar have been confirmed. These pharmacological effects are considered to be mainly brought about by the flavonoids contained in Wogon. The anti-allergic effect of Wogon is said to be developed by the chromone, thus being the base nucleus of baicalein, and is the source for developing anti-allergic drugs (Koda 1973, 1987).

1.3 A Fundamental Study of the Cultivation of *S. baicalensis* Georgi

Detailed study of seasonal variations of growth and flavonoid content has been reported by Tomimori et al. (1986b). The dry weight and flavonoids of the root gradually increased over 3 years, although it decreased during the resting periods. During budding, some flavone glucuronides were hydrolyzed to form aglycone and glucosidation was accelerated in the flowering and fruit-bearing periods. The baicalin content was highest in the latter half of fruit-bearing, and was almost the same as in the root of the 3-year-old plant; ca. 40% of the roots were delayed after fruit-bearing. It is therefore considered that from the viewpoint of both yield and quality, the root of *S. baicalensis* should be harvested in the second year in the latter half of the fruit-bearing period (in late autumn) (Tomimori et al. 1986b).

2 In Vitro Approach

Studies of tissue culture of *Scutellaria* plants are reported only in *S. baicalensis* by Trofimova (1982) and Yamamoto et al. (1986a,b,c,d, 1987, 1989a,b). Parsley (*Petroselinum hortense*) callus produces flavonoids under light (Hahlbrock et al. 1971; Krenzaler and Hahlbrock 1973), but callus cultures of *S. baicalensis* produced flavonoids in the dark (Yamamoto et al. 1986a,b,c,d).

2.1 Culture Methods of *S. baicalensis* Stem Callus

2.1.1 Callus Induction and Subculture

The stems of *S. baicalensis* Georgi are washed with a stream of water for about 1 h, dipped in 70% alcohol for 30 s, and then sterilized in 0.5% hypochlorite solution for 30 min. After sterilization, they are thoroughly washed with sterilized water and tissue is inoculated on LS basal medium (Linsmaier and Skoog 1965) containing 10^{-6}M IAA and 10^{-6}M kin. For static cultures, the callus is subcultured and incubated in the dark at 25 ± 1 °C. A part of the culture is transferred to a new medium every 30 days for repeated subculture.

2.1.2 Measurement of Callus Growth and Flavonoid Content

Callus growth was expressed by fresh weight per tube containing 10 ml medium, as measured at the end of each experiment. Samples for determination of flavonoid contents were prepared in the same way as in the previous reports (Yamamoto et al. 1986a,b,c,d), i.e., about 200 mg of dried callus was extracted with 50% EtOH and EtOH respectively, and analysis of flavonoids was carried out by high pressure liquid chromatography (HPLC) (Tomimori et al. 1985) and reversed phase ODS columns. As in previous work (Yamamoto et al. 1986b,c,d), quantitative determinations were made on 11 flavonoids (Fig. 2) including nine flavone (chrysin, baicalein, oroxylin A, wogonin, skullcupflavone II, baicalein-glucoside, baicalin, oroxylin A-glucuronide, and wogonin-glucuronide) and two flavanone (dichloroxylin A and 5,7,2′,6′-tetra hydroxyflavone).

2.1.3 Identification of Flavones

One hundred grams (dry weight) of callus tissue (culture condition; LS medium – 3% sucrose + 5% maltose + 10^{-6}M IAA + 10^{-5}M kin) was extracted with MeOH

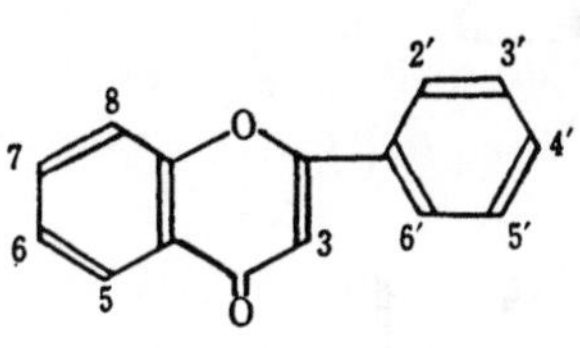

Flavone

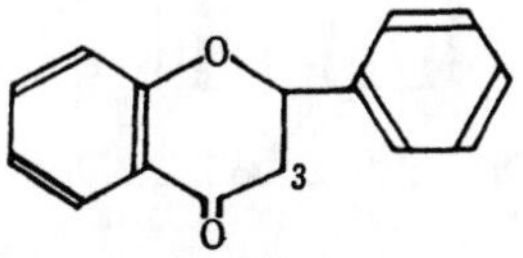

Flavanone

flavone	3	5	6	7	8	2′	6′
chrysin		OH		OH			
baicalein		OH	OH	OH			
oroxylin A		OH	OMe	OH			
wogonin		OH		OH	OMe		
skullcapflavone II		OH	OMe	OMe	OMe	OH	OMe
baicalein-glc.		OH	OH	O-glc.			
baicalin		OH	OH	O-glcuA			
oroxylinA-glcuA		OH	OMe	O-glcuA			
wogonin-glcuA		OH		O-glcuA	OMe		
flavanone							
dihydrooroxylinA		OH	OMe	OH			
5,7,2′,6′-tetra-hydroxyflavanone		OH		OH		OH	OH

Fig. 2. Structures of *Scutellaria* flavonoids (minor flavonoids are omitted)

and silica gel chromatography was repeated with respect to the MeOH extract according to the method of Tomimori et al. (1982, 1983, 1984a,b) and 11 kinds of flavonoids were isolated to be identified by direct comparison with their respective authentic specimen (melting point, TLC, UV, IR).

2.2 Callus Growth and Flavonoid Contents in *Scutellaria* Callus Culture

2.2.1 *Effect of Plant Growth Regulator*

Of the St-20 line and the St-7 line 35 ± 2 mg fresh weight were transferred to agar medium supplemented with various combinations of 0–10^{-5}M IAA, 2,4-D, NAA or kin, and were then incubated in the dark for 6 weeks (25 ± 1°C). The growth of the St-20 line was stimulated by increasing the kin concentration. The combination of 10^{-5}M kin, 10^{-7}M to 10^{-5}M NAA resulted in the largest fresh weight increases (Fig. 3A). This line had been routinely maintained on 10^{-5}M kin and 10^{-6}M IAA. Cell line St-7 did not show the requirement for high kin concentration, as growth was

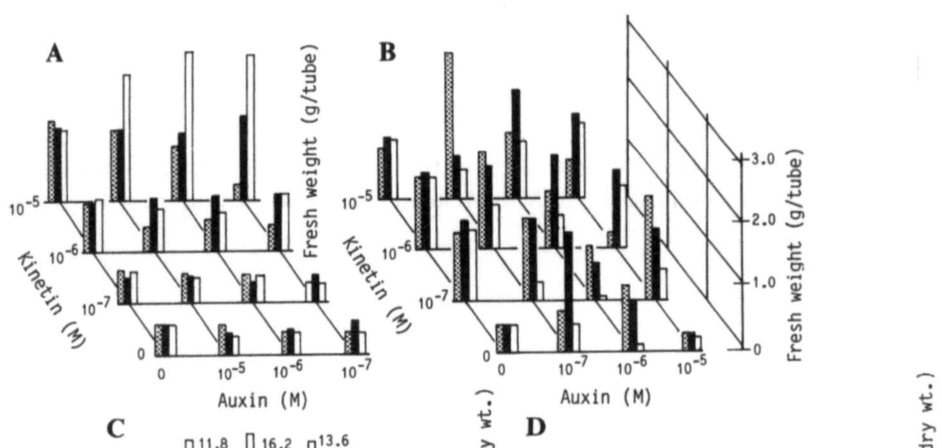

Fig. 3A-D. Effects of various auxins and kinetin on the growth and baicalin content of the *Scutellaria baicalensis* callus lines (10-month-old). Callus growth: **A** (St-20 line). **B** (St-7 line) baicalin content. **C** (St-20 line). **D** (St-7 line). Static culture: LS medium in the dark at 25 ± 1°C for 42 days. The value shows the mean average (n = 5). Various auxins: ▨ 2,4-D; ■ IAA; ▭ NAA. (Yamamoto et al. 1986a)

achieved on many different combinations of plant growth regulator (Fig. 3B). In general, baicalin was the main constituent of both lines. Under standard growth conditions the content of line St-20 was three times that of St-7 (Fig. 3C). Only in St-20 callus did those media promoting accumulation of high-level baicalin also contain a high level of baicalin, wogonin, and wogonin 7-O-glucuronide. Higher formation of flavonoids is thus correlated to good growth, which also holds true for line St-7 (Fig. 3D). Wogonin and wogonin 7-O-glucuronide were rarely found in St-20. The St-20 line gave a better callus for flavonoid production than the St-7 line and the glucuronide content was higher than that of aglycone. The flavonoid pattern and content of the St-20 line after the long culture period of 70 days were similar to those of *Scutellaria* root (Wogon), baicalin content was 5–8% dry weight in St-20 line callus cultures.

2.2.2 *Effect of Carbon Sources*

The callus tissue (St-20 line) was inoculated in LS basal medium containing 10^{-6}M IAA and 10^{-5}M kin and 1–8% of 11 carbon sources, namely monosaccharide: D-glucose, D-fructose, D-galactose; disaccharide: maltose, cellobiose, lactose, sucrose; trisaccharide: raffinose; sugar alcohol: D-sorbitol, D-mannitol, soluble starch, and incubated in the dark at 25 ± 1 °C for 42 days, when callus growth and flavonoid content were determined. Figure 4 and Table 1 show the effect of carbon

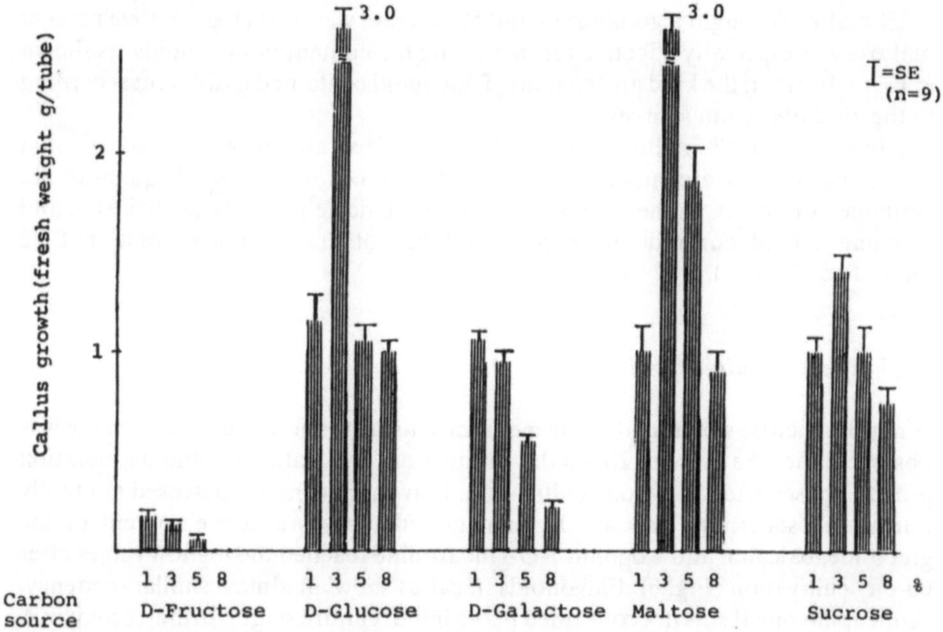

Fig. 4. Effects of carbon sources of callus growth in *Scutellaria baicalensis* callus cultures (St-20, 15-month-old). LS basal medium minus sucrose plus 10^{-6} M IAA and 10^{-5} M kin. Static culture in the dark at 25 ± 1 °C for 40 days. (Yamamoto et al. 1986b)

Table 1. Effects of carbon sources on callus growth and flavonoid contents in *Scutellaria baicalensis* callus cultures (St-20, 23-month-old). (Yamamoto et al. 1986b)

Carbon source	%	Flavonoid content[a] Baicalin	Flavonoid content[a] Wogonin-glcu A	Growth[b] fresh wt.	Carbon source	%	Flavonoid content[a] Baicalin	Flavonoid content[a] Wogonin-glcu A	Growth[b] fresh wt.
Lactose	1	–	–	0.33	Starch	1	0.33	0.06	0.58
	3	0.04	–	0.06	(soluble)	3	0.04	0.01	1.10
	5	0.13	–	1.35		5	0.01	–	1.43
	8	–	–	0.57		8	0.08	0.02	1.03
Cellobiose	1	0.05	0.01	1.00	D-Mannitol	1	0.09	0.03	0.16
	3	0.16	0.06	1.00		3	0.18	0.05	0.13
	5	0.10	0.20	0.84		5	0.19	0.04	0.11
	8	0.19	0.39	0.73		8	0.15	0.05	0.15
Raffinose	1	0.03	0.01	0.64	D-Sorbitol	1	0.14	0.04	0.21
	3	0.16	0.06	1.83		3	0.13	0.05	0.81
	5	0.98	0.20	2.00		5	0.15	0.03	0.20
	8	1.96	0.39	1.14		8	0.18	0.03	0.20

LS basal medium minus sucrose plus 10^{-6}M IAA and 10^{-5}M kin. Static culture in the dark at 25 ± 1°C for 40 days.
[a] mg/tube.
[b] g/tube.

sources on the callus growth. The callus growth is enhanced by the addition of 3% D-glucose, 3–5% maltose, and 3% sucrose, and suppressed by the addition of 1–8% D-fructose, D-mannitol, and D-sorbitol. With respect to the content of flavonoids, 5–8% maltose brought a good result and 5% sucrose was next (Fig. 5). Five percent maltose was especially effective for increasing the content of flavonoids, as shown in Fig. 5. Further, the kind and content of flavonoid contained is different according to the kind of carbon sources.

In the case of 5% maltose, the large amount of baicalin, wogonin, and wogonin 7-O-glucuronide are contained, but other flavonoids in very small quantity are contained (Table 2). In the case of 8% raffinose, baicalein-glucoside, baicalin, and wogonin 7-O-glucuronide were produced but other flavonoids could not be quantified (Table 2).

2.2.3 *Effect of Culture Period*

When 3% sucrose was added to the medium, a logarithmic callus growth phase was observed after the lapse of 20–40-day cultivation, and stationary and deceleration growth phases after 50–60 day cultivation. Flavonoid content increased gradually during the stationary and deceleration growth phases and the content of the glucoside, baicalin, and wogonin 7-O-glucuronide reached the highest values after 60-day cultivation (Fig. 6). Flavonoids in callus show an almost similar tendency, skullcupflavone II was not confirmed in the initial culture stage; further, oroxylin A was not confirmed but it was shown that the content of crysin in the St-20 line was high as compared with Wogon (Table 3). On the addition of 5% sugars, the growth of callus was in the order of 5% maltose > 5% glucose ≧ 5% sucrose. On the 5%

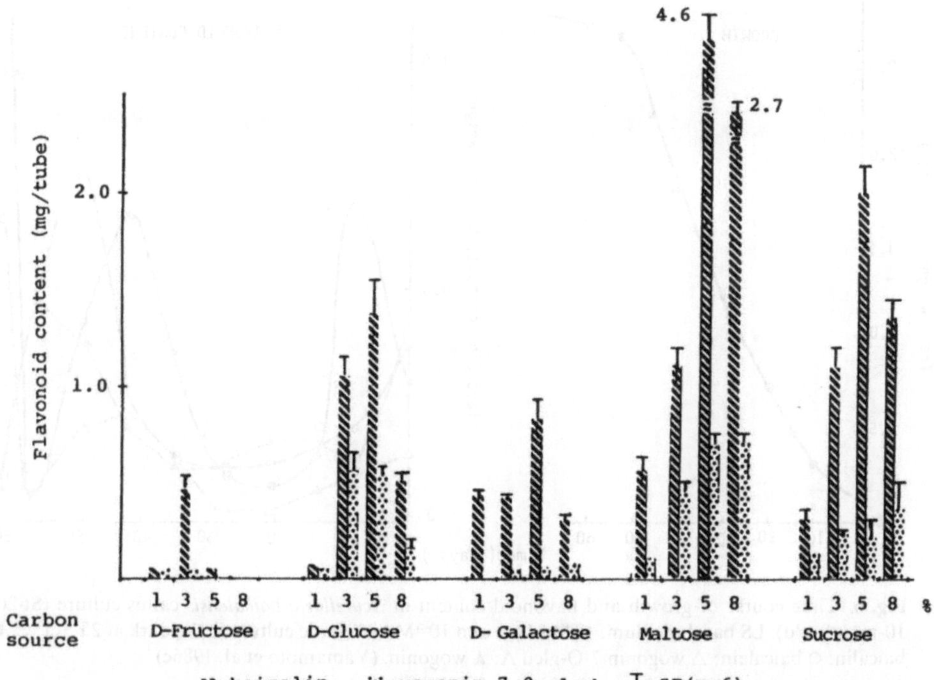

Fig. 5. Effects of carbon sources on flavonoid contents in *Scutellaria baicalensis* callus cultures (St-20, 25-month-old). LS basal medium minus sucrose plus 10^{-6} M IAA and 10^{-5} M kin. Static culture in the dark at 25 ± 1 °C for 40 days. (Yamamoto et al. 1986b)

Table 2. Effects of carbon sources on flavonoid contents of *Scutellaria baicalensis* callus culture (St-20, 23-month-old). (Yamamoto et al. 1986b)

Flavonoid content	Added carbon source				
	8% D-Glucose L[a]	5% Maltose D	8% Raffinose D	5% D-Mannitol D	5% D-Galactose D
Baicalein glc.	0.08[b]	–	0.07	–	–
Baicalin	1.57	4.30	1.96	0.19	1.40
Oroxylin A-glcu A	–	–	–	–	–
Wogonin-glcu A	0.70	1.20	0.39	0.04	0.84
5,7,2′,6′-Tetrahydroxyflavone	–	±	–	±	–
Oroxylin A	–	–	–	–	–
Dihydrooroxylin A	–	–	–	–	–
Baicalein	–	0.08	–	0.001	–
Skullcupflavone II	0.01	±	–	0.02	–
Wogonin	0.01	0.09	–	0.003	–
Chrysin	±	±	–	0.003	±

LS basal medium minus 3% sucrose plus 10^{-6} M IAA and 10^{-5} M kin. Static culture at 25 ± 1 °C for 42 days.
[a] D: dark, L: light (2500 lx).
[b] Flavonoid content (mg/tube).
–: undetected.

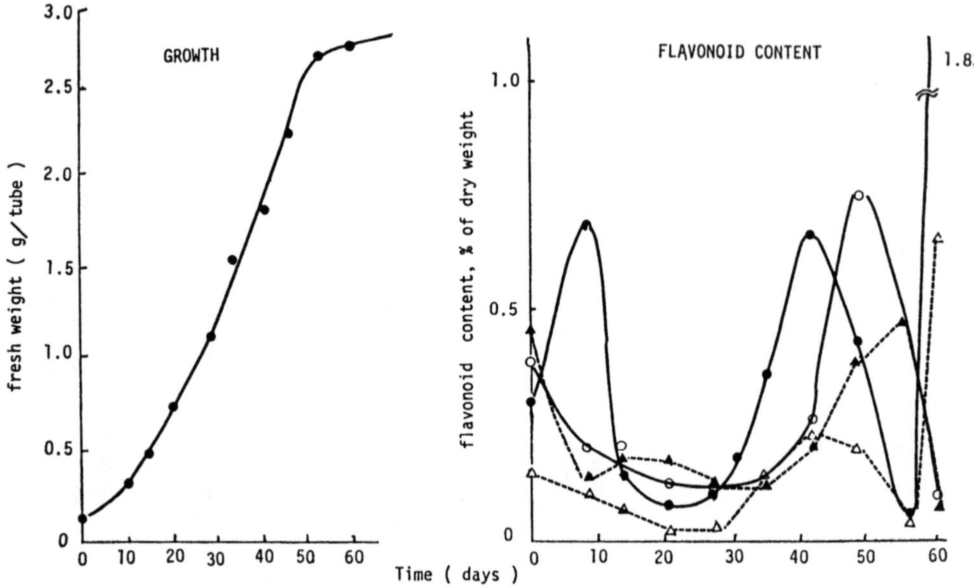

Fig. 6. Time course of growth and flavonoid content in *Scutellaria baicalensis* callus culture (St-20, 10-month-old). LS basal medium, 10^{-6}M IAA and 10^{-5}M kin. Static culture in the dark at 25 ± 1°C. ● baicalin; ○ baicalein; △ wogonin 7-O-glcu A; ▲ wogonin. (Yamamoto et al. 1986c)

Table 3. Flavonoid content (% of dry wt.) of *Scutellaria baicalensis* callus culture (St-20, 10-month-old). (Yamamoto et al. 1986c)

	Culture period (days)									
Flavonoid	0	9	13	20	27	34	41	48	55	60
Baicalin	0.26	0.68	0.14	0.07	0.08	0.33	0.67	0.42	0.04	1.83
Oroxylin-A-glucuronide	0.09	0.03	0.05	0.02	0.03	0.09	0.17	0.15	0.03	0.37
Wogonin-7-O-glucuronide	0.14	0.25	0.07	0.02	0.03	0.14	0.22	0.19	0.01	0.67
Baicalein	0.39	0.21	0.21	0.12	0.11	0.12	0.26	0.75	0.54	0.09
Skullcupflavone II	0.01	—	—	—	—	0.003	0.02	0.01	0.03	0.02
Wogonin	0.47	0.12	0.17	0.16	0.12	0.12	0.20	0.39	0.45	0.05
Chrysin	0.11	0.02	0.04	0.04	0.03	0.03	0.29	0.08	0.12	0.15

LS basal medium, 10^{-6}M IAA and 10^{-5}M kin.
Static culture in the dark at 25°C.

maltose-containing medium, a logarithmic growth phase was observed after a lapse of 10 days and a stationary growth phase after 50-day cultivation. Baicalin began to increase from about the 10th day on the addition of 5% maltose, and reached 16 mg/10 ml on the 70th day (Fig. 7, Table 4). On addition of 5% glucose, the flavonoid content is half that compared with the addition of 5% maltose, and on the addition of 5% sucrose, the flavonoid content is low. In every case, the crysin content is low (Table 5). On the addition of 5% sucrose the content of the aglycones baicalein and wogonin was relatively high, and baicalein and wogonin or the like were detected

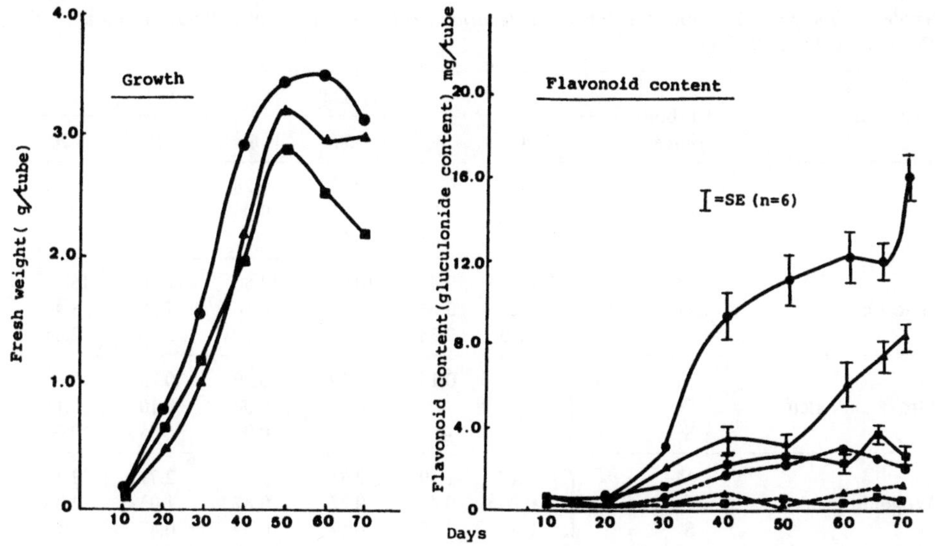

Fig. 7. Effects of carbon sources on callus growth and flavonoid content in *Scutellaria baicalensis* callus culture (St-20, 19-month-old). LS basal medium minus sucrose plus 10^{-6}M IAA and 10^{-5}M kin. Static culture in the dark at 25 ± 1°C for 70 days. ● 5% maltose; ▲ 5% D-glucose; ■ 5% sucrose; – baicalin; --- wogonin 7-O-glcu A. (Yamamoto et al. 1986c)

on the 40th day (Table 5). Brief light irradiation of the callus which had been cultured for 20 days in the dark effectively increased the rate of callus growth and the flavonoid content (Fig. 8).

2.2.4 Effect of 5% Maltose and Plant Growth Regulators

The effect of 5% maltose instead of 3% sucrose and plant growth regulators on the callus growth and flavonoid formation of some *S. baicalensis* stem callus lines was studied. The optimum combination of plant growth regulators in the St-20 line was found to be 10^{-7}M NAA and 10^{-5}M kin, 10^{-5}M NAA and 10^{-5}M kin for callus growth and 10^{-6}M IAA and 10^{-5}M kin, and 10^{-7}M NAA and 10^{-5}M kin for flavonoid content (Figs. 9, 10). When 5% maltose was added to the medium, an increase in flavonoid content was observed only in the St-20 line among the four lines tested (Fig. 11). By the addition of 2,4-D, none of the callus lines formed chlorophyll under light irradiation. The St-20 line is high in baicalein 7-O-glucuronide (baicalin) content as compared with other lines (Fig. 1x). The St-13 line is high in content of wogonin and wogonin 7-O-glucuronide as compared with other lines and the callus thereof is light yellow and wet, but the contents of baicalein and baicalin in this line are low. The formation of flavonoids is largely affected by various factors such as the line of callus, plant growth substances, sugars, light, and others.

Table 4. Change in flavonoid contents of *Scutellaria baicalensis* callus culture (St-20, 17-month-old). (Yamamoto et al. 1986c)

Flavonoid	Carbon source	Days						
		0	20	40	50	60	65	70
Baicalein-glc.	M	–[a]	–	0.10	0.23	0.28	0.23	0.44
	G		–	0.04	0.05	0.06	0.07	0.15
	S		–	0.01	0.03	±	±	±
Baicalin	M	1.87	0.18	9.42	10.90	12.80	12.00	16.1
	G		0.23	3.64	3.12	6.16	7.20	8.46
	S		0.73	2.38	2.77	2.20	3.90	2.65
Oroxylin A-glcu A	M	–	–	0.15	0.19	0.19	0.23	0.31
	G		–	–	0.05	0.08	0.10	0.11
	S		–	–	–	0.04	0.05	0.06
Wogonin-glcu A	M	0.32	0.17	1.99	2.37	3.00	2.71	2.10
	G		0.04	0.47	0.22	0.87	1.03	1.03
	S		0.15	0.49	0.54	0.59	0.87	0.68
5, 7, 2′, 6′-Tetra-hydroxyflavanone	M	0.24	–	0.22	–	000 –		–
	G		–	–	0.01	–	–	–
	S		–	–	0.03	0.22	–	–
Baicalein	M	0.54	–	0.22	0.02	0.14	0.05	–
	G		–	–	–	–	–	–
	S		–	0.13	0.10	0.35	0.35	0.54
Skullcupflavone II	M	0.25	–	0.25	–	–	–	–
	G		–	–	0.01	–	–	0.01
	S		–	–	–	–	–	–
Wogonin	M	0.46	–	–	0.08	0.06	0.03	–
	G		–	–	0.03	–	–	0.01
	S		0.01	0.02	0.05	0.10	0.05	0.06
Chrysin	M	0.03	–	–	–	–	–	–
	G		–	–	–	–	–	–
	S		–	0.01	–	0.03	0.02	–

LS medium minus sucrose plus 10^{-6} M IAA and 10^{-5} M kin.
Static culture in the dark at 25 ± 1 °C for 70 days
M: 5% maltose, G: 5% glucose, S: 5% sucrose.
[a] Flavonoid content, mg/tube (10 ml medium).

Table 5. Effects of carbon sources and growth regulators on flavonoid content (mg/tube) in the *Scutellaria baicalensis* callus culture (St-20, 19-month-old). (Yamamoto et al. 1986d)

Flavonoid	Carbon source									
	5% D-Glucose			5% Maltose				5% Sucrose		
	Growth regulators									
	I[a]	II	III	I	II	III	IV	I	II	III
Baicalein-glc.	–[b]	0.1	–	–	0.12	–	–	–	0.07	0.12
Baicalin	2.60	1.10	0.94	7.27	0.79	1.90	7.69	2.40	0.80	0.90
Oroxylin A-glcu A	0.04	0.02	–	0.11	0.01	–	0.02	0.04	0.01	0.03
Wogonin-glcu A	0.33	0.24	0.13	1.36	0.19	0.29	1.56	0.46	0.20	0.10
5, 7, 2′, 6′-Tetra-hydroxyflavanone	–	–	–	–	0.01	–	–	–	0.04	–
Baicalein	–	0.01	0.02	–	0.03	0.01	–	–	0.02	0.03
Wogonin	–	0.02	–	–	0.04	0.02	–	–	0.03	0.02
Chrysin	–	–	0.01	–	–	–	–	–	0.01	–

LS basal medium minus sucrose. Static culture in the dark at 25 ± 1°C for 40 days

[a] I: 10^{-6}M IAA + 10^{-5}M kin, II: 10^{-6}M 2,4-D + 10^{-6}M kinetin, III: 10^{-4}M NAA + 10^{-5}M kin, IV: 10^{-5}M NAA + 10^{-5}M kin.

[b] Flavonoid contents (mg/tube).

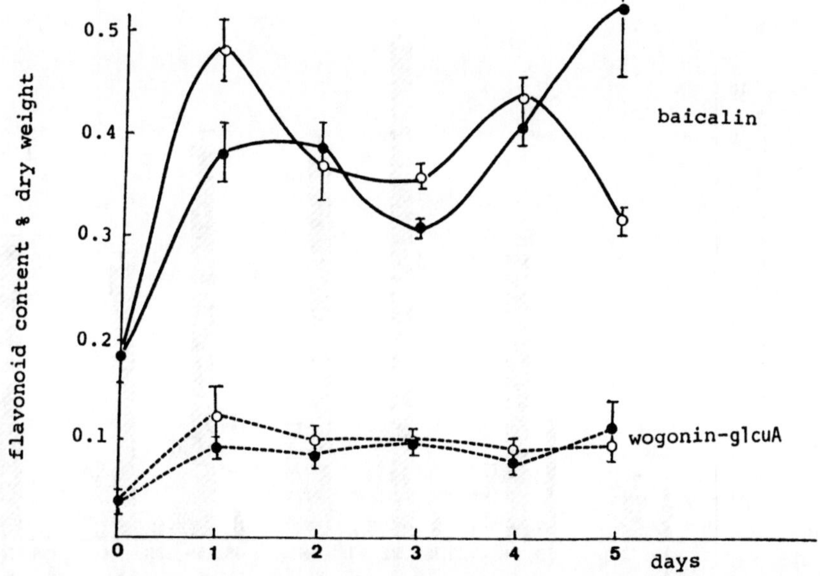

Fig. 8. Time course of the baicalin and wogonin-glcu A content of *Scutellaria baicalensis* callus culture (St-20, 25-month-old). LS basal medium, 10^{-6}M IAA and 10^{-5}M kin. ○ light; ● dark. (Yamamoto et al. 1986c)

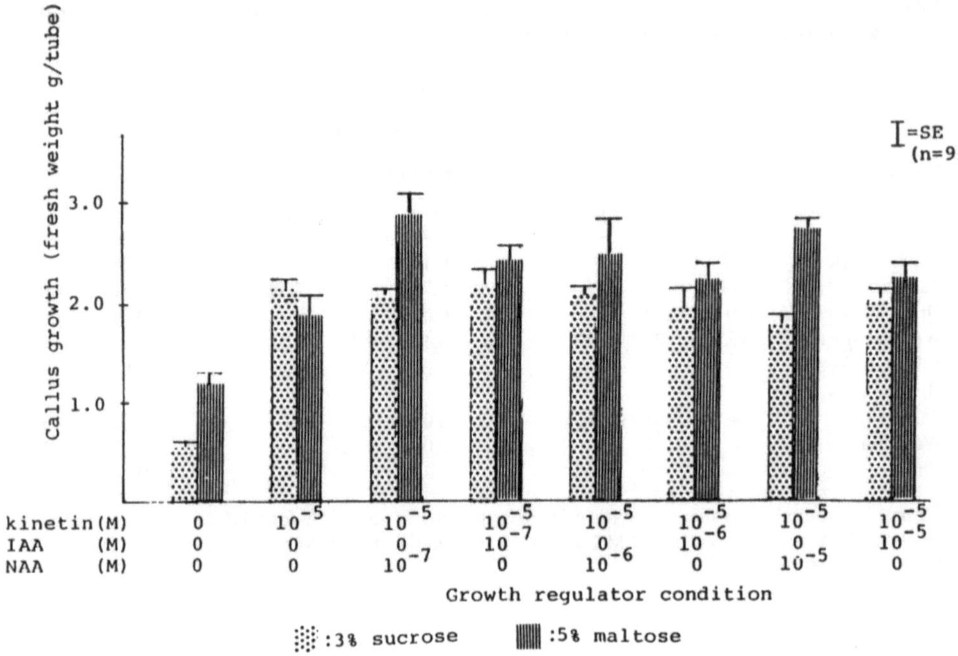

Fig. 9. Effects of carbon sources and growth regulators on callus growth in *Scutellaria baicalensis* callus culture (St-20, 25-month-old). LS basal medium minus sucrose. (Yamamoto et al. 1986d)

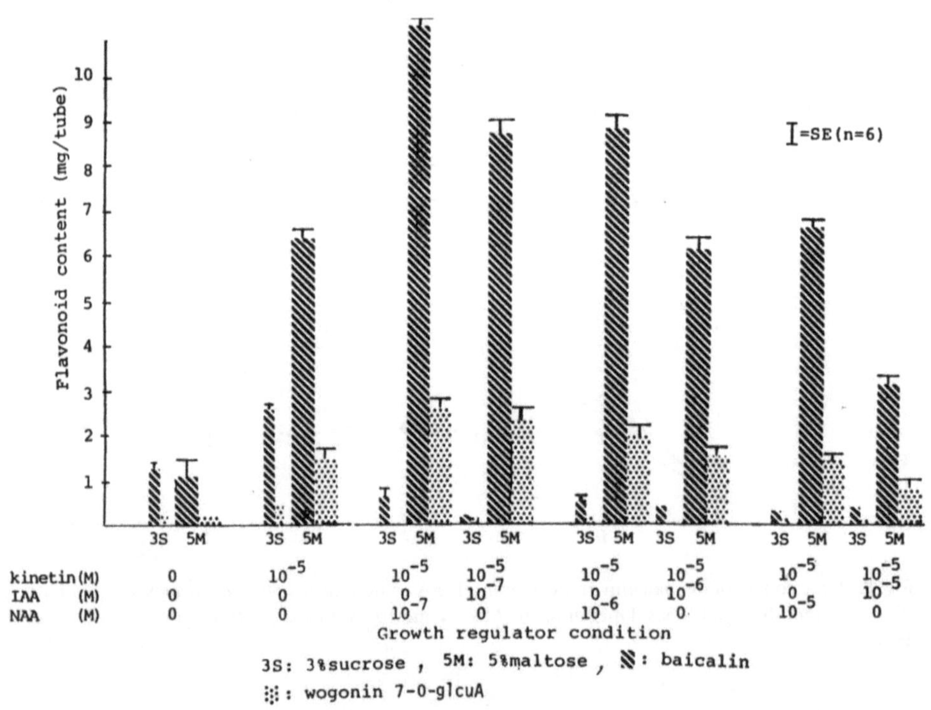

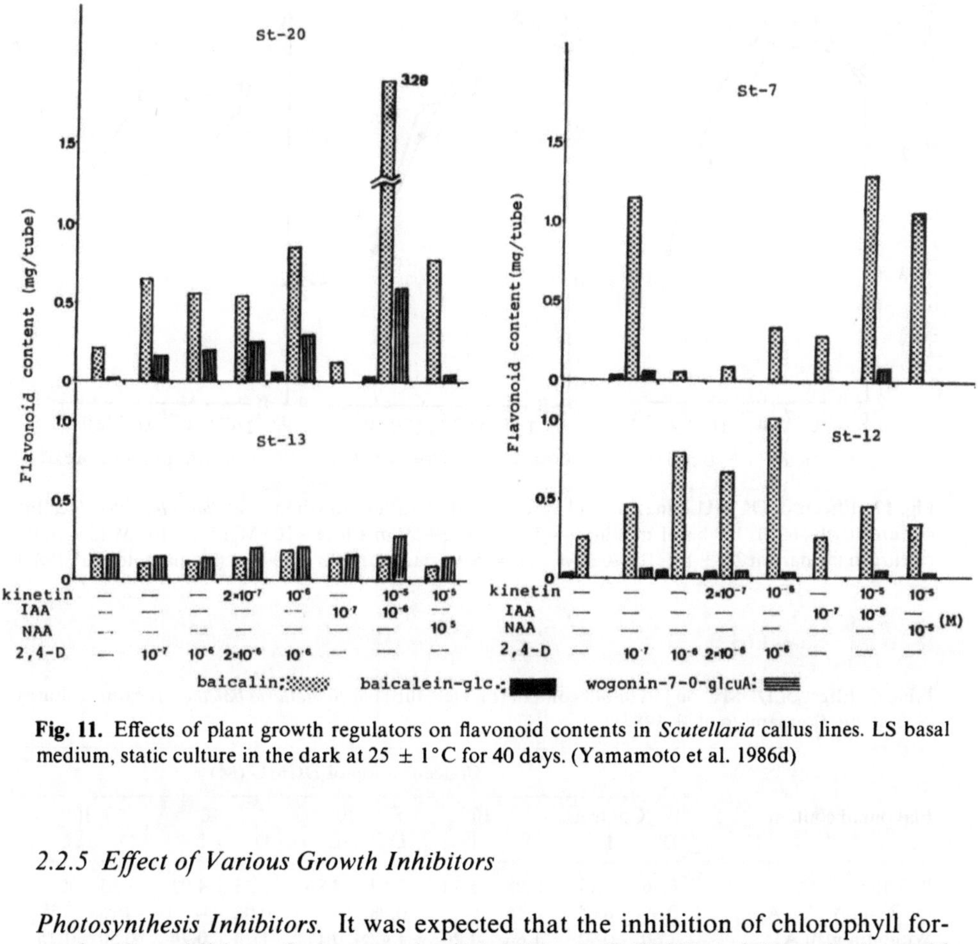

Fig. 11. Effects of plant growth regulators on flavonoid contents in *Scutellaria* callus lines. LS basal medium, static culture in the dark at 25 ± 1 °C for 40 days. (Yamamoto et al. 1986d)

2.2.5 *Effect of Various Growth Inhibitors*

Photosynthesis Inhibitors. It was expected that the inhibition of chlorophyll formation by addition of photosynthesis inhibitors (DCMU and Simazine) would result in an indirect stimulation of secondary metabolism. As shown in Fig. 12, callus growth was promoted by the addition of a low DCMU (a urea herbicide), but was inhibited by DCMU at a higher dose. This tendency was more obvious when cultured in the dark. These findings are in good agreement with the reported fact that on addition of DCMU the growth-promoting effect of light disappears and growth decreases to the level of growth in the dark. However, two aglycones, baicalein and wogonin, which usually appear at a low concentration level, were produced in relatively large quantities (Table 6). Simazine, a triazine herbicide, slightly promoted callus growth at low concentration, but markedly inhibited the growth at high concentration. The main flavonoids produced were baicalin (9.7 mg/10 ml) and wogonin 7-O-glucuronide (1.6 mg/10 ml) (Table 7).

◄

Fig. 10. Effects of carbon sources and growth regulators on flavonoid contents in *Scutellaria baicalensis* callus culture (St-20, 25-month-old). LS basal medium minus sucrose, static culture in the dark at 25 ± 1 °C for 40 days. (Yamamoto et al. 1986d)

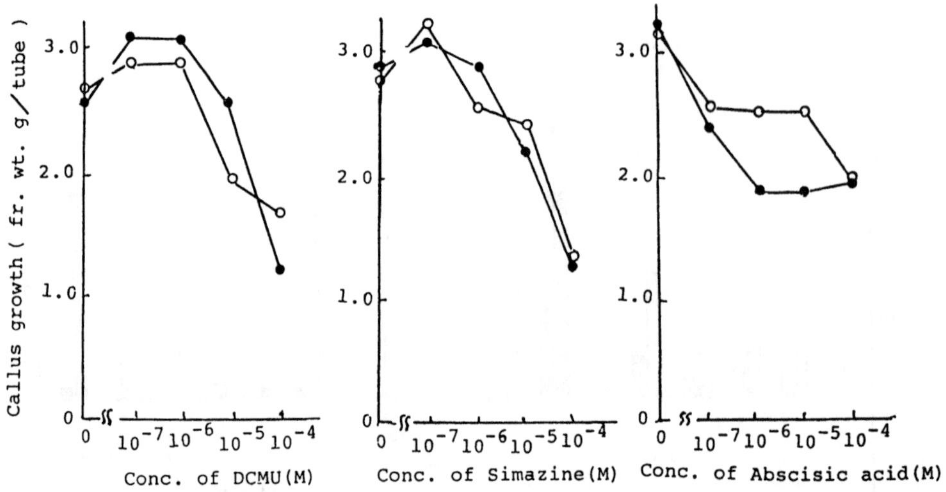

Fig. 12. Effects of DCMU, simazine and abscisic acid on callus growth in *Scutellaria baicalensis* callus cultures (Scb-st-20). LS basal medium – 3% sucrose + 5% maltose + 10^{-6}M IAA + 10^{-5}M kin, static culture in the dark at 25 ± 1°C for 40 days. $\bar{x}$: n = 5. ●: dark, ○: light (2500 lx). (Yamamoto et al. 1987)

Table 6. Effect of DCMU on flavonoid content (mg/test tube) of *Scutellaria baicalensis* callus cultures (Scb-st-20). (Yamamoto et al. 1987)

Flavonoid content	Concentration of DCMU (M)									
	Control		10^{-7}		10^{-6}		10^{-5}		10^{-4}	
	D[a]	L	D	L	D	L	D	L	D	L
Baicalin	8.86	5.17	8.98	5.74	7.52	4.86	6.25	4.03	1.33	4.44
Oroxylin A-glcu A	0.09	0.05	0.16	–	0.08	–	0.08	–	0.16	–
Wogonin-glcu A	1.60	0.72	1.66	1.20	1.42	0.11	1.28	0.74	0.50	1.15
5, 7, 2′, 6′-Tetra-hydroxyflavanone	–	–	0.10	–	0.07	–	–	–	0.01	–
Baicalein	0.02	–	0.24	0.04	0.11	0.04	0.01	–	0.02	0.01
Skullcupflavone II	–	–	–	–	–	–	–	–	0.01	–
Wogonin	0.02	–	0.10	0.02	0.04	0.01	0.01	–	0.02	0.01
Chrysin	–	–	0.02	–	–	–	–	–	–	–

LS basal medium – 3% sucrose + 5% maltose + 10^{-6}M IAA + 10^{-5}M kin. Static culture at 25 ± 1°C for 40 days.

[a] D, dark; L, light (2400 lx).

Table 7. Effect of simazine on flavonoid content (mg/test tube) of *Scutellaria baicalensis* callus cultures (Scb-st-20). (Yamamoto et al. 1987)

Flavonoid	Concentration of Simazine (M)									
	Control		10^{-7}		10^{-6}		10^{-5}		10^{-4}	
	D[a]	L	D	L	D	L	D	L	D	L
Baicalein-glc.	–	–	0.22	–	–	–	–	–	–	–
Baicalin	2.67	1.28	8.61	5.62	9.73	5.45	5.81	4.90	0.20	1.92
Oroxylin A-glcu A	0.02	–	0.11	0.51	0.08	0.51	0.04	0.03	–	–
Wogonin-glcu A	0.04	0.25	1.28	1.08	1.55	1.08	0.71	0.97	0.40	0.33
5, 7, 2′, 6′-Tetra-hydroxyflavanone	0.18	–	–	–	–	–	–	1.45	–	0.13
Baicalein	–	–	0.11	–	–	–	–	–	–	–
Wogonin	–	–	1.28	–	–	–	–	–	–	–

LS basal medium – 3% sucrose + 5% maltose + 10^{-6}M IAA + 10^{-5}M kin. Static culture at 25 ± 1°C for 40 days.
[a] D, dark; L, light (2500 lx)

Abscisic Acid (An Antagonist to the Growth-Promoting Activity of Auxin). Under both light and dark conditions, callus growth decreased by about 70% at 10^{-7}M of abscisic acid and by about 50% at 10^{-4}M (Fig. 12). The flavonoid content showed a general tendency to decrease with increasing concentration of abscisic acid either in the light or in the dark (Table 8).

L-Phenylalanine and p-Fluorophenylalanine (PFP; An Amino Acid Analog). As Fig. 13 shows, the callus growth was hardly affected by the addition of phenylalanine, whereas it was strongly inhibited by the addition of PFP. However, the production of flavonoids, especially baicalin, increased markedly when a low concentration of either PFP or phenylalanine was added to the cultures.

Table 8. Effect of abscisic acid on flavonoid content/(mg/test tube) of *Scutellaria baicalensis* callus cultures (Scb-st-20). (Yamamoto et al. 1987)

Flavonoid	Concentration of Abscisic acid (M)											
	0		10^{-7}		5×10^{-7}		10^{-6}		10^{-5}		10^{-4}	
	D[a]	L	D	L	D	L	D	L	D	L	D	L
Baicalin	9.42	3.23	4.80	2.60	1.89	1.95	0.86	1.14	5.40	2.80	2.83	1.70
Oroxylin A-glcu A	0.15	–	0.06	–	–	–	–	–	–	–	–	–
Wogonin-glcu A	1.99	0.58	1.10	0.66	0.61	0.47	0.20	0.27	0.80	0.48	0.38	0.29
5, 7, 2′, 6′-Tetrahydroxy-flavanone	0.02	–	–	–	0.02	–	0.01	–	–	–	–	–
Baicalein	0.05	0.04	0.06	–	0.03	0.07	–	–	0.01	0.01	0.01	–
Wogonin	–	–	0.02	–	–	0.07	–	0.03	–	–	–	–

LS basal medium – 3% sucrose + 5% maltose + 10^{-6}M IAA + 10^{-5}M kin. Static culture at 25 ± 1°C for 40 days.
[a] D, dark; L, light (2500 lx).

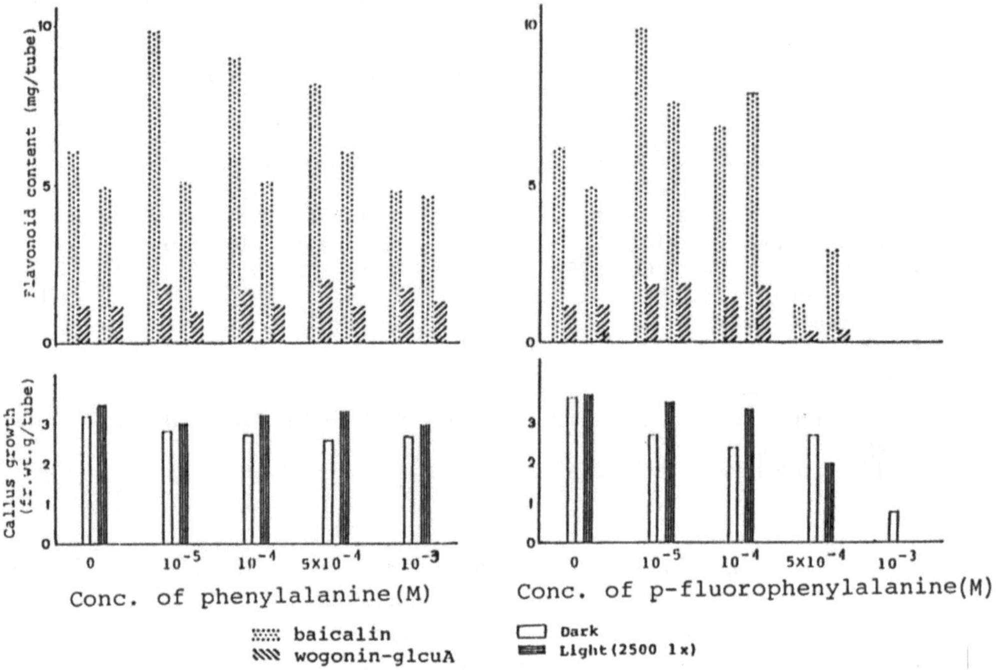

Fig. 13. Effects of phenylalanine and p-DL-fluorophenylalanine (PFP) on callus growth and flavonoid content in *Scutellaria baicalensis* callus cultures (Scb-st-20). LS basal medium – 3% sucrose + 5% maltose + 10^{-6}M IAA + 10^{-5}M kin, static culture at 25 ± 1 °C for 30 days. (Yamamoto et al. 1987)

Inhibition of Protein and Nucleic Acid Synthesis. Both callus growth and flavonoid production were significantly inhibited in proportion to the concentration of cycloheximide added (Yamamoto et al. 1987). Streptomycin and chloramphenicol apparently had no effect on callus growth and flavonoid production.

2.2.6 The Establishment of Culture and the Formation of Flavonoids

S. baicalensis stem callus (St-20 line) is inoculated in LS agar media containing 5% maltose instead of sucrose, 10^{-6}M IAA, and 10^{-5}M kin, and the tissue is cultured in the dark at 25 ± 1 °C for 40 days. The contents of baicalin and wogonin 7-O-glucuronide are consistently 5.92%, 1.52% per dry weight (yield 0.73 g/l, 0.14 g/l) respectively; the flavonoid content of the St-20 line for 70 days culture period gives a baicalin yield 1.6 g/l (Yamamoto et al. 1986b). Further, studies were made on the effects of basal media, carbon sources and plant growth regulation on the growth and flavonoid formation in callus culture of *S. baicalensis* in liquid medium by Yamamoto et al. (1989a). The best callus growth was obtained in the LS basal medium containing 10^{-7}M IAA, whereas the highest flavonoid content was in White's medium containing 10^{-6}M IAA and 10^{-5}M kin. Addition of 5% maltose not only increased cell growth by 52% but also the flavonoid content (baicalin, wogonin 7-O-glucuronide) by 92%.

The two-stage liquid culture system of *S. baicalensis* callus cultures (St-20 line) was studied for the purpose of promoting flavonoid production (Yamamoto et al. 1989b). Callus cultures were first grown on LS medium for 20 days and subsequently on White's medium containing 10^{-7}M IAA and 5% maltose for 20 days. Consequently, the contents of baicalin and wogonin 7-O-glucuronide increased to 9.07 and 6.33% per dry weight (yield 1.17 g/l, 0.82 g/l) as compared with 4.0 and 1.5% respectively in the ordinary batch cultures. These results showed that the two-stage cultures yielded as many flavonoids as the root (Wogon) of the intact plant. The flavonoid content of callus obtained by the two-stage liquid culture method was higher by 40 days than that of the roots differentiated from callus culture.

3 Conclusion

In the basic research on the formation of flavonoids in liquid culture of *S. baicalensis* (Yamamoto et al. 1989a), two-stage liquid cultures and the culture of differentiated roots (Yamamoto et al. 1989b) are still being studied for the purpose of mass proliferation of flavonoids (baicalein, baicalin) having anti-allergic action. The experiments on cultured tissue of *S. baicalensis* were performed from the following two viewpoints to obtain a definite result.

1. Various culture conditions are investigated using cultured cells of *S. baicalensis* and a useful substance is specifically formed, that is, flavonoids are isolated to be used as medicines.
2. The flavonoid pattern and content of *S. baicalensis* callus cultures are allowed to approach a crude drug Wogon, and *S. baicalensis* is made utilizable in a herb medicine recipe as a substitute medicine for the so-called Wogon.

Abbreviations: baicalein glc.: baicalein-glucoside: baicalein 7-O-glucoside, oroxylin A-glcu A: oroxylin A-glucuronide: oroxylin A 7-O-glucuronide, wogonin-glcu A: wogonin-glucuronide: wogonin 7-O-glucuronide

Acknowledgments. The author is grateful to Professor T. Tomimori for providing the facilities.

References

Aonuma S, Miura T, Tarutani M (1957) Effects of *Coptis, Scutellaria, Rhubarb* and *Bupleurum* on the serum cholesterol and phospholipid. Yakugaku Zashii 77:1303–1306

Hahlbrock K, Ebel J, Ortamann R, Sutter A (1971) Regulation of enzyme activities related on the biosynthesis of flavone glycosides in cell suspension cultures of parsley (*Petroselinum hortense*). Biochim Biophys Acta 244:7–15

Hirai Y, Takase H, Kobayashi H, Yamamoto M, Fujioka N, Kohda H, Yamasaki K, Yasuhara T, Nakajima T (1983) Screening test for anti-inflammatory crude drugs bases on inhibition effect of histamin release from mast cell. Shoyakugaku Zasshi 37:347–380

Institute of Pharmacy, Chinese Ministry of Sciences (ed) (1979) Chinese pharmacopoeia I. Peoples Hygienic Publications, Beijing, pp 546–555

Izawa B (1966) Coloured illustrations of medicinal plant (materia medica) of Japan. 2. Seibundo-Shincosha, Tokyo, pp 125–127

Kimura Y, Kubo M, Tani T, Arichi S, Ohminami H, Okuda H (1981a) Studies on Scutellariae Radix III. Effect on lipid metabolism in serum, liver, fat cell of rat. Chem Pharm Bull 29:2308–2312

Kimura Y, Kubo M, Tani T, Arichi S, Okuda H (1981b) Studies on Scutellariae Radix IV. Effects on lipid peroxidation in rat liver. Chem Pharm Bull 29:2610–2617

Kimura Y, Okuda H, Tani H, Arichi S (1982a) Studies on Scutellariae Radix V. Effects of flavone compounds on lipid peroxidation in rat liver. Chem Pharm Bull 30:1792–1795

Kimura Y, Kubo M, Kusaka K, Tani T, Higashino M, Arichi S, Okuda H (1982b) Studies on Scutellariae Radix VI. Effects on ethanol-induced hyperlipemia and lipolysis in isolated fat cell. Chem Pharm Bull 30:219–222

Kimura Y, Okuda H, Arichi S (1985) Studies on Scutellariae Radix. Effects of various flavonoids on arachidonate metabolism in leukucytes. Planta Med 47:132–136

Kitamura S, Murata G, Hori M (1975) Coloured illustrations of herbaceous plants of Japan (Sympetalae), rev edn, Hoikusha, Tokyo, pp 186–187

Koda A (1973) Pharmacological action of Scutellariae radix, principally baicalin and baicalein. Metab Disease (J Wakanyaku) 10:268–277

Koda A (1987) The relationship between crude drugs and allergic reaction. In: 34 Annu Meet Jpn Soc Pharmacog, Osaka, October, pp 9–12

Koda A, Sakai K (1966) Pharmacological action of baicalin and baicalein. Folia Pharmacol Jpn 62:39–40

Koda A, Nagai H, Wada H (1970a) The pharmacological action of baicalin and baicalein(I). Effects of active anaphylactic reactions. Folia Pharmcol Jpn 66:194–213

Koda A, Nagai H, Wada H (1970b) The pharmacological action of baicalin and baicalein(II). Effect upon dependent anaphylactic reaction. Folia Pharmacol Jpn 66:237–247

Koda A, Nagai H, Yoshida Y, Ro Hon Cat (1970c) The pharmacological action of baicalin and baicalein(III). Effect upon experimental asthma. Folia Pharmacol Jpn 66:471–486

Koda A, Katsuta E, Watanabe S, Mizuno M (1970d) Anti-allergic action of crude drugs. Folia Pharmacol Jpn 66:366–378

Koda A, Nishi K, Nagai H, Matsuura N, Tsutiya H (1982) Anti-allergic actions of crude drugs and blended chinese traditional medicines. Effects on Type I and Type IV. Folia Pharmacol Jpn 80:31–41

Koso New Medical Institute (ed) (1978) Dictionary of Chinese medicines, vol A. Sci Tech Publ, Shanghai, pp 2017–2021

Krenzaler E, Hahlbrock K (1973) Flavonoid glycosides from illuminated cell suspension culture of *Petroselium hortense.* Phytochemistry 12:1149–1152

Kubo M, Kimura Y, Odani T, Tani T, Namba T (1981) Studies on Scutellariae Radix. Planta Med 43:194–201

Kubo M, Matsuda H, Tanaka M, Kimura Y, Okuda H, Higashino M, Tani T, Namba K, Arichi S (1984) Studies on Scutellariae Radix VII. Anti-arthritic and anti-inflammatory action of methanolic extract and flavonoid components from Scutellariae Radix. Chem Pharm Bull 32:2724–2729

Kubo M, Matsuda H, Tani T, Arichi S, Kimura Y (1985) Studies on Scutellariae Radix XII. Anti-thrombic action various flavonoid from Scutellariae Radix. Chem Pharm Bull 33:2411–2415

Kumazaki H (1958) On the pharmacological action of the Scutellaria, a chinese crude drug. Gifu Pharm Univ Bull 6:94–111; 153–163; 164–168; 352–359; 372–376

Linsmaier EM, Skoog F (1965) Organic growth factor requirements of tobacco tissue culture. Physiol Plant 18:100–127

Liu Y-L, Li N-W, Sung W-C, Wu C (1980) Study on the flavonoid constituents in *Scutallaria amoena* Wright. Chung Tsáo Yao 11:337–340

Miyaichi Y, Imoto Y, Tomimori T, Chun-Ching Lin (1987) Studies on the constituents of *Scutellaria* species IX. On the flavonoid constituents of the root of *Scutellaria indica* L. Chem Pharm Bull 35:3720–3725

Miyaichi Y, Imoto Y, Kizu H, Tomimori T (1988) Studies on the Nepalese crude drugs (X). On the flavonoid and stilbene constituents of the leaves of *Scutellaria scandens* Buch, Ham. ex D. Don. Shoyakugaku Zasshi 42:204–207

Nanking Pharmaceutical Institute (ed) (1959) Materials for medicine. Nanking Pharm Inst, Shoka, pp 637–640

Niphonkoteisho Kyokai (ed) (1986) *Scutellaria* root, Pharmacopoeia japonica editio Undecima, D. Hirokawa, Tokyo, pp 113–116

Nishiyori T, Koda A, Tani T, Arichi S (1982) Antiallergic action of crude drugs and blended Chinese traditional medicine. Proc Symp Wakan-Yaku 15:187–191

Satake Y, Ohwi J, Kitamura S, Watari S, Tominari T (1981) Wild flowers of Japan herbaceous plants. Heibonsha, Tokyo, pp 75–78

Shibata S, Hattori S (1931) Über die konstitution des baicalein und des wogonin. Yakugaku Zasshi 51:15–17

Takagi S, Yamaki M, Inoue K (1980) Studies on the water-soluble constituents of roots of *Scutellaria baicalensis* Georgi (Wogon). Yakugaku Zasshi 100:1220–1224

Takagi S, Yamaki M, Inoue K (1981a) Flavone DI-*C*-glycoside from *Scutellaria baicalensis*. Phytochemistry 20:2443–2444

Takagi S, Yamaki M, Inoue K (1981b) On the minor constituents of the root of *Scutellaria baicalensis* Georgi (Wogon). Yakugaku Zasshi 101:899–903

Takido M (1973) On the chemical constituents of the root of *Scutellaria baicalensis* (Wogon) and possibility of its clinical applications. Metab Disease (J Wakanyaku) 10:261–267

Takido M, Aimi M, Takahashi S, Yamanouchi S, Torii H, Doi S (1975) Studies on the constituents in the water extracts of crude drug I. On the roots of *Scutellaria baicalensis* Georgi (Wogon) (1). Yakugaku Zasshi 96:108–113

Takido M, Yasukawa K, Matsuura S, Iinuma M (1979) On the relative structure of skullcapflavone II, a flavone compound in the roots of *Scutellaria baicalensis* Georgi (Wogon). Yakugaku Zasshi 99:443–444

Tomimori T, Miyaichi Y, Kizu H (1982) On the flavonoids constituents from the roots of *Scutellaria baicalensis* Georgi I. Yakugaku Zasshi 104:338–341

Tomimori T, Miyaichi Y, Imoto Y, Kizu H, Tanabe Y (1983) Studies on the constituents of *Scutellaria* species. II. On the flavonoid constituents of the root of *Scutellaria baicalensis* Georgi (2). Yakugaku Zasshi 103:607–611

Tomimori T, Miyaichi Y, Imoto Y, Kizu H, Tanabe Y (1984a) Studies on the constituents of *Scutellaria* species. III. On the flavonoid constituents of the root of *Scutellaria baicalensis* Georgi (3). Yakugaku Zasshi 104:524–528

Tomimori T, Miyaichi Y, Imoto Y, Kizu H, Suzuki C (1984b) Studies on the constituents of *Scutellaria* species. IV. On the flavonoid constituents of the root of *Scutellaria baicalensis* Georgi (4). Yakugaku Zasshi 104:529–534

Tomimori T, Miyaichi Y, Imoto Y, Kizu H (1984c) Studies on the constituents of *Scutellaria* species. V. On the flavonoid constituents of "Ban Zhi Lian", the whole herb of *Scutellaria rivularis* Wall (1). Shoyakugaku Zasshi 38:249–252

Tomimori T, Jin H, Miyaichi Y, Toyofuku S, Namba T (1985) Studies on the constituents of *Scutellaria* species. VI. On the constituents of the root of *Scutellaria baicalensis* Georgi (5). Quantitative analysis of flavonoids in *Scutellaria* roots by high-performance liquid chromatography. Yakugaku Zasshi 105:148–155

Tomimori T, Miyaichi Y, Imoto Y, Kizu H, Namba T (1986a) Studies on the Nepalese crude drugs. VI. On the flavonoid constituents of the root of *Scutellaria discolor* Colebr. (2) Chem Pharm Bull 34:406–408

Tomimori T, Miyaichi Y, Jin H, Toyofuku S, Yamamoto M (1986b) Studies on the constituents of *Scutellaria* species. VII. Seasonal variations of growth and flavonoid content in the root of *Scutellaria baicalensis* Georgi. Shoyakugaku Zasshi 40:381–389

Tomimori T, Miyaichi Y, Imoto Y, Kizu H (1986c) Studies on the constituents of *Scutellaria* species (VIII). On the flavonoid constituents of "Ban Zhi Lian", the whole herb of *Scutellaria rivularis* Wall (2). Shoyakugaku Zasshi 40:432–433

Trofimova NA (1982) *Scutellaria baicalensis* in isolated tissue culture (USSR). Rastit Resurs Yuzh Sib Ratsion Isp Okhrana, Tomsk, pp 80–82 (in Russian) [cf CA 99:155378z (1983)]

Yamahara J, Yamada T, Nakanishi H, Sawada T, Fujimura H (1981) Inhibitory effect of crude drugs on the denaturation of human γ-globurin induced by heat and Cu^{2+}. Shoyakugaku Zasshi 35:103–107

Yamamoto H, Chatani N, Kitayama Z, Tomimori T (1986a) Flavonoid production in *Scutellaria baicalensis* callus cultures. Plant Cell Tissue Org Cult 5:219–222

Yamamoto H, Chatani N, Watanabe K, Tomimori T (1986b) Effect of carbon sources on the growth and flavonoid formation of *Scutellaria baicalensis* stem callus culture. Shoyakugaku Zasshi 40:19–25

Yamamoto H, Chatani N, Watanabe K, Tomimori T (1986c) Effect of culture period on the growth and flavonoid formation of *Scutellaria baicalensis* stem callus culture. Shoyakugaku Zasshi 40:26–32
Yamamoto H, Chatani N, Watanabe K, Tomimori T (1986d) Effects of 5% maltose and plant growth regulators on the callus growth and flavonoid formation of some *Scutellaria baicalensis* stem callus lines. Shoyakugaku Zasshi 40:33–39
Yamamoto H, Watanabe K, Tomimori T (1987) Effects of various growth inhibitors on the callus growth and flavonoid production of *Scutellaria baicalensis* callus cultures. Shoyakugaku Zasshi 41:97–101
Yamamoto H, Sano T, Tomimori T (1989a) Growth and flavonoid formation of *Scutellaria baicalensis* callus culture in liquid medium. Shoyakugaku Zasshi 43:87–92
Yamamoto H, Sano T, Takeuchi S, Tanaka M, Tomimori T (1989b) Flavonoid production by two-stage culture and differentiated roots of *Scutellaria baicalensis* callus in liquid medium. Shoyakugaku Zasshi 43:188–191
Yu L-R, Liu M-L, Wang X-T (1984) Studies on the flavonoids of *Scutellaria viscidula* Bunge. Acta Pharm Sin 19:397–398

XXV *Securinega suffruticosa*: In Vitro Culture and the Formation of Securinega Alkaloids

AKIO IDE[1]

1 Introduction

1.1 Botany and Morphology

Securinega suffruticosa Rhed (*S. fluggeoides* Muell. Arg.) belonging to the family Euphorbiaceae, is a deciduous shrub up to 2 m high with greenish branches (Fig. 1). The leaves are alternate, short-petioled, oval or ovate, ovate-lanceolate, glabrous thin, acute or obtuse, and from 3 to 5 cm long. The flowers are dioecious, inconspicuous greenish white, in axillary clusters or solitary with small and subglobose greenish capsules, with five small sepals. Petals are absent. The male flowers are in fascicle, usually with five stamens. Two to eight female flowers form clusters with slightly longer peduncles and have a tristigmata style. The ovary is composed of three carpels, and one or two seeds are borne in a three-lobed dehiscent capsule. This plant is found in Japan, Korea, China, and some areas of the Eurasian continent.

1.2 Distribution and Alkaloids of the Plant

Securinega alkaloids are found in the genera *Securinega* and *Phyllanthus*, which are most widespread on the Eurasian continent, but not in the American.

The securinega alkaloids possess a securinane skeleton having a tetracyclic structure (Fig. 2). The following compounds have been found in the plant.

Securinine (*1*) (Muraveva and Bankovskii 1956a,b; Mukherjee et al. 1963), allosecurinine (*3*) (Parello et al. 1963), dihydrosecurinine (*5*) (Saito et al. 1963c), norsecurinine (*6*) (Saito et al. 1965a), securitinine (*8*) (Horii et al. 1967), securinol A (*10*) (Horii et al. 1965), securinol B (*11*), securinol C(*12*) (Horii et al. 1965, 1970), and phyllantidine (*13*) (Parello and Muravalli 1965) have been isolated from *S. suffruticosa*. Virosecurinine (*2*) (Nakano et al. 1963a,b), viroallosecurinine (*4*) (Saito et al. 1964a,b), (*6*) (Iketubosin and Mathieson 1963), virosine (dihydronorsecurinine, *7*) (Saito et al. 1964a,b,c) and hordenine (*14*) (Iketubosin and Mathieson 1963) were found in *S. virosa* Pax. et Hoffm. Phyllanthine (*9*) (Parello et al. 1965), nirurin (*15*) (Niu and Linag 1986), norvirosecurinine (enantiomer of 6) (*16*) (Rouffiac and Parello 1969), and 4-methoxynorsecurinine (*17*) (Mulchandani and Hassarajani 1984) have been isolated from *Phyllanthus*

[1]Faculty of Agriculture, Yamaguchi University, 1677-1 Yoshida, Yamaguchi 753, Japan

Biotechnology in Agriculture and Forestry, Vol. 15
Medicinal and Aromatic Plants III (ed. by Y.P.S. Bajaj)

Fig. 1A-G. *Securinega suffruticosa*. **A** A general view; **B** the male flower; **C** fruit; **D** – **G** callus culture, callus grown in MS medium containing 10^{-5} M NAA (**D**) or 10^{-5} M 2,4-D (**E**) and 10^{-7} M Kin in the dark, and 10^{-5} M NAA (**F**) or 10^{-6} 2,4-D (**G**) and 10^{-6} M Kin in the light, respectively

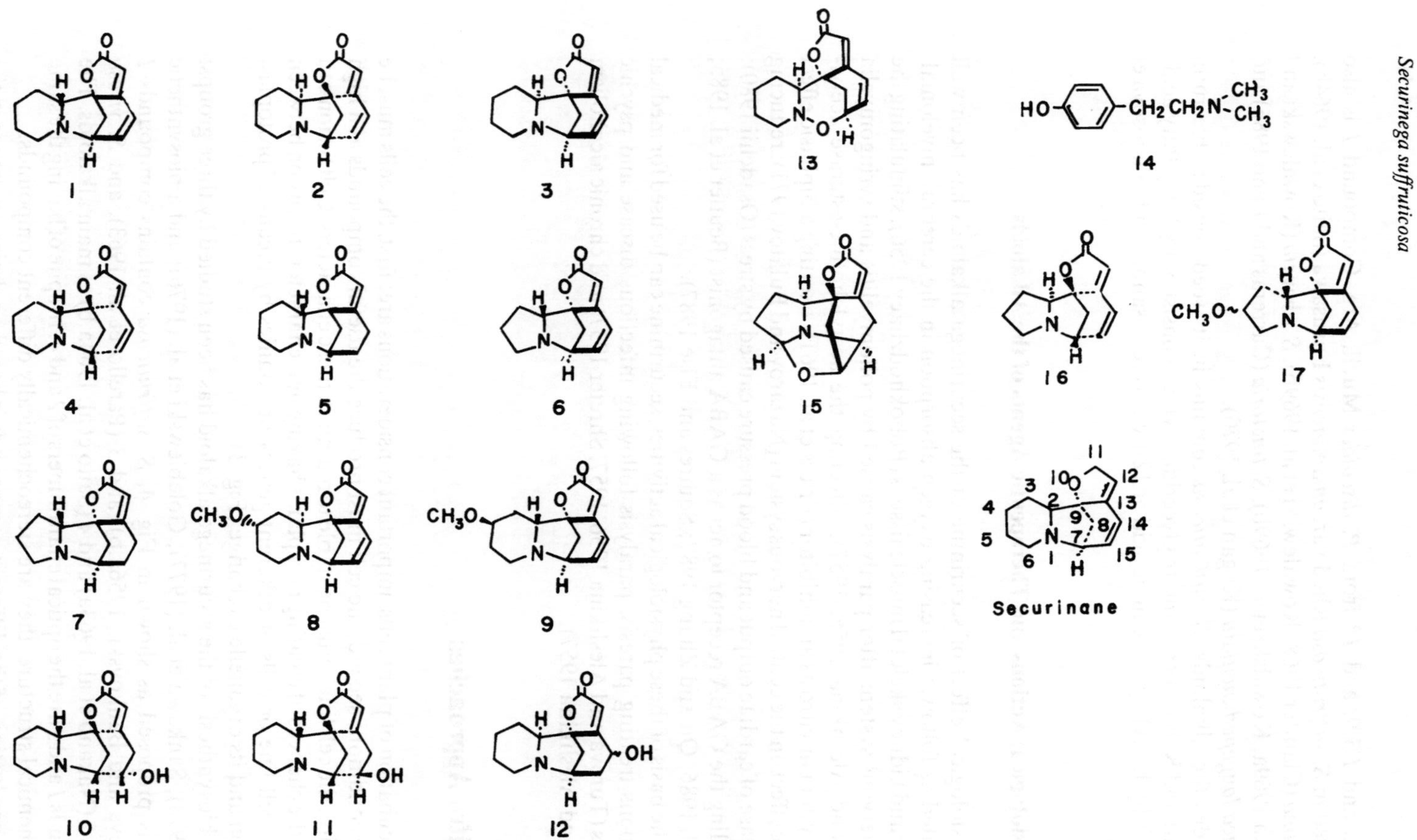

Fig. 2. Chemical structures of the securinega alkaloids *1* securinine; *2* virosecurinine; *3* allosecurinine; *4* viroallosecurinine; *5* dihydrosecurinine; *6* norsecurinine; *7* virosine; *8* securitinine; *9* phyllanthine; *10* securinol A; *11* securinol B; *12* securinol C; *13* phyllantidine; *14* hordenine; *15* nirurin; *16* norvirosecurinine; *17* 4-methoxy-norsecurinine. Securinane is the basal structure of the securinega alkaloids. The compound name of securinane is [6S-(6α, 11aβ, 11bα)]-2,6,9,10,11,11a-hexahydro-8H-6,11b-methanofuro[2,3-c]-pyrido[1,2-a]azepine

niruri L., and *1,3,9*, and *13* from *P. discoides* Muell. Arg. Compound *1* is also recognized in *S. suffruticosa* Rhed var. *amaminensis* Furusawa (Saito et al. 1964b), *S. durissima* (Clauder 1968; Kowalewski et al. 1969b), *S. obovata* (Kowalewski and Kowalska 1969a; Kowalski et al. 1969b), *S. tinctoria* (Carreras and Fano 1986) and *Securidaca longepedunculata* (Kogan et al. 1970).

Besides the alkaloids, *S. suffruticosa* contains in its seed considerable rutin (Nasudari and Krivut 1981) and triglycerides, which consist of hydroxy fatty acids such as 12-hydroxyheptadecanoic acid or 12-hydroxyeicosanoic acid (Yunosova et al. 1986).

1.3 Physiological Actions and Therapeutic Agents of the Alkaloids

The physiological effect of securinine of the securinega alkaloids has been well investigated as follows: increasing oxygen absorption in the cerebral, myelomal, and liver and kidney skeletal muscle tissues (Bobokhodzhaev 1956); stimulating the central nervous system after paralysis caused by poliomyelitis and syringomyelia (Trowa and Aleshkina 1956, 1957); affecting the acetylcholine-esterase acetylcholine system in vitro as an inhibitor (Friess et al. 1961); exerting a biphasic effect and tonic effect on the central nervous system (Nazarov and Bulatov 1971); reducing the increase of cardiac output and blood pressure caused by stress (Osadchii 1980); and binding the GABA receptor to act as a GABA antagonist (Beutler et al. 1985; Tao et al. 1986; Qu and Zhang 1987; Squires and Else 1987).

On the basis of these physiological activities, securinine can be used for medical preparations treating paresis, paralysis following infectious disease and psychic disorders (Turova and Aleshkina 1956, 1957, Shreter 1957) and chronic alcoholism (Trova and Aleshkina 1957).

2 In Vitro Approaches

In the incubation of plant cells, important considerations are: first, the cells must be in prime condition; second, the cell must produce the useful compounds at a high level; third, the cell lines must be stable to maintain these characters. Cell fusion can increase the chances of creating new plants having new characters by fusion between the same cell line or cells of different genera or families by means of protoplast formation and its organelle exchange (Fig. 3).

The biosynthesis of the securinega alkaloid has been studied by three groups: Parry (1978); Sankawa et al. (1977); Golebiewski et al. (1976); and a biosynthetic scheme is proposed as shown in Fig. 4. *S. suffruticosa* contains compounds *1* (Muraveva and Bankovskii 1956a,b), and *3* (Parello et al. 1963), and *S. virosa* contains *2* (Nakano et al. 1963a) and *4* (Saito et al. 1964a,b) as main alkaloids. The compounds *1* and *3* are the optical enantiomers of *2* and *4*. In spite of having the same planar chemical structure, they are stereochemically different compounds.

The production of the different compounds is based on their occurrence in the chiral center formation at the 9th carbon of the securinane skeleton rather than the

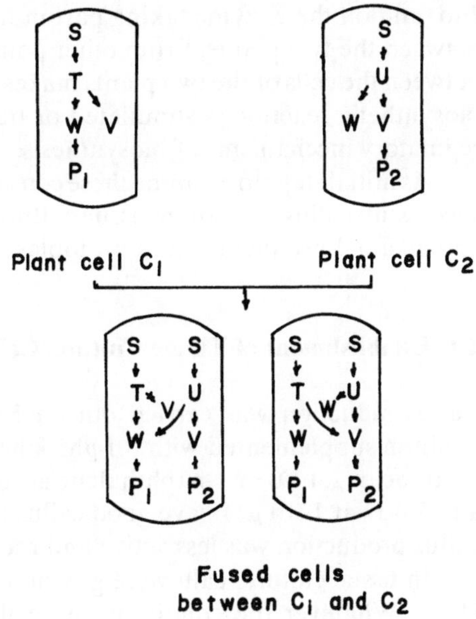

Fig. 3. Effects of transduction in the cell. Plant cell C1 has biosynthetic system P_1, and C_2 has P_2

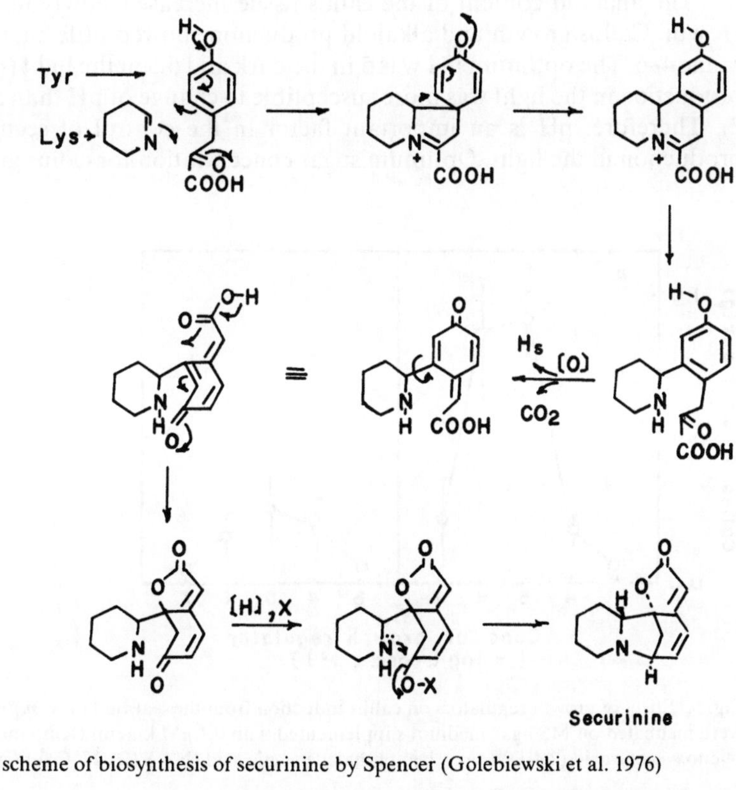

Fig. 4. A proposed scheme of biosynthesis of securinine by Spencer (Golebiewski et al. 1976)

2nd, and on the enzyme taking part in this reaction, which is essentially different between the two plants. From other points of view, transduction of cell character between the cells of the two plants makes it possible to create new cells in which the biosynthetic reaction is stimulated or transformed artificially, and to resolve the regulatory mechanisms of biosynthesis.

As initial steps to examine these considerations, the following experiments were successful: callus formation, tissue culture, optimum conditions for growth of tissue and alkaloid production, and protoplast formation (Ide et al. 1986).

2.1 Establishment of Tissue Culture/Cell Suspension

Callus induction was carried out on Murashige and Skoogs (MS) agar (1.3%) medium supplemented with 0.1 μM kinetin (Kin) and with 2,4-dichlorophenoxy-acetic acid (2,4-D) or 1-naphthalene acetic acid (NAA) in the dark. NAA at 10 μM and 2,4-D at 1–10 μM gave good callus formation (Fig. 5), but in the former case callus production was less active and root-like tissue was produced.

In tissue culture, calli were grown on a medium supplemented with 2,4-D or NAA. Whatever the condition, the callus showed a significant lag phase after inoculation (Fig. 6). They reached the stationary stage after 10–12 weeks through a logarithmic phase. Callus growth in the medium containing NAA in the dark was slow compared with that under other conditions.

The alkaloid content of the callus tissue increased slowly in step with callus growth. Callus growth and alkaloid production showed little change over the pH range 4–6. The optimum pH was 5 in the dark and 5.5 in the light (Fig. 7). Alkaloid production in the light was more susceptible to change of pH than in the dark (Fig. 7). Therefore, pH is an important factor in the control of securinega alkaloid production in the light. Optimum sugar concentration for callus growth was 5% in

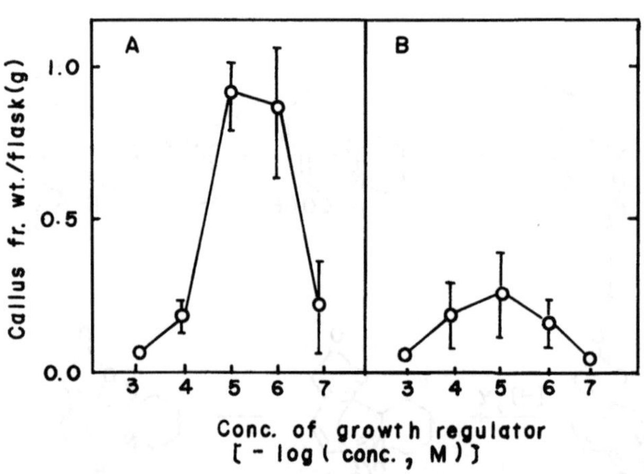

Fig. 5. Effect of growth regulators on callus induction from the leaf bud of *S. suffruticosa*. The tissues were incubated on MS agar medium supplemented with 0.1 μM kinetin (Kin) and with 2,4-dichloro-phenoxyacetic acid (2,4-D) (**A**) or 1-naphthalene acetic acid (NAA (**B**) 27°C for 6 weeks in the dark

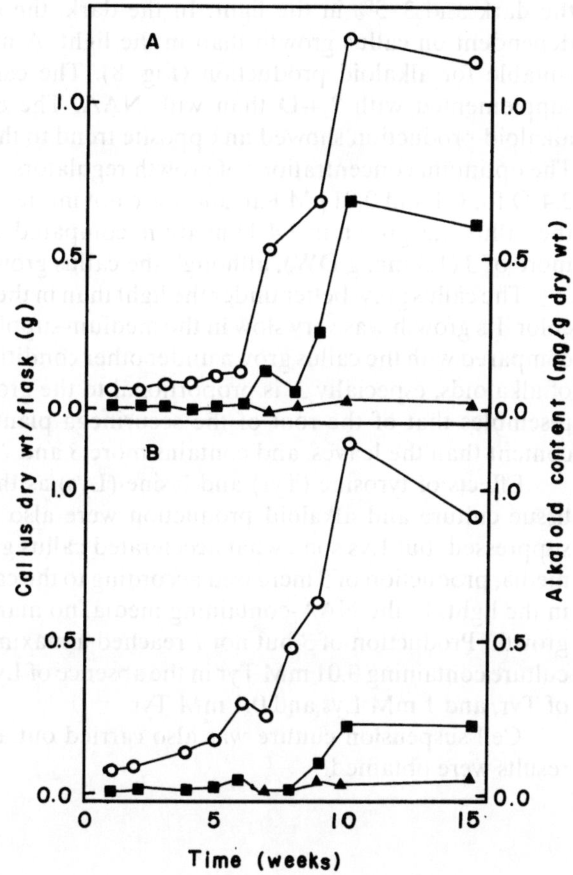

Fig. 6A,B. Growth as measured by dry weight (○) and alkaloid contents (▲—▲, *1*; ■—■, *3*) in callus of *S. suffruticosa*. The callus was grown on MS agar medium supplemented with 0.1 μM Kin and 1 μM 2,4-D at 27°C in the dark (**A**) or in the light (**B**)

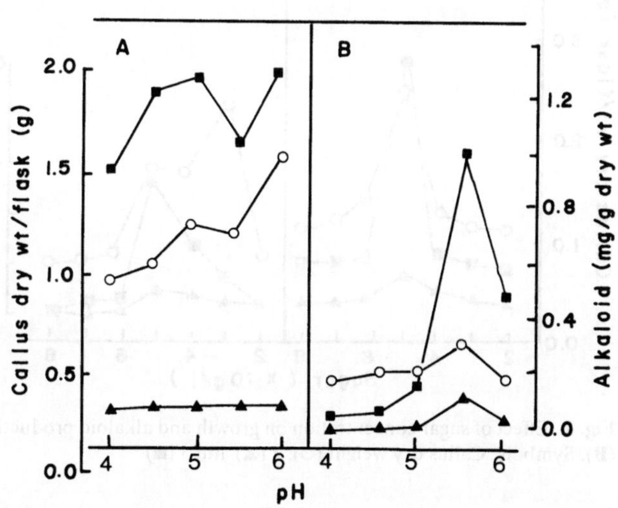

Fig. 7. Effect of pH on growth and production of alkaloids. Here and in Figs. 8 calli were grown for 10 weeks on MS agar medium supplemented with 2,4-D (1 μM) and Kin (0.1 μM) containing 5% sugar in the dark (**A**) or in the light (**B**). Symbols: Callus dry weight (○), *1* (▲) and *3* (■)

the dark and 3–5% in the light. In the dark, the alkaloid production was more dependent on callus growth than in the light. A medium containing 5% sugar is suitable for alkaloid production (Fig. 8). The callus grew better in a medium supplemented with 2,4-D than with NAA. The effect of growth regulation on alkaloid production showed an opposite trend to that of callus growth in the dark. The optimum concentrations of growth regulators for the callus growth were 1 μM 2,4-D for 0.1 and 0.01 μM Kin and were not influenced by light (Figs. 9, 10). When the callus was grown in 2,4-D medium compared with NAA, it produced slightly more of *3* (1.3 mg/g DW), although the callus growth was not so good.

The callus grew better under the light than in the dark and assumed a pale green color. Its growth was very slow in the medium supplemented with NAA in the dark compared with the callus grown under other conditions. In contrast, the production of alkaloids, especially *3*, is proportional to the growth of the callus. This pattern resembles that of the root of the securinega plant, which has a higher alkaloid content than the leaves, and contains more *3* and *1* than the leaves.

Effects of tyrosine (Tyr) and lysine (Lys) as the precursor of the alkaloid on tissue culture and alkaloid production were also examined (unpubl. data). Tyr suppressed, but Lys somewhat accelerated callus growth. In the 2,4-D-containing media, production of *3* increased according to the callus growth in the dark, but not in the light. In the NAA-containing media, no marked effects were seen on callus growth. Production of *3* but not *1* reached a maximum (0.7–0.8 mg/g DW) in the culture containing 0.01 mM Tyr in the absence of Lys, 0.1–1 mM Lys in the absence of Tyr, and 1 mM Lys and 0.1 mM Tyr.

Cell suspension culture was also carried out as in tissue culture, and similar results were obtained.

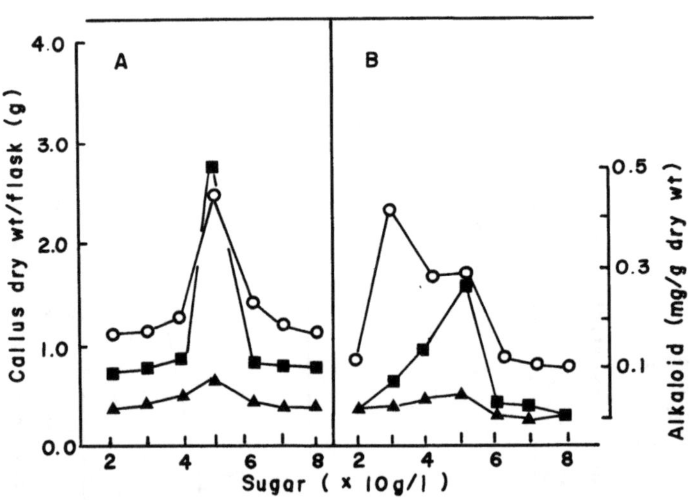

Fig. 8. Effect of sugar concentration on growth and alkaloid production in the dark (**A**) and in the light (**B**). Symbols: Callus dry weight (○), *1* (▲) and *3* (■)

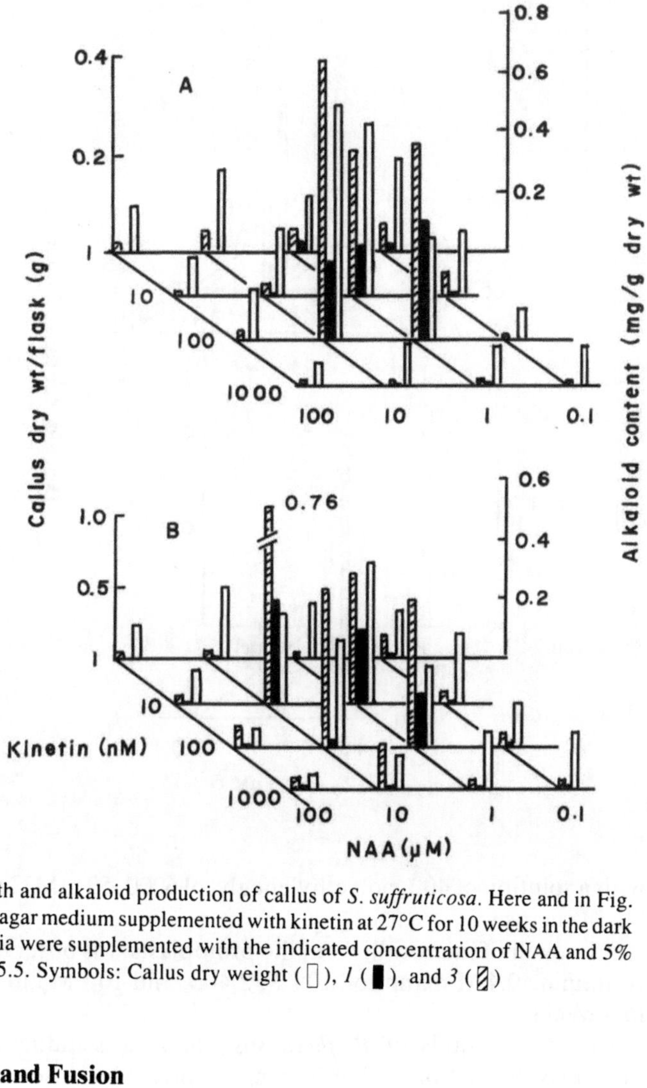

Fig. 9. Effect of NAA on growth and alkaloid production of callus of *S. suffruticosa*. Here and in Fig. 10, callus was incubated on MS agar medium supplemented with kinetin at 27°C for 10 weeks in the dark (**A**) or in the light (**B**). The media were supplemented with the indicated concentration of NAA and 5% sugar, and pH was adjusted to 5.5. Symbols: Callus dry weight (▯), *1* (▮), and *3* (▨)

2.2 Protoplast Isolation and Fusion

Protoplasts could be prepared from the young leaves or calli of *S. suffruticosa* and *P. flexuosus*, the following techniques being suitable. Calli or young leaves were cut into small pieces and the sliced pieces were incubated in a mixed enzyme solution of 1.0% cellulase (Onozuka R-10) and 0.05% pectoriase (Y-23) in 0.6 M mannitol (pH 5.5) for 6 h. The protoplast initiated well from the 8-week-old callus which was subcultured several times. The incubated mixture was filtered through a G-1 glass filter and the filtrate was centrifuged. The precipitate was washed with 0.6 M mannitol three times.

The precipitate was suspended in 0.6 M mannitol until it reached 10^6 cell/ml. Fifteen percent of the protoplasts fused when the protoplast suspension was mixed

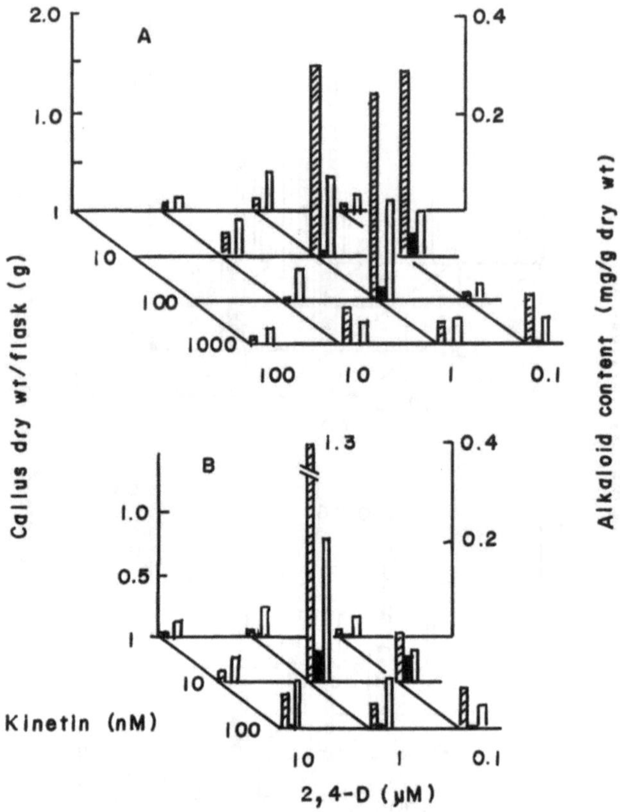

;. **10A,B.** Effect of 2,4-D on ›wth and alkaloid produc- n of callus of *S. suffruticosa*. mbols as in Fig. 9

with a solution of 40% polyethyleneglycol 6000, 50 mM $CaCl_2$, 50 mM glycine, and 0.1 M sucrose.

Growth of microcalli from the protoplasts was observed in MS liquid medium containing 0.4 M mannitol, 10^{-5} M 2,4-D, and 10^{-7} M kin at 23 °C in the dark for 3 to 4 weeks.

From the buds of *P. flexuosus*, callus was induced by the same method. Protoplast formation and growth of the protoplast also succeeded under the same conditions as for *S. suffruticosa* plants.

2.3 Extraction and Structure of the Securinega Alkaloids

Extraction (Ide et al. 1986) of the securinega alkaloids from callus tissue can be carried out as from the plant (Horii et al. 1963a). Calli were homogenized in a mixed solution of 2% NH_3 aq. and dichloroethane. The organic layer was extracted with 1 M H_2SO_4. The aqueous layer was basified with 10% NH_4OH, and extracted with $CHCl_3$. The $CHCl_3$ layer gave crude alkaloids.

From the plant, dry matter above ground and the roots were crushed with a hammer and extracted in the same way as the calli. The crude extracts were crystallized from ethanol for *1*, and the mother liquid was submitted to the chromatograph of SiO_2 or treated with organic acid, such as d-camphorsulfonic acid or oxalic acid to prepare organic salts (Horii et al. 1963a).

The chemical structures of the securinega alkaloids isolated from plants are shown in the Fig. 2. The norsecurinine groups seem to be synthesized from ornithine in the plant cell instead of lysine as in the case of the securinine groups. The structure of nirurin looks strange, however, its formation is not considered to be impossible when dopa is substituted for instead of tyrosine on the biosynthetic scheme.

3 Conclusions and Prospects

Securine improves muscle tone, stimulates respiration, strengthens cardiac contraction, and raises blood pressure (Bobokhodzhaev 1956; Muraveva and Bankovskii 1956b; Turova and Aleskina 1956, 1957; Golebiewski et al. 1976) and is therefore useful for therapeutic purposes. For this reason, it is important not only to establish cultivation methods for a good strain, but also to create new cells by artificial regulation. With these aims, we have experimented with *S. suffruticosa*, and studied callus derived from budding stem explants of the germinated seeds on MS agar medium supplemented with 0.1 μM kinetin and 1-10 μM 2,4-D. The major alkaloids, securinine and allosecurinine, were produced in tissue culture at pH 5.5, sugar 5% in the medium, and 1 μM 2,4-D and 0.01 μM kinetin in the light, cell suspension cultures were raised, and protoplasts were formed from the callus tissue. However, it is necessary to investigate the relationship between growth of tissue and production of alkaloids at the gene level, as there is a limitation to improving the tissue culture method and selecting the best strain.

References

Beutler JA, Karbon EW, Brubaker AN, Malik R, Curtis DR, Enna SJ (1985) Securinine alkaloids: a new class of GABA receptor antagonist. Brain Res 330(1):135-140

Bobokhodzhaev IY (1956) Pharmacology of securinine alkaloid. Farmakol Tokisikol 19 suppl: 3-5

Carreras LM, Fano B (1986) "Tamujo": a new medicinal plant in the Iberic flora. Ann R Acad Farm 52(2):319-329

Clauder O (1968) *Securinega fluggeoides* and *S. durissima* alkaloids. Acta Pharm Hung 38(2-3):126-132

Friess SL, Durent RC, Whitcomb ER, Reber LT, Tommesen WC (1961) Some toxicologic properties of the alkaloids galanthamine and securinine. Toxicol Appl Pharmacol 3:347-357

Golebiewski WM, Horsewood P, Spencer ID (1976) Biosynthesis of securinine: the mode of incorporation of lysine. J Chem Soc Chem Commun 1976(6):217-218

Horii Z, Tanaka T, Tamura Y, Saito S, Matsumura C, Sugimoto N (1963a) Securinine 1. Isolation and properties of securinega alkaloid. Yakugaku Zasshi 83:602-605

Horii Z, Ikeda M, Yamawaki Y, Tamura Y, Saito S, Kodera K (1963b) Absolute stereochemistry of securinine. Tetrahedron 19(12):2101-2110

Horii Z, Ikeda M, Tamura Y, Saito S, Kodera K, Iwamoto T (1965) Isolation of securinol A, B and C from *Securinega suffruticosa* and the structures of securinol A and B. Chem Pharm Bull 13(11):1307-1311

Horii Z, Ikeda M, Hanaoka M, Yamauchi M, Tamura Y, Saito S, Tanaka T, Kodera K, Sugimoto N (1966) Structure of securitinine. Chem Pharm Bull 14(8):917-918

Horii Z, Ikeda M, Hanaoka M, Yamauchi M, Tamura Y, Saito S, Tanaka T, Kodera K, Sugimoto N (1967) Structure of securitinine. Chem Pharm Bull 15(11):1633–1640

Horii Z, Yamauchi M, Ikeda M, Momose T (1970) Structure of securinol B and C. Chem Pharm Bull 18(10):2009–2012

Horii Z, Imanishi T, Yamauchi M, Honda M, Perello J, Munavalli S (1972) Structure of phyllanthidine. Tetrahedron Lett 1972:1877

Ide A, Nagano K, Tanaka N, Iwasaki K, Yamane Y, Koga D, Yagishita K, Nakao K, Kurisu Y, Yamasaki K (1986) Growth and alkaloid production of callus culture induced from *Securinega suffruticosa*. Phytochemistry 26(1):145–148

Iketubosin GO, Mathieson DW (1963) Isolation of hordenine and norsecurinine from *Securinega virosa*. Structure of norsecurinine. J Pharm Pharmacol 15(12):810–815

Kogan VI, Gerbunev VD, Rostostskii BK (1970) Securinine from bark of *Securidaca longepedunculata*. USSR 2277796

Kowalewski Z, Kowalska M (1969a) Rutoside levels in selected species of the genus *Securinega*. Ann Pharm (Poznan) 7:93–98

Kowalewski Z, Frencel I, Urszulak I, Filarowska A (1969b) Securinine content in some species of the genus *Securinega* cultivated in Poland. Ann Pharm (Poznan) 7:99–103

Mukherjee R, Das B, Arya VP, Chatterjee A (1963) Chemistry of securinine, an alkaloid of the root of *Securinega suffruticosa* (Pall) Rhed. Naturwissenschaften 50(5):55

Mulchandani NB, Hassarajani SA (1984) 4-Methoxynorsecurinine, a new alkaloid from *Phyllanthus niruri*. Planta Med 50(1):104–105

Muraveva VI, Bankovskii AI (1956a) Chemical study of alkaloids of *Securinega suffruticosa*. Doklady Akad Nauk SSSR 110:998–1000

Muraveva VI, Bankovskii AI (1956b) Securinine – a new alkaloid of strychinine-like action. Med Prom 10(2):27–28

Nakano T, Yang TH, Terao S (1963a) Alkaloids of *Securinega virosa* 1. Structure of virosecurinine. Tetrahedron 19:609–619

Nakano T, Yang TH, Terao S (1963b) Studies on the alkaloids of *Securinega virosa* 2. The absolute configuration of C-6 in virosecurinine and the stereochemical interrelationship of virosecurinine, securinine and allosecurinine. J Org Chem 28(10):2619–2621

Nasudari AA, Krivut BA (1981) Chromatographic-spectrometric method for the quantitative determination of rutin in *Securinega suffruticosa* Rhed. Dokl Akad Nauk Az SSR 37(9):66–70

Nazarov GS, Bulatov AA (1971) Pharmacotoxicological action of securinine. Veterinariya (Moscow) 48(8):78–80

Niu C, Liang X (1986) Rearrangement of an alpha-ketol. Hauxue Xuebao 44(7):746–749

Oboimakova AP, Malykhina SA (1956) New medical preparations. Proc Acad Sci USSR Sect Chem 110:631–633

Osadchii LI (1980) Pharmacological analysis of physiological mechanisms of orthostatic hemodynamic stability. Kosm Biol Aviakosm 14(2):45–49

Parello J, Meleva A, Goutarel (1963) Phyllochrisine and securinine, alkaloids of *Phyllanthus discoides*. Bull Soc Chim France 1963(4):898–910

Parello J, Muravalli S (1965) Phyllantine and phyllantidine, alkaloids of *Phyllanthus discoides*. Compt Rend 260(1):337–340

Parello J (1968) Structure of phyllanthine, a major alkaloid from *Phyllanthus discoides*. Bull Soc Chim Fr 1968(3):1117–1129

Parry RJ (1978) Biosynthesis of the securinega alkaloids. Investigations of the early and late stages of the pathway. Bioorg Chem 34(4):408–413

Petchenaree P, Bunyapraphatsara N, Cordell G, Cowe HJ, Cox PJ, Howie RA, Patt SL (1986) X-ray crystal and molecular structure of nirurine, a novel alkaloid related to the securinega alkaloid skeleton, from *Phyllanthus niruri* (Euphorbiaceae). J Chem Soc Perkin Trans 1:1551–1556

Qu ZW, Zhang JT (1987) Influence of securinine on 5 kinds of neurotransmitter receptors and membrane potential of spinal cord. Acta Pharm Sin 22(4):305–307

Rouffiac R, Parello J (1969) *Phyllanthus niruri* alkaloids. Presence of the optical antipode of norsecurinine. Plant Med Phytother 3(3):220–223

Saito S, Kodera K, Shigematsu N, Ide A, Sugimoto N, Horii Z, Hanaoka M, Yamawaki Y, Tamura Y (1963a) Structure of securinine. Tetrahedron 19(12):2085–2099

Saito S, Kodera K, Shigematsu N, Ide A, Horii Z, Tamura Y (1963b) Stereochemistry of securinine and virosecurinine. Chem Ind 1963:689–690

Saito S, Shigematsu N, Horii Z (1963c) Securinine 2. Isolation of dihydrosecurinine from *Securinega suffruticosa*. Yakugaku Zasshi 83:800–801

Saito S, Iwamoto T, Tanaka T, Matsumura C, Sugimoto N, Horii Z, Tamura Y (1964a) Two new alkaloids, viroallosecurinine and virosine, isolated from *Securinega virosa*. Chem Ind 1964(28):1263–1264

Saito S, Tanaka T, Iwamoto T, Matsumura C, Sugimoto N, Horii Z, Makita M, Ikeda M, Tamura Y (1964b) Determination of securinine and its stereoisomers in plants of the Securinega species, and isolation of viroallosecurinine and virosine. Yakugaku Zasshi 84(11):1126–1133

Saito S, Tanaka T, Kodera K, Nakai H, Sugimoto N, Horii Z, Ikeda M, Tamura Y (1964c) Structure and stereochemistry of norsecurinine and dihydrosecurinine. Chem Pharm Bull 12(12):1520–1523

Saito S, Tanaka T, Kodera K, Nakai H, Sugimoto N, Horii Z (1965a) Structure of norsecurinine and dihydronorsecurinine. Chem Pharm Bull 13:786

Saito S, Tanaka T, Kodera K, Nakai H, Sugimoto N, Horii Z, Ikeda M, Tamura Y (1965b) Absolute stereochemistry at C9a of norsecurinine and dihydronorsecurinine. Chem Pharm Bull 13:614–616

Sankawa U, Ebizuka Y, Yamasaki K (1977) Biosynthesis of natural products. Part 2. Biosynthesis of securinine, the main alkaloid of *Securinega suffruticosa*. Phytochemistry 16(5):561–563

Shreter AI (1957) A new medical plant in the Soviet flora. Botani Zhur 42:925–928

Squires RF, Else S (1987) GABA receptor blockers reverse the inhibitory effect of GABA on brain-specific sulfur 35 TBPS binding. Brain Res 414(2):357–364

Tao S, Peng J, Lu M (1986) Central convulsive action of l-securinine. Zhongguo Yaoli Xuebao 7(1):9–12

Turova AD, Aleshkina YAS (1957) Securinine – a new therapeutic agent.Med Prom S S S R 11(1):54–55

Turova AD, Aleshkina YAS (1956) Pharmacology of the new alkaloid, securinine. Farmakol Tokisikol 19(4):11–17

Yunusova SG, Gusakova SD, Glushenkova AI (1986) Lipids of *Securinega suffruticosa* seeds. Khim Prir Soedin 1986:277–281

XXVI *Solanum eleagnifolium* Cav. (Silverleaf Nightshade): In Vitro Culture and the Production of Solasodine

A.M. GIULIETTI[1], H.M. NIGRA[1], and O. CASO[2]

1 General Account

Although diosgenin is the most important source for the production of steroids (Applezweig 1969), alternative raw materials for replacing *Dioscorea* spp. (Mexican yams) have been sought. In this regard solasodine, aglycone of the glycoalkaloids solamargine and solasonine, has been shown to be a promising replacement for diosgenin. About 100 *Solanum* species have been tested for their glycoalkaloid content (Mann 1978). The same author considered that *S. aviculare*, *S. khasianum*, and *S. laciniatum* were the main candidates for solasodine production.

During the 1970's, the Organic Chemistry Department of the Universidad Nacional de San Luis (Argentina) started the screening of Argentinian species of the genus *Solanum*, in order to explore the possibility of local production of the steroids (Guerreiro et al. 1971; Guerreiro 1972; Kavka et al. 1973). Table 1 shows the results of this screening, *S. eleagnifolium* Cav. var. *leprosum* (Ortega) Dunal has a good concentration of glycoalkaloid solamargine.

This chapter deals with the agronomic and ecological aspects related to the variety of *S. eleagnifolium* Cav. that grows in Argentina, the establishment of tissue culture, and also solasodine production from solamargine.

1.1 Botanical Description

Solanum eleagnifolium Cav. var. *leprosum* (Ortega) Dunal has been described by several authors (Parodi 1974; Cabrera 1965; Morton 1976). A member of the *Solanaceae* it is a perennial shrubby herb, 30–50 cm high, with very branched stems, especially near the base, that are covered by hairs and spines. Leaves are numerous, alternate, with both sides of the leaf blade covered by hairs and with spines along petiole and veins. The leaf shape is oval-oblong, sometimes linear-oblong, and with entire or lobed margins. The root system is dimorphic, composed of two kinds of roots. Each plant has a strong vertical root, clearly geotropic, which can reach depths of 2 m; from this root others grow that are more or less horizontal, with diageotropic

[1]Biotecnología – Facultad de Farmacia y Bioquímica, Universidad de Buenos Aires, (1113) Junín 956, Buenos Aires, Argentina
[2]Departamento de Ciencias Biológicas, Facultad de Farmacia y Bioquímica, Universidad de Buenos Aires, (1113) Junín 956, Buenos Aires, Argentina

Biotechnology in Agriculture and Forestry, Vol. 15
Medicinal and Aromatic Plants III (ed. by Y.P.S. Bajaj)

Table 1. Glycoalkaloids content of Argentinian *Solanum* species, screened by members of the Department of Organic Chemistry of the Universidad de San Luis, Argentina. (Compiled by E. Guerreiro)

Species	Organs	Total Glycoalkaloids (g/kg dry matter)	Glycoalkaloids Found	Diosgen-in kg/dry matter
S. eleagnifolium Cav. var. *leprosum* (Ortega) Dun.	Unripe fruits	72.70	Solamargine	1.95
S. lorentzii Bitt.	Leaves and stems	37.10	Solamargine Solasonine	–
S. calophyllum Phil.	Leaves and stems	5.76	Solamargine	0.25
S. juncallense Reiche	Leaves and stems	5.05	Solamargine Solasonine	1.10
S. sublobatum Willd. ex Roem. et Schult	Fruits	11.60	Solamargine	0.90
S. atriplicifolium Gill. ex Mey.	Fruits	–	–	8.05
S. euacanthum Phil.	Leaves and stems	3.50	Solamargine	0.25
S. pyretrifolium Griseb.	Leaves and stems	12.50	Solamargine	0.90
S. juvenale Thell.	Leaves and stems	–	–	1.90

growth, that can extend to a distance of ca. 2.5–3 m. Along the horizontal roots and creeping rhizomes, adventitious buds usually originate (Vigna et al. 1981).

The flowers are blue, rarely white, either single or in groups, subterminals or opposite to leaves. Fruits are berries, orange-yellow when ripe, with diameters of 10–13 mm; 40–50 large seeds, agglutinated by a substance rich in saponins. Due to this, the fruits are used for soap, while Indians used them for leather tanning (Ruiz Leal 1972; Vigna et al. 1981). It is also considered a toxic plant for cattle, horses, and mules (Vigna et al. 1981).

S. eleagnifolium has different common names: quillo, meloncillo, revienta-caballo in Argentina; silverleaf nightshade in the U.S.A. It is a plant of American origin, with widespread distribution in Argentina, Chile, Uruguay, SW Brazil, Mexico, and the United States. According to Morton (1976), the South American plants are different from those of North America, and as a consequence he assigned the var. *leprosum* to the species with Argentinian distribution.

It is a very noxious weed, due to the presence of the bud-forming roots (Fernández and Brevedan 1972; Boyd and Murray 1982). It has been reported in India, Australia, South Africa, Egypt, Italy, and Greece (Boyd and Murray 1982).

1.2 Ecology

S. eleagnifolium is a very ubiquitous species, easily prospering in environments with different types of climates and soils. It can be found in arid and semi-arid zones (200–600 mm rainfall), as well as in humid areas (about 1000 mm). Although it is more common in sandy soils, it can also be found in saline ones. As a consequence of this growth in such diverse environments, different phenotypic characteristics can be found in the species. These characteristics are related to the growth habit, the size and shape of leaves, and number and size of flowers and fruits (Vigna et al. 1981).

It is a hemicryptophyte, with spring-summer growth of shoots. Senescence of the aerial parts occurs after the first frosts. The regrowth of shoots is produced by buds located at the base of stems, near the soil surface or barely beneath it (Boyd and Murray 1982). Flowering starts in early December and continue until March. Fruiting occurs from January on, and the first ripe fruits can be seen at the end of February.

Reproduction is by seeds, that can germinate from September to December, with another peak in February. These phenological data correspond to the situation in Argentina.

Its ability for invasion is increased by the formation of adventitious buds along the horizontal roots (Fernández and Brevedan 1972; Boyd and Murray 1982). The formation of these buds is favored by soil management prior to the sowing of either a crop or a pasture. Small pieces of root may form several buds, that can emerge from depths of about 20–40 cm.

Due to the danger as a weed, numerous trials for controlling this plant have been conducted in the United States (Cooley and Smith 1973; Jackson et al. 1977), Australia (Cuthbertson et al. 1976), and Argentina (Vigna et al. 1981).

1.3 Agricultural Possibilities

As has been pointed out, *S. eleagnifolium* is a plant with a wide distribution in different countries. However, it has to be introduced as a crop in order to be a reliable source for solasodine production. Since 1980, in the Instituto Argentino de Investigaciones de las Zonas Aridas (Mendoza-Argentina) a project is developing with the aim of investigating the potentialities for cultivation of this plant. A germplasm bank has already been started, with seeds collected in different areas of Argentina and surrounding countries. Ecological and physiological studies are also under way, headed by S.O. Trione.

1.4 Chemistry

The glycoalkaloid solamargine is present in the leaves and fruits. Although the leaves must be the organs where the alkaloid is biosynthesized, its highest concentrations occur in the unripe fruits. In the literature, concentrations higher than 3% expressed as dry weight (DW) are cited. Thus, 3.2% has been determined in India

(Maiti and Mathew 1967), 7.27% in Argentina (Guerreiro 1972), and from 0.4 to 2% in Australia (Bradley et al. 1978).

Leaves and fruits contain saponins. The mucilage that agglutinates the seeds is rich in these substances

Solasodine is the common aglycone of solamargine and solasonine, glycosides which differ only in one sugar (Fig. 1). Solamargine was the glycoside isolated from the whole plant materials as well as from in vitro culture. The biosynthesis of steroidal compounds in *Solanum* species has been reviewed by Mann (1978), see also Macek (1989).

The importance of solasodine lies in the fact that it has, in the same way as diosgenin, the Δ^5 linkage that can be easily transformed into 20 keto pregnane compounds, which are the key for steroid drugs production. Production of solasodine from dried, fresh plants or plant cell culture includes extraction and hydrolysis of the glycosides. The hydrolysis can be chemical or microbial. The chemical procedures have been reviewed by Weston (1976) for solasodine production from *S. aviculare* and *S. laciniatum*, but there are no objections to using *S. eleagnifolium* for that purpose (Rodriguez 1984). In this respect, glycoalkaloids from *S. eleagnifolium* were converted into solasodine by microbial hydrolysis using *Aspergillus niger* NRRL 3 (Rodriguez et al. 1979). The solasodine obtained was identified by its physical constants such as melting point and $[\alpha]_D$, which values, 199–202 °C and 102 °C (Cl_3CH) respectively, are in agreement with those mentioned in the literature. In addition, the infrared, NMR, and mass spectrum tests strengthen this identification.

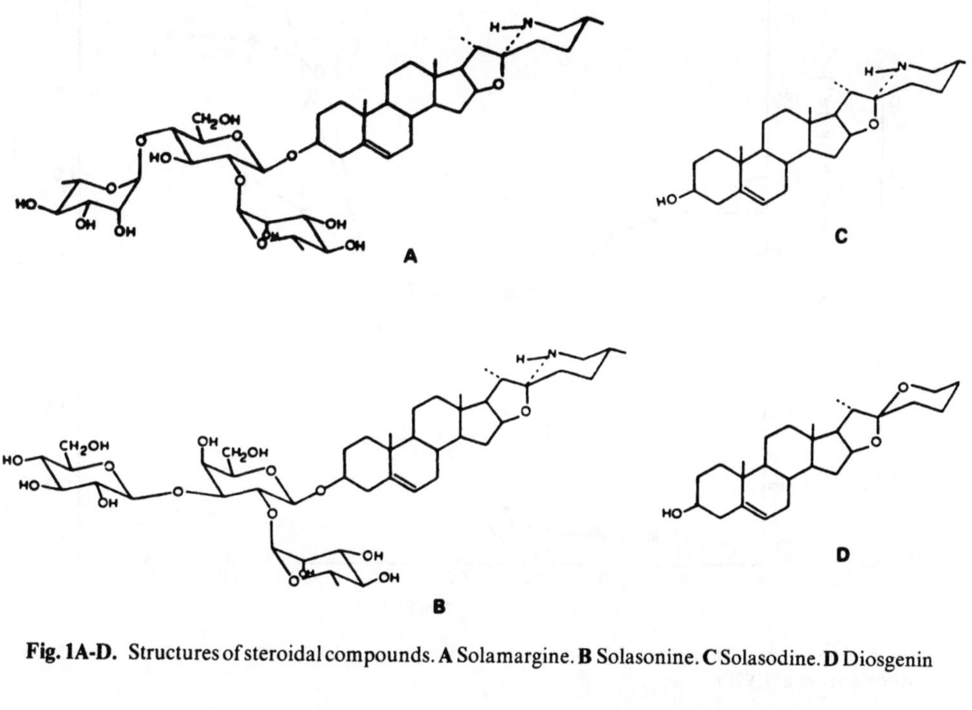

Fig. 1A-D. Structures of steroidal compounds. **A** Solamargine. **B** Solasonine. **C** Solasodine. **D** Diosgenin

The optimum conditions for achieving high microbial conversion of glycoalkaloids into solasodine were: initial pH 4.5, temperature 30°C, concentration of glycoalkaloids 0.4–0.6%, time of fermentation: 60–70 h. The use of higher concentration of glycoalkaloids was not convenient because a certain inhibition of growth was observed. Solasodine formation was a growth-associated phenomenon, as is shown in Fig. 2. The conversion reached the maximum value (near 100%) during the stationary phase of growth.

These yields are in agreement with those obtained with other sapogenins. The yields of diosgenin by microbial hydrolysis of *Dioscorea* ranged between 90 and 100% of the amounts expected by acid hydrolysis. It is interesting to note that no side-products were detected (Rodriguez et al. 1979). Solasodiene, frequently reported as another substance produced during chemical hydrolysis in aqueous medium, was absent in microbial hydrolysis. Although the yields obtained (1.5 g of solasodine for 100 g of dry material) are comparable to those attained by chemical hydrolysis (0.9–2.18 g of solasodine for 100 g of dry material), the mycelium maintained its hydrolase activity at the end of the process. This allows its re-use for further enzymatic conversion of glycoalkaloids. The re-use of the mycelium was

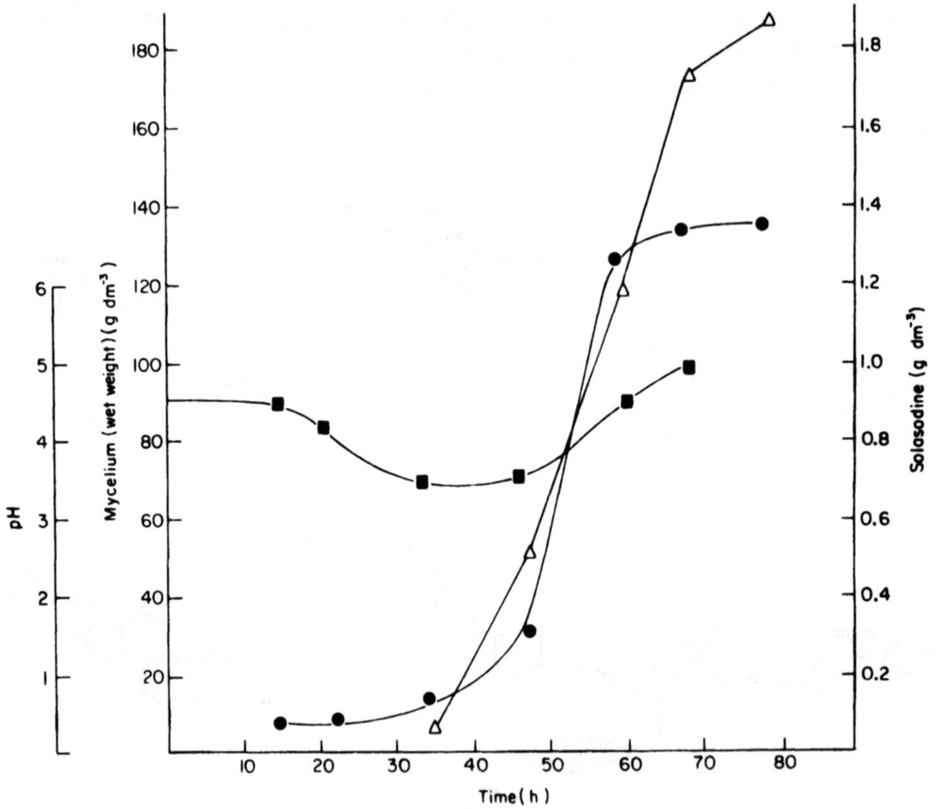

Fig. 2. The course of changes in solasodine concentration (△), pH value (■), and mycelium weight (●). (Rodriguez et al. 1979)

carried out at 45°C for 10–12 h. The crude solasodine has a purity degree of 80% determined by gas chromatography, and was used directly for synthesis of 16-dehydropregnenolone acetate (16 DHPA) with 48–53% yield. 16 DHPA is the key compound for the pharmaceutical industry for production of steroidal compounds.

The synthesis of 16 DHPA from solasodine takes place in four steps: acetylation of solasodine into O,N-diacetylsolasodine; transformation by isomerization into O,N-diacetylpseudosolasodine; oxidation of this compounds to give 16 (δ acetylamino-γ-methyl-valeroxy)-pregnenolone acetate; hydrolysis of the latter into 16 DHPA. The DHPA was used for progesterone production by means of *Nocardia* sp. (Rodriguez 1984).

2 In Vitro Approach

2.1 Callus Initiation

In vitro culture of several species of *Solanum* for solasodine production has been reported by several workers (Khanna et al. 1976; Khanna and Manot 1976; Hosoda and Yatazawa 1979; Hosoda et al. 1979; Chandler and Dodds 1983a,b; Emke and Eilert 1986; Cooner 1987; see also Macek 1989). The only reference to *S. eleagnifolium* is the one reported by Khanna et al. (1976), with no specifications of the explant origin. Fruits were collected from wild plants, sterilized for 2–3 min in 70% ethanol followed by 10 min in sodium hypochlorite (1% Cl_2 active) containing 0.1% Triton X-100 and rinsed with sterile distilled water. Seeds were also sterilized and germinated on MS medium (Murashige and Skoog 1962) supplemented with 5% sucrose and solidified with 0.8% agar. The cultures were incubated at 25 ± 2°C and exposed to photoperiods of 16 h.

Explants from seedlings (cotyledons, hypocotyls, roots, and leaves) and fruits were cultured in cylindrical flasks (70 mm diameter) on MS medium with the addition of myo-inositol (100 mg/l), sucrose (30 g/l) and RT vitamin complex (20 ml/l) (Khanna and Staba 1968). The pH was adjusted to 5.6–5.8 and the culture solidified with 8 g/l agar. This medium will be cited as MSRT.

Growth and solasodine production in calli originating from the different explants on MSRT medium supplemented with 2,4-D (4.5 mM) is shown in Fig. 3 (Nigra et al. 1987). Although growth was similar, with growth rates from 0.080 to 0.086/day, the callus formed from fruit explants showed longer lag periods. Moreover, callus initiation from fruit explants was more difficult to obtain than those from seedling sources. It is possible to relate the longer induction period to different growth potential according to the origin of the explants (Aitchinson et al. 1977).

Solasodine production was associated with growth (Fig. 3). These results differ from those of Emke and Eilert (1986) on *S. dulcamara*. The pattern of total solasodine production was similar in all the cases analyzed. First, there was an increase till the stationary growth phase was reached and then the production decreased. The reduction in solasodine concentration at this stage could be ex-

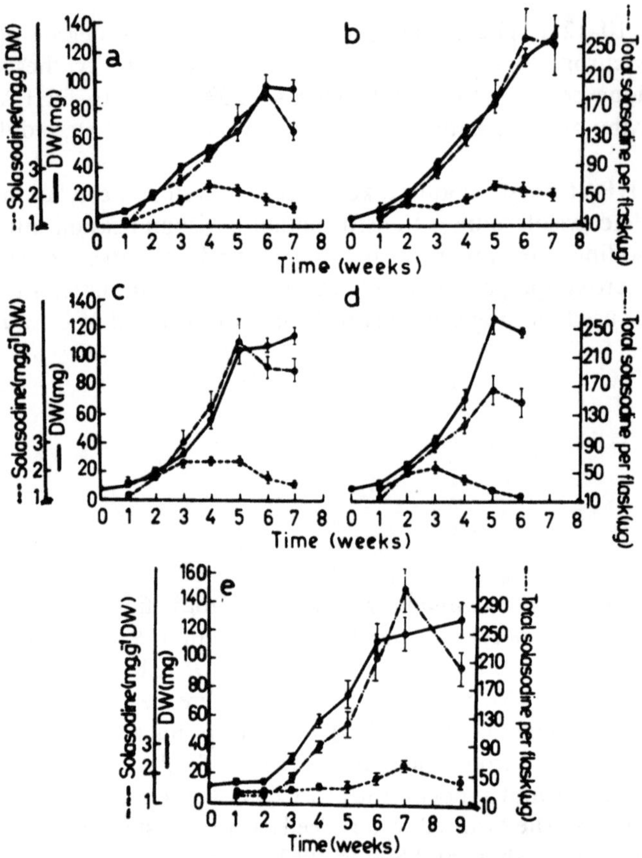

Fig. 3a-e. Time course of growth and solasodine accumulation in *S. eleagnifolium* calli. **a** Cotyledon calli. **b** Leaf calli. **c** Hypocotyl calli. **d** Root calli. **e** Fruit calli. Mean ± Se from 9 to 15 replicates. (Nigra et al. 1987)

plained either by a lower production rate or by the initiation of a degradation process.

The values of solasodine obtained (1–2.15 mg/g DW) were higher than those reported by Khanna et al. (1976), who found a value of 0.32 mg/g DW for the same species in plant cell culture. Compared with the values obtained for the whole plant, the solasodine concentrations in calli cultures were higher than those determined in stems and leaves (0.3–0.6 mg/g DW), but lower than those in fruits (30–50 mg/g DW) (Nigra et al. 1985).

2.2 Effect of Auxins

In order to examine the effect of different auxins on calli initiation, MSRT medium was supplemented with the following auxins (4.5 mM): IAA, 2,4-D, NAA, IBA, and 2,4,5-T. In all cases explants from hypocotyls were used.

The effect of different auxins on calli initiation is shown in Fig. 4. Those obtained with 2,4-D presented cellular hyperplasia after 1 week of culture and a friable callus after 4 weeks. When NAA or 2,4,5-T was used, callus formation and abundant roots were observed after 8 weeks. If calli were separated from the roots and transferred to the same medium, they showed profuse proliferation.

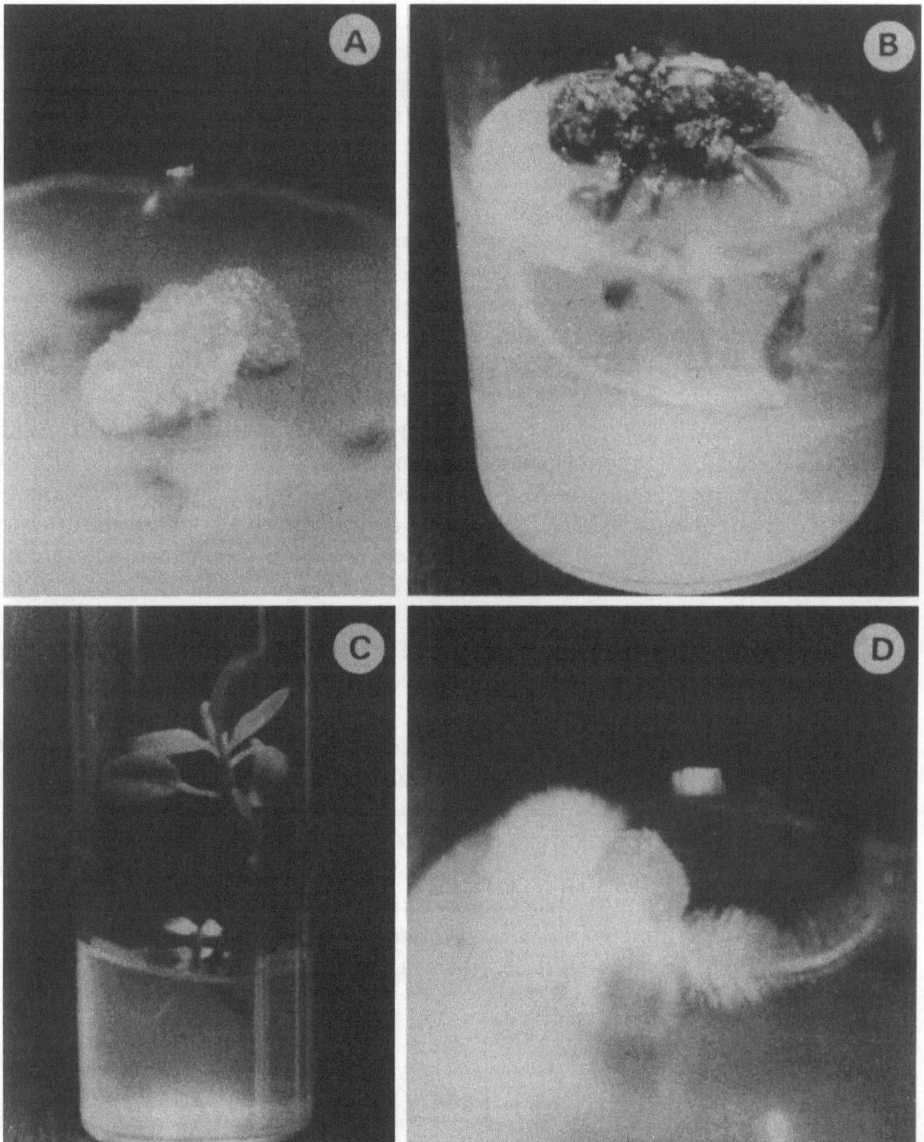

Fig. 4A-D. Effect of different auxins on calli initiation of *S. eleagnifolium* **A** 2,4-D. **B** 2,4,5-T. **C** IAA. **D** NAA. (Alvarez et al. unpubl.)

Regeneration of the whole plant with no callus formation was observed with IAA after 8 weeks of culture. However, when IBA was employed no callus was produced; some buds occurred but they died after a few days. These experiments suggest that 2,4-D is the best growth regulator for callus initiation, while NAA and 2,4,5-T are the auxins preferred as inducers of root formation. Although untransformed root cultures may be exploited for secondary metabolite production, their establishment and maintenance is difficult in some species. On the other hand, transformed roots induced by *Agrobacterium rhizogenes* are stable and grow faster than untransformed ones (Flores et al. 1987), even without decreasing metabolite production. However, hyoscyamine and scopolamine obtained from Solanaceae species were produced at the same levels in untransformed root culture as in transformed root (Flores and Filner 1985).

2.3 Influence of Auxins on Growth and Solasodine Production

The effect of auxins was analyzed in calli initiated in MSRT medium with 2,4-D and then transferred to the same basal medium, supplemented with 4.5 mM of 2,4-D, IBA, NAA, or IAA. The effect of different auxins on growth and solasodine production through five successive subcultures is shown in Fig. 5. The highest values, both in growth and solasodine production, were obtained using 2,4-D. With this auxin, the solasodine concentration obtained was about 1.4 mg/g while with the other auxins it was about 0.5–0.8 mg/g DW. The relatively high content of solasodine obtained with IBA, IAA, and NAA in the first and second subculture may be the result of a residual effect of 2,4-D, because from the third subculture no significant differences, either in growth or solasodine production, were obtained using these auxins.

These results differ from those reported for *S. laciniatum*, which produces the same level of solasodine in tissue cultures irrespective of the auxin present in the culture medium (Hosoda et al. 1979).

Concerning the influence of 2,4-D, it has generally been found to be less suitable for triggering secondary metabolism in plant cell cultures than either IAA or NAA (Mantell and Smith 1983). In this respect and according to Misawa (1985), 2,4-D stimulates the dedifferentiation of cells and consequently diminishes the level of secondary metabolites. However, this is not always the case, for example the production of L-DOPA by *Macuma pruriensis*, ubiquinona 10 by *Nicotiana tabacum*, and diosgenin by *Dioscorea deltoidea* were stimulated by high levels of 2,4-D (Brain 1976; Ikeda et al. 1976; Misawa 1985).

2.4 Influence of Light

To study the influence of light, calli from hypocotyls were started and subcultured three times during 25–27 days, in MSRT medium plus 2,4-D (4.5 mM). Afterwards, the calli were divided into two groups. One was exposed to photoperiods of 16 h and the other was kept in darkness. The effect of light on growth index and solasodine concentration in calli is shown in Fig. 6. From the second subculture the solasodine

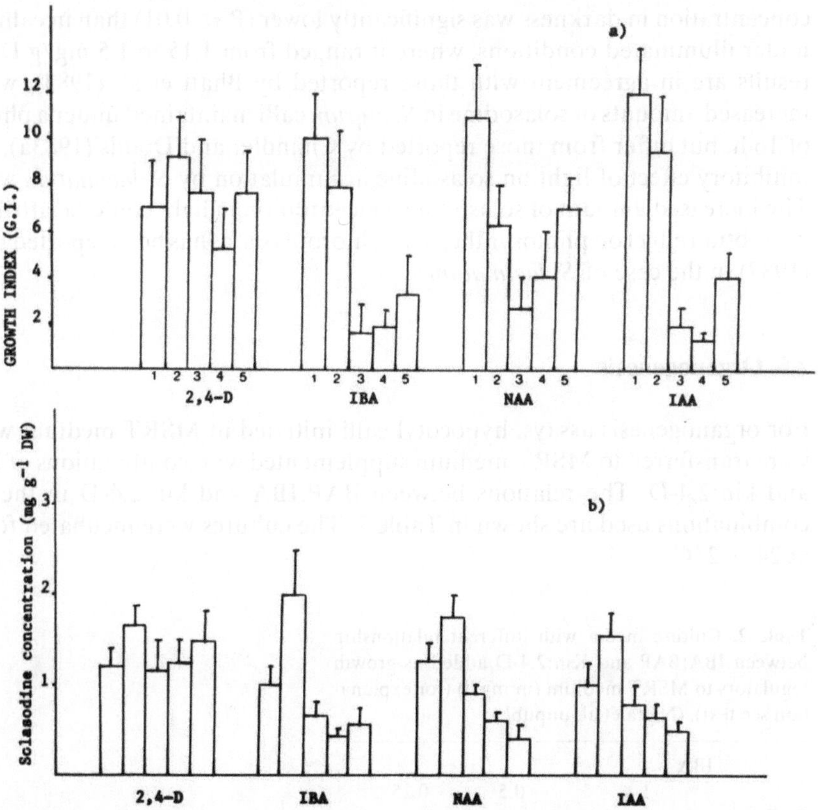

Fig. 5a,b. Effect of different auxins (4.5 mM) on **a** growth index. **b** solasodine concentration. (Nigra et al. unpubl.)

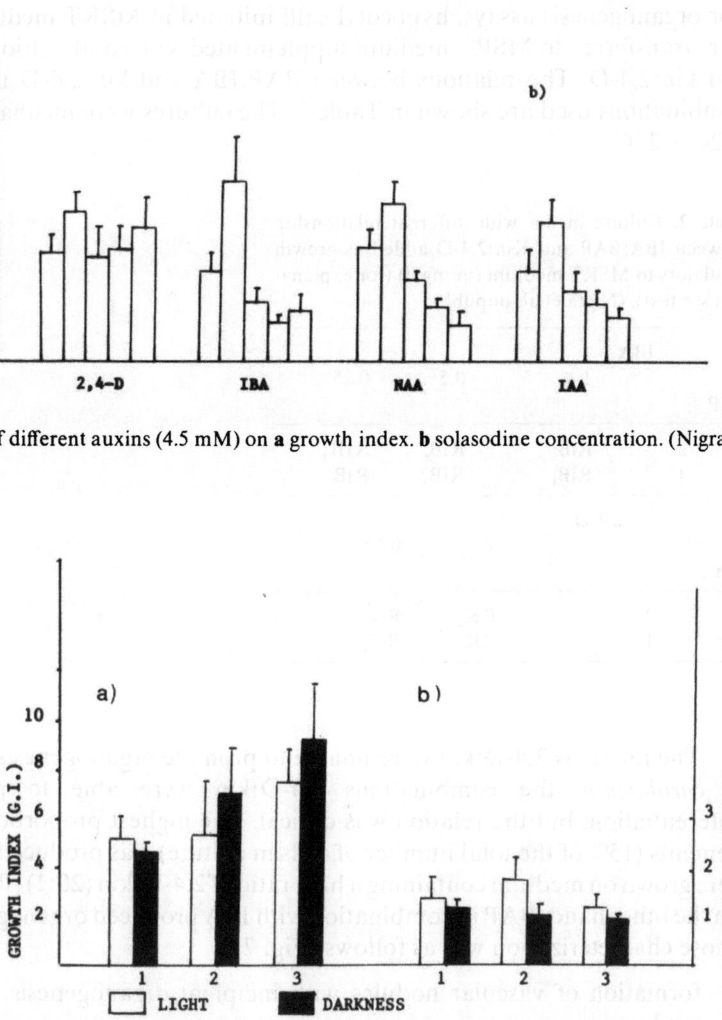

Fig. 6a,b. Effect of light on **a** growth index. **b** solasodine concentration. (Nigra et al. unpubl.)

concentration in darkness was significantly lower ($P < 0.01$) than in calli obtained under illuminated conditions, where it ranged from 1.15 to 1.5 mg/g DW. These results are in agreement with those reported by Bhatt et al. (1983), who found increased amounts of solasodine in *S. nigrum* calli maintained under a photoperiod of 16 h, but differ from those reported by Chandler and Dodds (1983a), where an inhibitory effect of light on solasodine accumulation by *S. laciniatum* was found. The increased amount of solasodine concentration in light could be attributed to a promotion effect on photosynthesizing chloroplasts, as has been reported by Cooner (1987) in the case of *S. laciniatum*.

2.5 Organogenesis

For organogenesis assays, hypocotyl calli initiated in MSRT medium with 2,4-D were transferred to MSRT medium supplemented with combinations of BAP:IBA and kin:2,4-D. The relations between BAP:IBA and kin:2,4-D in the different combinations used are shown in Table 2. The cultures were incubated for 30 days at 24 ± 2°C.

Table 2. Culture media with different relationship between IBA:BAP and Kin:2,4-D added as growth regulators to MSRT medium (in mg/l) (for explanation see text). (Nigra et al. unpubl.)

BAP \ IBA	1	0.5	0.25
2	RiB_1	RiB_2	RiB_3
1	RiB_4	RiB_5	RiB_6

Kin \ 2,4-D	1	0.5
2	RK_1	RK_3
1	RK_2	RK_4

The relations 2,4-D:kin were unable to promote organogenesis. In the case of *S. carolinense*, the combinations 2,4-D:kin were able to promote cytodifferentiation, but the relation was critical. The highest proportion of tracheary elements (15% of the total number of cells in culture) was produced when cultures were grown on medium containing a high ratio of 2,4-D:kin (20:1) (Reynolds 1987). On the other hand, BAP in combination with IBA produced organogenetic changes whose characterization was as follows (Fig. 7):

1. formation of vascular nodules with incipient organogenesis, in RiB_5, RiB_6 media,
2. formation of buds in RiB_1, RiB_2, RiB_4 media,
3. internode elongation and foliar expansion in some buds in RiB_2 medium.

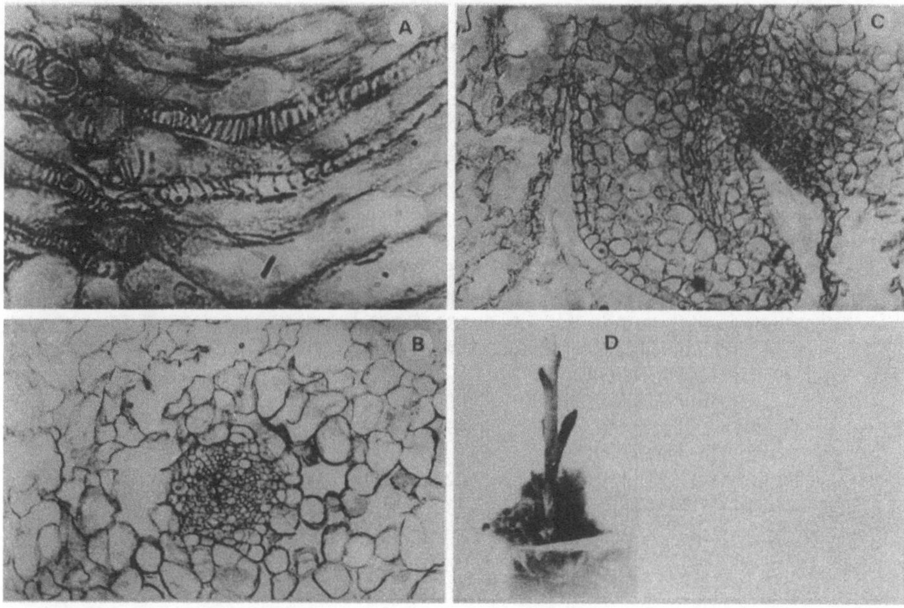

Fig. 7A-D. Organogenesis in calli of *S. eleagnifolium*. **A,B** Vascular nodules. **C** Bud and foliar primordium. Micrographs: **A** (400 x) **B,C** (100 x). **D** Internode elongation and foliar expansion in bud. (Nigra et al. unpubl.)

Additionally, when calli maintained in MSRT medium with 2,4-D for 6 months were transferred to the above BAP:IBA media, no organogenetic changes were observed. This behavior can be attributed to the irreversible dedifferentiation effect induced by 2,4-D during prolonged subcultures (Narayanaswamy 1977; Chandler and Dodds 1983a).

Solasodine productivity in nondifferentiated and differentiated tissues shown in Table 3 is in agreement with the results reported by Kamimura et al. (1976), Chandler and Dodds (1983a), Misawa (1985), and Emke and Eilert (1986) for secondary metabolite production in other plant species. The beneficial effect of cellular differentiation can be due to an increased capacity of vacuolar cell accumulation of metabolites of the differentiated tissues (Aitchinson et al. 1977).

2.6 Batch Suspension Cultures

2.6.1 Effect of Auxins on Establishment of Suspension Cultures, Growth Rate, and Solasodine Productivity

Cell suspensions were initiated in liquid MSRT medium and then transferred to four media, each supplemented with 4.5 mM of 2,4-D, NAA, IAA, and 2,4-5-T. Incubation was carried out in a rotary shaker at 120 r.p.m. 24 $\pm 2\,^{\circ}$C and with a 16-h photoperiod.

Table 3. Solasodine production in differentiated and undifferentiated cells of *Solanum eleagnifolium* Cav. (for explanation see text). (Nigra et al. unpubl.)

Medium	Plant cell material	Solasodine mg/g DW
RiB_1	Undifferentiated calli	0.66 ± 0.12
	Buds	1.00 ± 0.16
RiB_2	Undifferentiated calli	0.86 ± 0.04
	Buds	1.68 ± 0.12
RiB_3	Undifferentiated calli	0.96 ± 0.25
	Buds	1.31 ± 0.07
RiB_5	Undifferentiated calli	0.47 ± 0.16
	Nodule formation with incipient organogenesis	1.02 ± 0.85
RiB_6	Undifferentiated calli	0.79 ± 0.06
	Nodule formation with incipient organogenesis	1.27 ± 0.47

The characteristics of the cell suspension and of the individual cell varied according to the growth regulator used. 2,4-D and 2,4,5-T produced fine, homogeneous suspensions with numerous tubular cells with a little dense cytoplasm. On the other hand, NAA and IAA produced suspensions with thick clumps in which free cells as well as cell aggregates were observed. The majority of cells in this case are tubular, presenting cytoplasmic granules which give a negative reaction with Sudan and Millon tests and positive with Lugol, which suggests the presence of plastids containing high-grade hydrated carbohydrates. The auxins have different effects on specific growth rate and productivity. Thus, NAA and 2,4,5-T do not affect the specific growth rate (μ) in the range of the concentration tested, whereas 2,4-D and IAA have an inhibitory effect at increasing auxin concentrations. Concerning productivity, no significant changes are observed with different concentrations of IAA and 2,4,5-T, whereas it is enhanced with the increase of NAA, and decreased when 2,4-D increased (Alvarez et al. unpubl.).

2.6.2 *Effect of Inoculum Size*

Two cell suspensions of different concentrations (4×10^3 cells/ml and 2×10^4 cells/ml) were inoculated in MRST medium plus 2,4-D (4.5 mM), and the effect of inoculum size on growth and solasodine production was studied (Fig. 8). Longer lag periods occurred in the growth of *S. eleagnifolium* with the small inoculum. This could be due to the occurrence of the "nursery effect" induced by the large inoculum. The specific yield of solasodine production was similar (180 mg/g) with all the inocula used. These values were reached earlier when large inocula were utilized.

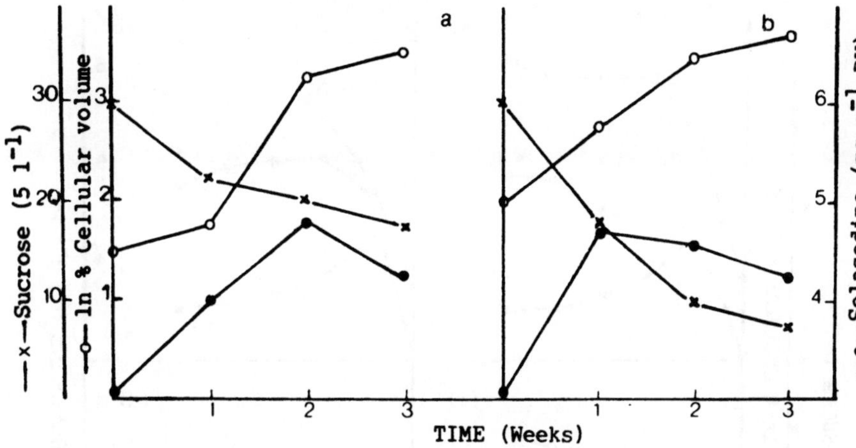

Fig. 8a,b. Effect of inoculum size on growth and solasodine production. **a** 4.10^3 (cell/ml). **b** 2.10^4 (cell/ml). Medium MSRT + sucrose 30 g/l. (Nigra et al. unpubl.)

2.6.3 Carbohydrate Source

The effect of several carbohydrates as carbon sources on growth and solasodine production were analyzed (Fig. 9). The culture medium (MSRT without sucrose) was supplemented with the respective sugars (glucose, sucrose, fructose, and maltose) at concentrations ranging from 90 to 180 mM. The specific growth rate of suspension cultures was approximately 0.15–0.20/day.

The consumption rate of all sugars was similar in the first 6 days of culture. Over this period, 10–12% of sugar was consumed, but after 24 days, the consumption was as follows: glucose: 38%, fructose: 54%, sucrose: 27%, and maltose: 20%. The highest biomass values were obtained with sucrose and glucose while the lowest one corresponded to maltose (Fig. 10) but the specific yield ($Y_{p/x}$) was similar. For this reason, maltose-fed batch cultures could be a possible alternative when this sugar is used as a sole carbon source. These results show that sucrose and glucose are superior to the other sugars tested for biomass production and solasodine productivity.

With the exception of fructose, improved values of biomass concentrations were obtained when the sugar concentration increased from 90 to 180 mM. The same behavior was shown in diosgenin production by *D. deltoidea* (Rokem et al. 1985) and in rosmarinic acid production by *Coleus blumei* (Misawa 1985) when the concentration of sucrose increased from 15 to 30 g/l. Fructose appeared to have an inhibitory effect on the growth of the suspensions.

2.6.4 Effect of the Relationship Between Nitrate and Ammonium Source

The effect of the relative concentration of nitrate and ammonium on solasodine production was tested using different culture media, whose compositions were

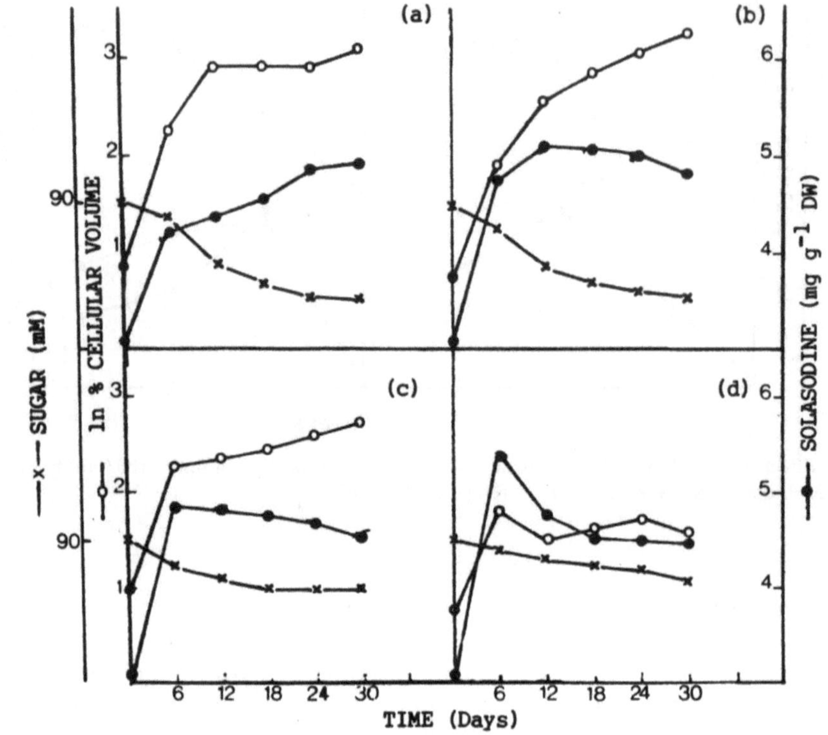

Fig. 9a-d. Growth and solasodine production on different carbon source (90 mM). **a** Fructose. **b** Glucose. **c** Sucrose. **d** Maltose. (Nigra et al. unpubl.)

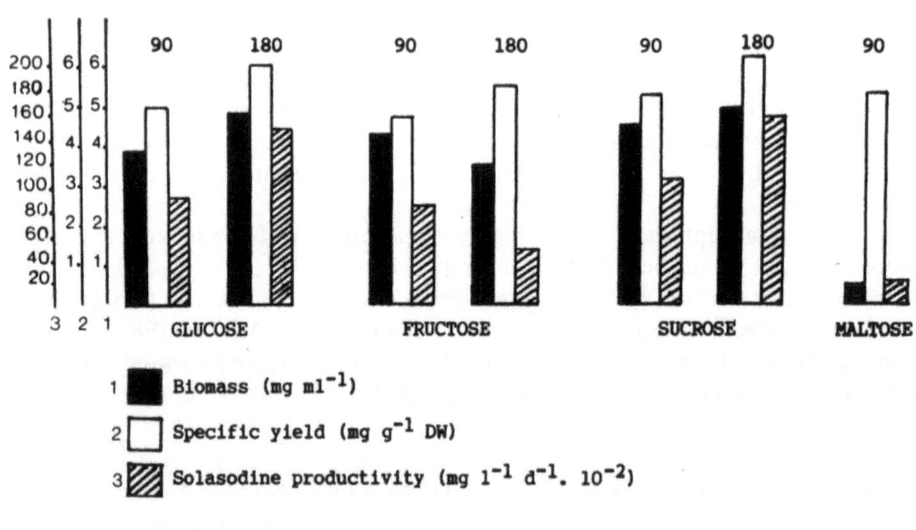

Fig. 10. Effect of several carbohydrates on specific yield and productivity of solasodine in batch suspension culture. (Nigra et al. unpubl.)

Table 4. Influence of the relationships between NO_3^- : NH_4^+ sources on biomass and solasodine specific yield (for explanation see text). (Nigra et al. unpubl.)

Time		4 days		16 days		24 days	
N_2 Source		DW	Solasodine	DW	Solasodine	DW	Solasodine
NO_3^+	NH_4^-	mg/ml	mg/g DW	mg/ml	mg/g DW	mg/ml	mg/g DW
1		0.60	3.3	2.94	1.3	4.69	1.03
	1	0.33	0.6	0.12	0.83	0.21	0.41
1	1	0.32	1.58	3.74	1.96	4.89	2.50
2	1	0.83	3.83	3.38	3.80	6.88	4.08
1	2	0.60	2.19	3.58	3.00	5.56	3.20

similar to the MSRT medium except for the nitrogen source (Table 4). The influence of nitrate, ammonium, and the relationship of NO_3^-: NH_4^+ on the growth and solasodine production is shown in Table. 4. As can be seen, when ammonium was the main nitrogen source, the biomass obtained was lower. Dougall (1977) suggested that plant cells can grow with NH_4^+ as the sole nitrogen source provided that a dicarboxylic acid of the Krebs cycle is added to the culture medium. The cultures presented a long lag period when nitrate was the main nitrogen source, suggesting that nitrate is more slowly assimilated than ammonium. As was expected, the relationship between nitrate and ammonium used in the culture media had a marked effect on growth and solasodine production. The highest values of biomass and yield of solasodine were produced for the relation NO_3^-: NH_4^+ 2:1, results which are in agreement with those found in *D. deltoidea* for diosgenin production by Rokem et al. (1985).

In general, the effect of nitrogen on secondary metabolite production should be viewed as a delicate interaction with the carbon source available to the cell where even small alterations of the C:N ratio may have a marked influence on growth and solasodine production.

3 Summary and Conclusions

The in vitro production of solasodine from *S. eleagnifolium* was studied. Callus culture was initiated from hypocotyl sections of 21-day-old seedling in a MS medium with the addition of myo-inositol (100 mg/l), sucrose (30 g/l), and RT vitamin complex (20 ml/l). The pH was adjusted to 5.6–5.8 and solidified with 8 g/l agar (MSRT medium). The peak of solasodine production was achieved after 21 days of culture (3 mg/g DW).

Solasodine content (as mg/g DW) was higher than that found in stems and leaves of the plant, but lower than in fruits. Some organogenesis (bud formation) was obtained when the MSRT medium was supplemented with IBA (0.5 mg/l), plus BAP (2 mg/l).

Cell suspension cultures were produced when calli were transferred to liquid conditions in a rotary shaker (120 r.p.m.) in MSRT plus 2,4-D (4.5 mM). After 11 days of culture, the maximum values of solasodine achieved were ca. 6 mg/g DW. Concerning the future of solasodine production, it is evident that the in vitro cell culture (including immobilization techniques, use of elicitors, etc.) should be economically compared with production from the whole plant. The results obtained so far have indicated that it is an attractive possibility which deserves further investigation, although no data for estimating the cost of production are yet available.

4 Protocol
Media and Culture Conditions for the Establishment of Cultures and Solasodine Production from *Solanum eleagnifolium*

1. Explants: hypocotyl sections of 21-day-old seedlings, grown on MSRT medium, 25 ± 2°C, 16 h photoperiod (1.8 W/m²/s)
2. Callus: same medium + 2,4-D (4.5 mM), same conditions. Culture periods: 4 weeks
3. Differentiation: MSRT medium supplemented with 2 mg/l BAP + 0.5 mg/l IBA
4. Cell suspension: MSRT medium plus 2,4-D (4.5 mM), rotary shaker (120 r.p.m.), 24 ± 2°C, 16 h photoperiod
5. Solasodine production: same medium and conditions as for cell suspension after 11 days of culture

Acknowledgments. The authors wish to acknowledge the useful information provided by Dr. O. Giordano, Dr. E. Guerreiro (Universidad Nacional de San Luis, Argentina), Sinibaldo Trione (IADIZA, Mendoza, Argentina) and Osvaldo Fernández (CERZOS, Bahia Blanca, Pcia. de Buenos Aires, Argentina.

References

Aitchinson PA, McLeod AJ, Yeoman MM (1977) Tissue (callus) culture. In: Street HE (ed) Plant tissue and cell culture. Blackwell, Oxford, pp 267–306

Applezweig N (1969) Oral medication for fertility control. Chem Week 17:58–72

Bhatt PN, Bhatt DP, Sussex I (1983) Studies on some factors affecting solasodine contents in tissue cultures of *Solanum nigrum*. Physiol Plant 57:159–162

Boyd JW, Murray DS (1982) Growth and development of silverleaf nightshade (*Solanum eleagnifolium*). Weed Sci 30:238–243

Bradley V, Collins DJ, Crabbe PG, Eastwood FM, Irvine MC, Swan JM, Symon DE (1978) A survey of Australian *Solanum* plants for potentially useful sources of solasodine. Aust J Bot 26:723–754

Brain KR (1976) Accumulation of L-DOPA in cultures from *Mucuma pruriens*. Plant Sci Lett 7:157

Cabrera A (1965) *Solanaceae*. In: Cabrera A (ed) Flora de la Provincia de Buenos Aires. Colección Cientifica INTA, vol 4, parte 5; pp 190–250

Chandler S, Dodds J (1983a) Solasodine production in rapidly proliferating tissue cultures on *Solanum laciniatum* Ait. Plant Cell Rep 2:69–72

Chandler S, Dodds JH (1983b) The effect of phosphate, nitrogen and sucrose on the production of phenols and solasodine in callus culture of *Solanum laciniatum*. Plant Cell Rep 2:205–208

Cooley AW, Smith DT (1973) Silverleaf nightshade response to glyphosate. Proc 26th Annu Meet South. Weed Sci Soc, p 59. Cited in: Trione S, Cony M (1988) *Solanum eleagnifolium* Cav, fuente de precursores esteroidales. Analisis y perspectiva de su cultivo. Interciencia 13:303–310

Cooner A (1987) Differential solasodine accumulation in photoautotrophic and heterotrophic tissue cultures of *Solanum laciniatum*. Phytochemistry 26(10):2744–2750

Cuthbertson EG, Leys AR, McMaster J (1976) Silverleaf nightshade: A potential threat to agriculture. Agric Gaz NSW 87:11–13

Dougall DK (1977) Current problems in the regulation of nitrogen metabolism in plant cell culture. In: Barz W, Reinhard E, Zenk M (eds) Plant tissue culture and its bio-technological application. Springer, Berlin Heidelberg New York, pp 76–84

Emke A, Eilert T (1986) Steroidal alkaloids in tissue cultures and regenerated plants of *Solanum dulcamara*. Plant Cell Rep 5:31–34

Fernández OA, Brevedan RE (1972) Regeneración de *Solanum eleagnifolium* Cav. a partir de fragmentos de sus raíces. Darwiniana 17:433–442

Flores HE, Filner P (1985) Metabolic relationship of putrescine. Gaba and alkaloids in cell and roots cultures of Solanaceae. In: Neumann KH, Barz W, Reinhard E (eds) Primary and secondary metabolism in plant cell culture. Springer, Berlin Heidelberg New York Tokyo, pp 174–186

Flores HE, Hoy MW, Picard JJ (1987) Secondary metabolites from roots cultures. Trends Biotechnol 5:64–69

Guerreiro E (1972) Glicoalcaloides esteroidales en especies del género *Solanum* (Solanaceae). Doctoral thesis, Universidad Nacional de San Luis, Argentina

Guerreiro E, Giordano OS, Kavka J, D'Arcangelo AT (1971) Aislamiento de compuestos esteroidales del *Solanum eleagnifolium* Cav. An Quím 67:789–794

Hosoda N, Yatazawa M (1979) Sterols, steroidal sapogenin and steroidal alkaloid in callus culture of *Solanum laciniatum* Ait. Agric Biol Chem 43:821–825

Hosoda N, Ito H, Yatazawa M (1979) Some accounts on culture conditions of callus tissues of *Solanum laciniatum* Ait for producing solasodine. Agric Biol Chem 43:1745–1748

Ikeda T, Matsumoto T, Nagouchi M (1976) Formation of ubiquinone by tobacco plant cells in suspension culture. Phytochemistry 15:568–569

Jackson DW, Abermathy JR, Keeling JW (1977) Sorghum and silverleaf nightshade response to incorporated triazine herbicides. Proc 30th Annu Meet South. Weed Sci, p 32. Cited in: Trione S, Cony M (1988) *Solanum eleagnifolium* Cav, fuente de precursores esteroidales. Analisis y perspectiva de su cultivo. Interciencia 13:303–310

Kamimura S, Akutsu M, Nishikawa M (1976) Tissue culture of *Papaver bracteatum*. Agric Biol Chem 40:913–919

Kavka J, Guerreiro E, Gianello JC, Giordano OS, D'Arcangelo AT (1973) Glicoalcaloides esteroidales en cuatro especies del género *Solanum* (Solanaceae). An R Soc Esp Fís Quím 69:929–932

Khanna P, Manot SK (1976) Effect of cholesterol on suspension culture of *Solanum xanthocarpum* Schrad and Wendl. Indian J Exp Biol 14:631–633

Khanna P, Uddin A, Sharma GL, Manot SK, Rathore AK (1976) Isolation and characterization of sapogenin and solasodine from in vitro cultures of some *Solanaceous* Plants. Indian J Exp Biol 14:694–696

Macek TE (1989) *Solanum aviculare* and *Solanum laciniatum*: In vitro culture and the production of solasodine. In: Bajaj YPS (ed) Biotechnology in agriculture and forestry, vol 7. Medicinal and aromatic plants II. Springer, Berlin Heidelberg New York Tokyo, pp 443–467

Maiti PC, Mathew R (1967) Rich sources of solasodine. Curr Sci 36:126

Mann JD (1978) Production of solasodine for the pharmaceutical industry. Adv Agron 30:207–245

Misawa M (1985) Production of useful plant metabolites. In: Fiechter A (ed) Plant Cell Culture. Springer, Berlin Heidelberg New York Tokyo, pp 59–88

Morton CV (1976) A revision of the Argentine species of *Solanum*. Acad Nac Córdoba, Argentina, pp 260

Murashige T, Skoog F (1962) A revised medium for rapid growth and bioassays with tobacco tissue culture. Physiol Plant 15:473–497

Narayanaswamy S (1977) Regeneration of plants from tissue cultures. In: Reinert J, Bajaj YPS (eds) Applied and fundamental aspects of plant cell, tissue, and organ culture. Springer, Berlin Heidelberg New York, pp 179–209

Nigra HM, Caso O, Giulietti AM (1985) Producción de solasodina en callos de *Solanum eleagnifolium* Cav. Resúmenes XVI Reunión Nacional de Fisiología Vegetal. La Plata,, Argentina, pp 76

Nigra HM, Caso O, Giulietti AM (1987) Production of solasodine by calli from different parts of *Solanum eleagnifolium* Cav. plant. Plant Cell Rep 6:135–137

Parodi L (1974) Enciclopedia Argentina de Agricultura y Jardinería. Vol. 2, II parte. Ed. Acme, Buenos Aires, Argentina

Reynolds TH (1987) Cytodifferentiation in cell suspension cultures of *Solanum carolinense*: Effect of 2,4-D and kinetin. J Plant Physiol 130:373–383

Rodriguez J (1984) Producción de solasodina por hidrólisis microbiana de los gliacoalcaloides de *Solanum eleagnifolium* Cav.; Doctoral Thesis. Universidad Nacional de San Luis, Argentina

Rodriguez J, Segovia R, Guerreiro E, Ferretti F, Zamarbide de Sosa G, Ertola R (1979) Production of solasodine by microbial hydrolysis of glycoalkaloids of *Solanum eleagnifolium* Cav. and its application to the synthesis of 16 dehidropregnenolone acetate. J Chem Technol Biotechnol 29:525–530

Rokem JS, Tal B. Goldberg I (1985) Methods for increasing diosgenin production by *Dioscorea* cells in suspension culture. J Nat Prod 48:210–222

Ruiz Leal A (1972) Flora Popular Mendocina. Deserta 3:9–296

Vigna MR, Fernandez OA, Brevedan RE (1981) Biologia y control de *Solanum eleagnifolium* Cav. (revisión bibliográfica). Rev Fac Agron B Aires 2:79–89

Weston RJ (1976) An investigation of methods for isolation of the steroidal alkaloid solasodine from native New Zealand *Solanum* species. J Appl Chem Biotechnol 26:657

XXVII *Swertia* spp.: In Vitro Culture, Regeneration, and the Production of Secondary Metabolites

H. Miura[1]

1 Introduction

Swertia japonica Makino (Fig. 1) and *S. pseudochinensis* Hara (Fig. 2), biennial herbs of the family Gentianaceae, are found wild on sunny slopes of hills and mountains. *S. japonica*, 5–40 cm high, growing wild in Japan, Korea, and China, bears white five-petaled flowers with purple streaks in the autumn. *S. pseudochinensis*, growing wild in Japan, Korea, and the temperate regions of Southeast Asia, is much like *S. japonica* in external appearance, although it is a little larger and bears bright bluish purple flowers with purple streaks (Shibata 1944; Okuyama 1960; Namba 1980). All the parts of these herbs collected at flowering have been used as Japanese folk remedies as stomach bitters. *S. japonica* is especially important and officially listed in the Japanese pharmacopoeia, because it has the bitterest taste in the genus *Swertia*; it is now being tried as a hair tonic. On the other hand, *S. pseudochinensis* is inferior in the quality of its bitterness, which is explained by the chemical fact that this herb contains swertiamarin, which is less bitter than amarogentin and amaroswerin, which *S. japonica* contains, together with sweroside as major bitter constituents (Fig. 3) (Inouye and Nakamura 1971; Hayashi et al. 1976; Hayashi and Higashino 1976; Inouye et al. 1967; Sakamoto et al. 1983). Other species of this genus found wild in Japan are not used as medicinal plants. *Chiretta, S. chirata* Hamilton an annual herb of a height of about 1 m indigenous to mountainous districts of northern India, has been used as a tonic rather than a stomach bitter in India (Wallis 1960). The only countries which use the genus *Swertia* for medicinal purposes are Japan, China (Tibet), and India. Thus there are almost no *Swertia* species in use as bitters except *S. japonica* and *S. pseudochinensis*. In spite of the great demand for these herbs, their supplies still depend on wild sources which are becoming critical on account of the overharvesting of the plants and the progressive clearance of their habitats. It would therefore certainly be helpful to utilize in vitro culture techniques to ensure the supply of the crude drugs from these herbs and the production of secondary metabolites. Clonal propagation of plants for agricultural and horticultural uses has been carried out for a long time, and it is already being put to practical use in plants of various kinds; but also studied on the micropropagation of medicinal plants and the production of secondary metabolites are now increasing from year to year in spite of a belated start (see Bajaj et al. 1988).

[1] Faculty of Pharmaceutical Sciences, Nagasaki University, 1-14, Bunkyo-machi, Nagasaki 852, Japan

Biotechnology in Agriculture and Forestry, Vol. 15
Medicinal and Aromatic Plants III (ed. by Y.P.S. Bajaj)

Fig. 1. Two-year-old plant of *S. japonica* with flowers, growing wild in the field

Fig. 2. Two-year-old plant of *S. pseudochinensis* with flowers

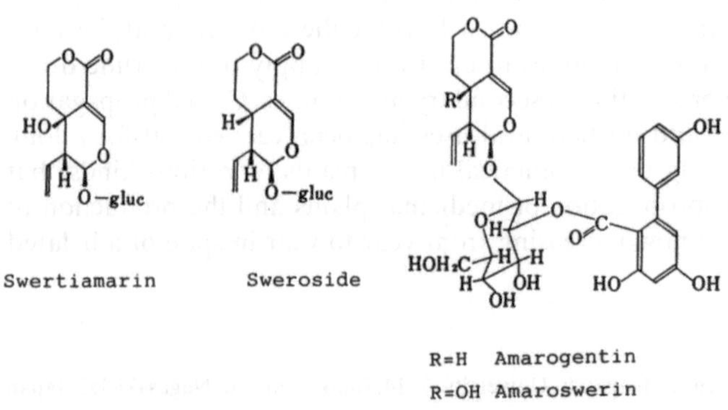

Fig. 3. Major bitter glucosides in *S. japonica*

This chapter deals mainly with in vitro culture and the examination of secondary metabolites of the callus in *S. japonica* as well as callus and organ cultures, plant regeneration, and bitter glucoside production by cultured tissues of *S. pseudochinensis*.

2 In Vitro Approaches

2.1 Review of Tissue Culture Studies

In spite of the great demand of *Swertia* plants, the general method for their cultivation in the field has not been established (Nishiyama et al. 1978; Yamada et al. 1980). Such a situation has remained practically unchanged since our first study on in vitro culture of *S. japonica* (Miura et al. 1978a,b). There are so far no publications on in vitro culture of the genus *Swertia* except our studies on *S. japonica* and *S. pseudochinensis*, summarized in Table 1.

Table 1. In vitro culture studies in the genus *Swertia*: tissue culture and secondary metabolite production

Species[a]	Tissue culture established	Secondary products identified	Reference
1. *Sj*	Callus from seedlings		Miura et al. (1978a)
2. *Sj*	Callus culture	Scopoletin, scopoletin glucoside, phytosterols (β-sitosterol, campesterol and stigmasterol)	Miura et al. (1978b)
3. *Sj*	Cell suspension from callus	Glucosylation of naringenin (naringenin 7-O-glucoside and gentiobioside)	Miura et al. (1986a)
4. *Sp*	Excised root culture from seedling		Kitamura et al. (1987)
5. *Sp*	Callus and excised shoot cultures, whole plant via embryogenesis from callus	Swertiamarin (in regenerated plant)	Kitamura et al. (1988)
6. *Sj*	Shoot and root from callus		Miura and Kitamura (unpubl.)[b]

[a] *Sj* = *S. japonica*, *Sp* = *S. pseudochinensis*.
[b] This result will be reported at the meeting of Kyusyu Branch, Pharmaceutical Society of Japan, Dec. 1988.

2.2 Callus Culture and Organogenesis in *S. japonica*

Callus tissues were successfully derived from young roots of sterile seedlings on modified White's agar medium (White's medium) (Steward 1964) containing 10% coconut milk (CM) and 1 mg/l 2,4-D in the dark at 25 ± 1 °C. The callus tissues, subcultured on the same fresh medium every month for over 2 years, were used to determine the requirements needed for rapid growth, and gave the following results: without the supplement of CM, with a suitable combination of auxin and kinetin (kin), particularly by the addition of 2,4-dichlorophenoxyacetic acid (2,4-D) 1 mg/l, indole-3-acetic acid (IAA) 10 mg/l or 1-naphthaleneacetic acid (NAA) 10 mg/l to the medium containing kin 0.01 mg/l, successful callus growth was obtained. Callus cultures supplied with either 10% CM or 1% yeast extract as natural extracts showed vigorous growth; some varieties grew best in yeast extract. The optimum pH and sucrose concentration were found to be pH 5.5–7.5 and 1–2%, respectively. Illumination markedly inhibited the growth of callus on the medium containing IAA. The callus tissues in early stages of subculturing were unsuitable for the determination because of their heterogeneity and poor growth. The long-term cultured callus tissues, which were selected to obtain only the white and well-growing ones from every subculturing, showed a tendency to change their cultural requirement for rapid growth. For example, such callus tissues grew rapidly on the medium containing 2,4-D 1 mg/l and kin 0.01 mg/l, unless supplied with CM, and on MS (Murashige and Skoog 1962) or LS (Linsmaier and Skoog 1965) basal agar medium, in any of which satisfactory growth of short-term cultures could not be expected.

As it was found that there was little hope of success in the production of bitter glucosides in callus tissues as described in Section 3.1, much effort has been directed to organogenesis, since the method of callus culture of *S. japonica* had been established as a preliminary step (unpublished results given below). Shoots were induced from callus on White's medium with the supplement of 1 mg/l of 6-benzyladenine (BA) and 2 g/l of casein hydrolysate (CH) under light (Fig. 4A). They kept growing on a hormone-free MS medium, but without success in rooting, in spite of many trials. Root formation (Fig. 4B) from the callus of 2-year-old plant leaf origin was observed in the early stages after callus induction.

2.3 Callus Formation and Organogenesis in *S. pseudochinensis*

Seeds were collected from 2-year-old wild plants, and treated with gibberellic acid (GA) 400 mg/l for 24 h for germination (Inoue et al. 1977) and sterilized. The seeds germinated on agar medium without nutrients in the light. After 5–7 days of germination, the seedlings were placed on MS medium supplemented with BA and NAA combined at various concentrations and cultured at 25 ± 1 °C in continuous light to give results as summarized in Table 2. The pH of the media was adjusted to 5.8 before autoclaving. Root and leaf developments of seedlings were observed at NAA 0.01 mg/l + BA 0–0.1 mg/l and at NAA 0–0.01 mg/l + BA 1–10 mg/l, respectively. Further experiments showed that NAA 1 μg/l was suitable for root development. Buds were maintained on MS medium containing BA 1 mg/l without

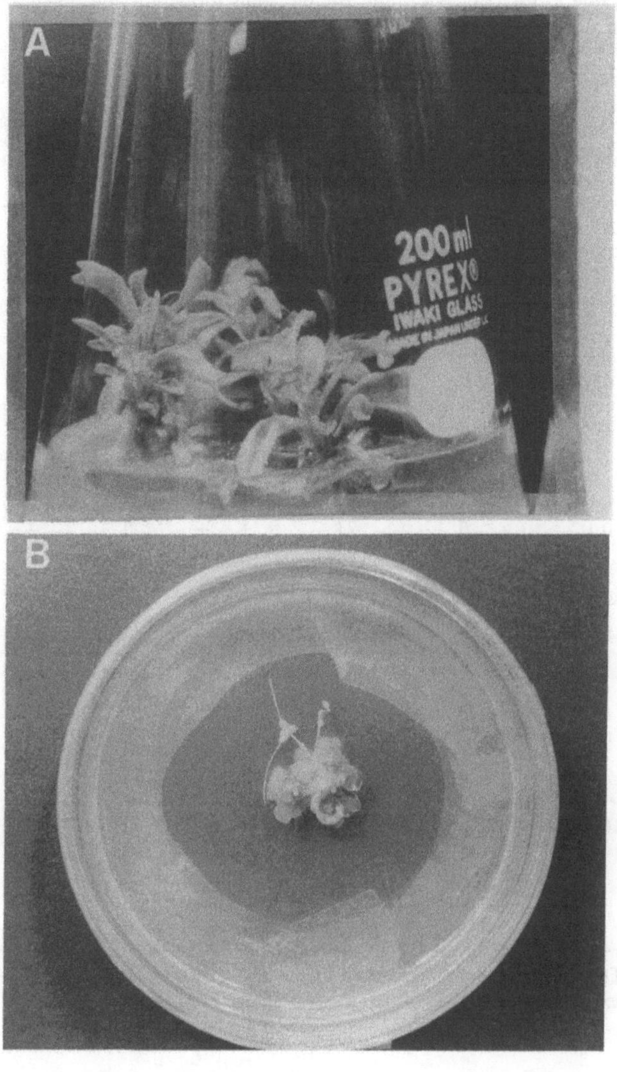

Fig. 4A,B. Organogenesis from *S. japonica* callus. **A** Shoots growing on a hormone-free MS medium after induction from callus on White's medium containing BA (1 mg/l) and casein hydrolysate (2 g/l) under light. **B** Adventitious roots from callus of 2-year-old plant leaf origin formed in early stage after callus induction on White's medium containing 2,4-D (1 mg/l) and coconut milk (10%)

rooting. Callus was formed at NAA 0.1–1 mg/l + BA 0–1 mg/l from hypocotyls and roots, and they were subcultured every 6–8 weeks on the medium containing NAA 1 mg/l in continuous light, which was better for successive callus growth than in the dark. Throughout 6 months' culturing, three callus lines were established according to their morphogenesis; root differentiation, embryogenesis, and no morphogenesis (Table 3). It was found that the NAA concentration used for callus induction was the most important factor in morphogenesis. Embryoids were observed in the calli which were induced at high level (10 mg/l) of NAA. The color and hardness of callus were also useful makers for morphogenesis. These cell lines were maintained for 2 years.

Table 2. Effect of NAA and BA on callus formation and organ development of *S. pseudochinensis* seedlings

Hormone		Response					
		Callus formation		Bud development		Root development	
NAA(mg/l)	BA(mg/l)	Frequency (%)	Intensity	Frequency (%)	Intensity	Frequency (%)	Intensity
0	0	0	–	88	±	88	±
	0.01	0	–	88	+	88	±
	0.1	0	–	77	+	70	±
	1	0	–	88	+ +	75	±
	10	0	–	78	+ +	0	–
0.01	0	58	+	92	±	33	+ + +
	0.01	9	+	91	+	91	+ + +
	0.1	0	–	92	+	92	+ +
	1	0	–	83	+ +	63	±
	10	0	–	67	+ +	0	–
0.1	0	73	+ + +	0	–	0	–
	0.01	88	+ + +	0	–	0	–
	0.1	55	+ + +	64	±	36	±
	1	92	+ + +	15	±	0	–
	10	50	+	40	±	0	–
1	0	100	+ + +	0	–	0	–
	0.01	100	+ + +	0	–	0	–
	0.1	90	+ + +	0	–	0	–
	1	82	+ + +	0	–	0	–
	10	33	+	11	±	0	–
10	0	0	+	0	–	0	–
	0.01	44	+	0	–	0	–
	0.1	38	+	0	–	0	–
	1	22	+	0	–	0	–
	10	38	+	0	–	0	–

Determination was performed at the end of 6 weeks in culture; 8–12 seedlings were provided per experiment; –, ±, +, + +, + + +, intensity of growth.

Table 3. Morphogenesis of different callus lines of *S. pseudochinensis*

Callus line	Morphogenesis	NAA concentration used for callus induction (mg/l)	Phenotype
1	Root differentiation	0.1–1	Pale yellow, compact
2	Embryo formation	10	Blight yellow, granular
3	No morphogenesis	1–10	White or pale yellow soft, moisten

2.4 Excised Root Cultures of *S. pseudochinensis*

When the seedlings were 1 month old, the root tips (2 cm) were isolated. Five root tips were put into an Erlenmeyer flask containing 25 ml B5 basal medium (Gamborg et al. 1968) supplemented with 10 μg/l NAA and 2% sucrose and cultured on a rotary shaker in dim light for 6 weeks. Root tips of the cultured roots were transferred to half-strength B5 basal medium containing 1 μg/l of NAA and 3% of sucrose at 6–8 weeks intervals, and subcultured successfully for more than 2 years. The effects of various auxins (Figs. 5 and 6), basal media, and sucrose concentrations on root growth were studied, and it was shown that a half-strength B5 basal medium containing 1 μg/l NAA and 3% sucrose was most suitable (Kitamura et al. 1987).

2.5 Plant Regeneration from Callus Cultures of *S. pseudochinensis*

Typical granular, bright yellow embryogenic calli were subcultured every 6 weeks. After 3–4 weeks in culture on a fresh MS medium containing NAA 1 mg/l, numerous embryoids (Fig. 7A) were found in the callus under the microscope. Further, many visual green spots appeared on the surface of the callus after 4–5 weeks in culture. When the embryoids or embryogenic calli were transferred to a hormone-free medium at earlier stages of culture, intact plantlets could not be obtained. When regenerated plantlets at the cotyledonary stage (Fig. 7B) were cultured on MS medium without hormones for 2 months, they developed to plants (Fig. 7C). The plants were transferred to pots containing a mixture of sand and peat moss and covered with glass beakers to maintain high humidity for 2–3 weeks. After taking the cover away, the plant continued to grow (Fig. 7D).

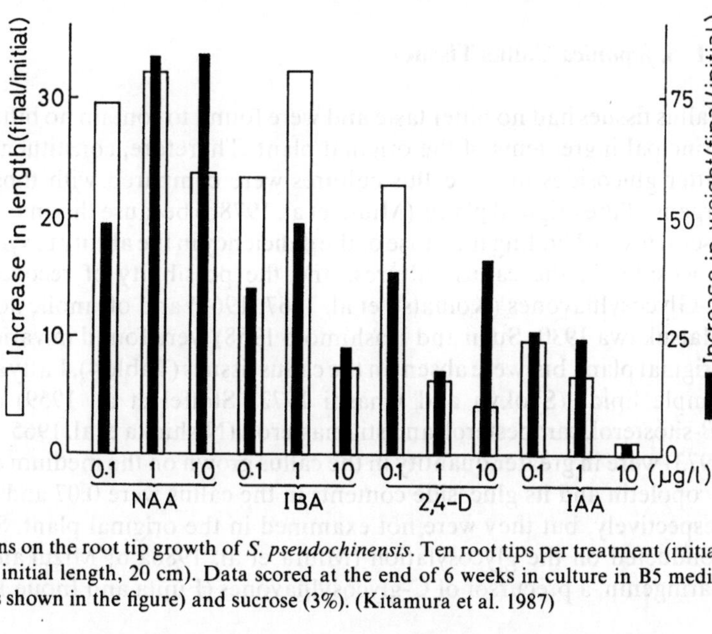

Fig. 5. Effect of auxins on the root tip growth of *S. pseudochinensis*. Ten root tips per treatment (initial fresh weight, 0.02 g; initial length, 20 cm). Data scored at the end of 6 weeks in culture in B5 media containing auxins (as shown in the figure) and sucrose (3%). (Kitamura et al. 1987)

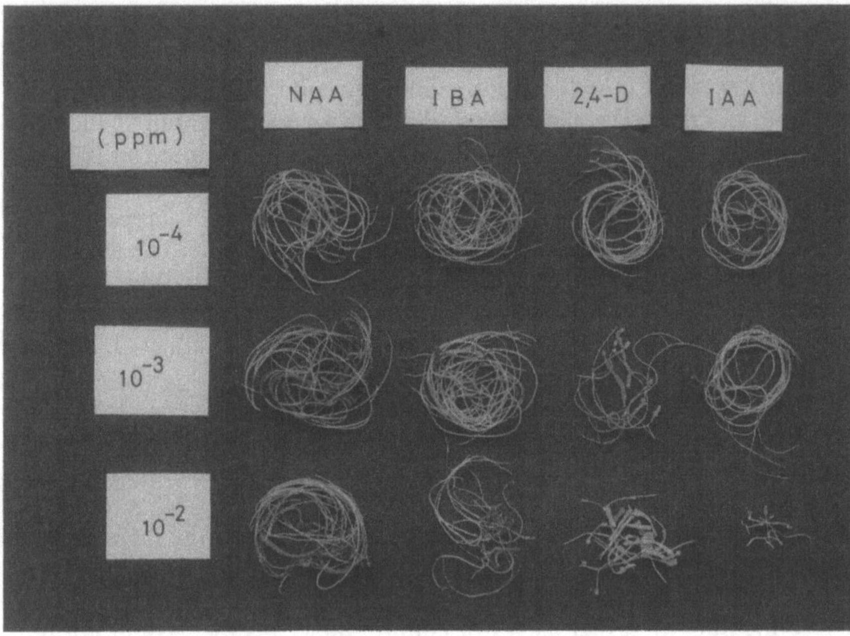

Fig. 6. Roots of *S. pseudochinensis* cultured for 6 weeks in B5 media containing auxins as shown in the figure and sucrose (3%). Five root tips per treatment (initial fresh weight), 0.01 g; initial length, 10 cm). (Kitamura et al. 1987)

3 Secondary Metabolites

3.1 *S. japonica* Callus Tissues

Callus tissues had no bitter taste and were found to contain no bitter glucosides, the principal ingredients of the original plant. Therefore, constituents other than the bitter glucosides in the callus cultures were compared with those in the various organs of the original plants (Miura et al. 1978b) because this may give information useful toward finding the cause of the deficiency in the ability to synthesize the bitter glucosides in the callus cultures, and the possibility of recovering this ability. C-Glycosylflavones (Komatsu et al. 1967, 1968) and oleanolic acid (Kuwada and Matsukawa 1930; Sugii and Hashimoto 1958) were found in various organs of the original plant, but were absent in the callus tissues (Table 4). Fatty acids constituting simple lipids (Syokyu and Ohashi 1972; Stoffel et al. 1959) and phytosterols (β-sitosterol, campesterol, and stigmasterol) (Nishioka et al. 1965; Song and Tattrie 1973) were in greater quantity in the callus grown on the medium containing NAA. Scopoletin and its glucoside contents in the callus were 0.07 and 0.39 μM /g DW, respectively, but they were not examined in the original plant. Studies were also conducted on the glycosylation (Miura et al. 1986a,b; Mizukami et al. 1983) of naringenin, a precursor of C-glycosylflavones (Fujita and Inoue 1979), in cultured

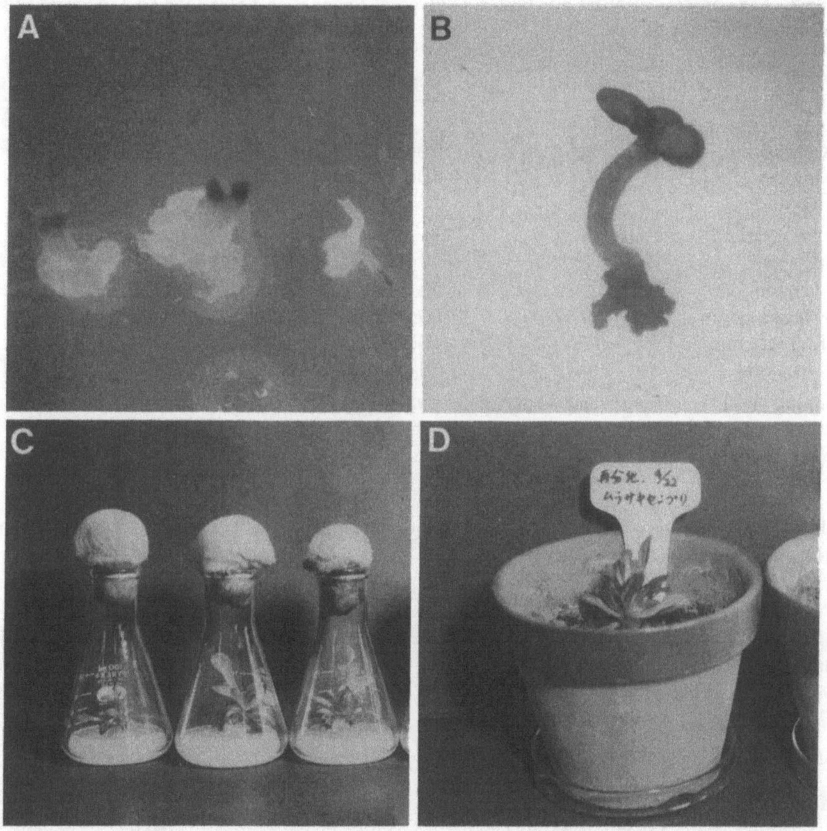

Fig. 7A-D. Plant regeneration of *S. pseudochinensis*. **A** Embryoids from calli cultured for 3–4 weeks after transferring onto MS medium containing NAA (1 mg/l). **B** Plantlet with cotyledonary leaves cultured on MS medium containing NAA (1 mg/l). **C** Regenerated plants from plantlets with cotyledonary leaves cultured for 2 months after transferring onto MS medium without hormone. **D** 6-month-old regenerated plant cultivated in a pot

cells and the glycosylation products isolated were identified as naringenin 7-O-β-D-glucoside and naringenin 7-O-β-gentiobioside, which were absent in the original plant. From this result it was found that there were differences in the glycosylation pattern between the callus cultures and the original plants.

3.2 Swertiamarin Production in Cultured Tissues of *S. pseudochinensis*

Production of swertiamarin, a bitter glucoside, in various cultured tissues was investigated and compared with that of the original plant (Table 5) (Kitamura et al. 1988). Lyophilized plant materials were extracted according to a procedure similar to that of Inouye et al. (1970a) and used for the analysis of swertiamarin. Identification and quantitative estimation of swertiamarin were performed by TLC

Table 4. The contents of bitter glycosides, C-glycosyl-flavones and oleanolic acid in various tissues of *S. japonica*. (Miura et al. 1978b)

	Flowers	Leaves	Stems	Roots	Callus
Bitter glycosides					
Amarogentin	0.51	0.30	0.20	0.21	–
Amaroswerin	0.68	0.30	0.20	–	–
Sweroside	7.1	1.5	0.9	–	–
Swertiamarin	14.2	9.0	0.6	–	–
C-glycosylflavones					
Swertisin	2.15	0.57	0.26	–	–
Swertiajaponin	0.49	0.25	0.16	–	–
Homoorientin	0.50	0.32	0.24	–	–
Isovitexin	0.43	0.11	0.16	–	–
Oleanolic Acid	1.43	1.72	0.42	0.16	–

Results are given as % of dry weight.

Table 5. Swertiamarin content in callus, cultured organs regenerated plants and the original plant of *S. pseudochinensis*. (Kitamura et al. 1988)

Tissue examined		Swertiamarin content(%)	Tissue examined		Swertiamarin content(%)
Cultured tissue			Original plant tissue		
Callus		n.d.	20-day-old seedling	Root	0.02
Organ culture	Root	n.d.		Leaf	2.70
	Leaf	n.d.	2-year-old wild plant	Root	0.29
5-Month-old RP[a]	Root	n.d.		Stem	0.34
	Leaf	4.17 ±2.44		Leaf	0.49
8-Month-old RP[b]	Root	0.87 ±0.52		Flower	3.33
	Leaf	2.42 ±0.93		Seed	0.02

RP = regenerated plant. n.d. = not detected.
[a]n = 15.
[b]n = 5.

in comparison with an authentic sample and by dual-wavelength densitometry according to the procedure of Hayashi et al. (1976). Swertiamarin was not detected in callus, cultured root, or leaf tissues, whereas it was present in 5- and 8-month-old regenerated plants, but absent in roots of 5-month-old regenerated plants. The contents in the leaves of 5-month-old regenerated plants varied greatly, and one of the plants contained 11% swertiamarin.

4 Conclusion and Prospects

4.1 *S. japonica*

On White's medium containing 10% CM and 1 mg/l 2,4-D, callus tissues were successfully derived from seedling root. The calli maintained on the medium for over 2 years were used to determine the cultural requirements of callus. By the addition of 2,4-D 1 mg/l, IAA 10 mg/l, or NAA 10 mg/l to the medium containing kin 0.01 mg/l, successful growth of callus was obtained without the supplement of CM. Shoots were induced from callus on the medium containing BA 1 mg/l and CH 2 g/l under continuous light. They continued to grow on a hormone-free MS medium without rooting.

The parent plant of *S. japonica* contains bitter glucosides having a strong, bitter taste, whereas it was found that callus tissues had no bitter taste and no bitter glucosides. Therefore, the speedy establishment of an in vitro method for cultured tissues and/or regenerated plants with medicinal qualities is greatly needed for *S. japonica*. To accomplish this purpose, results successfully obtained from tissue culture studies of *S. pseudochinensis* may be a great help.

4.2 *S. pseudochinensis*

Root and leaf developments of seedlings were observed at NAA 0.01 mg/l + BA 0–0.1 mg/l and NAA 0–0.1 mg/l + BA 1–10 mg/l, respectively. Callus was induced at NAA 0.1–1 mg/l + BA 0–1 mg/l from hypocotyls and roots and subcultured on the medium containing NAA 1 mg/l. After 3–4 weeks in culture on the same medium, numerous embryoids were formed in the embryogenic calli, which were developed to plants when cultured for 2 months. The plants obtained were transferred to pots containing a mixture of sand and peat moss and continued to grow.

Excised root cultures were obtained from root tips of the seedlings. Successive cultures were maintained by subculturing root tips at 6–8-week intervals for 2 years. Their best growth was obtained when cultured in half-strength B5 liquid medium containing NAA 1 μg/l and sucrose 3% in dim light.

Swertiamarin was present in 5- and 8-month-old regenerated plants, but not in roots of 5-month-old ones. It is particularly interesting from the viewpoint of bitter glucoside biosynthesis that swertiamarin exists in regenerated plants but in neither roots or leaves cultured separately. The regenerated plants obtained may be useful not only for medicinal source plants, but also for a simple experimental system for studies on the bitter glucoside biosynthesis (Inouye et al. 1967, 1970b) because they contain predominantly swertiamarin, the same as the original plants.

Acknowledgments. The author is grateful to Dr. T. Ikenaga for the photograph of wild-growing *S. japonica*.

References

Bajaj YPS, Furmanowa M, Olszowska O (1988) Biotechnology of the micropropagation of medicinal and aromatic plants. In: Bajaj YPS (ed) Biotechnology in agriculture and forestry, vol 4. Springer, Berlin Heidelberg New York Tokyo, pp 60–103

Fujita M, Inoue T (1979) Biosynthesis of C-glucosylflavones in *Swertia japonica*. Yakugaku Zasshi 99:165–171

Gamborg OL, Miller RA, Ojima K (1968) Nutrient requirements of suspension cultures of soybean root cells. Exp Cell Res 50:151–158

Hayashi T, Higashino M (1976) Studies on crude drugs from gentianaceous plants. IV. Discrimination of *Swertiae* herbs from some other bitter crude drugs. Yakugaku Zasshi 96:366–369

Hayashi T, Tsuji Y, Matuda T (1976) Studies on crude drugs originated from gentianaceous plants. II. Determination of swertiamarin, the major bitter principle of *Swertia* herbs and related crude drugs. Yakugaku Zasshi 96:498–502

Inoue K, Ikenaga T, Okubo S, Ohashi H (1977) Ecology and cultivation of *Swertia*-plant. IV. Effect of low temperature and gibberellin treatment on germination of seed in *Swertia japonica* Makino. Shoyakugaku Zasshi 31:75–81

Inouye H, Nakamura Y (1971) Studies on monoterpene glucosides and related natural products. XVI. Occurrence of Secoiridoid glucosides in *Gentiana* and *Swertia*. Yakugaku Zasshi 91:755–759

Inouye H, Ueda S, Nakamura Y (1967) Biosynthesis of bitter glucosides of the *Gentianaceae*, gentiopicroside, swertiamarin, and sweroside. Tetrahedron Lett 1967:3221–3226

Inouye H, Ueda S, Nakamura Y (1970a) Monoterpene D-glucosides. X. Secoiridoid D-glucosides from *Swertia japonica*. Isolation of five secoiridoid glucosides as well as the structural clarification of sweroside, swertiamarin, and gentiopicroside. Chem Pharm Bull 18:1856–1865

Inouye H, Ueda S, Nakamura Y (1970b) Studies on monoterpene glucosides. XII. Biosynthesis of Gentianaceous secoiridoid glucosides. Chem Pharm Bull 18:2043–2049

Kitamura Y, Nakano M, Miura H, Sugii M (1987) Root cultures of *Swertia pseudochinensis*. Shoyakugaku Zasshi 41:58–60

Kitamura Y, Dono M, Miura H, Sugii M (1988) Production of swertiamarin in cultured tissues of *Swertia pseudochinensis*. Chem Pharm Bull 36: 1575–1576

Komatsu M, Tomimori T, Makiguchi Y (1967) Studies on the constituents of *Swertia japonica* spp. Isolation and structure of new flavonoid, swertiajaponin. Chem Pharm Bull 15:1567–1572

Komatsu M, Tominori T, Makiguchi Y, Asano K (1968) Studies on the constituents of *Swertia japonica*. III. On the flavonoid constituents of the plant of *Swertia* spp. Yakugaku Zasshi 88:832–837

Kuwada S, Matsukawa M (1930) Identification of swertic acid and oleanolic acid. Yakugaku Zasshi 53:680–686

Linsmaier EM, Skoog F (1965) Organic growth factor and bioassays with tobacco tissue cultures. Physiol Plant 18:100–127

Miura Y, Hirata K (1988) Formation of Vinblastine in multiple shoot culture of *Catharanthus roseus*. Planta Med 54:18–20

Miura H, Kondo Y, Kitamura Y, Sugii M (1978a) Studies on the tissue culture of *Swertia japonica* Makino (I). On the cultural requirement of callus. Shoyakugaku Zasshi 32:59–65

Miura H, Ida M, Kitamura Y, Sugii M (1978b) Studies on the tissue culture of *Swertia japonica* Makino (II). A comparison of constituents between callus cultures of various organs of original plant. Shoyakugaku Zasshi 32:90–95

Miura H, Kawashima K, Kitamura Y, Sugii M (1986a) Studies on the tissue culture of *Swertia japonica* Makino (III). Glycosylation of naringenin in cultured cells. Shoyakugaku Zasshi 40:40–43

Miura H, Kitamura Y, Sugii M (1986b) Glycosylation of naringenin in *Duboisia myoporoides* cultured cell. Shoyakugaku Zasshi 40:113–115

Mizukami H, Terao T, Miura H, Ohashi H (1983) Glucosylation of salicyl alcohol in cultured plant cells. Phytochemistry 22:679–680

Murashige T, Skoog F (1962) A revised medium for rapid growth and bioassays with tobacco tissue cultures. Physiol Plant 15:473–497

Namba T (1980) Coloured illustrations of Wakanyaku. Vol 2. Hoikusya Publishing, Osaka, Japan, pp 46–47

Nishioka I, Ikekawa N, Yagi Y, Kawasaki T (1965) Studies on the plant sterols and triterpenes. II. Separation of stigmasterol, β-sitosterol and campesterol, and about so-called "γ-sitosterol". Chem Pharm Bull 13:379–384

Nishiyama H, Ikenaga T, Ohashi H (1978) Synthetic discussion of the biological studies on *Swertia japonica*. Jpn J Michurin Biol 14:2–9

Okuyama S (1960) Coloured illustrations of wild plants of Japan. Seibundo-Shinkosha Publishing, Tokyo, Japan, vol 4, pp 169–170; vol 5, p 126

Sakamoto I, Morimoto K, Tanaka O, Inouye H (1983) Application of high performance liquid chromatography and field desorption mass spectroscopy to separative analysis of bitter secoiridoid glucosides of *Swertia* herba. Chem Pharm Bull 31:25–30

Shibata K (ed) (1944) A cyclopedia of useful plants and plant products. The Hokuryu Kan, Tokyo, Japan, pp 393–394

Song M, Tattrie NH (1973) Lipid composition of morning glory (*Ipomoea* species) cells grown in suspension cultures. Can J Bot 51:1893–1897

Steward FC (1964) Advanced laboratory course in plant physiology. A print for laboratory exercise. Cornell Univ, New York

Stoffel W, Chu F, Ahrens Jr EH (1959) Analysis of long-chain fatty acids by gas-liquid chromatography micromethod for preparation of methyl esters. Anal Chem 31:307–308

Sugii M, Hashimoto Y (1958) The utilization of sodium acetate-carboxyl-C by *Swertia japonica* Mak. Bull Inst Chem Res 36:127–133

Syokyu N, Ohashi H (1972) Cultivation and breeding of *Oenothera* plants. I. Geographical variation of *Oenothera* plants. Shoyakugaku Zasshi 26:25–35

Wallis TE (1960) Text book of pharmacognocy, 4th edn. J & A Churchill, London, pp 315–316

Yamada H, Ikenaga T, Ohashi H (1980) The culture method of *Swertia japonica* Makino. J Jpn Soc Hortic Sci 1980:19–23

XXVIII *Symphytum officinale* (Comfrey): In Vitro Culture, Regeneration, and Biogenesis of Pyrrolizidine Alkaloids

H.J. Huizing[1] and J.H. Sietsma[2]

1 Introduction

Symphytum officinale L. (Fig. 1) and *Symphytum asperum* L. (Boraginaceae) are allopatric taxa, which are able to intercross and to form interspecific hybrids with different chromosome numbers. The species differ not only in a number of morphological characters but also ecologically, *S. asperum* being a species of higher elevations (upper montane zone), *S. officinale* of lowland and the lower montane zone. *S. asperum* is a Caucasian species, which has the sporophytic chromosome number $2n = 32$. It was introduced from the Caucasus into Europe as a fodder plant. *S. officinale* is variable, containing cytotypes with $2n = 24, 48, 56, 40$, and occurs throughout Europe.

The most common chromosome number of *S. officinale* is $2n = 48$. Scattered diploid populations of $2n = 24$ occur in Western, Central, and Eastern Europe; they are white-flowered throughout. The tetraploids are white- or purple-flowered in western Europe and purple in eastern Europe. Populations in which purple- and white-flowered individuals occur intermingled are very common in western Europe. In eastern Europe, mixed populations are very rare and consist of white-flowered diploid plants with purple-flowered tetraploid plants (Huizing 1985).

The cytotype $2n = 56$ has been found at only one location in the high Tatra mountains of Czechoslovakia and the cytotype $2n = 40$ is very common in low-lying peat regions in The Netherlands. There exists, however, some controversy whether the latter represents a true subspecies, *S. uplandicum* Nym., or a hybrid between *S. asperum* ($2n = 32$) and *S. officinale* (Hills 1976; Huizing 1985).

There has been renewed interest in *S. officinale* regarding its use in the treatment of several diseases, as well as for treatment of neuralgia and paradontosis (Röder 1979). Also for the treatment of burns, a preparation containing an extract of *S. officinale* has been used recently (Huizing 1985). In the DAC (Deutscher Arzneimittel Codex) *S. officinale* has been included in the monograph "Symphyti radix". Also a number of other described physiological activities of alcoholic extracts of *S. officinale* led to a strongly increased interest for this plant in phytotherapeutics (Stengl et al. 1982).

It is not only for phytotherapeutic reasons that *Symphytum* (comfrey) taxa have regained interest. Hills (1975, 1976) reported that comfrey is a plant which should

[1] Agrotechnological Research Institute ATO, P.O. Box 17, 6700 AA Wageningen, The Netherlands
[2] Department of Molecular Plant Biology, University of Groningen, Biological Centre, 9751 NN Haren, The Netherlands

Biotechnology in Agriculture and Forestry, Vol. 15
Medicinal and Aromatic Plants III (ed. by Y.P.S. Bajaj)

Fig. 1. Flowering plant of *Symphytum officinale*

be able to produce vegetable protein at the highest yield, per unit of time and culturing area, known for plants. He also claimed that comfrey should be well suited not only for fodder, but also human consumption was considered. Furthermore, Farnsworth (1979) has pointed out that comfrey is used in the most common herbal teas sold to the American public, which points to its widespread use. However, application should be considered with caution for its possible hepatotoxic and carginogenic properties.

In 1900, Greiner (see Huizing 1985) demonstrated the presence of small amounts of alkaloids in roots of *S. officinale*, which had been purchased under the name "Radix consolidae". It was in 1968 that Furuya and Araki (1968) isolated symphytine and echimidine from *S. officinale*. These alkaloids belong to the di-ester pyrrolizidine alkaloids with an allylic-ester structure (Fig. 2). This latter configuration determines the hepatotoxic action of pyrrolizidine alkaloids (Bull et al. 1968); it also has been shown that symphytine is carcinogenic to rats (Hirono et al. 1979).

Indicine-N-oxide (INO), a mono-ester pyrrolizidine alkaloid, apparently seems to be a compound of interest for its low hepatotoxic properties and its high anti-tumor activity (Kovach et al. 1979). Therefore the production of these pyrrolizidine alkaloids by biotechnical techniques might be of interest. In vitro-grown cells of comfrey were tested for their feasibility as a model system for the production of these alkaloids. For this purpose biogenetic properties of cell suspension cultures

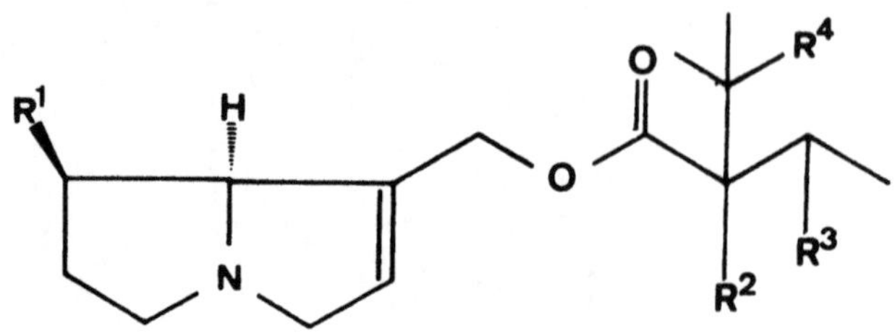

	hydroxyls:	erythro	threo
R^1 = hydroxyl , R^2 = R^3 = hydroxyl, R^4 = H:		lycopsamine	intermedine
R^1 = acetoxy , R^2 = R^3 = hydroxyl, R^4 = H:		ac-lycopsamine	ac-intermedine
R^1 = tigloxy , R^2 = R^3 = hydroxyl, R^4 = H:		symphytine	myoscorpine
R^1 = angeloxy , R^2 = R^3 = hydroxyl, R^4 = H:		symlandine	echiumine
R^1 = tigloxy , R^2 = R^3 = R^4 = hydroxyl :		echimidine	-

Fig. 2. Summary of structures of the main pyrrolizidine alkaloids found in *Symphytum officinale*, *S. uplandicum*, and *S. asperum*

and regenerating callus of a *S. officinale* cytotype have been surveyed, with emphasis on the analysis of precursors in the pyrrolizidine alkaloid biogenesis.

2 Initiation, Maintenance, and Regeneration of in Vitro Cultures

Primary calli were induced from different organs of *Symphytum officinale* L. (2n = 24) on MS medium with various growth regulator ratios (Table 1). Callus tissue from petioles and peduncles were obtained in 2 weeks.

At most plant growth regulator ratios, leaflets and/or shoots emerged from the primary callus tissue. Others have also observed that primary calli differentiated easily (Tanaka et al. 1974). Roots were formed less frequently than shoots, but they could be induced by transferring shoots from differentiating calli to basal MS medium supplemented with IAA only (0.1, 1 or 2 mg/l) (Fig. 3).

To avoid differentiation, callus was induced on solid B5 medium supplemented with 2,4-D in combination with BA or kinetin. Omitting the cytokinin resulted in white callus tissue with white rootlike protuberances, but the combination of 2,4-D with BA at 2 or 1 mg/l resulted in greyish calli showing no differentiation. These calli have been kept in culture for more than 3 years (with more than 45 times of subculturing). Cell suspension cultures were obtained from callus tissue grown on solid B5 medium, containing 2,4-D (2 mg/ml) and BA (1 mg/ml) by omitting the agar or agar substitute. Suspension cultures were transferred every week, by diluting in a 1:3 ratio with fresh medium, containing 4% sucrose as a carbon source.

Table 1. Effect of plant growth regulators (PGR) on development of primary callus (c), leaflets (l) and rootlets (r) from *Symphytum officinale* explants on MS medium with 2% sucrose after 20 days of incubation. Explants originated from roots (R), leaves (L), petioles (P), peduncles (FS) and stems (S). Order of increasing response: (c), c, cc and *cc* (same designations hold for l and r)

PGR in mg/l	Explant origin				
	R	L	P	FS	S
BA:1	(c)	(c)	ccll	*cc*l	cc*ll*
BA:1+IAA:10^{-2}	ccl	cll	*cc*ll	*ccll*r	cc*ll*
BA:1+IAA:10^{-1}	*cc*l	ccll	*cc*ll	*cc*ll	ccllrr
BA:1+IAA:1	*cc*	(c)llrr	*ccll*	*ccll*rr	ccll

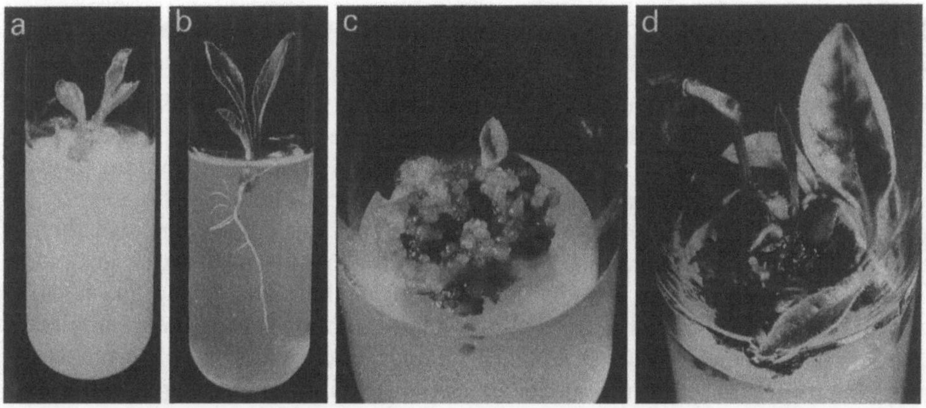

Fig. 3. **a** Shoots emerging from primary callus from a peduncle explant of *Symphytum officinale*, induced on MS medium supplemented with BA and IAA at 1 mg/l each. **b** Rooting of an excised regenerated shoot from a primary callus of *S. officinale* on MS medium supplemented with IAA at 1 mg/l. **c** Differentiation of a callus from a plated cell suspension culture of *S. officinale* on MS medium supplemented with BA at 1 mg/l and IAA at 0.1 mg/l. **d** Formation of plantlets on callus derived from a plated cell suspension culture of *S. officinale* under the same conditions as designated in **c**

When cells from a cell suspension culture were transferred to solidified MS medium containing 1 mg/ml BA and 0.1 mg/ml IAA and incubated at room temperature with a 16-h photoperiod, developing calli turned green and started to differentiate. After about 2 months, plantlets from these cultures could be transferred to soil.

Cell suspension cultures of 14 months old were still able to regenerate into normal plants. Chromosome numbers in root tips and the phenotype of the regenerated plants did not show any alteration in comparison with the explant source.

Ethanolic extracts from 4-week-old cell suspension cultures reacted still positive with the Dragendorff reagent (Huizing 1985), indicating the presence of pyrrolizidine alkaloids. After preparative thin-layer chromatography of these extracts, a major Dragendorff positive spot was isolated and examined by mass spectrometry. The spectrum obtained was identical with the spectrum of a mixture of symphytine isomers (see Fig. 4). These isomers show the same fragmentation patterns in mass spectrometry. No other pyrrolizidine alkaloids common to *S. offcinale* could be recognized on TLC by their R_f values. When cell suspension cultures were grown for a period longer than 2 months, no pyrrolizidine alkaloids were detectable, even after application of extracted-ion current profile EI GC-MS (Budde and Eichelberger 1977) or positive-ion CI GC-MS to silylated extracts.

However, when plants were regenerated from these cell suspension cultures, they contained all the characteristic alkaloids (Fig. 4). Thus the loss of expression of pyrrolizidine alkaloid biogenesis appears to be connected with the in vitro culture conditions.

3 Investigations of Ploidy Levels of in Vitro Cultures of a Cytotype

When considering the usefulness of plant cell cultures for biogenetic studies, it is of interest to know whether these cultures remain stable in their characteristics, e.g., their DNA content. From the observation that regenerated plants produced pyrrolizidine alkaloids again, one could conclude that a cessation of the expression of the alkaloid biogenesis during in vitro culture is not caused by progressive alterations in the genetic information of the cells, e.g., due to mutations, exchange or loss of parts of chromosomes. The dedifferentiation of the plant during the in vitro cultivation could also be the cause of an alteration in the secondary metabolism.

Microdensitometry of fixed, Feulgen-stained nuclei of *S. officinale* protoplasts, prepared from callus cells, revealed a bimodel distribution pattern representing G1

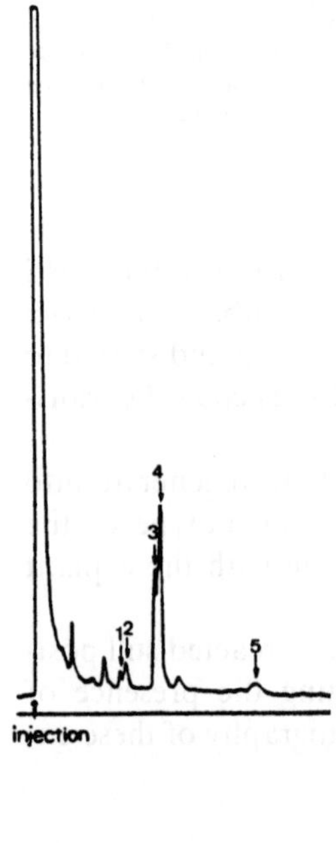

Fig. 4. Separation of TMS derivatives of pyrrolizidine alkaloids from roots of a *S. officinale* plant, regenerated from a cell suspension culture, by GLC on a glass column packed with 4% OV-17. *1* lycopsamine; *2* intermediate; *3* acetyl-lycop-samine; *4* acetyl-intermedine; *5* symphytine and/or isomers. *Arrow* indicates the start of the run

and S + G2 phase cells, respectively, and four of 100 nuclei also showed a higher ploidy level (Fig. 6). Because the latter nuclei showed a staining intensity four times greater than the G1-phase nuclei and twice of that of S + G2-phase nuclei, it could be concluded that these nuclei represent a tetraploid population in the callus culture. In protoplasts derived from leaf material also a bimodel distribution was observed, but only one ploidy level could be detected (Fig. 5).

Judging from the apparently low tendency to form polyploid cells (4%) and the high regenerating ability of callus as well as cell suspension cultures and subsequent formation of plantlets producing again pyrrolozidine alkaloids, a high genetic instability of the in vitro cultures even after a 3-year growing period, does not seem very likely. Therefore an inhibition in expression of pyrrolizidine alkaloid biogenesis due to dedifferentiation of the used plant cells in in vitro cultures seems more likely.

4 Biogenetic Properties of Cell Suspension Cultures and Regenerating Calli

In 1964, Bottomley and Geissmann found that labeled 2- and 5-^{14}C-ornithine as well as 1,4-^{14}C-putrescine when fed to *Senecio douglasii* plants were incorporated into the retronecine moiety of the alkaloids retrorsine, riddeline, senecionine, and seneciphylline. They therefore proposed a possible pathway for retronecine biosynthesis, as indicated in Fig. 7.

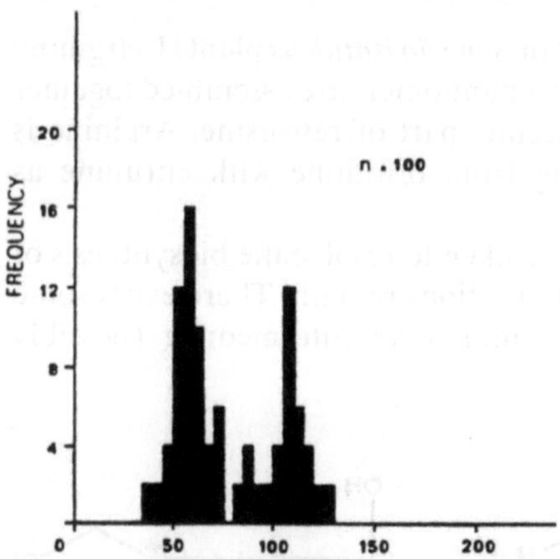

Fig. 5.

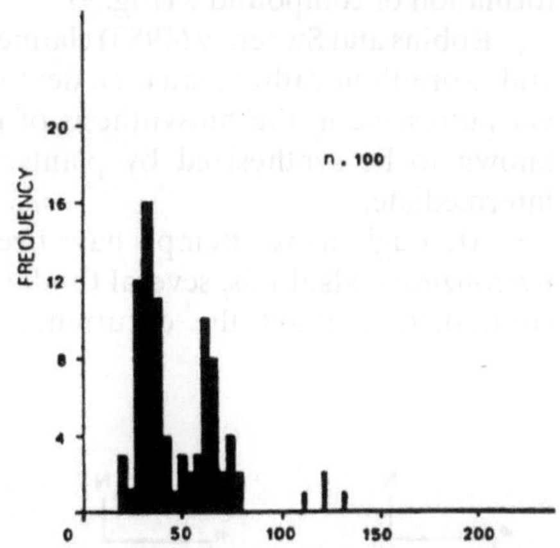

Fig. 6.

Fig. 5. Frequency distribution of the relative DNA content of nuclei originating from leaf material of *Symphytum officinale*

Fig. 6. Frequency distribution of the relative DNA content of nuclei in callus cells of *Symphytum officinale*

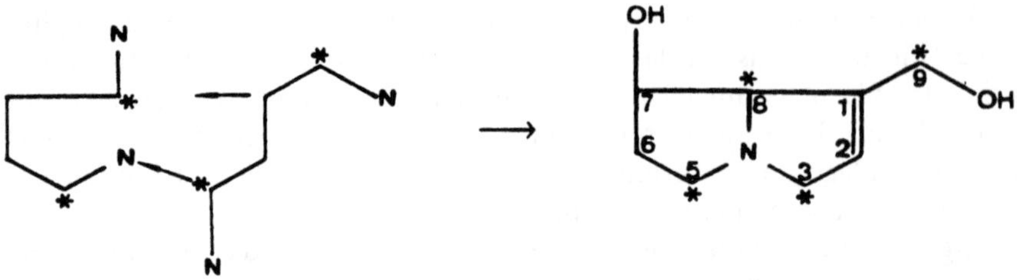

1 2

Fig. 7. Incorporation scheme of 1,4-^{14}C-labeled putrescine (*1*) into retronecine (*2*). Labeled C-atoms are marked with *asterisks*

By feeding *Senecio isatideus* plants 1,4-^{14}C or 2,3-^{13}C-putrescine (Fig. 8) and subsequently isolating and hydrolyzing the pyrrolizidine alkaloid retrorsine, a C_4-N-C_4 intermediate in the biosynthesis of its amino-alcohol moiety, i.e., retronecine, was suggested by Khan and Robins (1981) from the recorded ^{13}C-NMR data.

Grue-Sørensen and Spencer (1981) also suggested a symmetrical dimeric intermediate (possibly analogous to sym-homo-spermidine: *3* in Fig. 8) to be crucial in the biosynthesis of retronecine in *Senecio vulgaris* plants (Fig. 9). They based this suggestion on extensive ^{13}C-NMR studies on derived retronecine after administration of double labeled 1-^{13}C, 1-^{15}N-putrescine also in admixture with 1-^{14}C-putrescine. It could not be excluded, however, that the intermediate *6* (Fig. 9), which shows symmetry, in fact is absent and that rather an asymmetrical compound is present which can undergo a sigmatropic reaction. Further, two different pathways with nonsymmetrical dimeric intermediates might give rise in addition to the formation of compound *7* (Fig. 9).

Robins and Sweeney (1983) claimed that in *Senecio isatideus* plants L-arginine and L-ornithine rather than their dextrorotary enantiomers are assembled together via putrescine in the biosynthesis of retronecine, part of retrorsine. Arginine is known to be synthesized by plants starting from ornithine with citrulline as intermediate.

Although many attempts have been undertaken to resolve the biosynthesis of pyrrolizidine alkaloids, several fundamental questions remain. There exists some contradiction about the occurrence of a symmetrical intermediate (possibly

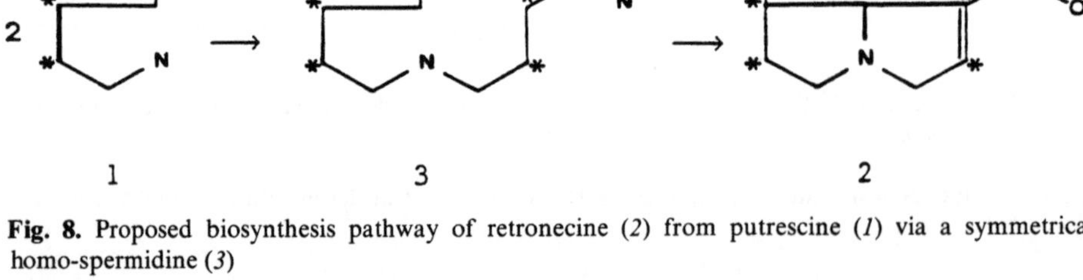

1 3 2

Fig. 8. Proposed biosynthesis pathway of retronecine (*2*) from putrescine (*1*) via a symmetrical homo-spermidine (*3*)

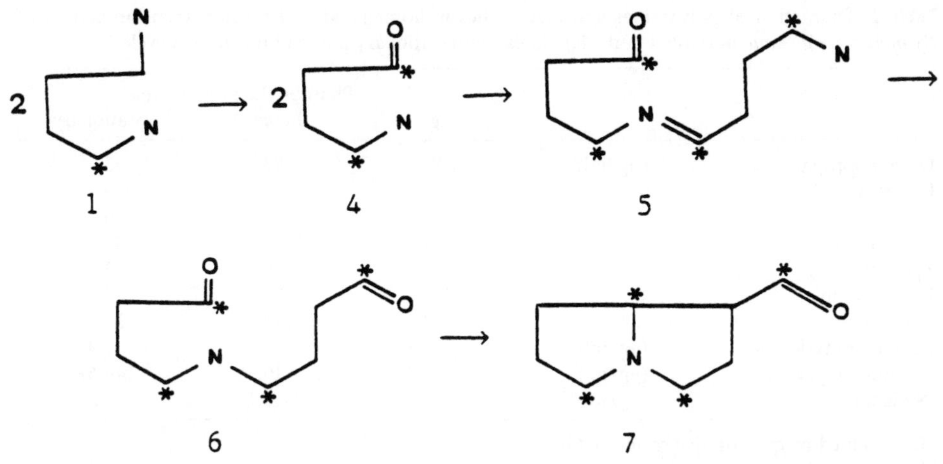

Fig. 9. Biosynthesis of the pyrrolizidine alkaloid amino-alcohol moiety from two single-C labeled putrescine (*1*) molecules. Structure *4* gives the average picture constructed from the sum of all possible labeling patterns, which may occur after the a-specific oxidation of single-labeled putrescine

homologous to sym-homo-spermidine) or the existence of a condensation of different metabolites. Apart from the description of the incorporation of the amino acids ornithine, arginine, and proline as precursors, putrescine, spermidine, and spermine have also been shown to act as specific precursors of retronecine (Robins and Sweeney 1979). However, the main problem for the construction of a biogenetic pathway for pyrrolizidine alkaloids comes from the fact that none of the possible intermediates has been isolated and characterized. Therefore, the pathways constructed in fact still have to be considered as rather hypothetical and need further investigation.

As described, in vitro-grown *Symphytum officinale* cells are not able to produce pyrrolizidine alkaloids. Genetic instability, rendering permanent distortion of the biogenesis, seems to be unlikely and therefore the loss of expression must be caused by the in vitro culture system. In principal though, pyrrolizidine alkaloid biogenesis seems to be possible in callus cultures, and has been described for *Crotalaria juncea* cells (Khanna and Manot 1977).

For this reason we tried to detect what went wrong in the biogenesis of the pyrrolizidine alkaloid in cell suspension cultures and regenerating calli of *Symphytum officinale* by searching for possible precursors and intermediates.

5 Polyamines, Precursors of Pyrrolizidine Alkaloid Biosynthesis

Extraction of a homogenized cell suspension culture revealed the presence of putrescine, spermine, and some minor unknown compounds. The presence and identity of the polyamines were further confirmed by monitoring of the extracted ion current profiles (EICP) on capillary GC-MS of heptafluorobutyrylated (HFB) extracts (Table 2).

Table 2. Estimation of polyamines and metabolites in homogenates of a cell suspension culture of *Symphytum officinale* harvested in the lag, linear, and stationary phase of the growth cycle

		Phase in the growth cycle		
		Lag	Linear	Stationary
Diaminopropane	(ng/ml)[a]	85	93	83
Putrescine	–	724	646	2686
Cadaverine	–	51	16	16
Spermidine	–	959	1933	2794
Spermine	–	52	49	32
Isoputreanine	–	2296	2328	2234
Putreanine	–	64	179	123
γ-Aminobutyric acid	(μg/ml)[a]	2.3	24.7	147.9
γ-Aminobutyric acid	(μg/ml)[b]	1.35	19.36	144.61
β-Alanine	(ng/ml)[a]	2.3	3.1	4.2

[a] Determined by quantitative capillary GLC.
[b] Determined by amino acid analyzer.

Cadaverine was present only in relatively low amounts in the cell suspension cultures of *S. officinale*. This compound was more generally found in several leguminous plants, where it is derived from lysine and then may become incorporated into quinolizidine alkaloids. However, when cadaverine (3–10 mM) was fed to *S. officinale* cell suspension cultures, no quinolizidines were produced (Wink et al. 1980). Lysine, as the presumed precursor for cadaverine, was also present in cell suspension cultures of *S. officinale*, as can be seen in Table 3.

Putrescine and spermidine were quite abundant in cell suspension cultures of *S. officinale*, as was suggested before from the identification procedures. Putrescine (pu) and spermidine (spd) in particular showed an increase in abundance on further progression of the cells through the growth cycle.

In members of the Solanaceae, putrescine may be synthesized from ornithine through mediation of ornithine decarboxylase, an enzyme which indeed is expressed at high levels in tobacco cell suspension cultures (Heimer and Mizrahi 1982). Berlin (1981) showed that in tobacco cell lines, ornithine as well as arginine can also act as putrescine precursors with about equal efficiency, although it is more generally accepted that in plants putrescine is formed from L-arginine via agmatine by the enzyme L-arginine decarboxylase. Heimer et al. (1979) demonstrated an apparent association of L-ornithine decarboxylase activity with rapidly proliferating cells (e.g., cell suspension cultures). Arginine was demonstrated in *S. officinale* cell suspension cultures (Table 3) and therefore ornithine and citrulline should be present as precursors (although possibly short-lived) considering the well-known biosynthetic pathway of arginine. Agmatine could not be demonstrated in *S. officinale* cell suspension cultures by the TLC techniques used. Putrescine subsequently can act as the precursor for spermidine and spermine by addition of propylamino residues originating from decarboxylated SAM to putrescine or spermidine, respectively (Smith 1975; Hirasawa and Suzuki 1983). The data presented show that the pathways for biosynthesis of several presumed precursors for pyrrolizidine alkaloids, i.e., putrescine, spermine, spermidine (Robins and Sweeney

Table 3. Estimation of amino acids (μg/ml) in homogenates of a cell suspension culture of *Symphytum officinale* harvested in the lag, linear, and stationary phase of the growth cycle

Amino acid	Growth phase		
	Lag	Linear	Stationary
Glycine	0.66	3.55	7.66
Alanine	2.99	27.04	38.71
Valine	1.12	6.54	10.96
Leucine	1.01	6.88	13.23
Isoleucine	0.28	4.69	7.66
Serine	1.44	19.34	17.45
Threonine	8.93	6.90	10.35
Lysine	1.56	6.35	15.21
Arginine	0.78	6.04	10.34
Histidine	0.25	1.19	2.88
Fenylalanine	0.94	8.51	14.36
Tyrosine	0.18	2.30	5.47
Methionine	0.30	1.37	3.19
Glutamic acid	4.48	49.20	14.89
Glutamine	18.37	36.16	43.98
Aspartic acid	0.65	9.91	0.55
Asparagine	0.41	4.40	7.33
γ-amino-butyric acid	1.35	19.36	144.61

1979, Grue-Sørensen and Spencer 1981), or ornithine and arginine (Robins and Sweeney 1983), are not blocked in the cell suspension cultures of *S. officinale*. Furthermore, addition of these precursors, separate or in combination, to cell suspension cultures of *S. officinale* did not give rise to the production of the usual pyrrolizidine alkaloids found in mature plants. Also, when in precursor studies, putrescine, cadaverine, spermidine, or spermine were added to cell suspension cultures together with the polyamino oxidase inhibitors diethyldithiocarbamate, β-hydroxyethylhydrazine or δ-hydroxyquinoline, no pyrrolizidine alkaloids could be detected. The polyamino oxidase inhibitors were added to prevent the breakdown of the precursors to 1-pyrroline or 1-(3-aminopropyl)pyrroline (Smith 1972).

On the other hand, other metabolites of spermidine, i.e., isoputreanine (or its lactam) and putreanine were detected. Their identity was confirmed by comparison with authentic samples employing capillary GLC of their HFB-derivatives. Putreanine (Fig. 10: *22*) was isolated before from mammalian brain tissue (Huizing

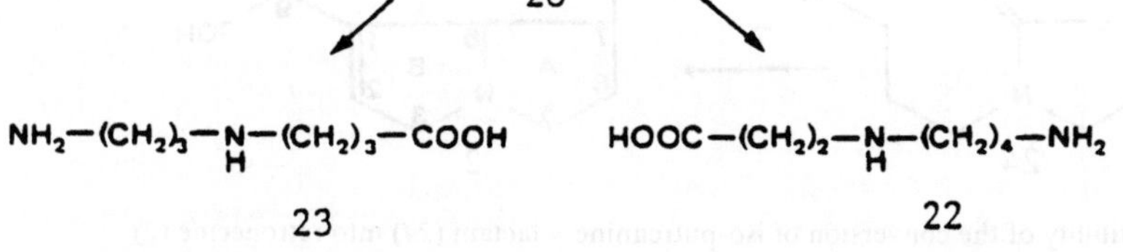

Fig. 10. Formation of isoputreanine (*23*) and putreanine (*22*) from spermidine (*20*)

1985) and might be a condensation product of putrescine and β-alanine, compounds both found in cell suspension cultures of *S. officinale*, as can be seen in Table 2. More recently, it has been suggested that putreanine, but also isoputreanine, are metabolites of spermidine (Asatoor 1979) and/or spermine (Muskiet et al. 1982).

Isoputreanine is apparently present as a γ-lactam (unpubl.) and might be conceived as a pyrrolizidine alkaloid precursor. However, a further methylene group (C-9) has to be added to the compound making the construction of a retronecine moiety possible (Fig. 11).

When it is assumed that the isoputreanine-lactam is synthesized from putrescine via spermidine and that a theoretical labeling pattern is constructed starting from 1,4-labeled-putrescine, then ultimately only C_5, C_8 or C_5, C_8, C_1 sequences should become labeled if the lactam forms the A-ring in Fig. 11. If the lactam gives rise to the B-ring, then C_3, C_8 or C_3, C_8, C_7 in retronecine should become labeled. C_9 could be linked afterwards to the isoputreanine-lactam moiety and then also has to originate from putrescine or a related metabolite which became labeled at the appropriate places. As pointed out in the beginning, a labeling at C_7 or C_1 in the respective sequences has not been found in labeling studies carried out with 1,4-^{14}C-putrescine. It therefore seemed more likely to consider isoputreanine rather as a spermidine metabolite than as a pyrrolizidine alkaloid precursor.

Another metabolite of putrescine that occurs in large amounts in *S. officinale* cell suspension cultures is γ-aminobutyric acid (GABA). In brain tissue it is apparently formed by the action of glutamic acid decarboxylase on glutamic acid (Holdiness et al. 1981). In the fungus *Panus tigrinus*, GABA is formed by oxidation of putrescine (Boldt et al. 1971). Putrescine feeding to sandalwood tree caused a production of GABA (Smith 1975). It might be most likely to assume that in *S. officinale* cultures GABA is also formed through oxidation of putrescine.

In the sandalwood tree also the presence of symmetrical homo-spermidine has been confirmed. It could be synthesized by a condensation of putrescine and an intermediate in GABA biosynthesis, i.e., γ-aminobutyraldehyde (see Fig. 12). This sym-homospermidine was also suggested as a possible intermediate for retronecine (Khan and Robins 1981), but the compound could not be found in extracts of cell suspension cultures and whole plants of *S. officinale*. After addition of GABA together with putrescine to cell suspension cultures, neither sym-homospermidine nor pyrrolizidine alkaloids could be identified in extracts. The absence of sym-homospermidine may either be seen as a crucial step in the biogenesis of pyr-

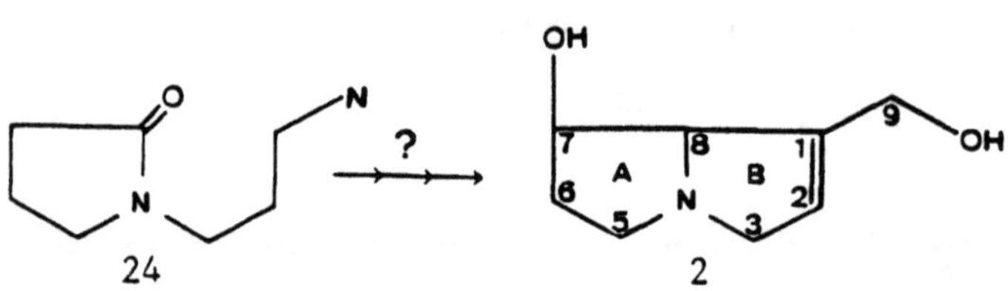

Fig. 11. Impractibility of the conversion of iso-putreanine γ-lactam (*24*) into retronecine (*2*)

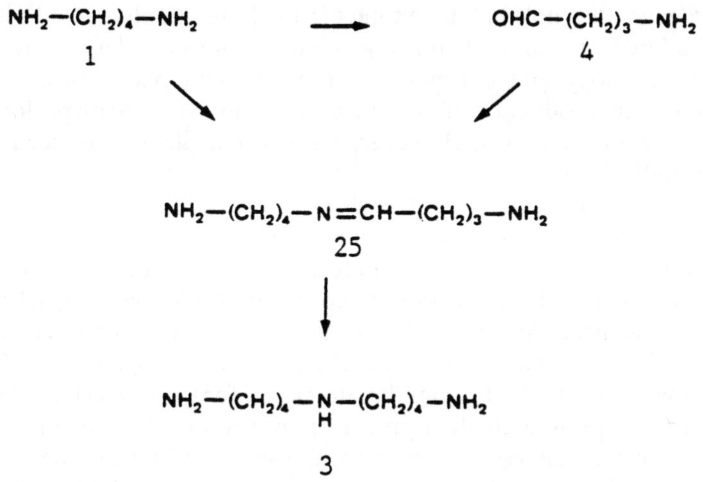

Fig. 12. Formation of sym-homospermidine (*3*) from putrescine (*1*) and γ-aminobutyraldehyde (*4*)

rolizidine alkaloids in in vitro cultures of *S. officinale*, or it may be a short-lived intermediate in the biogenesis of these alkaloids in whole plants.

Although precursors for the obviously initial steps in the biosynthesis of pyrrolizidine alkaloids were found (i.e., arginine, polyamines) or may not be excluded as being present as a short-lived precursor (i.e., ornithine), no further possible precursors were detected in cell suspension cultures of *S. officinale*. Metabolites of polyamines apparently caused no negative effect on the alkaloid biosynthesis, because addition of polyamine or amino acids failed to initiate alkaloid production. Another reason for the absence of alkaloids might be that catabolism of these alkaloid moieties to other, so far unknown, derivatives could occur in the cell suspension cultures. Information concerning this aspect might be gained from a study of the biotransformation properties of a cell suspension culture of *S. officinale* after addition of pyrrolizidine alkaloids.

6 Summary and Conclusions

Because some of the pyrrolizidine alkaloids from Boraginaceae have useful medical applications, including cytostatic effects, an attempt has been made to investigate the usefulness of in vitro-grown cells from *Symphytum officinale* for the production of related alkaloids.

The induction of primary calli from various organs of *S. officinale* (2n = 24) plants on solid MS and B5 medium supplemented with appropriate growth regulators is described. The callus cultures were subcultured on B5 medium. Cell suspension cultures were obtained from B5-grown calli by transfer to liquid B5 medium. Cells from calli as well as from cell suspension cultures were induced to regenerate whole plants on solid MS medium, supplemented with indole-acetic acid and benzyladenine. Plants regenerated from short-term cultures were identical in morphology and chromosome number with plants from which cultures were initiated. Production of pyrrolizidine alkaloids ceased on prolonged subculturing of suspension cultures. However, regenerated plants produced again the alkaloids usually found.

Microdensitometry of Feulgen-stained fixed protoplasts from callus cells of *S. officinale* (2n = 24 cytotype) did not show great deviation in nuclear DNA content from protoplasts derived from leaf material from the same cytotype. Therefore it was concluded that the loss of expression of alkaloid biogenesis was not a result of genetic instability, but rather an effect of an alteration of cell metabolism.

It was demonstrated that the polyamine biogenesis in *S. officinale* did not deviate substantially from that described for tobacco cell cultures. The abundance of the supposed alkaloid precursors was not likely to be the limiting factor in the alkaloid biogenesis. In addition, the supply of polyamines and inhibitors of the polyamine catabolism did not give rise to the alkaloid synthesis. Isoputreanine-lactam, a metabolite of spermidine and spermine, did not appear to be an alkaloid precursor either. Cell suspension cultures of *S. officinale* were able to acetylate lycopsamine. There was evidence that an oxygenation of the necine moiety of the alkaloid might also occur (in prep.).

N-(4-amino-butan-1-one)-2-pyrrolidone, a compound found in shoot regenerating calli, was conceived as a precursor in pyrrolizidine alkaloid biogenesis (unpubl.). It may be formed via a condensation of two γ-aminobutyric acid (GABA) molecules; GABA has been detected in cell suspension of *S. officinale*. A hypothetical pathway for the pyrrolizidine alkaloid biogenesis in *S. officinale* is given. So far, it was not possible to initiate alkaloid synthesis in cell suspension cultures of *S. officinale*. There are indications, however, that photomixotrophic cultures may be essential for the induction of the alkaloid biogenesis.

References

Asatoor AM (1979) Isolation and characterisation of a new urinary metabolite derived from spermidine. Biochim Biophys Acta 586:55–62

Berlin J (1981) Formation of putrescine and cinnamoyl putrescine in tobacco cell cultures. Phytochemistry 20:53–55

Boldt A, Miersch J, Reinbothe H (1971) Metabolism of agmatine in fruit-bodies of the fungus *Panus tigrinus*. Phytochemistry 10:731–738

Bottomley W, Geissmann TA (1964) Pyrrolizidine alkaloids. The biosynthesis of retronecine. Phytochemistry 3:357–360

Budde WL, Eichelberger JW (1977) The mass spectrometer as a substance-selective detector in chromatography. J Chromatogr 134:147–158

Bull LB, Culvenor CCJ, Dick AT (1968) The pyrrolizidine alkaloids. In: Neuberger A, Tatum EL (eds) Frontiers of biology 9. North-Holland Publishing, Amsterdam

Farnsworth NR (1979) The present and future of pharmacognosy. Am Soc Pharmacog News Lett 16:4–12

Furuya T, Araki K (1968) Studies on constituents of crude drugs I. Alkaloids of *Symphytum officinale* L. Chem Pharm Bull 16:2512–2516

Grue-Sørensen G, Spenser JD (1981) Biosynthesis of retronecine. J Am Chem Soc 103:3208–3210

Heimer YM, Mizrahi Y (1982) Characterization of ornithine decarboxylase of tobacco cells and tomato ovaries. Biochem J 201:373-376
Heimer YM, Mizrahi Y, Bachrach U (1979) Ornithine decarboxylase activity in rapidly proliferating plant cells. FEBS Lett 104:146-148
Hills LD (1975) Comfrey report. The Rateavers, California, USA
Hills LD (1976) Comfrey: past, present and future. Faber & Faber, London
Hirasawa E, Suzuki Y (1983) Biosynthesis of spermidine in maize seedlings. Phytochemistry 22:102-106
Hirono I, Haga M, Fujii M, Matsubara N, Nakayama M, Furuya T, Hikichi M, Takanashi H, Uchida E, Hosaka S, Ueno I (1979) Induction of hepatic tumors in rats by senkirkine and symphytine. J Natl Cancer Inst 63:469-472
Holdiness MR, Justice JB, Salamone JD, Neill DB (1981) Gas chromatographic-mass spectrometric determination of glutamic acid decarboxylase activity in subregions of rat brain. J Chromatogr Biomed Appl 225:283-290
Huizing HJ (1985) Phytochemistry, systematics and biogenesis of pyrrolizidine alkaloids of *Symphytum* taxa. Thesis Univ Groningen, Groningen, The Netherlands
Huizing HJ, Pfauth EC, Malingré ThM, Sietsma JH (1983) Regeneration of plants from tissue and cell suspension cultures of *Symphytum officinale* and effect of in vitro culture on pyrrolizidine alkaloid production. Plant Cell Tissue Organ Cult 2:227-238
Khan HA, Robins DJ (1981) Pyrrolizidine alkaloid biosynthesis. Incorporation of carbon-13 labeled putrescines into retronecine. J Chem Soc Chem Commun 1981:146-147
Khanna P, Manot SK (1977) Production of pyrrolizidine alkaloid from in vitro tissue culture of *Crotalaria juncea* L. Indian J Exp Biol 15:807-808
Kovach JS, Ames MM, Powis G, Moertel CG, Hahn RG, Creagan ET (1979) Toxicity and pharmacokinetics of a pyrrolizidine alkaloid, indicine N-oxide, in humans. Cancer Res 39:4540-4544
Muskiet FAJ, Stratingh CM, Fremouw-Ottevangers DC, Halie MR (1982) J Chromatogr Biomed Appl 230:142-147
Robins DJ, Sweeney JR (1979) Pyrrolizidine alkaloids: Evidence for the involvement of spermidine and spermine in the biosynthesis of retronecine. J Chem Soc Chem Commun 1979:120-121
Robins DJ, Sweeney JR (1983) Pyrrolizidine alkaloid biosynthesis: derivation of retronecine from L-arginine and L-ornithine. Phytochemistry 22:457-459
Röder E (1979) Zur möglichen Toxizität und Kanzerogenität handelsüblicher Drogen. Dtsch Apoth Ztg 119:391-392
Smith TA (1972) Purification and properties of the polyamine oxidase of barley plants. Phytochemistry 11:899-910
Smith TA (1975) Recent advances in the biochemistry of plant amines. Phytochemistry 14:865-890
Stengl P, Wiedenfeld H, Röder E (1982) Lebertoxische Pyrrolizidine Alkaloide in Symphytum-Präparaten. Dtsch Apoth Ztg 122:851-855
Tanaka K, Machida Y, Tanaka H, Mukai N, Misawa M (1974) Accumulation of glutamine by suspension cultures of *Symphytum officinale*. Agric Biol Chem 38:987-992
Wichers HJ, Kate J ten, Buys CHCM, Huizing HJ (1984) A simple and rapid procedure to obtain nucleated protoplasts from plant material. Cytologia 49:529-535
Wink M, Hartmann T, Witte L (1980) Biotransformation of cadaverine and potential biogenetic intermediates of lupanine biosynthesis by plant cell suspension cultures. Planta Med 40:31-39

XXIX *Syringa vulgaris* L. (Common Lilac): In Vitro Culture and the Occurrence and Biosynthesis of Phenylpropanoid Glycosides

C.C.S. CHAPPLE[1] and B.E. ELLIS[2]

1 Introduction

Syringa vulgaris L., or common lilac is a horticulturally important member of the Oleaceae, a family in which other economically significant genera (*Olea*, *Fraxinus*, *Jasminum*, *Forsythia*) also occur. Thirty species of *Syringa* are found distributed across the temperate and south temperate zones of Europe and Asia, while the common lilac has been introduced even more widely as an ornamental. A woody shrub (or small tree) in habit (Fig. 1), lilac is typical of the Oleaceae in having opposite leaves, a calyx of four fused sepals and a corolla of four united petals. The highly scented flowers occur in thyrses, each branch bearing a terminal flower.

The high demand for common lilac and lilac hybrids in the woody ornamental trade has led to several studies of in vitro propagation of this and related species. New shoot formation could be induced from explants of actively growing shoot tips on MS medium (Murashige and Skoog 1962) supplemented with 0.1 mg/l 6-benzylaminopurine (BA) and 0.1–0.5 mg/l indoleacetic acid (Hildebrandt and Harney 1983). These were multiplied on MS medium containing high (7.5 mg/l) levels of BA and low (0.1 mg/l) levels of auxin (NAA). Low cytokinin levels led to callus formation, from which regeneration has yet to be reported in lilac (Uhring 1980). Successful in vitro multiplication of *S. vulgaris* has also been reported by Einset and Alexander (1984) and Pierik et al. (1988). In *Syringa chinensis*, 0.1 mg/l NAA and 3.0 mg/l BA yielded the highest number of shoots and nodes per explant (Welander 1987). Under conditions of high auxin and low cytokinin levels, explants of *Syringa vulgaris* and related species readily form fast-growing callus tissue.

Phytochemically, the Oleaceae are probably best known for the occurrence of iridoid glycosides and sugar alcohols. More recently, however, numerous reports have appeared concerning the presence, within this family and related families in the Asteridae, of a remarkably diverse family of aromatic metabolites generally referred to as phenylpropanoid glycosides. Since the original identification of echinacoside in roots of *Echinacea angustifolia* in 1950 (Stoll et al. 1950), many related phenylpropanoid glycosides have been isolated. The glucoside salidroside [1-O-β-D-glucosyl-2-(4′-hydroxyphenyl)ethanol] is often isolated from plants containing the more complex derivatives. Other related deacyl compounds have also been isolated (Endo and Hikino 1982; Shimomura et al. 1987b) and may

[1]MSU-DOE Plant Research Laboratory, Michigan State University, East Lansing, MI 48824–1312, USA
[2]The University of British Columbia, Department of Plant Science, Suite 248, 2357 Main Mall, Vancouver, B.C., Canada V6T 2A2

Biotechnology in Agriculture and Forestry, Vol. 15
Medicinal and Aromatic Plants III (ed. by Y.P.S. Bajaj)

Fig. 1. *Syringa vulgaris* L. (common lilac)

represent either biosynthetic intermediates or degradation products. Table 1 illustrates the structures of the phenylpropanoid glycosides characterized to date, as well as the structure of salidroside.

The apparently diverse array of phenylpropanoid glycosides in Table 1 can actually be simplified into a relatively small number of structural classes. The verbascoside class is characterized by a central 3-rhamnosyl-, 4-caffeoyl-glucose component. The core units of many of the other classes are similar: purpureaside A class (3-glucosyl-, 4-caffeoyl-glucose), conandroside class (3-xylosyl-, 4-caffeoyl-glucose), calceolarioside C class (6-xylosyl-, 4-caffeoyl-glucose), forsythoside class (6-rhamnosyl-, 4-caffeoyl-glucose). Others such as the eukovoside class (3-rhamnosyl-, 4-isoferuloyl-glucose), and the leucosceptoside A class (3-rhamnosyl-, 4-feruloyl-glucose) are distinguished from the verbascoside class by the type of acyl unit esterified at the 4 position. Classes are not rigidly defined and presumably reflect the spectrum of acyltransferases and glucosyltransferases expressed within a given plant species. In some cases, the glucosyltransferases required for formation of two different classes of phenylpropanoid glycosides may be expressed in the same plant, giving rise to compounds such as arenarioside, pheliposide, poliumoside, and purpureaside B, which can be classified as belonging to either of two classes (Table 1). Further common substitution patterns such as 6-glycosylation, acetylation, and O-methylation can be seen in a number of classes. The result is a complex array of compounds, whose interrelationships could only be fully represented by a multidimensional matrix. This array must represent both the inheritance and expression of the genes required for phenylpropanoid glycoside biosynthesis.

Table 1. Phenylpropanoid glycosides known in the plant kingdom

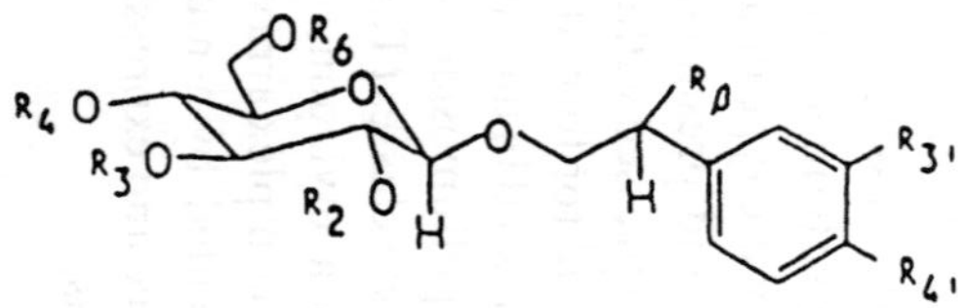

Glucoside	$R_{3'}$	$R_{4'}$	R_β	R_2	R_3	R_4	R_6	Class	Reference
Angoroside A	OH	OH	H	H	rha	caff	ara	V	Calis et al. (1987)
Angoroside B	OH	OCH_3	H	H	rha	caff	ara	V	Calis et al. (1988)
Angoroside C	OH	OCH_3	H	H	rha	fer	ara	LA	Calis et al. (1988)
Arenarioside	OH	OH	H	H	rha	caff	xyl	V,CC	Andary et al. (1985a)
Calceolarioside C	OH	OH	H	H	H	caff	xyl	CC	Nicoletti et al. (1988)
Campneoside I	OH	OH	OCH_3	H	rha	caff	H	V	Imakura et al. (1985)
Cistanoside A	OCH_3	OH	H	H	rha	caff	gluc	V	Kobayashi et al. (1984a)
Cistanoside B	OCH_3	OH	H	H	rha	fer	gluc	LA	Kobayashi et al. (1984a)
Cistanoside C	OCH_3	OH	H	H	rha	caff	H	V	Kobayashi et al. (1984b)
Cistanoside D	OCH_3	OH	H	H	rha	fer	H	LA	Kobayashi et al. (1984b)
Conandroside	OH	OH	H	H	xyl	caff	H	Co	Nonaka and Nishioka (1977)
Echinacoside	OH	OH	H	H	rha	caff	gluc	V	Becker et al. (1982)
Ehrenoside	OH	OH	H	ara	rha	caff	H	V	Lahloub et al. (1986)
Eukovoside	OH	OH	H	H	rha	isofer	H	E	Sticher et al. (1982)
Forsythoside A[1]	OH	OH	H	H	H	caff	rha	F	Endo et al. (1981)
Forsythoside B	OH	OH	H	H	rha	caff	apio	V	Endo et al. (1982)
Grayanoside A	H	OH	H	H	H	H	fer	S	Shimomura et al. (1987b)
Grayanoside B	OH	OH	H	H	H	H	fer	–	Shimomura et al. (1987b)
Isomartynoside	OH	OCH_3	H	H	rha	H	fer	LA	Calis et al. (1984)
Isoverbascoside[2]	OH	OH	H	H	rha	H	caff	V?	Schilling et al. (1982)
Leucosceptoside A	OH	OH	H	H	rha	fer	H	LA	Miyase et al. (1982)
Leucosceptoside B	OH	OCH_3	H	H	rha	fer	apio	LA	Miyase et al. (1982)
Magnolidin[3]	OH	OH	H	H,caff?	rha	H,caff?	rha	–	Rao and Juneau (1975)
Martynoside	OH	OCH_3	H	H	rha	fer	H	LA	Sasaki et al. (1978)
Mussatioside	H	OH	H	H	H	H	xyl-rha	M	Jiminez et al. (1988)

Mussatioside, (4-cinnamoyl)-[4]	H	OH	H	H	H	H	cinn/xyl-rha	M	Jiminez et al. (1987)
Mussatioside, (4-cinnamoyldesxylosyl)-	H	OH	H	H	H	H	cinn-rha	M	Jiminez et al. (1988)
Mussatioside, (4-p-coumaroyl)-	H	OH	H	H	H	H	p-coum/xyl-rha	M	Jiminez et al. (1988)
Mussatioside, (4-dimethoxycinnamoyl)-[5]	H	OH	H	H	H	H	dimethoxycinn/ xyl-rha	M	Jiminez et al. (1987)
Mussatioside, (4-feruloyl)-	H	OH	H	H	H	H	fer/xyl-rha	M	Jiminez et al. (1988)
Mussatioside, (4-p-methoxycinnamoyl)-[6]	H	OH	H	H	H	H	4-methoxycinn/ xyl-rha	M	Jiminez et al. (1987)
Mussatioside, (4-vanilloyl)-	H	OH	H	H	H	H	van/xyl-rha	M	Jiminez et al. (1988)
Myricoside	OH	OH	H	H	apio-rha	caff	H	V	Cooper et al. (1980)
Neoverbascoside	OH	OH	H	caff	rha	H	gluc	N	Birkofer et al. (1968)
Orobanchoside	OH	OH	OH	rha	H	caff	H	O	Andary et al. (1982)
Osmanthuside B	H	OH	H	H	rha	p-coum	H	S	Kobayashi et al. (1984b)
Pheliposide	OH	OH	H	acetyl	rha	caff	xyl	V,CC	Andary et al. (1985)
Poliumoside	OH	OH	H	H	rha	caff	rha	V,F	Andary et al. (1985)
Purpureaside A	OH	OH	H	H	gluc	caff	H	PA	Shoyama et al. (1986)
Purpureaside B	OH	OH	H	H	gluc	caff	rha	PA,F	Shoyama et al. (1986)
Purpureaside C	OH	OH	H	H	rha	caff	gal	V	Matsumoto et al. (1987)
Rossicaside A	OH	OH	H	H	gluc-rha	caff	H	V	Konishi et al. (1987)
Stachyoside[7]	OH	OH	H	H	gluc-rha	caff	H	–	Andary and Ibrahim (1986)
Suspensaside[8]	OH	OH	OH	H	H	caff	rha	F	Nishibe et al. (1982)
Teucrioside	OH	OH	H	H	lyx-rha	caff	H	V	Gross et al. (1988)
Tubuloside A	OH	OH	H	acetyl	rha	caff	gluc	V	Kobayashi et al. (1987)
Tubuloside B	OH	OH	H	acetyl	rha	H	caff	V?	Kobayashi et al. (1987)
Tubuloside C	OH	OH	H	acetyl	triacetyl-rha	caff	gluc	V	Kobayashi et al. (1987)
Tubuloside D	OH	OH	H	acetyl	triacetyl-rha	p-coum	gluc	S	Kobayashi et al. (1987)
Unnamed 1	H	OH	H	H	H	H	caff	–	Shimomura et al. (1987a)
Unnamed 2[9]	OH	OH	H	H	H	H	caff	V	Shimomura et al. (1987a)
Verbascoside[2,10]	OH	OH	H	H	rha	caff	H	V	Andary et al. (1982)
Verbascoside, (2′-acetyl)-[2]	OH	OH	H	acetyl	rha	caff	H	V	Kobayashi et al. (1984b)

(*to be continued*)

Table 1. *(Continued)*

Glucoside	$R_{3'}$	$R_{4'}$	R_β	R_2	R_3	R_4	R_6	Class	Reference
Verbascoside, (desrhamnosyl)-[2,11]	OH	OH	H	H	H	caff	H	V	Matsumoto et al. (1987)
Verbascoside, (β-hydroxy)-[2,12]	OH	OH	OH	H	rha	caff	H	V	Kitigawa et al. (1984)
Salidroside	H	OH	H	H	H	H	H	S	Thieme (1969)

Abbreviations: apio, apiose; ara, arabinose; rha, rhamnose; caff, caffeoyl; cinn, cinnamoyl; p-coum, p-coumaroyl; fer, feruloyl; gluc, glucose; isofer, isoferuloyl; lyx, lyxose; xyl, xylose; van, vanilloyl.
Classes: CC: calceolarioside C; Co: conandroside; E: eukovoside; F: forsythoside A; LA: leucosceptoside A; M: mussatioside; N: neoverbascoside; O: orobanchoside; PA: purpureaside A; S: salidroside; V: verbascoside; –: questionable.

[1] Synonym: forsythiaside (Nishibe et al. 1982a)
[2] Nomenclature of verbascoside and derivatives reflects priority of original isolation (Scarpati and Del Monache 1963)
[3] Acylation site not determined
[4] Synonym: mussatioside I (Jiminez et al. 1987)
[5] Synonym: mussatioside II (Jiminez et al. 1987)
[6] Synonym: mussatioside III (Jiminez et al. 1987)
[7] Linkage of glucose to rhamnose was not determined; may be identical to Rossicaside A.
[8] Synonym: forsythoside C (Endo and Hikino 1982)
[9] Subsequently named calceolarioside B (Nicoletti et al. 1988)
[10] Synonyms: acteoside (Birkofer et al. 1968), kusaginin (Sakurai and Kato 1983)
[11] Synonym: calceolarioside A (Nicoletti et al. 1988)
[12] Synonym: campneoside II (Imakura et al. 1985)

There is little information concerning the physiological role of these compounds in the plant. Myricoside, extracted from roots of *Clerodendrum myricoides*, acts as a potent antifeedant against the African armyworm (*Spodoptera exempta*) at concentrations as low as 10 ppm (Cooper et al. 1980). A more general role for these compounds may be inferred from the high UV-absorbtivity conferred by the caffeoyl moiety, which may enable these compounds to protect the plant from UV-induced mutation (Caldwell et al. 1983). In addition, the two ortho-dihydroxy ring substituents are readily oxidized following tissue damage, and may participate in the response to plant pathogenic interactions. Consistent with this hypothesis is the report that verbascoside is a phytoalexin in root tissue of *Rehmannia glutinosa* (Shoyama et al. 1987).

Most investigations of the occurrence of these compounds have arisen from interests in chemotaxonomy (Calis et al. 1984, 1987; Andary et al. 1985a), bitter principles (Nonaka and Nishioka 1977; Imakura et al. 1985; Shimomura et al. 1987a), or components of plant materials used as folk medicines (Endo et al. 1981; Nishibe et al. 1982b; Kitigawa et al. 1984; Kobayashi et al. 1984a,b, 1987; Lahloub et al. 1986; Shoyama et al. 1986; Burger et al. 1987; Henry et al. 1987; Jiminez et al. 1987; Konishi et al. 1987; Gross et al. 1988; Potterat et al. 1988). As the result of studies of the potential medicinal properties of these compounds, a number of them [forsythoside A (Endo et al. 1981; Nishibe et al. 1982b; Shoyama et al. 1986), forsythoside B (Endo et al. 1982), suspensaside (Endo and Hikino 1982), deshamnosyl verbascoside, purpureaside A (Shoyama et al. 1986), purpureaside C, and verbascoside (Shoyama et al. 1987)] have been shown to possess antibacterial properties. Verbascoside and orobanchoside have been investigated as analgesic and antihypertensive compounds, and as potential DOPA agonists (Andary et al. 1982 and references cited therein; Sticher and Lahloub 1982). The biological activity of analogous compounds has been documented as well, such as the cytotoxicity activity of caffeic acid phenylethyl ester on human tumor cell lines (Grunberger et al. 1988). The potential of these compounds as pharmacologic agents, the ease with which cell cultures can be established from many of the plants which produce them, and the yield of the phenylpropanoid glycosides by cultured cells, make these systems of particular interest both for the study of the diversity and control of plant secondary metabolism, and for possible commercial exploitation.

2 Distribution of Phenylpropanoid Glycosides

Phenylpropanoid glycosides have most commonly been identified in members of the Asteridae (Table 2). One example is known from the Magnoliidae (Rao and Juneau 1975), and compounds lacking a second sugar moiety (Table 1: unnamed 1,2, grayanoside A,B) have been isolated from *Prunus* in the Subclass Rosidae (Shimomura et al. 1987a; Shoyama et al. 1986). Whether this accurately summarizes the taxonomic distribution of this class of compounds is unknown, since few systematic attempts have been made to determine how widely the compounds occur in the plant kingdom (Kvist and Pedersen 1986). A recent paper has summarized the

Table 2. Taxonomic Distribution of Phenylpropanoid Glycosides

Compound	Source	Family	Reference
Angoroside A	*Scrophularia scopolii*	Scrophulariaceae	Calis et al. (1987)
Angoroside B	*Scrophularia scopolii*	Scrophulariaceae	Calis et al. (1988)
Angoroside C	*Scrophularia scopolii*	Scrophulariaceae	Calis et al. (1988)
Arenarioside	*Orobanche arenaria*	Orobanchaceae	Andary et al. (1985a)
Calceolarioside C	*Calceolaria hypericina*	Scrophulariaceae	Nicoletti et al. (1988)
Campneoside I	*Campsis chinensis*	Bignoniaceae	Imakura et al. (1985)
Cistanoside A	*Cistanche salsa*	Orobanchaceae	Kobayashi et al. (1984a)
Cistanoside B	*Cistanche salsa*	Orobanchaceae	Kobayashi et al. (1984a)
Cistanoside C	*Cistanche salsa*	Orobanchaceae	Kobayashi et al. (1984b)
Cistanoside D	*Cistanche salsa*	Orobanchaceae	Kobayashi et al. (1984b)
Conandroside	*Conandron ramoidioides*	Gesneriaceae	Nonaka and Nishioka (1977)
Echinacoside	*Cistanche salsa*	Orobanchaceae	Kobayashi et al. (1984a)
	Cistanche tubulosa	Orobanchaceae	Kobayashi et al. (1987)
	Echinacea angustifolia	Asteraceae	Becker et al. (1982)
Ehrenoside	*Veronica bellidioides*	Scrophulariaceae	Lahloub et al. (1986)
Eukovoside	*Euphrasia rostkoviana*	Scrophulariaceae	Sticher et al. (1982)
Forsythoside A	*Digitalis purpurea*	Scrophulariaceae	Matsumoto et al. (1987)
	Fosythia suspensa	Oleaceae	Endo et al. (1981) Nishibe et al. (1982a)
	Rehmannia glutinosa[a]	Scrophulariaceae	Shoyama et al. (1986)
Forsythoside B	*Forsythia koreana*	Oleaceae	Endo et al. (1982)
Grayanoside A	*Prunus grayana*	Rosaceae	Shimomura et al. (1987b)
Grayanoside B	*Prunus grayana*	Rosaceae	Shimomura et al. (1987b)
Isomartynoside	*Galeopsis pubescens*	Lamniaceae	Calis et al. (1984)
Isoverbascoside (neoacteoside)	*Cistanche tubulosa*	Orobanchaceae	Kobayashi et al. (1987)
	Clerodendrum myricoides	Verbenaceae	Cooper et al. (1980)
	Harpagophytum procumbens	Pedaliaceae	Burger et al. (1987)
	Leucosceptrum japonicum	Lamniaceae	Miyase et al. (1982)
	Paulownia tomentosa	Scrophulariaceae	Schilling et al. (1982)
	Syringa vulgaris	Oleaceae	Birkofer et al. (1968)
	Syringa vulgaris[a]	Oleaceae	Chapple et al. (1984)
Leucosceptoside A	*Leucosceptrum japonicum*	Lamniaceae	Miyase et al. (1982)
Leucosceptoside B	*Leocosceptrum japonicum*	Lamniaceae	Miyase et al. (1982)
Magnolidin	*Magnolia grandiflora*	Magnoliaceae	Rao and Juneau (1975)
Martynoside	*Galeopsis pubescens*	Lamniaceae	Calis et al. (1984)
	Leucosceptrum japonicum	Lamniaceae	Miyase et al. (1982)
	Martynia louisiana	Martyniaceae	Sasaki et al. (1978)
Mussatioside	*Mussatia* sp.	Bignoniaceae	Jiminez et al. (1988)
Mussatioside,	*Mussatia* sp.	Bignoniaceae	Jiminez et al. (1987)
(4-cinnamoyl)-	*Mussatia hyacintha*	Bignoniaceae	Jiminez et al. (1987)
Mussatioside, (4-cinnamoyl desxylosyl)-	*Mussatia* sp.	Bignoniaceae	Jiminez et al. (1988)
Mussatioside, (4-p-coumaroyl)-	*Mussatia* sp.	Bignoniaceae	Jiminez et al. (1988)
Mussatioside, (4-dimethyl caffeoyl)-	*Mussatia* sp.	Bignoniaceae	Jiminez et al. (1987)
Mussatioside, (4-feruloyl)-	*Mussatia* sp.	Bignoniaceae	Jiminez et al. (1988)
Mussatioside, (4-p-methoxy cinnamoyl)-	*Mussatia* sp.	Bignoniaceae	Jiminez et al. (1987)

Table 2. (*Continued*)

Compound	Source	Family	Reference
Mussatioside, (4-vanilloyl)-	*Mussatia* sp.	Bignoniaceae	Jiminez et al. (1988)
Myricoside	*Clerodendrum myricoides*	Verbenaceae	Cooper et al. (1980)
	Clerodendrum johnstonii	Verbenaceae	Cooper et al. (1980)
Orobanchoside	*Orobanche rapum-genistae*	Orobanchaceae	Andary et al. (1982)
Osmanthuside B	*Cistanche salsa*	Orobanchaceae	Kobayashi et al. (1984b)
Pheliposide	*Orobanche arenaria*	Orobanchaceae	Andary et al. (1985a)
Poliumoside	*Teucrium belion*	Lamniaceae	Andary et al. (1985b)
	Verbascum sp.	Scrophulariaceae	Andary et al. (1985b)
Purpureaside A	*Rehmannia glutinosa*[a]	Scrophulariaceae	Shoyama et al. (1986)
Purpureaside B	*Digitalis purpurea*	Scrophulariaceae	Matsumoto et al. (1987)
	Digitalis purpurea[a]	Scrophulariaceae	Matsumoto et al. (1987)
	Rehmania glutinosa[a]	Scrophulariaceae	Shoyama et al. (1986)
Purpureaside C	*Digitalis purpurea*[a]	Scrophulariaceae	Matsumoto et al. (1987)
	Rehmannia glutinosa	Scrophulariaceae	Shoyama et al. (1987)
Rossicaside A	*Boschniakia rossica*	Orobanchaceae	Konishi et al. (1987)
Stachyoside	*Stachys albens*	Lamniaceae	Andary and Ibrahim (1986)
	Stachys rigida	Lamniaceae	Andary and Ibrahim (1986)
	Stachys lanata	Lamniaceae	Andary and Ibrahim (1986)
Suspensaside	*Forsythia suspensa*	Oleaceae	Nishibe et al. (1982b) Endo and Hikino (1982)
Teucrioside	*Teucrium chamaedrys*	Lamniaceae	Gross et al. (1988)
Tubuloside A	*Cistanche tubulosa*	Orobanchaceae	Kobayashi et al. (1987)
Tubuloside B	*Cistanche tubulosa*	Orobanchaceae	Kobayashi et al. (1987)
Tubuloside C	*Cistanche tubulosa*	Orobanchaceae	Kobayashi et al. (1987)
Tubuloside D	*Cistanche tubulosa*	Orobanchaceae	Kobayashi et al. (1987)
Un-named 1	*Prunus grayana*	Rosaceae	Shimomura et al. (1987a)
Un-named 2	*Calceolaria hypericina*	Scrophulariaceae	Nicoletti et al. (1988)
	Digitalis purpurea	Scrophulariaceae	Matsumoto et al. (1987)
	Prunus grayana	Rosaceae	Shimomura et al. (1987a)
Verbascoside	*Ajuga incisa*	Lamniaceae	Shimomura et al. (1987c)
	Ajuga japonica	Lamniaceae	Shimomura et al. (1987c)
	Ajuga nipponensis	Lamniaceae	Shimomura et al. (1987c)
	Calceolaria hypericina	Scrophulariaceae	Nicoletti et al. (1988)
	Campsis chinensis	Bignoniaceae	Imakura et al. (1985)
	Cistanche salsa	Orobanchaceae	Kobayashi et al. (1984a)
	Cistanche tubulosa	Orobanchaceae	Kobayashi et al. (1987)
	Clerodendron trichotomum	Verbenaceae	Sakurai and Kato (1983)
	Clerodendrum myricoides	Verbenaceae	Cooper et al. (1980)
	Clerodendrum johnstonii	Verbenaceae	Cooper et al. (1980)
	Conandron ramoidioides	Gesneriaceae	Nonaka and Nishioka (1977)
	Digitalis purpurea[a]	Scrophulariaceae	Matsumoto et al. (1987)
	Forsythia × intermedia[a]	Oleaceae	Chapple and Ellis (1984)
	Forsythia koreana	Oleaceae	Endo et al. (1982)
	Forsythia viridissima	Oleaceae	Kitigawa et al. (1984)
	Harpagophytum procumbens	Pedaliaceae	Burger et al. (1987)
	Hygrophila erecta	Acanthaceae	Henry et al. (1987)
	Hygrophila erecta[a]	Acanthaceae	Henry et al. (1987)
	Leucosceptrum japonicum	Lamniaceae	Miyase et al. (1982)
	Martynia louisiana	Martyniaceae	Sasaki et al. (1978)
	Olea europea	Oleaceae	Amiot et al. (1986)
	Orobanche rapum-genistae	Orobanchaceae	Andary et al. (1982)

(*to be continued*)

Table 2. (*Continued*)

Compound	Source	Family	Reference
	Paulownia tomentosa	Scrophulariaceae	Schilling et al. (1982) Sticher and Lahloub (1982)
	Rehmania glutinosa[a]	Scrophulariaceae	Shoyama et al. (1986)
	Scrophularia scopolii	Scrophulariaceae	Calis et al. (1987)
	Sesamum angolense	Pedaliaceae	Potterat et al. (1988)
	Syringa vulgaris	Oleaceae	Birkofer et al. (1968)
	Syringa vulgaris[a]	Oleaceae	Ellis (1983)
	Verbascum sinuatum	Scrophulariaceae	Scarpati and Delle Monache (1963)
Verbascoside	*Cistanche salsa*	Orobanchaceae	Kobayashi et al. (1984b)
(2′-acetyl)-	*Cistanche tubulosa*	Orobanchaceae	Kobayashi et al. (1987)
Verbascoside	*Ajuga nipponensis*	Lamniaceae	Shimomura et al. (1987c)
(desrhammosyl)-	*Calceolaria hypericina*	Scrophulariaceae	Nicoletti et al. (1988)
	Digitalis purpurea	Scrophulariaceae	Matsumoto et al. (1987)
Verbascoside	*Campsis chinensis*	Bignoniaceae	Imakura et al. (1985)
(β-hydroxy)-	*Forsythia viridissima*	Oleaceae	Kitigawa et al. (1984)

[a] Denotes isolation from a cell culture.

taxonomic distribution of phenylpropanoid glycosides and many other caffeic acid esters in plants (Molgaard and Ravn 1988).

Phenylpropanoid glycosides have been isolated from plant cell cultures (Table 2). Overall, biosynthesis of phenylpropanoid glycosides is maintained in culture, but some differences in the spectrum of the compounds accumulated have been observed. Verbascoside has been isolated from callus cultures of *Rehmannia glutinosa* (Shoyama et al. 1986), *Hygrophila erecta* (Henry et al. 1987), and *Digitalis purpurea* (Matsumoto et al. 1987), but in *Syringa vulgaris* suspension cultures it is accumulated to very high levels (Ellis 1983). Callus and suspension cultures of *Forsythia* × *intermedia* were reported to accumulate substantial amounts of verbascoside but none of the forsythoside typical of the source tissue (Chapple and Ellis 1984). Tissue of *D. purpurea* also ceased to produce forsythoside when induced to form callus, while the accumulation of verbascoside and purpureasides A, B, and C continued (Matsumoto et al. 1987).

3 Establishment of Cultures and Growth Characteristics

Sections of immature stem tissue of *Syringa vulgaris* L. (obtained from the University of Guelph Arboretum) were treated with 70% ethanol for 1 min, followed by surface sterilization in 20% commercial bleach (final concentration of sodium hypochlorite approximately 1.2%) for 30 min. The tissue was rinsed with sterile 0.01% HCl, and then several times with water. The tissue was then cut aseptically into 0.5-cm segments and transferred to Gamborg's B5 medium (Gamborg and Eveleigh 1968), containing 0.1 mg/l 2,4-D, 0.1 mg/l kinetin, and 0.8% agar.The callus tissue which formed was maintained on the same medium. Suspension cultures were initiated by transfer of callus to 250 ml flasks containing 50 ml of liquid B5 medium supplemented with the same growth regulators. Once well established, suspension cultures were

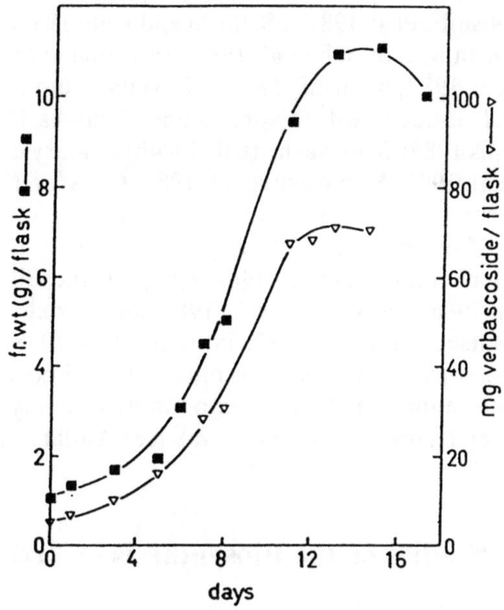

Fig. 2. Tissue growth and verbascoside production in cell suspension cultures of *Syringa vulgaris* (Ellis 1983)

subcultured every 10 days into fresh medium, and were maintained under conditions of continuous light.

The kinetics of *Syringa* cell growth and verbascoside accumulation are shown in Fig. 2. Although verbascoside accumulation reaches a maximum level of approximately 70 mg/flask between days 10 and 12 of the culture cycle, the level of verbascoside on a fresh weight basis is relatively constant throughout the culture cycle. A similar accumulation pattern has been reported for cultures of *Hygrophila erecta*, but the verbascoside content in these cultures was between 0.01% and 1% of the tissue dry weight. In contrast, verbascoside comprises up to 16% of the dry weight of *Syringa* cell cultures, while salidroside is accumulated to about 25% of the level of verbascoside. This represents a very high commitment to secondary metabolism.

4 Extraction and Identification of Phenylpropanoid Glycosides

The extraction of any of the compounds listed in Table 1 can be accomplished using hot methanol, ethanol, or acetone, and after concentration of the organic phase, the compounds are readily extractable into n-butanol. Purification of the phenylethanol glycosides has been achieved using many techniques including paper chromatography (Ellis 1983; Chapple et al. 1986; Calis et al. 1987), thin layer chromatography (Nishibe et al. 1982a; Imakura et al. 1985; Andary and Ibrahim 1986; Burger et al. 1987), column chromatography on silica gel (Nonaka and Nishioka 1977; Sasaki et al. 1978; Miyase et al. 1982; Nishibe et al. 1982a; Sticher et al. 1982; Sticher and Lahloub 1982; Kobayashi et al. 1984a,b, 1987; Lahloub et al. 1986; Shoyama et al. 1986; Konishi et al. 1987; Shimomura et al. 1987a; Calis et al. 1988), cellulose (Miyase et al. 1982; Konishi et al. 1987), Sephadex LH-20

(Nishibe et al. 1982b; Sakurai and Kato 1983; Kitigawa et al. 1984; Kobayashi et al. 1984a,b; Lahloub et al. 1986; Shoyama et al. 1986, 1987; Burger et al. 1987; Calis et al. 1987; Konishi et al. 1987; Matsumoto et al. 1987; Shimomura et al. 1987a), or polyamide powder (Nonaka and Nishioka 1977; Cooper et al. 1980; Chapple and Ellis 1984; Kobayashi et al. 1984b; Andary et al. 1985b; Lahloub et al. 1986; Calis et al. 1987; Kobayashi et al. 1987; Gross et al. 1988), as well as more sophisticated techniques such as counter-current distribution (Cooper et al. 1980; Andary et al. 1985a; Jiminez et al. 1987, 1988; Nicoletti et al. 1988; Potterat et al. 1988), and HPLC (Sticher and Lahloub 1982; Chapple and Ellis 1984; Fleuriet et al. 1984; Amiot et al. 1986; Calis et al. 1987, 1988; Jiminez et al. 1987, 1988; Kobayashi et al. 1987; Matsumoto et al. 1987; Potterat et al. 1988). Nuclear magnetic resonance spectroscopy (NMR) has become the method of choice for the accurate structural determination of these compounds. Virtually all of the compounds of Table 1 have been characterized in this way (see Andary et al. 1982 for an excellent example).

5 Studies on the Biosynthesis of Verbascoside

The high levels of verbascoside biosynthesis in *Syringa* cultures and the rapid uptake of radiolabeled precursors by cell cultures in general make *Syringa vulgaris* suspension cultures an excellent system in which to study the verbascoside biosynthetic pathway. In an effort to explore this possibility, various compounds were administered to *Syringa* cultures as potential precursors of the aromatic moieties of verbascoside and salidroside (Ellis 1983). ^{14}C-phenylalanine and ^{14}C-tyrosine were efficient precursors of the caffeoyl and dihydroxyphenylethanol units of verbascoside respectively, while ^{14}C-tyrosine also strongly labeled the aromatic moiety of salidroside. ^{14}C-salidroside was incorporated at low levels into verbascoside. This pattern of labeling is consistent with parallel pathways incorporating each of the two amino acid precursors into the aromatic moieties of verbascoside (Fig. 3). The incorporation of labeled salidroside into verbascoside may indicate that salidroside serves as an intermediate in verbascoside synthesis, but the large quantities of salidroside accumulated in the cultures make it unlikely that it serves solely as an intermediate. Salidroside is probably an endproduct in itself, representing a branching of the tyrosine pathway.

To determine the sequence of reactions that transform the precursor amino acids into verbascoside and salidroside, labeled tyramine, tyrosol, dopamine, and DOPA were administered to *Syringa* cell cultures. ^{14}C-tyramine was incorporated into salidroside and the dihydroxyphenylethanol moiety of verbascoside at a level comparable to that of tyrosine, while labeled tyrosol was incorporated efficiently only into salidroside. This preferential incorporation into salidroside may reflect a metabolic branching of the verbascoside/salidroside pathways or an artifactual diversion of tyrosol away from the normal in vivo pathway. The dihydroxy compounds tested, DOPA and dopamine, were poor precursors for verbascoside and salidroside. This may indicate that ring hydroxylation normally occurs after modification of the side chain or that the dihydroxy compounds have only limited access to the verbascoside/salidroside biosynthetic site.

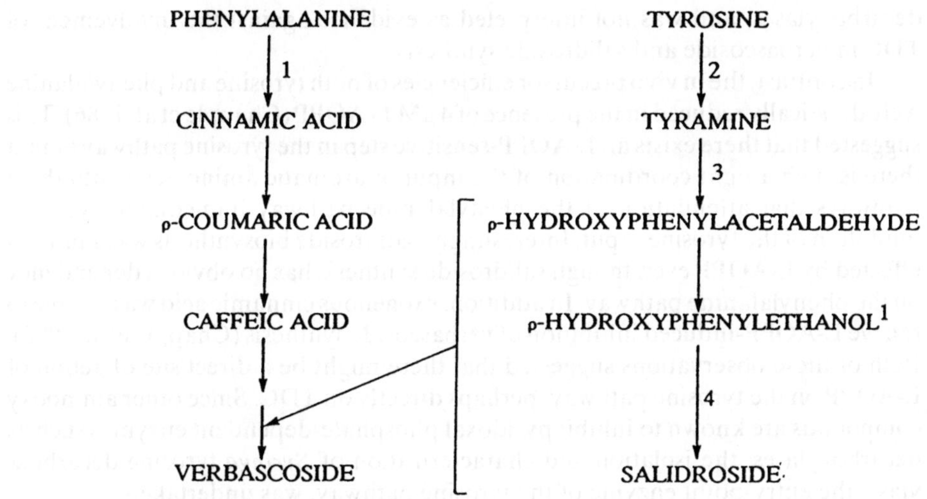

Fig. 3. Biosynthetic pathway leading to the production of verbascoside and salidroside. Potential catalysts identified in *Syringa vulgaris* suspension culture extracts include *1* Phenylalanine ammonia lyase; *2* Tyrosine decarboxylase; *3* Tyramine oxidase; *4* UDP-glucose: tyrosol glucosyltransferase. ([1] ρ-hydroxyphenylethanol = tyrosol)

5.1 In Vivo Inhibition Studies

The pattern of labeling seen in feeding experiments led to the hypothesis that phenylalanine serves as the precursor for the caffeic acid moiety of verbascoside, initially chaneled into the pathway by phenylalanine ammonia lyase (PAL). In parallel, tyrosine is diverted into salidroside and verbascoside biosynthesis through decarboxylation to tyramine catalyzed by tyrosine decarboxylase (TDC), and subsequent oxidation and reduction to tyrosol. To examine this hypothesis, and to determine what metabolic coordination might exist between these two pathways, in vivo inhibition studies were employed (Chapple et al. 1986).

The potent PAL inhibitor L-α-aminooxy-β-phenylpropionic acid (L-AOPP), and inhibitors of mammalian aromatic amino acid decarboxylase, α-fluoromethyltyrosine and α-fluoromethylDOPA (Maycock et al. 1980), were employed to inhibit the putative entry point enzymes in verbascoside and salidroside synthesis, PAL and TDC. The incorporation of label from 2,6-^{3}H-phenylalanine and U-^{14}C-tyrosine was unaffected by the α-fluoromethyl amino acids (data not shown). This lack of effect of the α-fluoromethyl amino acid might be due to differences in active site architecture between the mammalian and *Syringa* decarboxylases, and was not interpreted as evidence against the involvement of TDC in verbascoside and salidroside synthesis.

In contrast, the in vivo precursor efficiencies of both tyrosine and phenylalanine were drastically reduced in the presence of 4 μM L-AOPP (Chapple et al. 1986). This suggested that there exists an L-AOPP-sensitive step in the tyrosine pathway or that there is such a tight coordination of the input of aromatic amino acids into these pathways that attenuation of the phenylalanine pathway immediately leads to inhibition of the tyrosine input. Interestingly, salidroside biosynthesis was similarly effected by L-AOPP even though salidroside synthesis has no obvious dependency on the phenylalanine pathway. In addition, exogenous cinnamic acid was unable to relieve L-AOPP-induced inhibition of verbascoside synthesis (Chapple et al. 1986). Both of these observations suggested that there might be a direct site of action of L-AOPP on the tyrosine pathway, perhaps directly on TDC. Since other aminooxy compounds are known to inhibit pyridoxal phosphate-dependent enzymes such as decarboxylases, the isolation and characterization of *Syringa* tyrosine decarboxylase, the entry-point enzyme of the tyrosine pathway, was undertaken.

5.2 Characterization of *Syringa* Tyrosine Decarboxylase and Phenylalanine Ammonia Lyase

Tyrosine decarboxylase from *Syringa vulgaris* suspension culture was extracted and assayed as described previously (Chapple et al. 1986). The inclusion of β-mercaptoethanol increased TDC activity 100% over control extracts. The pH optimum for *Syringa* TDC was found to be between 7.5 and 8.0, and in this respect is similar to aromatic amino acid decarboxylases from *Hordeum* (pH optimum 7.3) (Hosoi 1974), and mammalian sources (pH optima 8.0 to 8.5) (Christenson et al. 1970; Bowsher and Henry 1983). In contrast, the pH optimum for bacterial TDC is 5.5 (Epps 1944).

The inclusion of 7.5 μM pyridoxal phosphate (PLP) in assays of crude enzyme preparations increased TDC activity 50%, and increased activity in ammonium sulfate precipitated enzyme up to three fold. Extensive dialysis of extracts failed to completely eliminate TDC activity, indicating that there may be a PLP component which is very tightly bound to the TDC apoenzyme. Activation of *Syringa* TDC by PLP suggest that it is a PLP-dependent enzyme similar to those from other plant, animal, and bacterial sources.

The K_M for tyrosine of the *Syringa* TDC was determined to be 0.3 mM. When assayed at substrate concentrations of 0.4 mM, the pattern of relative substrate preference was found to be tyrosine: DOPA: phenylalanine [1.0:0.63:0.04].

The extraction and assay of PAL from *Syringa vulgaris* suspension cultures was performed essentially as described previously (Razzaque and Ellis 1977; Ramsey 1984). In the case of PAL, the inclusion of 5 mM β-mercaptoethanol in the extraction buffer increased PAL activity eight fold. The pH optimum for the *Syringa* enzyme was found to be between 8.0 and 8.3, similar to the pH optima previously reported from PAL from other sources (Camm and Towers 1973). The K_M of the enzyme for L-phenylalanine was 0.2 mM. The most highly purified preparation had a native molecular mass of 290 kDa and a subunit mass of ca. 79 kDa (Ramsey 1984).

5.2.1 *Effect of L-AOPP on TDC and PAL in Vitro*

In view of the dramatic effects of L-AOPP on verbascoside synthesis in vivo, the effect of this inhibitor on PAL and TDC was examined in vitro. Concentrations of L-AOPP between 10 and 100 nM caused a striking decrease in TDC activity in partially purified preparations (Chapple et al. 1986). Analysis of this inhibition using a double reciprocal plot indicated that the K_i for L-AOPP was 0.01 μM. This value was approximately tenfold higher than that previously reported for buckwheat PAL (Amrhein and Godeke 1977), but is close to the value determined for *Syringa* PAL (K_i approximately 0.02 μM). This indicates that the L-AOPP-induced inhibition of ^{14}C-tyrosine incorporation into verbascoside and salidroside in vivo can be accounted for by a direct action of L-AOPP on TDC, and provides supporting evidence that TDC catalyzes the first step in the pathway. This finding should also encourage a cautious interpretation of results from any experiments using L-AOPP as a probe of phenylpropanoid metabolism. It is clear that the specificity of inhibitors such as L-AOPP should not be inferred solely from their potency, since other metabolic targets may be similarly affected.

5.2.2 *Effects of Selected Metabolites on PAL and TDC*

The first enzyme in a metabolic sequence is often found to be regulated by pathway intermediates or endproducts. To determine whether the entry point enzymes in verbascoside and salidroside biosynthesis have any important interactions with intermediates of the pathway, a range of metabolites was tested for their ability to activate or inhibit PAL and TDC (Table 3). None of the compounds tested produced any enhancement of activity but several substances did have inhibitory effects. Inhibition of PAL by cinnamate has been reported in other cases (Camm and Towers 1973; Hahlbrock et al. 1982) and may be a method of enzyme activity modulation in vivo (Shields et al. 1982; Bolwell et al. 1986) but it is unclear what physiological significance the remainder of the results may have. PAL is prefer-

Table 3. Effect of selected metabolites on the in vitro activity of phenylalanine ammonia lyase (PAL) and tyrosine decarboxylase (TDC) from *Syringa vulgaris* suspension cultures. Data expressed as percent inhibition relative to control

	PAL		TDC	
Compound	0.1 mM	1.0 mM	0.1 mM	1.0 mM
Tyramine	10	7	0	0
Tyrosol	0	0	6	48
Cinnamate	19	65	0	0
p-Coumarate	4	2	43	91
Caffeate	2	26	31	80
Salidroside	5	5	0	17
Verbascoside	0	8	9	30
Phenylpyruvate	22	57	0	15
p-Hydroxyphenylpyruvate	0	0	50	91

entially inhibited by nonhydroxylated aromatic compounds while TDC is more strongly affected by compounds with one or two phenolic hydroxyls. This generalization is in keeping with the substrate preferences of the two enzymes and may reflect some aspect of their active site architecture. No significant feedback inhibition of PAL, and only a modest decrease in TDC activity, was caused by the pathway endproducts, verbascoside, and salidroside.

5.2.3 Induction of PAL and TDC Throughout the Culture Cycle

The fluctuations in extractable PAL and TDC activity throughout the culture cycle are shown in Fig. 4. As has been previously reported for other enzymes devoted to secondary metabolism in plant cell cultures (Hahlbrock et al. 1982; Dixon and Bolwell 1986), the activities of these two enzymes are induced shortly after subculture, perhaps by exposure of the cells to the higher nutrient concentrations of fresh medium. Both activities decline rapidly thereafter but they rise again later in the culture cycle between days 6 and 10. This second increase may be the result of

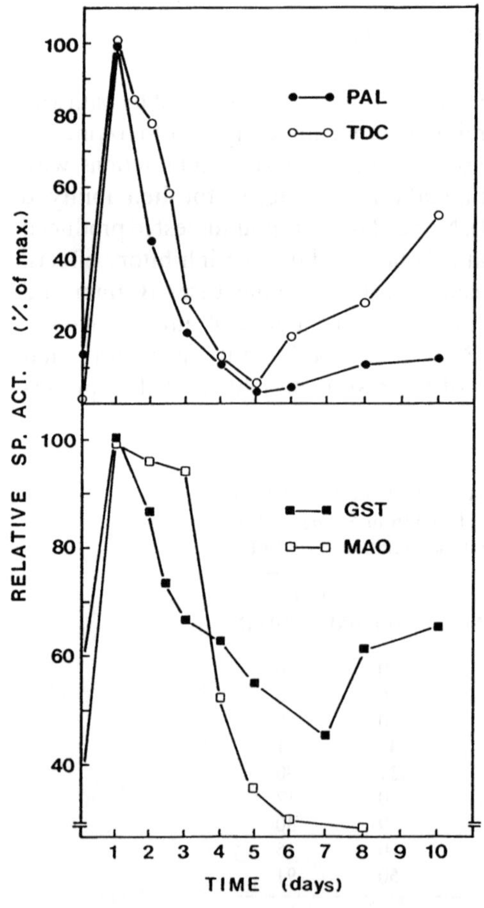

Fig. 4. Activity of enzymes associated with verbascoside and salidroside production in *Syringa vulgaris* suspension cultures throughout the culture cycle. *PAL* phenylalanine ammonia lyase; *TDC* tyrosine decarboxylase; *GST* UDP-glucose: tyrosol glucosyltransferase; *MAO* tyramine oxidase

a separate stimulus, perhaps the depletion of one or more nutrients in the culture medium. The relationship between these two induction phenomena was examined by subculturing *Syringa* cells into fresh medium before, during, and after the late rise in TDC activity, and evaluating extractable TDC activity 24 h later. These experiments revealed that the level of extractable TDC 24 h after subculture is strongly dependent upon the physiological state of the culture at the time of transfer to fresh medium, with maximum inducibility observed at day 7 of the culture cycle. A similar "inducibility window" has recently been reported for elicitor-induced TDC from *Thalictrum* cell cultures (Gugler et al. 1988).

5.3 Enzymes of Tyramine Metabolism

Tyramine had been shown to serve as an efficient precursor of both the 3′,4′-dihydroxyphenylethanol moiety of verbascoside and the 4′-hydroxyphenylethanol (tyrosol) moiety of salidroside, when the labeled compound was administered to *Syringa* cell cultures. However, the specific reactions involved, and their sequence, remained uncertain. To gain some insight into this metabolic pattern, a search was undertaken for the relevant enzyme activities.

When radioactive tyramine was incubated with desalted protein extracts from 3-day-old *Syringa* cells, a radioactive compound rapidly appeared which was extractable into organic solvent. Following reduction with sodium borohydride, this radioactive product co-chromatographed with authentic tyrosol, consistent with the behavior expected of 4′-hydroxyphenylacetaldehyde. Since the enzymic reaction showed no requirement for cofactors, it is probable that the enzyme involved in removal of the amino group is an amine oxidase rather than an aminotransferase. No reaction could be detected with putrescine as a substrate, which indicates that the *Syringa* enzyme is a monoamine oxidase, as opposed to the well-known diamine oxidase activities isolated from *Pisum* (Glatz et al. 1987) and *Lens* (Floris et al. 1983). The *Syringa* enzyme was 96% inhibited by 10 μM trans-phenylcyclopropylamine, a classical specific inhibitor of monoamine oxidases (S. Schweissing and B. Ellis unpubl.).

An indication of the possible specificity of this monoamine oxidase activity for hydroxyphenylethanol producing tissues was obtained when extracts from *Anchusa officinalis*, *Nicotiana tabacum*, *Canavalia ensiformis*, and *Hordeum vulgare* tissues were found to have no deaminating activity against tyramine. Other workers have recently shown that cell cultures of *Berberis* species, which are thought to require a similar tyramine oxidase activity for benzylisoquinoline alkaloid formation, displayed only very low levels of the enzyme by comparison with *Syringa* cultures (Rueffer and Zenk 1987).

The monoamine oxidase activity in *Syringa* cultures was induced by transfer to fresh medium, in a manner similar to the PAL and TDC activities (Fig. 4). In crude extracts, a broad pH optimum around pH 6.3 was observed. No requirement for metal cations, pyridoxal phosphate, or thiol agents could be demonstrated.

When *undesalted* crude protein extracts were assayed with radioactive tyramine as substrate, a second radioactive product was obtained in addition to the aldehyde. The new product was chromatographically identified as tyrosol, formed by enzymic reduction of the aldehyde with consumption of residual nicotinamide reductants present in the fresh cell extracts. Supplementation of the extracts with $NADPH_2$ greatly enhanced the reduction, while $NADH_2$ was ineffective. This oxidoreductase activity was not characterized further.

Synthesis of salidroside requires a specific attachment of glucose to the alipathic hydroxyl of tyrosol to form the beta-glucoside. A corresponding glucosyltransferase activity was readily detected in crude cell extracts, with highest activity appearing in the cultures 1 day after transfer to fresh medium (Fig. 4) (Parissenti and Ellis unpubl.). The product of the reaction was chromatographically identified as salidroside, with no indication of formation of the 4'-O-glucoside of tyrosol. The enzyme was markedly activated and stabilized by 5 mM dithioerythritol, and displayed a pH optimum of 7.3. UDPG was the preferred sugar donor substrate; TDPG supported some activity, while ADPG and GDPG were ineffective. These characteristics are similar to those of a number of other plant glycosyltransferases (Hosel 1981), but the *Syringa* enzyme is unusual in its specificity for the alipathic, rather than an aromatic, hydroxyl as the site of glucosylation.

6 Conclusions and Prospects

Plant species which synthesize hydroxyphenylethanol glycosides generally appear to retain this capability when their tissues are cultured in vitro, in the cases reported thus far. The quantities of glycosides accumulated by the cultures exceed the levels of most other secondary metabolites produced in culture systems, which offers a number of opportunities. If compounds of this class are eventually found to be of commercial value, their production in large-scale culture systems could be considered, especially if the native plant source is a slow growing or inaccessible woody species. The level of glycoside synthesis and accumulation in the cultured cells also makes these systems attractive experimental objects in which to explore mechanisms of metabolite storage and transport in plant cells. Finally, the inducibility of several of the key enzymes involved in glycoside biosynthesis, coupled with the high levels of activity attained, makes it feasible to consider purifying the enzymes and retrieving the corresponding genes. In particular, the extensive involvement of specific glycosyltransferases in the synthesis of the more elaborate glycosides offers an opportunity to determine the structural and mechanistic requirements of this important class of enzymes, which is also crucial to glycoprotein biosynthesis in plants and animals. Considering current advances in the generation of antibodies and genetic probes and the experimental versatility afforded by cell suspension cultures, the elucidation of the molecular basis of coordinate induction of the biosynthetic enzymes, and the organization of the pathway within the cells, should be possible. An understanding of the biochemistry behind these specific high-level production patterns will be essential if the prospect of successfully manipulating plant secondary metabolism for industrial and agricultural purposes is to be realized.

References

Amiot M-J, Fleuriet A, Macheix J-J (1986) Importance and evolution of phenolic compounds in Olive during growth and maturation. J Agric Food Chem 34:823–826

Amrhein N, Godeke KH (1977) α-Aminooxy-β-phenylpropionic acid – a potent inhibitor of L-phenylalanine ammonia-lyase in vitro and in vivo. Plant Sci Lett 8:313–317

Andary C, Ibrahim RK (1986) Biosynthetic capacity of *Stachys* seedlings for verbascoside and related caffeoyl derivatives. Z Naturforsch 41c:18–21

Andary C, Wylde R, Laffite C, Privat G, Winternitz F (1982) Structures of verbascoside and orobanchoside, caffeic acid sugar esters from *Orobanche rapum-genistae*. Phytochemistry 21:1123–1127

Andary C, Privat G, Wylde R, Heitz A (1985a) Pheliposide et arenarioside, deux nouveaux esters heterosidiques de l'acide cafeique isoles de *Orobanche arenaria*. J Nat Prod 48:778–783

Andary C, Wylde R, Heitz A, Rascol JP, Roussel JL, Laffite C (1985b) Poliumoside, a caffeic acid glycoside ester from *Teucrium belion*. Phytochemistry 24:362–364

Becker H, Hsieh WC, Wylde R, Laffite C, Andary C (1982) Structure of echinacoside. Z Naturforsch 37c:351–353

Birkofer L, Kaiser C, Thomas U (1968) Acteosid und Neoacteosid; Zuckerester aus *Syringa vulgaris* L. Z Naturforsch 23b:1051–1058

Bolwell G, Cramer C, Lamb C, Schuch W, Dixon R (1986) L-Phenylalanine ammonia-lyase from *Phaseolus vulgaris*: modulation of the levels of active enzyme by trans-cinnamic acid. Planta 169:97–107

Bowsher RR, Henry DP (1983) Decarboxylation of p-tyrosine: a potential source of p-tyramine in mammalian tissues. J Neurochem 40:992–1002

Burger JFW, Brandt EV, Ferreira D (1987) Iridoid and phenolic glycosides from *Harpagophytum procumbens*. Phytochemistry 26:1453–1457

Caldwell MM, Robberecht R, Flint SD (1983) Internal filters: prospects for UV-acclimation in higher plants. Physiol Plant 58:445–450

Calis I, Lahloub MF, Rogenmoser E, Sticher O (1984) Isomartynoside, a phenylpropanoid glycoside from *Galeopsis pubescens*. Phytochemistry 23:2313–2315

Calis I, Gross GA, Sticher O (1987) Phenylpropanoid glycosides isolated from *Scrophularia scopolii*. Phytochemistry 26:2057–2061

Calis I, Gross GA, Sticher O (1988) Two phenylpropanoid glycosides from *Scrophularia scopolii*. Phytochemistry 27:1465–1468

Camm EL, Towers GHN (1973) Phenylalanine ammonia lyase. Phytochemistry 12:961–975

Chapple CCS, Ellis BE (1984) Preparative and analytical liquid chromatography of complex caffeoyl esters. J Chromatogr 285:171–177

Chapple CCS, Walker MA, Ellis BE (1986) Plant tyrosine decarboxylase can be strongly inhibited by L-α-aminooxy-β-phenylpropionate. Planta 167:101–105

Christenson JG, Dairman W, Udenfriend S (1970) Preparation and properties of a homogeneous aromatic L-aromatic amino acid decarboxylase from hog kidney. Arch Biochem Biophys 141:356–367

Cooper R, Solomon PH, Kubo I, Nakanishi K, Shoolery JN, Occolowitz JL (1980) Myricoside, an african armyworm antifeedant: separation by droplet countercurrent chromatography. J Am Chem Soc 102:7953–9755

Dixon RA, Bolwell GP (1986) Modulation of the phenylpropanoid pathway in bean (*Phaseolus vulgaris*) cell suspension cultures. In: Morris P, Scragg A, Stafford A, Fowler M (eds) Secondary metabolism in plant cell cultures. Cambridge Univ Press, Cambridge

Einset JW, Alexander JH (1984) Multiplication of *Syringa vulgaris* species and cultivars in tissue culture. Comb Proc Plant Prop Soc 34:628–636

Ellis BE (1983) Production of hydroxyphenylethanol glycosides in suspension cultures of *Syringa vulgaris*. Phytochemistry 22:1941–1943

Endo K, Hikino H (1982) Structure of forsythoside C and D, antibacterial principles of *Forsythia suspensa* fruits. Heterocycles 19:2033–2036

Endo K, Takahashi K, Abe T, Hikino H (1981) Structure of forsythoside A, an antibacterial principle of *Forsythia suspensa* leaves. Heterocycles 16:1311–1314

Endo K, Takahashi K, Abe T, Hikino H (1982) Structure of forsythoside B, an antibacterial principle of *Forsythia koreana* stems. Heterocycles 19:261–264

Epps HMR (1944) Studies on bacterial amino-acid decarboxylases 2. l(−)-tyrosine decarboxylase from *Streptococcus faecalis*. Biochem J 38:242–249

Fleuriet A, Machiex JJ, Andary C, Villemur P (1984) Identification and quantitative determination of verbascoside by high performance liquid chromatography in the fruit of six cultivars of *Olea europaea* L. Physiol Vég 299:253–256

Floris G, Giartosio A, Rinaldi A (1983) Diamine oxidase from *Lens esculenta* seedlings: purification and properties. Phytochemistry 22:1871–1874

Gamborg OL, Eveleigh DE (1968) Culture methods and detection of glucanases in suspension cultures of wheat and barley. Can J Biochem 46:417–421

Glatz Z, Kovar J, Macholan L, Pec P (1987) Pea (*Pisum sativum*) diamine oxidase contains pyrroloquinoline quinone as a cofactor. Biochem J 242:603–606

Gross G-A, Lahloub MF, Anklin C, Schulten H-R, Sticher O (1988) Teucrioside, a phenylpropanoid glycoside from *Teucrium chamaedrys*. Phytochemistry 27:1459–1463

Grunberger D, Banerjee R, Eisinger K, Oltz EM, Efros L, Caldwell M, Estevez V, Nakanishi K (1988) Preferential cytotoxicity on tumor cells by caffeic acid phenethyl ester from propolis. Experientia 44:230–232

Gugler K, Funk C, Brodelius P (1988) Elicitor-induced tyrosine decarboxylase in berberine-synthesizing suspension cultures of *Thalictrum rugosum*. Eur J Biochem 170:661–666

Hahlbrock K, Boudet AM, Chappell J, Kreuzaler F, Kuhn DN, Ragg H (1982) Differential induction of mRNAs by light and elicitor in cultured plant cells. In: Cifferi O, Dure L (eds) Structure and function of plant genomes. Plenum, New York

Henry M, Roussel JL, Andary C (1987) Verbascoside production in callus and suspension cultures of *Hygrophila erecta*. Phytochemistry 26:1961–2963

Hildebrandt V, Harney PM (1983) In vitro propagation of *Syringa vulgaris* 'Vesper'. HortScience 18:432–434

Hosel W (1981) Glycosylation and glycosidases. In: Conn EE (ed) The biochemistry of plants. A Comprehensive Treatise. Vol 7. Academic Press, New York

Hosoi K (1974) Purification and some properties of L-tyrosine carboxylyase from barley roots. Plant Cell Physiol 15:429–440

Imakura Y, Kobayashi S, Mima A (1985) Bitter phenyl propanoid glycosides from *Campsis chinensis*. Phytochemistry 24:139–146

Jiminez C, Villaverde MC, Riguera R, Castedo L, Stermitz FR (1987) Three phenylpropanoid glycosides from *Mussatia*. Phytochemistry 26:1805–1810

Jiminez C, Villaverde MC, Riguera R, Castedo L, Stermitz FR (1988) Five phenylpropanoid glycosides from *Mussatia*. Phytochemistry 27:2947–2951

Kitigawa S, Tsukamoto H, Hisada S, Nishibe S (1984) Studies on the Chinese crude drug "Forsythiae Fructus". VII. A new caffeoyl glycoside from *Forsythia viridissima*. Chem Pharm Bull 32:1209–1213

Kobayashi H, Karasawa H, Miyase T, Fukushima S (1984a) Studies on the constituents of Cistanchis Herba. III. Isolation and structures of new phenylpropanoid glycosides, cistanosides A and B. Chem Pharm Bull 32:3009–3014

Kobayashi H, Karasawa H, Miyase T, Fukushima S (1984b) Studies on the constituents of Cistanchis Herba. IV. Isolation and structures of two new phenylpropanoid glycosides, cistanosides C and D. Chem Pharm Bull 32:3880–3885

Kobayashi H, Oguchi H, Takizawa N, Miyase T, Ueno A, Usmanghani K, Ahmad M (1987) New phenylethanoid glycosides from *Cistanche tubulosa* (Schrenk) Hook. f. I. Chem Pharm Bull 35:3309–3314

Konishi T, Narumi Y, Watanabe K, Kiyosawa S, Shoji J (1987) Comparative studies on the constituents of a parasitic plant and its host. III. On the constituents of *Boschniakia rossica* Fedtsch. et Flerov. (2). Chem Pharm Bull 35:4155–4161

Kvist LP, Pedersen JA (1986) Distribution and taxonomic implications of some phenolics in the family Gesneriaceae determined by EPR spectroscopy. Biochem Syst Ecol 14:385–405

Lahloub MF, Gross G-A, Sticher O, Winkler T, Schulten H-R (1986) Ehrenoside, a new phenylpropanoid glycoside from *Veronica bellidioides*. Planta Med 52:352–355

Matsumoto M, Koga S, Shoyama Y, Nishioka I (1987) Phenolic glycoside compositon of leaves and callus cultures of *Digitalis purpurea*. Phytochemistry 26:3225–3227

Maycock AL, Aster SD, Patchett AA (1980) Inactivation of 3-(3,4-dihydroxyphenyl)alanine decarboxylase by 2-fluoromethyl)-3-(3,4-dihydroxyphenyl)alanine. Biochemistry 19:709–718

Miyase T, Koizumi A, Ueno A, Noro T, Kuroyanagi M, Fukushima S, Akiyama Y, Takemoto T (1982) Studies on the acyl glycosides from *Leucoseptrum japonicum* (Miq) Kitamura et Murata. Chem Pharm Bull 30:2732–2737

Molgaard P, Ravn H (1988) Evolutionary aspects of caffeoyl ester distribution in dicotyledons. Phytochemistry 27:2411–2421

Murashige T, Skoog F (1962) A revised medium for rapid growth and bioassays with tobacco tissue cultures. Physiol Plant 15:473–497

Nicoletti M, Galeffi C, Messana I, Marini-Bettolo GB, Garbarino JA, Gambaro V (1988) Phenylpropanoid glycosides from *Calceolaria hypericina*. Phytochemistry 27:639–641

Nishibe S, Okabe K, Tsukamoto H, Sakushima A, Hisada S (1982a) The structure of forsythiaside isolated from *Forsythia suspensa*. Chem Pharm Bull 30:1048–1050

Nishibe S, Okabe K, Tsukamoto H, Sakushima A, Hisada S, Baba H, Akisada T (1982b) Studies on the Chinese crude drug "Forsythiae fructus." VI. The structure and antibacterial activity of suspensaside isolated from *Forsythia suspensa*. Chem Pharm Bull 30:4548–4553

Nonaka G, Nishioka I (1977) Bitter phenylpropanoid glycosides from *Conandron ramoidioides*. Phytochemistry 16:1265–1267

Pierik RLM, Steegmans HHM, Elias AA, Stiekema OTJ, van der Velde AJ (1988) Vegetative propagation of *Syringa vulgaris* L. in vitro. Acta Hortic 226:25–40

Potterat O, Msonthi JD, Hostettmann K (1988) Four iridoid glucosides and a phenylpropanoid glycoside from *Sesamum angolense*. Phytochemistry 27:2677–2679

Ramsey B (1984) MSc Thesis, Univ of Guelph, Canada

Rao KV, Juneau RJ (1975) Glycosides of *Magnolia*. II. Structural elucidation of Magnolidin. Lloydia 38:339–342

Razzaque A, Ellis BE (1977) Rosmarinic acid production in *Coleus* cell cultures. Planta 137:287–291

Rueffer M, Zenk M (1987) Distant precursors of benzylisoquinoline alkaloids and their enzymatic formation. Z Naturforsch 42c:319–332

Sakurai A, Kato T (1983) A new glycoside, kusaginin isolated from *Clerodendron trichotomum*. Bull Chem Soc Jpn 56:1573–1574

Sasaki H, Taguchi T, Endo T, Yosioka I, Higashiyama K, Otomasu H (1978) The glycosides of *Martynia louisiana* Mill. A new phenylpropanoid glycoside, martynoside. Chem Pharm Bull 26:2111–2121

Scarpati ML, Delle Monache F (1963) Isolamento dal *Verbascum sinuatum* di due nuovi glucosidi: il verbascoside e l'isoverbascoside. Ann Chim 53:356–367

Schilling G, Hugel M, Mayer W (1982) Verbascoside and isoverbascoside from *Paulownia tomentosa* Steud. Z Naturforsch 37b:1633–1635

Shields S, Wingate V, Lamb C (1982) Dual control of phenylalanine ammonia-lyase production and removal by its product cinnamic acid. Eur J Biochem 123:389–395

Shimomura H, Sashida Y, Adachi T (1987a) Phenolic glucosides from *Prunus grayana*. Phytochemistry 26:249–251

Shimomura H, Sashida Y, Adachi T (1987b) Cyanogenic and phenylpropanoid glucosides from *Prunus grayana*. Phytochemistry 26:2363–2366

Shimomura H, Sashida Y, Ogawa K (1987c) Iridoid glucosides and phenylpropanoid glycosides in *Ajuga* species of Japan. Phytochemistry 26:1981–1983

Shoyama Y, Matsumoto M, Nishioka I (1986) Four caffeoyl glycosides from callus tissue of *Rehmannia glutinosa*. Phytochemistry 25:1633–1636

Shoyama Y, Matsumoto M, Nishioka I (1987) Phenolic glycosides from diseased roots of *Rehmannia glutinosa* var. *purpurea*. Phytochemistry 26:983–986

Sticher O, Lahloub MF (1982) Phenolic glycosides of *Paulownia tomentosa* bark. Planta Med 46:145–148

Sticher O, Salama O, Chaudhuri RK, Winkler T (1982) Structural analysis of eukovoside, a new phenylpropanoid glycoside from *Euphrasia rostkoviana* Hayne. Helv Chim Acta 65:1538–1542

Stoll A, Renz J, Brack A (1950) Isolierung und Konstitution des Echinacosids eines Glykosids aus den Wurzeln von *Echinacea angustifolia* D.C. Helv Chim Acta 33:1877–1893

Thieme H (1969) Über die Identität der Glucoside Rhodiolosid und Salidrosid. Sonderdruck aus Die Pharmazie 24:118–119

Uhring J (1980) Organ and tissue culture of lilac. HortScience 15:418 (abstract)

Welander NT (1987) Propagation of *Syringa chinensis* cv. Saugeana by in vitro culture of nodal explants. J Hortic Sci 62:89–96

Subject Index

MIX
Papier aus verantwortungsvollen Quellen
Paper from responsible sources
FSC® C105338

If you have any concerns about our products,
you can contact us on
ProductSafety@springernature.com

In case Publisher is established outside the EU,
the EU authorized representative is:
Springer Nature Customer Service Center GmbH
Europaplatz 3, 69115 Heidelberg, Germany

Printed by Libri Plureos GmbH
in Hamburg, Germany